Differential Geometry of Curves and Surfaces

Manfredo P. do Carmo

Instituto de Matematica Pura e Aplicada (IMPA)
Rio de Janeiro, Brazil

Prentice-Hall, Inc., Englewood Cliffs, New Jersey

Carmo, Manfredo Perdigao do.
Differential geometry of curves and surfaces

"A free translation, with additional material, of a book and a set of notes, both published originally in Portuguese."
Bibliography: p.
Includes index.
1. Geometry, Differential. 2. Curves. 3. Surfaces.
I. Title
QA641.C33 75–22094
ISBN: 0–13–212589–7

To Leny

Current printing:

10 9 8 7 6 5 4 3 2 1

Printed in the United States of America

Prentice-Hall International, Inc., *London*
Prentice-Hall of Australia Pty., Limited, *Sydney*
Prentice-Hall of Canada, Ltd., *Toronto*
Prentice-Hall of India Private Ltd., *New Delhi*
Prentice-Hall of Japan, Inc., *Tokyo*
Prentice-Hall of Southeast Asia Private Limited, *Singapore*

Contents

Preface v

Some Remarks on Using this Book vii

1. Curves 1

1-1 Introduction *1*
1-2 Parametrized Curves *2*
1-3 Regular Curves; Arc Length *5*
1-4 The Vector Product in $\boldsymbol{R}^3$ *11*
1-5 The Local Theory of Curves Parametrized by Arc Length *16*
1-6 The Local Canonical Form *27*
1-7 Global Properties of Plane Curves *30*

2. Regular Surfaces 51

2-1 Introduction *51*
2-2 Regular Surfaces; Inverse Images of Regular Values *52*
2-3 Change of Parameters; Differential Functions on Surfaces *69*
2-4 The Tangent Plane; the Differential of a Map *83*
2-5 The First Fundamental Form; Area *92*
2-6 Orientation of Surfaces *102*
2-7 A Characterization of Compact Orientable Surfaces *109*
2-8 A Geometric Definition of Area *114*
Appendix: A Brief Review on Continuity
and Differentiability *118*

3. The Geometry of the Gauss Map 134

3-1 Introduction *134*
3-2 The Definition of the Gauss Map and Its Fundamental Properties *135*
3-3 The Gauss Map in Local Coordinates *153*
3-4 Vector Fields *175*
3-5 Ruled Surfaces and Minimal Surfaces *188*
Appendix: Self-Adjoint Linear Maps and Quadratic Forms *214*

4. The Intrinsic Geometry of Surfaces 217

4-1 Introduction *217*
4-2 Isometries; Conformal Maps *218*
4-3 The Gauss Theorem and the Equations of Compatibility *231*
4-4 Parallel Transport; Geodesics *238*
4-5 The Gauss-Bonnet Theorem and its Applications *264*
4-6 The Exponential Map. Geodesic Polar Coordinates *283*
4-7 Further Properties of Geodesics. Convex Neighborhoods *298*
Appendix: Proofs of the Fundamental Theorems of The Local Theory of Curves and Surfaces *309*

5. Global Differential Geometry 315

5-1 Introduction *315*
5-2 The Rigidity of the Sphere *317*
5-3 Complete Surfaces. Theorem of Hopf-Rinow *325*
5-4 First and Second Variations of the Arc Length; Bonnet's Theorem *339*
5-5 Jacobi Fields and Conjugate Points *357*
5-6 Covering Spaces; the Theorems of Hadamard *371*
5-7 Global Theorems for Curves; the Fary-Milnor Theorem *380*
5-8 Surfaces of Zero Gaussian Curvature *408*
5-9 Jacobi's Theorems *415*
5-10 Abstract Surfaces; Further Generalizations *425*
5-11 Hilbert's Theorem *446*
Appendix: Point-Set Topology of Euclidean Spaces *456*

Bibliography and Comments 471

Hints and Answers to Some Exercises 475

Index 497

Preface

This book is an introduction to the differential geometry of curves and surfaces, both in its local and global aspects. The presentation differs from the traditional ones by a more extensive use of elementary linear algebra and by a certain emphasis placed on basic geometrical facts, rather than on machinery or random details.

We have tried to build each chapter of the book around some simple and fundamental idea. Thus, Chapter 2 develops around the concept of a regular surface in R^3; when this concept is properly developed, it is probably the best model for differentiable manifolds. Chapter 3 is built on the Gauss normal map and contains a large amount of the local geometry of surfaces in R^3. Chapter 4 unifies the intrinsic geometry of surfaces around the concept of covariant derivative; again, our purpose was to prepare the reader for the basic notion of connection in Riemannian geometry. Finally, in Chapter 5, we use the first and second variations of arc length to derive some global properties of surfaces. Near the end of Chapter 5 (Sec. 5-10), we show how questions on surface theory, and the experience of Chapters 2 and 4, lead naturally to the consideration of differentiable manifolds and Riemannian metrics.

To maintain the proper balance between ideas and facts, we have presented a large number of examples that are computed in detail. Furthermore, a reasonable supply of exercises is provided. Some factual material of classical differential geometry found its place in these exercises. Hints or answers are given for the exercises that are starred.

The prerequisites for reading this book are linear algebra and calculus. From linear algebra, only the most basic concepts are needed, and a standard undergraduate course on the subject should suffice. From calculus, a certain familiarity with calculus of several variables (including the statement of the implicit function theorem) is expected. For the reader's convenience, we have tried to restrict our references to R. C. Buck, *Advancd Calculus,* New York: McGraw-Hill, 1965 (quoted as Buck, *Advanced Calculus*). A certain knowledge of differential equations will be useful but it is not required.

This book is a free translation, with additional material, of a book and a set of notes, both published originally in Portuguese. Were it not for the enthusiasm and enormous help of Blaine Lawson, this book would not have come into English. A large part of the translation was done by Leny Cavalcante. I am also indebted to my colleagues and students at IMPA for their comments and support. In particular, Elon Lima read part of the Portuguese version and made valuable comments.

Robert Gardner, Jürgen Kern, Blaine Lawson, and Nolan Wallach read critically the English manuscript and helped me to avoid several mistakes, both in English and Mathematics. Roy Ogawa prepared the computer programs for some beautiful drawings that appear in the book (Figs. 1-3, 1-8, 1-9, 1-10, 1-11, 3-45 and 4-4). Jerry Kazdan devoted his time generously and literally offered hundreds of suggestions for the improvement of the manuscript. This final form of the book has benefited greatly from his advice. To all these people—and to Arthur Wester, Editor of Mathematics at Prentice-Hall, and Wilson Góes at IMPA—I extend my sincere thanks.

Rio de Janeiro Manfredo P. do Carmo

Some Remarks on Using This Book

We tried to prepare this book so it could be used in more than one type of differential geometry course. Each chapter starts with an introduction that describes the material in the chapter and explains how this material will be used later in the book. For the reader's convenience, we have used footnotes to point out the sections (or parts thereof) that can be omitted on a first reading.

Although there is enough material in the book for a full-year course (or a topics course), we tried to make the book suitable for a first course on differential geometry for students with some background in linear algebra and advanced calculus.

For a short one-quarter course (10 weeks), we suggest the use of the following material: Chapter 1: Secs. 1-2, 1-3, 1-4, 1-5 and one topic of Sec. 1-7—2 weeks. Chapter 2: Secs. 2-2 and 2-3 (omit the proofs), Secs. 2-4 and 2-5—3 weeks. Chapter 3: Secs. 3-2 and 3-3—2 weeks. Chapter 4: Secs. 4-2 (omit conformal maps and Exercises 4, 13-18, 20), 4-3 (up to Gauss theorema egregium), 4-4 (up to Prop. 4; omit Exercises 12, 13, 16, 18-21), 4-5 (up to the local Gauss-Bonnet theorem; include applications (b) and (f))—3 weeks.

The 10-week program above is on a pretty tight schedule. A more relaxed alternative is to allow more time for the first three chapters and to present survey lectures, on the last week of the course, on geodesics, the Gauss theorema egregium, and the Gauss-Bonnet theorem (geodesics can then be defined as curves whose osculating planes contain the normals to the surface).

In a one-semester course, the first alternative could be taught more leisurely and the instructor could probably include additional material (for instance, Secs. 5-2 and 5-10 (partially), or Secs. 4-6, 5-3 and 5-4).

Please also note that an asterisk attached to an exercise does not mean the exercise is either easy or hard. It only means that a solution or hint is provided at the end of the book. Second, we have used for parametrization a bold-faced **x** and that might become clumsy when writing on the blackboard. Thus we have reserved the capital X as a suggested replacement.

Where letter symbols that would normally be italic appear in italic context, the letter symbols are set in roman. This has been done to distinguish these symbols from the surrounding text.

1 Curves

1-1. Introduction

The differential geometry of curves and surfaces has two aspects. One, which may be called classical differential geometry, started with the beginnings of calculus. Roughly speaking, classical differential geometry is the study of local properties of curves and surfaces. By local properties we mean those properties which depend only on the behavior of the curve or surface in the neighborhood of a point. The methods which have shown themselves to be adequate in the study of such properties are the methods of differential calculus. Because of this, the curves and surfaces considered in differential geometry will be defined by functions which can be differentiated a certain number of times.

The other aspect is the so-called global differential geometry. Here one studies the influence of the local properties on the behavior of the entire curve or surface. We shall come back to this aspect of differential geometry later in the book.

Perhaps the most interesting and representative part of classical differential geometry is the study of surfaces. However, some local properties of curves appear naturally while studying surfaces. We shall therefore use this first chapter for a brief treatment of curves.

The chapter has been organized in such a way that a reader interested mostly in surfaces can read only Secs. 1-2 through 1-5. Sections 1-2 through 1-4 contain essentially introductory material (parametrized curves, arc length, vector product), which will probably be known from other courses and is included here for completeness. Section 1-5 is the heart of the chapter and contains the material of curves needed for the study of surfaces. For those wishing to go a bit further on the subject of curves, we have included Secs. 1-6 and 1-7.

1-2. Parametrized Curves

We denote by R^3 the set of triples (x, y, z) of real numbers. Our goal is to characterize certain subsets of R^3 (to be called curves) that are, in a certain sense, one-dimensional and to which the methods of differential calculus can be applied. A natural way of defining such subsets is through differentiable functions. We say that a real function of a real variable is *differentiable* (or *smooth*) if it has, at all points, derivatives of all orders (which are automatically continuous). A first definition of curve, not entirely satisfactory but sufficient for the purposes of this chapter, is the following.

DEFINITION. *A* parametrized differentiable curve *is a differentiable map* α*:* $\mathrm{I} \longrightarrow \mathrm{R}^3$ *of an open interval* $\mathrm{I} = (\mathrm{a}, \mathrm{b})$ *of the real line* R *into* R^3.†

The word *differentiable* in this definition means that α is a correspondence which maps each $t \in I$ into a point $\alpha(t) = (x(t), y(t), z(t)) \in R^3$ in such a way that the functions $x(t)$, $y(t)$, $z(t)$ are differentiable. The variable t is called the *parameter* of the curve. The word *interval* is taken in a generalized sense, so that we do not exclude the cases $a = -\infty$, $b = +\infty$.

If we denote by $x'(t)$ the first derivative of x at the point t and use similar notations for the functions y and z, the vector $(x'(t), y'(t), z'(t)) = \alpha'(t) \in R^3$ is called the *tangent vector* (or *velocity vector*) of the curve α at t. The image set $\alpha(I) \subset R^3$ is called the *trace* of α. As illustrated by Example 5 below, one should carefully distinguish a parametrized curve, which is a map, from its trace, which is a subset of R^3.

A warning about terminology. Many people use the term "infinitely differentiable" for functions which have derivatives of all orders and reserve the word "differentiable" to mean that only the existence of the first derivative is required. We shall not follow this usage.

Example 1. The parametrized differentiable curve given by

$$\alpha(t) = (a \cos t, a \sin t, bt), \qquad t \in R,$$

has as its trace in R^3 a helix of pitch $2\pi b$ on the cylinder $x^2 + y^2 = a^2$. The parameter t here measures the angle which the x axis makes with the line joining the origin 0 to the projection of the point $\alpha(t)$ over the xy plane (see Fig. 1-1).

---

†In italic context, letter symbols will not be italicized so they will be clearly distinguished from the surrounding text.

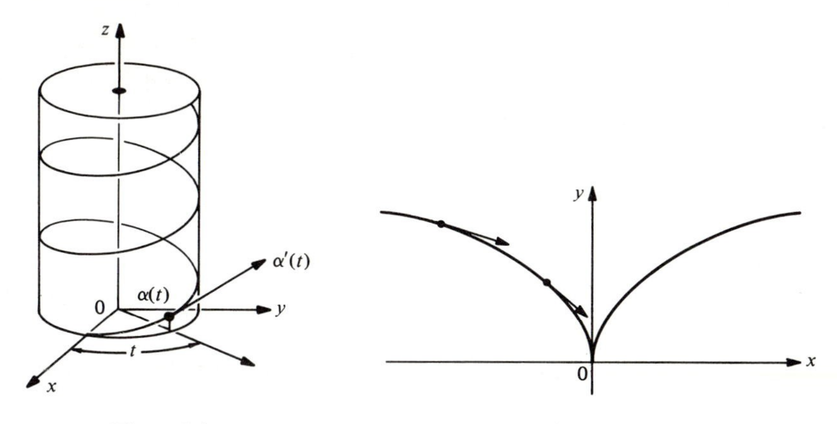

Figure 1-1 Figure 1-2

Example 2. The map $\alpha\colon R \longrightarrow R^2$ given by $\alpha(t) = (t^3, t^2)$, $t \in R$, is a parametrized differentiable curve which has Fig. 1-2 as its trace. Notice that $\alpha'(0) = (0, 0)$; that is, the velocity vector is zero for $t = 0$.

Example 3. The map $\alpha\colon R \longrightarrow R^2$ given by $\alpha(t) = (t^3 - 4t, t^2 - 4)$, $t \in R$, is a parametrized differentiable curve (see Fig. 1-3). Notice that $\alpha(2) = \alpha(-2) = (0, 0)$; that is, the map α is not one-to-one.

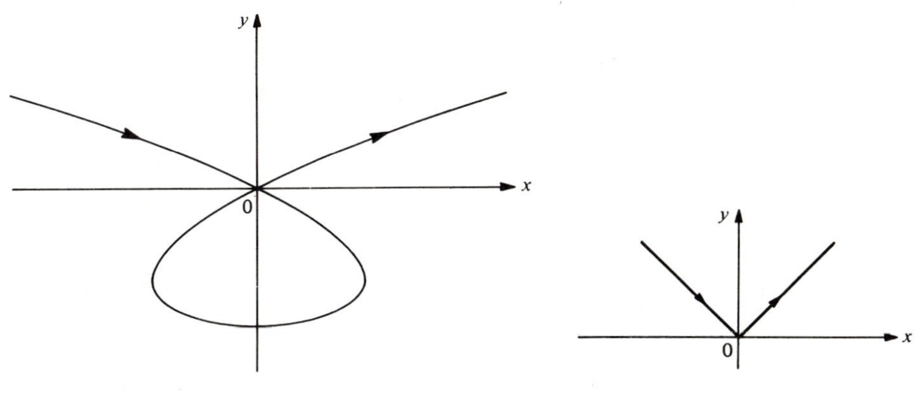

Figure 1-3 Figure 1-4

Example 4. The map $\alpha\colon R \longrightarrow R^2$ given by $\alpha(t) = (t, |t|)$, $t \in R$, is *not* a parametrized differentiable curve, since $|t|$ is not differentiable at $t = 0$ (Fig. 1-4).

Example 5. The two distinct parametrized curves

$$\alpha(t) = (\cos t, \sin t),$$
$$\beta(t) = (\cos 2t, \sin 2t),$$

where $t \in (0 - \epsilon, 2\pi + \epsilon)$, $\epsilon > 0$, have the same trace, namely, the circle $x^2 + y^2 = 1$. Notice that the velocity vector of the second curve is the double of the first one (Fig. 1-5).

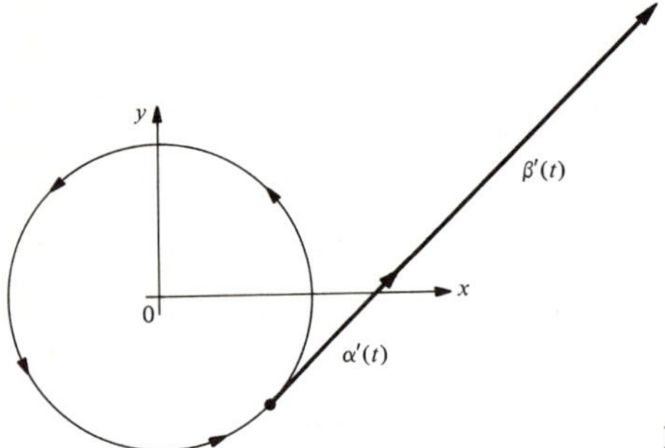

Figure 1-5

We shall now recall briefly some properties of the inner (or dot) product of vectors in R^3. Let $u = (u_1, u_2, u_3) \in R^3$ and define its *norm* (or *length*) by

$$|u| = \sqrt{u_1^2 + u_2^2 + u_3^2}.$$

Geometrically, $|u|$ is the distance from the point (u_1, u_2, u_3) to the origin $0 = (0, 0, 0)$. Now, let $u = (u_1, u_2, u_3)$ and $v = (v_1, v_2, v_3)$ belong to R^3, and let θ, $0 \leq \theta \leq \pi$, be the angle formed by the segments $0u$ and $0v$. The *inner product* $u \cdot v$ is defined by (Fig. 1-6)

$$u \cdot v = |u||v| \cos \theta.$$

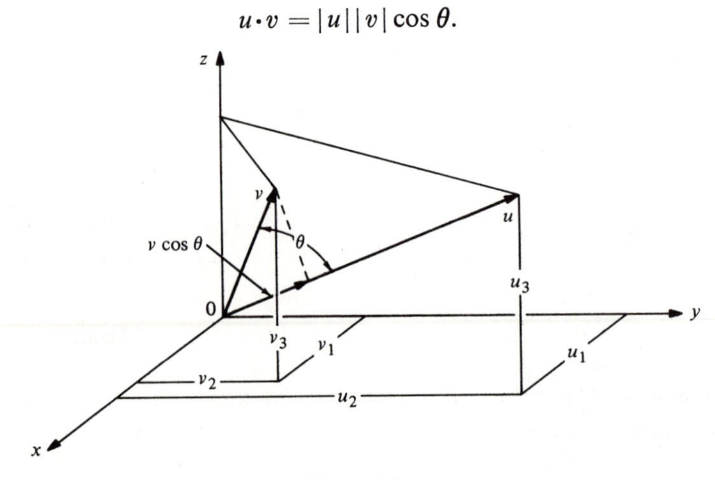

Figure 1-6

The following properties hold:

1. Assume that u and v are nonzero vectors. Then $u \cdot v = 0$ if and only if u is orthogonal to v.

2. $u \cdot v = v \cdot u$.
3. $\lambda(u \cdot v) = \lambda u \cdot v = u \cdot \lambda v$.
4. $u \cdot (v + w) = u \cdot v + u \cdot w$.

A useful expression for the inner product can be obtained as follows. Let $e_1 = (1, 0, 0)$, $e_2 = (0, 1, 0)$, and $e_3 = (0, 0, 1)$. It is easily checked that $e_i \cdot e_j = 1$ if $i = j$ and that $e_i \cdot e_j = 0$ if $i \neq j$, where $i, j = 1, 2, 3$. Thus, by writing

$$u = u_1 e_1 + u_2 e_2 + u_3 e_3, \qquad v = v_1 e_1 + v_2 e_2 + v_3 e_3,$$

and using properties 3 and 4, we obtain

$$u \cdot v = u_1 v_1 + u_2 v_2 + u_3 v_3.$$

From the above expression it follows that if $u(t)$ and $v(t)$, $t \in I$, are differentiable curves, then $u(t) \cdot v(t)$ is a differentiable function, and

$$\frac{d}{dt}(u(t) \cdot v(t)) = u'(t) \cdot v(t) + u(t) \cdot v'(t).$$

EXERCISES

1. Find a parametrized curve $\alpha(t)$ whose trace is the circle $x^2 + y^2 = 1$ such that $\alpha(t)$ runs clockwise around the circle with $\alpha(0) = (0, 1)$.

2. Let $\alpha(t)$ be a parametrized curve which does not pass through the origin. If $\alpha(t_0)$ is the point of the trace of α closest to the origin and $\alpha'(t_0) \neq 0$, show that the position vector $\alpha(t_0)$ is orthogonal to $\alpha'(t_0)$.

3. A parametrized curve $\alpha(t)$ has the property that its second derivative $\alpha''(t)$ is identically zero. What can be said about α?

4. Let $\alpha: I \longrightarrow R^3$ be a parametrized curve and let $v \in R^3$ be a fixed vector. Assume that $\alpha'(t)$ is orthogonal to v for all $t \in I$ and that $\alpha(0)$ is also orthogonal to v. Prove that $\alpha(t)$ is orthogonal to v for all $t \in I$.

5. Let $\alpha: I \longrightarrow R^3$ be a parametrized curve, with $\alpha'(t) \neq 0$ for all $t \in I$. Show that $|\alpha(t)|$ is a nonzero constant if and only if $\alpha(t)$ is orthogonal to $\alpha'(t)$ for all $t \in I$.

1-3. Regular Curves; Arc Length

Let $\alpha: I \longrightarrow R^3$ be a parametrized differentiable curve. For each $t \in I$ where $\alpha'(t) \neq 0$, there is a well-defined straight line, which contains the point $\alpha(t)$ and the vector $\alpha'(t)$. This line is called the *tangent line* to α at t. For the study of the differential geometry of a curve it is essential that there exists such a tangent line at every point. Therefore, we call any point t where $\alpha'(t) = 0$ a *singular point* of α and restrict our attention to curves without singular points. Notice that the point $t = 0$ in Example 2 of Sec. 1-2 is a singular point.

DEFINITION. *A parametrized differentiable curve* $\alpha: I \longrightarrow R^3$ *is said to be* regular *if* $\alpha'(t) \neq 0$ *for all* $t \in I$.

From now on we shall consider only regular parametrized differentiable curves (and, for convenience, shall usually omit the word differentiable).

Given $t \in I$, the *arc length* of a regular parametrized curve $\alpha: I \longrightarrow R^3$, from the point t_0, is by definition

$$s(t) = \int_{t_0}^{t} |\alpha'(t)|\, dt,$$

where

$$|\alpha'(t)| = \sqrt{(x'(t))^2 + (y'(t))^2 + (z'(t))^2}$$

is the length of the vector $\alpha'(t)$. Since $\alpha'(t) \neq 0$, the arc length s is a differentiable function of t and $ds/dt = |\alpha'(t)|$.

In Exercise 8 we shall present a geometric justification for the above definition of arc length.

It can happen that the parameter t is already the arc length measured from some point. In this case, $ds/dt = 1 = |\alpha'(t)|$; that is, the velocity vector has constant length equal to 1. Conversely, if $|\alpha'(t)| \equiv 1$, then

$$s = \int_{t_0}^{t} dt = t - t_0;$$

i.e., t is the arc length of α measured from some point.

To simplify our exposition, we shall restrict ourselves to curves parametrized by arc length; we shall see later (see Sec. 1-5) that this restriction is not essential. In general, it is not necessary to mention the origin of the arc length s, since most concepts are defined only in terms of the derivatives of $\alpha(s)$.

It is convenient to set still another convention. Given the curve α parametrized by arc length $s \in (a, b)$, we may consider the curve β defined in $(-b, -a)$ by $\beta(-s) = \alpha(s)$, which has the same trace as the first one but is described in the opposite direction. We say, then, that these two curves differ by a *change of orientation*.

EXERCISES

1. Show that the tangent lines to the regular parametrized curve $\alpha(t) = (3t, 2t^2, 2t^3)$ make a constant angle with the line $y = 0$, $z = x$.

2. A circular disk of radius 1 in the plane xy rolls without slipping along the x axis. The figure described by a point of the circumference of the disk is called a *cycloid* (Fig. 1-7).

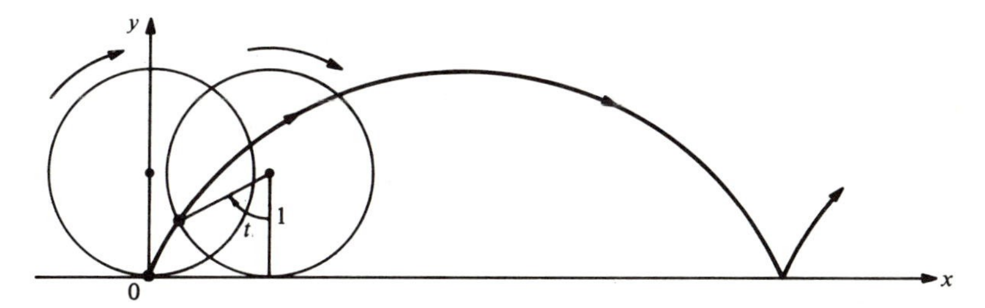

Figure 1-7. The cycloid.

***a.** Obtain a parametrized curve $\alpha: R \longrightarrow R^2$ the trace of which is the cycloid, and determine its singular points.

b. Compute the arc length of the cycloid corresponding to a complete rotation of the disk.

3. Let $0A = 2a$ be the diameter of a circle S^1 and $0Y$ and AV be the tangents to S^1 at 0 and A, respectively. A half-line r is drawn from 0 which meets the circle S^1 at C and the line AV at B. On $0B$ mark off the segment $0p = CB$. If we rotate r about 0, the point p will describe a curve called the *cissoid of Diocles*. By taking $0A$ as the x axis and $0Y$ as the y axis, prove that

a. The trace of

$$\alpha(t) = \left(\frac{2at^2}{1+t^2}, \frac{2at^3}{1+t^2}\right), \qquad t \in R,$$

is the cissoid of Diocles ($t = \tan\theta$; see Fig. 1-8).

b. The origin (0, 0) is a singular point of the cissoid.

c. As $t \longrightarrow \infty$, $\alpha(t)$ approaches the line $x = 2a$, and $\alpha'(t) \longrightarrow (2a, 0)$. Thus, as $t \longrightarrow \infty$, the curve and its tangent approach the line $x = 2a$; we say that $x = 2a$ is an *asymptote* to the cissoid.

4. Let $\alpha: (0, \pi) \longrightarrow R^2$ be given by

$$\alpha(t) = \left(\cos t, \cos t + \log \tan \frac{t}{2}\right),$$

where t is the angle that the y axis makes with the vector $\alpha(t)$. The trace of α is called the *tractrix* (Fig. 1-9). Show that

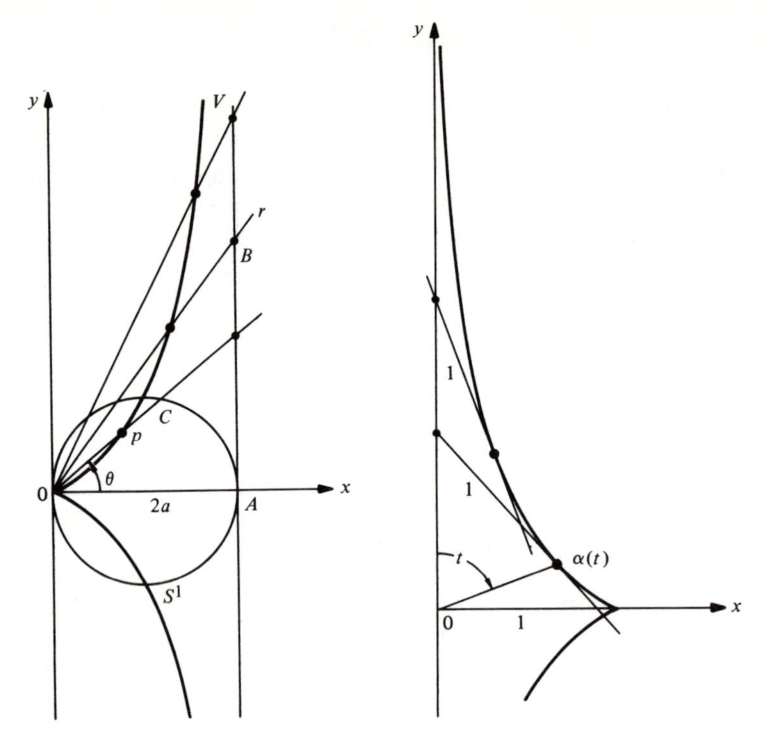

Figure 1-8. The cissoid of Diocles. **Figure 1-9.** The tractrix.

a. α is a differentiable parametrized curve, regular except at $t = \pi/2$.

b. The length of the segment of the tangent of the tractrix between the point of tangency and the y axis is constantly equal to 1.

5. Let $\alpha: (-1, +\infty) \longrightarrow R^2$ be given by

$$\alpha(t) = \left(\frac{3at}{1+t^3}, \frac{3at^2}{1+t^3}\right).$$

Prove that:

a. For $t = 0$, α is tangent to the x axis.

b. As $t \longrightarrow +\infty$, $\alpha(t) \longrightarrow (0, 0)$ and $\alpha'(t) \longrightarrow (0, 0)$.

c. Take the curve with the opposite orientation. Now, as $t \longrightarrow -1$, the curve and its tangent approach the line $x + y + a = 0$.

The figure obtained by completing the trace of α in such a way that it becomes symmetric relative to the line $y = x$ is called the *folium of Descartes* (see Fig. 1-10).

6. Let $\alpha(t) = (ae^{bt} \cos t, ae^{bt} \sin t)$, $t \in R$, a and b constants, $a > 0$, $b < 0$, be a parametrized curve.

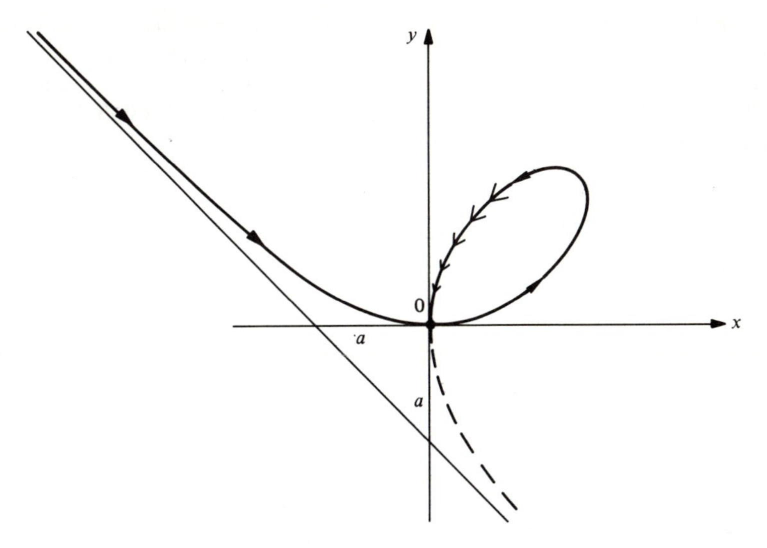

Figure 1-10. Folium of Descartes.

a. Show that as $t \longrightarrow +\infty$, $\alpha(t)$ approaches the origin 0, spiraling around it (because of this, the trace of α is called the *logarithmic spiral*; see Fig. 1-11).

b. Show that $\alpha'(t) \longrightarrow (0, 0)$ as $t \longrightarrow +\infty$ and that

$$\lim_{t \to +\infty} \int_{t_0}^{t} |\alpha'(t)|\, dt$$

is finite; that is, α has finite arc length in $[t_0, \infty)$.

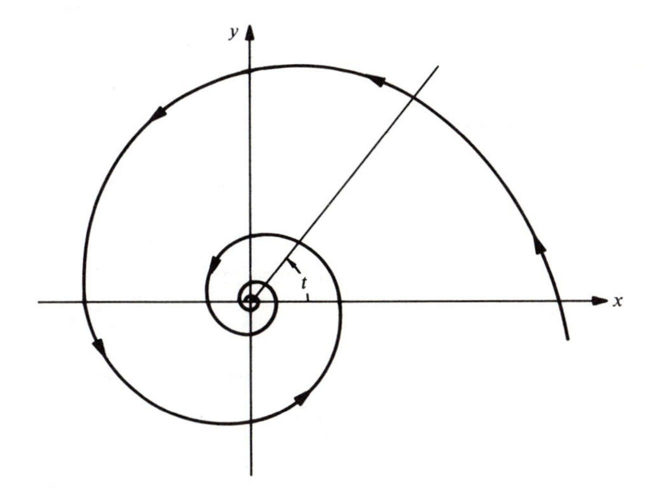

Figure 1-11. Logarithmic spiral.

7. A map $\alpha: I \longrightarrow R^3$ is *called a curve of class* C^k if each of the coordinate functions in the expression $\alpha(t) = (x(t), y(t), z(t))$ has continuous derivatives up to order k. If α is merely continuous, we say that α is of class C^0. A curve α is called *simple* if the map α is one-to-one. Thus, the curve in Example 3 of Sec. 1-2 is not simple.

Let $\alpha: I \longrightarrow R^3$ be a simple curve of class C^0. We say that α has a *weak tangent* at $t = t_0 \in I$ if the line determined by $\alpha(t_0 + h)$ and $\alpha(t_0)$ has a limit position when $h \longrightarrow 0$. We say that α has a *strong tangent* at $t = t_0$ if the line determined by $\alpha(t_0 + h)$ and $\alpha(t_0 + k)$ has a limit position when $h, k \longrightarrow 0$. Show that

a. $\alpha(t) = (t^3, t^2)$, $t \in R$, has a weak tangent but not a strong tangent at $t = 0$.

***b.** If $\alpha: I \longrightarrow R^3$ is of class C^1 and regular at $t = t_0$, then it has a strong tangent at $t = t_0$.

c. The curve given by

$$\alpha(t) = \begin{cases} (t^2, t^2), & t \geq 0, \\ (t^2, -t^2), & t \leq 0, \end{cases}$$

is of class C^1 but not of class C^2. Draw a sketch of the curve and its tangent vectors.

***8.** Let $\alpha: I \longrightarrow R^3$ be a differentiable curve and let $[a, b] \subset I$ be a closed interval. For every *partition*

$$a = t_0 < t_1 < \cdots < t_n = b$$

of $[a, b]$, consider the sum $\sum_{i=1}^{n} |\alpha(t_i) - \alpha(t_{i-1})| = l(\alpha, P)$, where P stands for the given partition. The norm $|P|$ of a partition P is defined as

$$|P| = \max(t_i - t_{i-1}),\ i = 1, \ldots, n.$$

Geometrically, $l(\alpha, P)$ is the length of a polygon inscribed in $\alpha([a, b])$ with vertices in $\alpha(t_i)$ (see Fig. 1-12). The point of the exercise is to show that the arc length of $\alpha([a, b])$ is, in some sense, a limit of lengths of inscribed polygons.

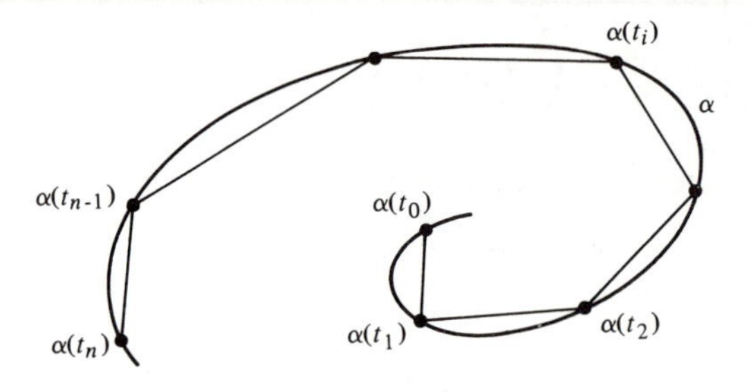

Figure 1-12

Prove that given $\epsilon > 0$ there exists $\delta > 0$ such that if $|P| < \delta$ then

$$\left| \int_a^b |\alpha'(t)|\, dt - l(\alpha, P) \right| < \epsilon.$$

9. a. Let $\alpha: I \longrightarrow R^3$ be a curve of class C^0 (cf. Exercise 7). Use the approximation by polygons described in Exercise 8 to give a reasonable definition of arc length of α.

b. (*A Nonrectifiable Curve.*) The following example shows that, with any reasonable definition, the arc length of a C^0 curve in a closed interval may be unbounded. Let α: $[0, 1] \longrightarrow R^2$ be given as $\alpha(t) = (t, t \sin(\pi/t))$ if $t \neq 0$, and $\alpha(0) = (0, 0)$. Show, geometrically, that the arc length of the portion of the curve corresponding to $1/(n+1) \leq t \leq 1/n$ is at least $2/(n + \frac{1}{2})$. Use this to show that the length of the curve in the interval $1/N \leq t \leq 1$ is greater than $2 \sum_{n=1}^{N} 1/(n+1)$, and thus it tends to infinity as $N \longrightarrow \infty$.

10. (*Straight Lines as Shortest.*) Let $\alpha: I \longrightarrow R^3$ be a parametrized curve. Let $[a, b] \subset I$ and set $\alpha(a) = p$, $\alpha(b) = q$.

a. Show that, for any constant vector v, $|v| = 1$,

$$(q - p) \cdot v = \int_a^b \alpha'(t) \cdot v\, dt \leq \int_a^b |\alpha'(t)|\, dt.$$

b. Set

$$v = \frac{q - p}{|q - p|}$$

and show that

$$|\alpha(b) - \alpha(a)| \leq \int_a^b |\alpha'(t)|\, dt;$$

that is, the curve of shortest length from $\alpha(a)$ to $\alpha(b)$ is the straight line joining these points.

1-4. The Vector Product in R^3

In this section, we shall present some properties of the vector product in R^3. They will be found useful in our later study of curves and surfaces.

It is convenient to begin by reviewing the notion of orientation of a vector space. Two ordered bases $e = \{e_i\}$ and $f = \{f_i\}$, $i = 1, \ldots, n$, of an n-dimensional vector space V have the *same orientation* if the matrix of change of basis has positive determinant. We denote this relation by $e \sim f$. From elementary properties of determinants, it follows that $e \sim f$ is an equivalence relation; i.e., it satisfies

1. $e \sim e$.
2. If $e \sim f$, then $f \sim e$.
3. If $e \sim f$, $f \sim g$, then $e \sim g$.

The set of all ordered bases of V is thus decomposed into equivalence classes (the elements of a given class are related by $\sim$) which by property 3 are disjoint. Since the determinant of a change of basis is either positive or negative, there are only two such classes.

Each of the equivalence classes determined by the above relation is called an *orientation* of V. Therefore, V has two orientations, and if we fix one of them arbitrarily, the other one is called the opposite orientation.

In the case $V = R^3$, there exists a natural ordered basis $e_1 = (1, 0, 0)$, $e_2 = (0, 1, 0)$, $e_3 = (0, 0, 1)$, and we shall call the orientation corresponding to this basis the *positive orientation* of R^3, the other one being the *negative orientation* (of course, this applies equally well to any R^n). We also say that a given ordered basis of R^3 is *positive* (or *negative*) if it belongs to the positive (or negative) orientation of R^3. Thus, the ordered basis e_1, e_3, e_2 is a negative basis, since the matrix which changes this basis into e_1, e_2, e_3 has determinant equal to -1.

We now come to the vector product. Let $u, v \in R^3$. The *vector product* of u and v (in that order) is the unique vector $u \wedge v \in R^3$ characterized by

$$(u \wedge v) \cdot w = \det(u, v, w) \qquad \text{for all } w \in R^3.$$

Here $\det(u, v, w)$ means that if we express u, v, and w in the natural basis $\{e_i\}$,

$$u = \sum u_i e_i, \qquad v = \sum v_i e_i,$$
$$w = \sum w_i e_i, \qquad i = 1, 2, 3,$$

then

$$\det(u, v, w) = \begin{vmatrix} u_1 & u_2 & u_3 \\ v_1 & v_2 & v_3 \\ w_1 & w_2 & w_3 \end{vmatrix},$$

where $|a_{ij}|$ denotes the determinant of the matrix (a_{ij}). It is immediate from the definition that

$$u \wedge v = \begin{vmatrix} u_2 & u_3 \\ v_2 & v_3 \end{vmatrix} e_1 - \begin{vmatrix} u_1 & u_3 \\ v_1 & v_3 \end{vmatrix} e_2 + \begin{vmatrix} u_1 & u_2 \\ v_1 & v_2 \end{vmatrix} e_3. \tag{1}$$

Remark. It is also very frequent to write $u \wedge v$ as $u \times v$ and refer to it as the *cross product*.

The following properties can easily be checked (actually they just express the usual properties of determinants):

1. $u \wedge v = -v \wedge u$ (anticommutativity).
2. $u \wedge v$ depends linearly on u and v; i.e., for any real numbers a, b, we have

$$(au + bw) \wedge v = au \wedge v + bw \wedge v.$$

3. $u \wedge v = 0$ if and only if u and v are linearly dependent.
4. $(u \wedge v) \cdot u = 0$, $(u \wedge v) \cdot v = 0$.

It follows from property 4 that the vector product $u \wedge v \neq 0$ is normal to a plane generated by u and v. To give a geometric interpretation of its norm and its direction, we proceed as follows.

First, we observe that $(u \wedge v) \cdot (u \wedge v) = |u \wedge v|^2 > 0$. This means that the determinant of the vectors $u, v, u \wedge v$ is positive; that is, $\{u, v, u \wedge v\}$ is a positive basis.

Next, we prove the relation

$$(u \wedge v) \cdot (x \wedge y) = \begin{vmatrix} u \cdot x & v \cdot x \\ u \cdot y & v \cdot y \end{vmatrix},$$

where u, v, x, y are arbitrary vectors. This can easily be done by observing that both sides are linear in u, v, x, y. Thus, it suffices to check that

$$(e_i \wedge e_j) \cdot (e_k \wedge e_l) = \begin{vmatrix} e_i \cdot e_k & e_j \cdot e_k \\ e_i \cdot e_l & e_j \cdot e_l \end{vmatrix}$$

for all $i, j, k, l = 1, 2, 3$. This is a straightforward verification.

It follows that

$$|u \wedge v|^2 = \begin{vmatrix} u \cdot u & u \cdot v \\ u \cdot v & v \cdot v \end{vmatrix} = |u|^2 |v|^2 (1 - \cos^2 \theta) = A^2,$$

where θ is the angle of u and v, and A is the area of a parallelogram generated by u and v.

In short, the vector product of u and v is a vector $u \wedge v$ perpendicular to a plane generated by u and v, with a norm equal to the area of a parallelogram generated by u and v and a direction such that $\{u, v, u \wedge v\}$ is a positive basis (Fig. 1-13).

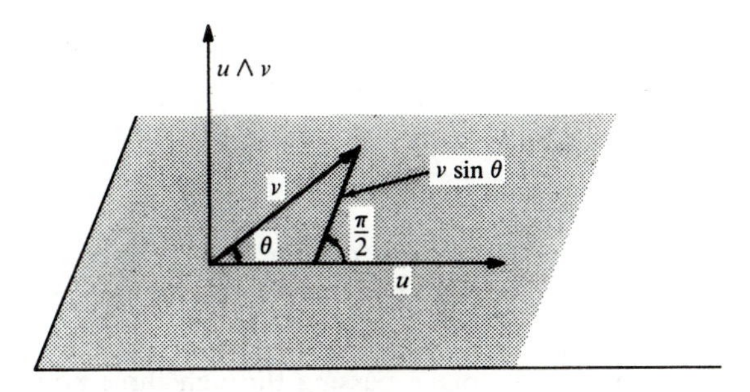

Figure 1-13

The vector product is not associative. In fact, we have the following identity:

$$(u \wedge v) \wedge w = (u \cdot w)v - (v \cdot w)u, \tag{2}$$

which can be proved as follows. First we observe that both sides are linear in u, v, w; thus, the identity will be true if it holds for all basis vectors. This last verification is, however, straightforward; for instance,

$$(e_1 \wedge e_2) \wedge e_1 = e_2 = (e_1 \cdot e_1)e_2 - (e_2 \cdot e_1)e_1.$$

Finally, let $u(t) = (u_1(t), u_2(t), u_3(t))$ and $v(t) = (v_1(t), v_2(t), v_3(t))$ be differentiable maps from the interval (a, b) to R^3, $t \in (a, b)$. It follows immediately from Eq. (1) that $u(t) \wedge v(t)$ is also differentiable and that

$$\frac{d}{dt}(u(t) \wedge v(t)) = \frac{du}{dt} \wedge v(t) + u(t) \wedge \frac{dv}{dt}.$$

Vector products appear naturally in many geometrical constructions. Actually, most of the geometry of planes and lines in R^3 can be neatly expressed in terms of vector products and determinants. We shall review some of this material in the following exercises.

EXERCISES

1. Check whether the following bases are positive:

a. The basis $\{(1, 3), (4, 2)\}$ in R^2.

b. The basis $\{(1, 3, 5), (2, 3, 7), (4, 8, 3)\}$ in R^3.

***2.** A plane P contained in R^3 is given by the equation $ax + by + cz + d = 0$. Show that the vector $v = (a, b, c)$ is perpendicular to the plane and that $|d|/\sqrt{a^2 + b^2 + c^2}$ measures the distance from the plane to the origin $(0, 0, 0)$.

***3.** Determine the angle of intersection of the two planes $5x + 3y + 2z - 4 = 0$ and $3x + 4y - 7z = 0$.

***4.** Given two planes $a_i x + b_i y + c_i z + d_i = 0$, $i = 1, 2$, prove that a necessary and sufficient condition for them to be parallel is

$$\frac{a_1}{a_2} = \frac{b_1}{b_2} = \frac{c_1}{c_2},$$

where the convention is made that if a denominator is zero, the corresponding numerator is also zero (we say that two planes are parallel if they either coincide or do not intersect).

5. Show that the equation of a plane passing through three noncolinear points $p_1 = (x_1, y_1, z_1)$, $p_2 = (x_2, y_2, z_2)$, $p_3 = (x_3, y_3, z_3)$ is given by

$$(p - p_1) \wedge (p - p_2) \cdot (p - p_3) = 0,$$

where $p = (x, y, z)$ is an arbitrary point of the plane and $p - p_1$, for instance, means the vector $(x - x_1, y - y_1, z - z_1)$.

***6.** Given two nonparallel planes $a_i x + b_i y + c_i z + d_i = 0$, $i = 1, 2$, show that their line of intersection may be parametrized as

$$x - x_0 = u_1 t, \qquad y - y_0 = u_2 t, \qquad z - z_0 = u_3 t,$$

where (x_0, y_0, z_0) belongs to the intersection and $u = (u_1, u_2, u_3)$ is the vector product $u = v_1 \wedge v_2$, $v_i = (a_i, b_i, c_i)$, $i = 1, 2$.

***7.** Prove that a necessary and sufficient condition for the plane

$$ax + by + cz + d = 0$$

and the line $x - x_0 = u_1 t$, $y - y_0 = u_2 t$, $z - z_0 = u_3 t$ to be parallel is

$$au_1 + bu_2 + cu_3 = 0.$$

***8.** Prove that the distance ρ between the nonparallel lines

$$x - x_0 = u_1 t, \qquad y - y_0 = u_2 t, \qquad z - z_0 = u_3 t,$$
$$x - x_1 = v_1 t, \qquad y - y_1 = v_2 t, \qquad z - z_1 = v_3 t$$

is given by

$$\rho = \frac{|(u \wedge v) \cdot r|}{|u \wedge v|},$$

where $u = (u_1, u_2, u_3)$, $v = (v_1, v_2, v_3)$, $r = (x_0 - x_1, y_0 - y_1, z_0 - z_1)$.

9. Determine the angle of intersection of the plane $3x + 4y + 7z + 8 = 0$ and the line $x - 2 = 3t$, $y - 3 = 5t$, $z - 5 = 9t$.

10. The natural orientation of R^2 makes it possible to associate a sign to the area A of a parallelogram generated by two linearly independent vectors $u, v \in R^2$. To do this, let $\{e_i\}$, $i = 1, 2$, be the natural ordered basis of R^2, and write $u = u_1 e_1 + u_2 e_2$, $v = v_1 e_1 + v_2 e_2$. Observe the matrix relation

$$\begin{pmatrix} u \cdot u & u \cdot v \\ v \cdot u & v \cdot v \end{pmatrix} = \begin{pmatrix} u_1 & u_2 \\ v_1 & v_2 \end{pmatrix} \begin{pmatrix} u_1 & v_1 \\ u_2 & v_2 \end{pmatrix}$$

and conclude that

$$A^2 = \begin{vmatrix} u_1 & u_2 \\ v_1 & v_2 \end{vmatrix}^2.$$

Since the last determinant has the same sign as the basis $\{u, v\}$, we can say that A is positive or negative according to whether the orientation of $\{u, v\}$ is positive or negative. This is called the *oriented area* in R^2.

11. a. Show that the volume V of a parallelepiped generated by three linearly independent vectors $u, v, w \in R^3$ is given by $V = |(u \wedge v) \cdot w|$, and introduce an *oriented volume* in R^3.

b. Prove that

$$V^2 = \begin{vmatrix} u \cdot u & u \cdot v & u \cdot w \\ v \cdot u & v \cdot v & v \cdot w \\ w \cdot u & w \cdot v & w \cdot w \end{vmatrix}.$$

12. Given the vectors $v \neq 0$ and w, show that there exists a vector u such that $u \wedge v = w$ if and only if v is perpendicular to w. Is this vector u uniquely determined? If not, what is the most general solution?

13. Let $u(t) = (u_1(t), u_2(t), u_3(t))$ and $v(t) = (v_1(t), v_2(t), v_3(t))$ be differentiable maps from the interval (a, b) into R^3. If the derivatives $u'(t)$ and $v'(t)$ satisfy the conditions

$$u'(t) = au(t) + bv(t), \qquad v'(t) = cu(t) - av(t),$$

where a, b, and c are constants, show that $u(t) \wedge v(t)$ is a constant vector.

14. Find all unit vectors which are perpendicular to the vector $(2, 2, 1)$ and parallel to the plane determined by the points $(0, 0, 0)$, $(1, -2, 1)$, $(-1, 1, 1)$.

1-5. The Local Theory of Curves Parametrized by Arc Length

This section contains the main results of curves which will be used in the later parts of the book.

Let $\alpha: I = (a, b) \to R^3$ be a curve parametrized by arc length s. Since the tangent vector $\alpha'(s)$ has unit length, the norm $|\alpha''(s)|$ of the second derivative measures the rate of change of the angle which neighboring tangents make with the tangent at s. $|\alpha''(s)|$ gives, therefore, a measure of how rapidly the curve pulls away from the tangent line at s, in a neighborhood of s (see Fig. 1-14). This suggests the following definition.

DEFINITION. *Let* $\alpha: I \to R^3$ *be a curve parametrized by arc length* $s \in I$. *The number* $|\alpha''(s)| = k(s)$ *is called the* curvature *of* α *at* s.

If α is a straight line, $\alpha(s) = us + v$, where u and v are constant vectors ($|u| = 1$), then $k \equiv 0$. Conversely, if $k = |\alpha''(s)| \equiv 0$, then by integration $\alpha(s) = us + v$, and the curve is a straight line.

Notice that by a change of orientation, the tangent vector changes its direction; that is, if $\beta(-s) = \alpha(s)$, then

$$\frac{d\beta}{d(-s)}(-s) = -\frac{d\alpha}{ds}(s).$$

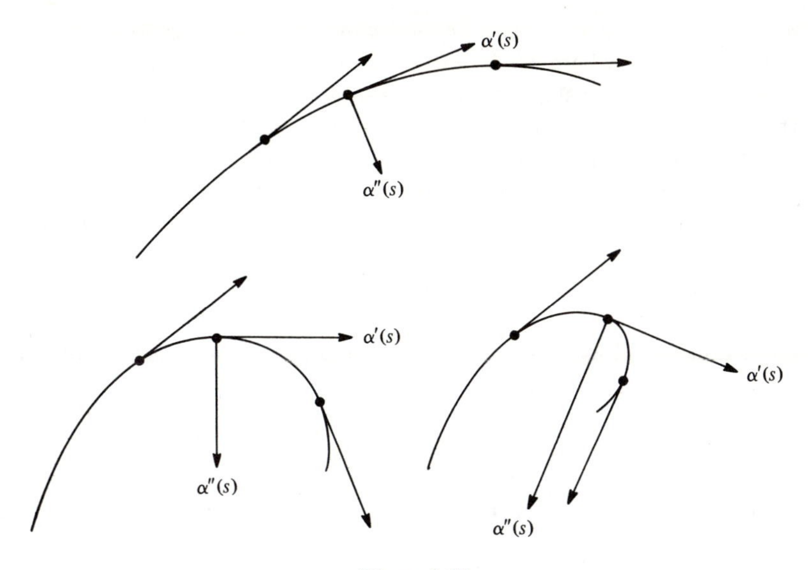

Figure 1-14

Therefore, $\alpha''(s)$ and the curvature remain invariant under a change of orientation.

At points where $k(s) \neq 0$, a unit vector $n(s)$ in the direction $\alpha''(s)$ is well defined by the equation $\alpha''(s) = k(s)n(s)$. Moreover, $\alpha''(s)$ is normal to $\alpha'(s)$, because by differentiating $\alpha'(s) \cdot \alpha'(s) = 1$ we obtain $\alpha''(s) \cdot \alpha'(s) = 0$. Thus, $n(s)$ is normal to $\alpha'(s)$ and is called the *normal vector* at s. The plane determined by the unit tangent and normal vectors, $\alpha'(s)$ and $n(s)$, is called the *osculating plane* at s. (See Fig. 1-15.)

At points where $k(s) = 0$, the normal vector (and therefore the osculating plane) is not defined (cf. Exercise 10). To proceed with the local analysis of curves, we need, in an essential way, the osculating plane. It is therefore

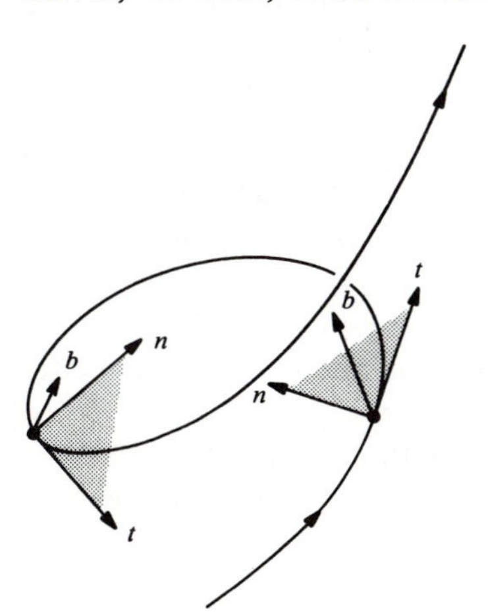

Figure 1-15

convenient to say that $s \in I$ is a *singular point of order* 1 if $\alpha''(s) = 0$ (in this context, the points where $\alpha'(s) = 0$ are called singular points of order 0).

In what follows, we shall restrict ourselves to curves parametrized by arc length without singular points of order 1. We shall denote by $t(s) = \alpha'(s)$ the unit tangent vector of α at s. Thus, $t'(s) = k(s)n(s)$.

The unit vector $b(s) = t(s) \wedge n(s)$ is normal to the osculating plane and will be called the *binormal vector* at s. Since $b(s)$ is a unit vector, the length $|b'(s)|$ measures the rate of change of the neighboring osculating planes with the osculating plane at s; that is, $b'(s)$ measures how rapidly the curve pulls away from the osculating plane at s, in a neighborhood of s (see Fig. 1-15).

To compute $b'(s)$ we observe that, on the one hand, $b'(s)$ is normal to $b(s)$ and that, on the other hand,

$$b'(s) = t'(s) \wedge n(s) + t(s) \wedge n'(s) = t(s) \wedge n'(s);$$

that is, $b'(s)$ is normal to $t(s)$. It follows that $b'(s)$ is parallel to $n(s)$, and we may write

$$b'(s) = \tau(s)n(s)$$

for some function $\tau(s)$. (*Warning*: Many authors write $-\tau(s)$ instead of our $\tau(s)$.)

DEFINITION. *Let* $\alpha: I \to R^3$ *be a curve parametrized by arc length* s *such that* $\alpha''(s) \neq 0$, $s \in I$. *The number* $\tau(s)$ *defined by* $b'(s) = \tau(s)n(s)$ *is called the* torsion *of* α *at* s.

If α is a plane curve (that is, $\alpha(I)$ is contained in a plane), then the plane of the curve agrees with the osculating plane; hence, $\tau \equiv 0$. Conversely, if $\tau \equiv 0$ (and $k \neq 0$), we have that $b(s) = b_0 =$ constant, and therefore

$$(\alpha(s) \cdot b_0)' = \alpha'(s) \cdot b_0 = 0.$$

It follows that $\alpha(s) \cdot b_0 =$ constant; hence, $\alpha(s)$ is contained in a plane normal to b_0. The condition that $k \neq 0$ everywhere is essential here. In Exercise 10 we shall give an example where τ can be defined to be identically zero and yet the curve is not a plane curve.

In contrast to the curvature, the torsion may be either positive or negative. The sign of the torsion has a geometric interpretation, to be given later (Sec. 1-6).

Notice that by changing orientation the binormal vector changes sign, since $b = t \wedge n$. It follows that $b'(s)$, and, therefore, the torsion, remains invariant under a change of orientation.

Let us summarize our position. To each value of the parameter s, we have associated three orthogonal unit vectors $t(s)$, $n(s)$, $b(s)$. The trihedron thus formed is referred to as the *Frenet trihedron* at s. The derivatives $t'(s) = kn$, $b'(s) = \tau n$ of the vectors $t(s)$ and $b(s)$, when expressed in the basis $\{t, n, b\}$, yield geometrical entities (curvature k and torsion τ) which give us information about the behavior of α in a neighborhood of s.

The search for other local geometrical entities would lead us to compute $n'(s)$. However, since $n = b \wedge t$, we have

$$n'(s) = b'(s) \wedge t(s) + b(s) \wedge t'(s) = -\tau b - kt,$$

and we obtain again the curvature and the torsion.

For later use, we shall call the equations

$$\begin{aligned} t' &= kn, \\ n' &= -kt - \tau b, \\ b' &= \tau n \end{aligned}$$

the *Frenet formulas* (we have ommited the s, for convenience). In this context, the following terminology is usual. The tb plane is called the *rectifying plane*, and the nb plane the *normal plane*. The lines which contain $n(s)$ and $b(s)$ and pass through $\alpha(s)$ are called the *principal normal* and the *binormal*, respectively. The inverse $R = 1/k$ of the curvature is called the *radius of curvature* at s. Of course, a circle of radius r has radius of curvature equal to r, as one can easily verify.

Physically, we can think of a curve in R^3 as being obtained from a straight line by bending (curvature) and twisting (torsion). After reflecting on this construction, we are led to conjecture the following statement, which, roughly speaking, shows that k and τ describe completely the local behavior of the curve.

FUNDAMENTAL THEOREM OF THE LOCAL THEORY OF CURVES. *Given differentiable functions* $k(s) > 0$ *and* $\tau(s)$, $s \in I$, *there exists a regular parametrized curve* $\alpha: I \longrightarrow R^3$ *such that* s *is the arc length,* $k(s)$ *is the curvature, and* $\tau(s)$ *is the torsion of* α. *Moreover, any other curve* $\bar{\alpha}$, *satisfying the same conditions, differs from* α *by a rigid motion; that is, there exists an orthogonal linear map* ρ *of* R^3, *with positive determinant, and a vector* c *such that* $\bar{\alpha} = \rho \circ \alpha + c$.

The above statement is true. A complete proof involves the theorem of existence and uniqueness of solutions of ordinary differential equations and will be given in the appendix to Chap. 4. A proof of the uniqueness, up to rigid motions, of curves having the same s, $k(s)$, and $\tau(s)$ is, however, simple and can be given here.

Proof of the Uniqueness Part of the Fundamental Theorem. We first remark that arc length, curvature, and torsion are invariant under rigid motions; that means, for instance, that if $M: R^3 \to R^3$ is a rigid motion and $\alpha = \alpha(t)$ is a parametrized curve, then

$$\int_a^b \left|\frac{d\alpha}{dt}\right| dt = \int_a^b \left|\frac{d(M \circ \alpha)}{dt}\right| dt.$$

That is plausible, since these concepts are defined by using inner or vector products of certain derivatives (the derivatives are invariant under translations, and the inner and vector products are expressed by means of lengths and angles of vectors, and thus also invariant under rigid motions). A careful checking can be left as an exercise (see Exercise 6).

Now, assume that two curves $\alpha = \alpha(s)$ and $\bar{\alpha} = \bar{\alpha}(s)$ satisfy the conditions $k(s) = \bar{k}(s)$ and $\tau(s) = \bar{\tau}(s)$, $s \in I$. Let t_0, n_0, b_0 and $\bar{t}_0, \bar{n}_0, \bar{b}_0$ be the Frenet trihedrons at $s = s_0 \in I$ of α and $\bar{\alpha}$, respectively. Clearly, there is a rigid motion which takes $\bar{\alpha}(s_0)$ into $\alpha(s_0)$ and $\bar{t}_0, \bar{n}_0, \bar{b}_0$ into t_0, n_0, b_0. Thus, after performing this rigid motion on $\bar{\alpha}$, we have that $\bar{\alpha}(s_0) = \alpha(s_0)$ and that the Frenet trihedrons $t(s), n(s), b(s)$ and $\bar{t}(s), \bar{n}(s), \bar{b}(s)$ of α and $\bar{\alpha}$, respectively, satisfy the Frenet equations:

$$\begin{aligned} \frac{dt}{ds} &= kn & \frac{d\bar{t}}{ds} &= k\bar{n} \\ \frac{dn}{ds} &= -kt - \tau b & \frac{d\bar{n}}{ds} &= -k\bar{t} - \tau\bar{n} \\ \frac{db}{ds} &= \tau n & \frac{d\bar{b}}{ds} &= \tau\bar{n}, \end{aligned}$$

with $t(s_0) = \bar{t}(s_0)$, $n(s_0) = \bar{n}(s_0)$, $b(s_0) = \bar{b}(s_0)$.

We now observe, by using the Frenet equations, that

$$\begin{aligned} \frac{1}{2}\frac{d}{ds}\{|t - \bar{t}|^2 + |n - \bar{n}|^2 + |b - \bar{b}|^2\} \\ &= \langle t - \bar{t}, t' - \bar{t}'\rangle + \langle b - \bar{b}, b' - \bar{b}'\rangle + \langle n - \bar{n}, n' - \bar{n}'\rangle \\ &= k\langle t - \bar{t}, n - \bar{n}\rangle + \tau\langle b - \bar{b}, n - \bar{n}\rangle - k\langle n - \bar{n}, t - \bar{t}\rangle \\ &\quad - \tau\langle n - \bar{n}, b - \bar{b}\rangle \\ &= 0 \end{aligned}$$

for all $s \in I$. Thus, the above expression is constant, and, since it is zero for

$s = s_0$, it is identically zero. It follows that $t(s) = \bar{t}(s)$, $n(s) = \bar{n}(s)$, $b(s) = \bar{b}(s)$ for all $s \in I$. Since

$$\frac{d\alpha}{ds} = t = \bar{t} = \frac{d\bar{\alpha}}{ds},$$

we obtain $(d/ds)(\alpha - \bar{\alpha}) = 0$. Thus, $\alpha(s) = \bar{\alpha}(s) + a$, where a is a constant vector. Since $\alpha(s_0) = \bar{\alpha}(s_0)$, we have $a = 0$; hence, $\alpha(s) = \bar{\alpha}(s)$ for all $s \in I$.

Q.E.D.

Remark 1. In the particular case of a plane curve $\alpha: I \longrightarrow R^2$, it is possible to give the curvature k a sign. For that, let $\{e_1, e_2\}$ be the natural basis (see Sec. 1-4) of R^2 and define the normal vector $n(s)$, $s \in I$, by requiring the basis $\{t(s), n(s)\}$ to have the same orientation as the basis $\{e_1, e_2\}$. The curvature k is then *defined* by

$$\frac{dt}{ds} = kn$$

and might be either positive or negative. It is clear that $|k|$ agrees with the previous definition and that k changes sign when we change either the orientation of α or the orientation of R^2 (Fig. 1-16).

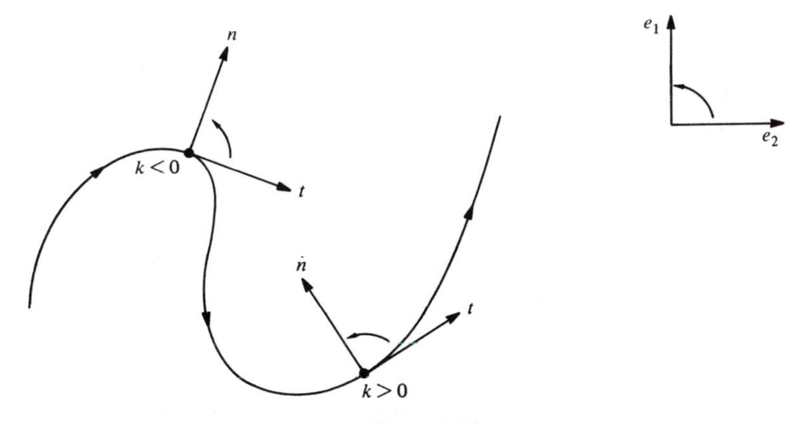

Figure 1-16

It should also be remarked that, in the case of plane curves ($\tau \equiv 0$), the proof of the fundamental theorem, refered to above, is actually very simple (see Exercise 9).

Remark 2. Given a regular parametrized curve $\alpha: I \longrightarrow R^3$ (not necessarily parametrized by arc length), it is possible to obtain a curve $\beta: J \longrightarrow R^3$ parametrized by arc length which has the same trace as α. In fact, let

$$s = s(t) = \int_{t_0}^{t} |\alpha'(t)|\, dt, \qquad t, t_0 \in I.$$

Since $ds/dt = |\alpha'(t)| \neq 0$, the function $s = s(t)$ has a differentiable inverse $t = t(s)$, $s \in s(I) = J$, where, by an abuse of notation, t also denotes the inverse function s^{-1} of s. Now set $\beta = \alpha \circ t \colon J \longrightarrow R^3$. Clearly, $\beta(J) = \alpha(I)$ and $|\beta'(s)| = |\alpha'(t) \cdot (dt/ds)| = 1$. This shows that β has the same trace as α and is parametrized by arc length. It is usual to say that β is a *reparametrization of $\alpha(I)$ by arc length.*

This fact allows us to extend all local concepts previously defined to regular curves with an arbitrary parameter. Thus, we say that the curvature $k(t)$ of $\alpha \colon I \longrightarrow R^3$ at $t \in I$ is the curvature of a reparametrization $\beta \colon J \longrightarrow R^3$ of $\alpha(I)$ by arc length at the corresponding point $s = s(t)$. This is clearly independent of the choice of β and shows that the restriction, made at the end of Sec. 1-3, of considering only curves parametrized by arc length is not essential.

In applications, it is often convenient to have explicit formulas for the geometrical entities in terms of an arbitrary parameter; we shall present some of them in Exercise 12.

EXERCISES

Unless explicity stated, $\alpha \colon I \longrightarrow R^3$ *is a curve parametrized by arc length* s, *with curvature* $k(s) \neq 0$, *for all* $s \in I$.

1. Given the parametrized curve (helix)

$$\alpha(s) = \left(a \cos \frac{s}{c}, a \sin \frac{s}{c}, b \frac{s}{c}\right), \qquad s \in R,$$

where $c^2 = a^2 + b^2$,

a. Show that the parameter s is the arc length.

b. Determine the curvature and the torsion of α.

c. Determine the osculating plane of α.

d. Show that the lines containing $n(s)$ and passing through $\alpha(s)$ meet the z axis under a constant angle equal to $\pi/2$.

e. Show that the tangent lines to α make a constant angle with the z axis.

***2.** Show that the torsion τ of α is given by

$$\tau(s) = -\frac{\alpha'(s) \wedge \alpha''(s) \cdot \alpha'''(s)}{|k(s)|^2}.$$

3. Assume that $\alpha(I) \subset R^2$ (i.e., α is a plane curve) and give k a sign as in the text. Transport the vectors $t(s)$ parallel to themselves in such a way that the origins of

$t(s)$ agree with the origin of R^2; the end points of $t(s)$ then describe a parametrized curve $s \longrightarrow t(s)$ called the *indicatrix of tangents* of α. Let $\theta(s)$ be the angle from e_1 to $t(s)$ in the orientation of R^2. Prove (a) and (b) (notice that we are assuming that $k \neq 0$).

a. The indicatrix of tangents is a regular parametrized curve.

b. $dt/ds = (d\theta/ds)n$, that is, $k = d\theta/ds$.

***4.** Assume that all normals of a parametrized curve pass through a fixed point. Prove that the trace of the curve is contained in a circle.

5. A regular parametrized curve α has the property that all its tangent lines pass through a fixed point.

a. Prove that the trace of α is a (segment of a) straight line.

b. Does the conclusion in part a still hold if α is not regular?

6. A *translation* by a vector v in R^3 is the map $A: R^3 \longrightarrow R^3$ that is given by $A(p) = p + v$, $p \in R^3$. A linear map $\rho: R^3 \longrightarrow R^3$ is an *orthogonal transformation* when $\rho u \cdot \rho v = u \cdot v$ for all vectors $u, v \in R^3$. A *rigid motion* in R^3 is the result of composing a translation with an orthogonal transformation with positive determinant (this last condition is included because we expect rigid motions to preserve orientation).

a. Demonstrate that the norm of a vector and the angle θ between two vectors, $0 \leq \theta \leq \pi$, are invariant under orthogonal transformations with positive determinant.

b. Show that the vector product of two vectors is invariant under orthogonal transformations with positive determinant. Is the assertion still true if we drop the condition on the determinant?

c. Show that the arc length, the curvature, and the torsion of a parametrized curve are (whenever defined) invariant under rigid motions.

***7.** Let $\alpha: I \longrightarrow R^2$ be a regular parametrized plane curve (arbitrary parameter), and define $n = n(t)$ and $k = k(t)$ as in Remark 1. Assume that $k(t) \neq 0$, $t \in I$. In this situation, the curve

$$\beta(t) = \alpha(t) + \frac{1}{k(t)} n(t), \qquad t \in I,$$

is called the *evolute* of α (Fig. 1-17).

a. Show that the tangent at t of the evolute of α is the normal to α at t.

b. Consider the normal lines of α at two neighboring points t_1, t_2, $t_1 \neq t_2$. Let t_1 approach t_2 and show that the intersection points of the normals converge to a point on the trace of the evolute of α.

8. The trace of the parametrized curve (arbitrary parameter)

$$\alpha(t) = (t, \cosh t), \qquad t \in R,$$

is called the *catenary*.

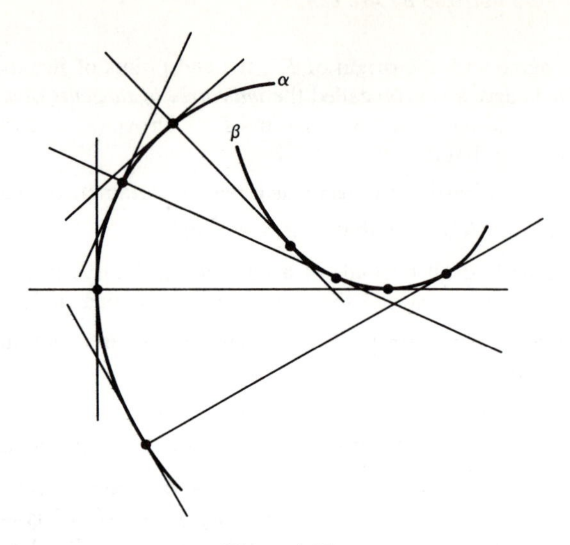

Figure 1-17

a. Show that the signed curvature (cf. Remark 1) of the catenary is

$$k(t) = \frac{1}{\cosh^2 t}.$$

b. Show that the evolute (cf. Exercise 7) of the catenary is

$$\beta(t) = (t - \sinh t \cosh t,\ 2 \cosh t).$$

9. Given a differentiable function $k(s)$, $s \in I$, show that the parametrized plane curve having $k(s) = k$ as curvature is given by

$$\alpha(s) = \left(\int \cos \theta(s)\, ds + a,\ \int \sin \theta(s)\, ds + b\right),$$

where

$$\theta(s) = \int k(s)\, ds + \varphi,$$

and that the curve is determined up to a translation of the vector (a, b) and a rotation of the angle φ.

10. Consider the map

$$\alpha(t) = \begin{cases} (t, 0, e^{-1/t^2}), & t > 0 \\ (t, e^{-1/t^2}, 0), & t < 0 \\ (0, 0, 0), & t = 0 \end{cases}$$

a. Prove that α is a differentiable curve.

b. Prove that α is regular for all t and that the curvature $k(t) \neq 0$, for $t \neq 0$, $t \neq \pm\sqrt{2/3}$, and $k(0) = 0$.

c. Show that the limit of the osculating planes as $t \to 0$, $t > 0$, is the plane $y = 0$ but that the limit of the osculating planes as $t \to 0$, $t < 0$, is the plane $z = 0$ (this implies that the normal vector is discontinuous at $t = 0$ and shows why we excluded points where $k = 0$).

d. Show that τ can be defined so that $\tau \equiv 0$, even though α is not a plane curve.

11. One often gives a plane curve in polar coordinates by $\rho = \rho(\theta)$, $a \leq \theta \leq b$.

a. Show that the arc length is

$$\int_a^b \sqrt{\rho^2 + (\rho')^2}\, d\theta,$$

where the prime denotes the derivative relative to θ.

b. Show that the curvature is

$$k(\theta) = \frac{2(\rho')^2 - \rho\rho'' + \rho^2}{\{(\rho')^2 + \rho^2\}^{3/2}}.$$

12. Let $\alpha\colon I \to R^3$ be a regular parametrized curve (not necessarily by arc length) and let $\beta\colon J \to R^3$ be a reparametrization of $\alpha(I)$ by the arc length $s = s(t)$, measured from $t_0 \in I$ (see Remark 2). Let $t = t(s)$ be the inverse function of s and set $d\alpha/dt = \alpha'$, $d^2\alpha/dt^2 = \alpha''$, etc. Prove that

a. $dt/ds = 1/|\alpha'|$, $d^2t/ds^2 = -(\alpha' \cdot \alpha''/|\alpha'|^4)$.

b. The curvature of α at $t \in I$ is

$$k(t) = \frac{|\alpha' \wedge \alpha''|}{|\alpha'|^3}.$$

c. The torsion of α at $t \in I$ is

$$\tau(t) = -\frac{(\alpha' \wedge \alpha'') \cdot \alpha'''}{|\alpha' \wedge \alpha''|^2}.$$

d. If $\alpha\colon I \to R^2$ is a plane curve $\alpha(t) = (x(t), y(t))$, the signed curvature (see Remark 1) of α at t is

$$k(t) = \frac{x'y'' - x''y'}{((x')^2 + (y')^2)^{3/2}}.$$

***13.** Assume that $\tau(s) \neq 0$ and $k'(s) \neq 0$ for all $s \in I$. Show that a necessary and sufficient condition for $\alpha(I)$ to lie on a sphere is that

$$R^2 + (R')^2T^2 = \text{const.},$$

where $R = 1/k$, $T = 1/\tau$, and R' is the derivative of R relative to s.

14. Let $\alpha\colon (a, b) \to R^2$ be a regular parametrized plane curve. Assume that there exists t_0, $a < t_0 < b$, such that the distance $|\alpha(t)|$ from the origin to the trace of

α will be a maximum at t_0. Prove that the curvature k of α at t_0 satisfies $|k(t_0)| \geq 1/|\alpha(t_0)|$.

***15.** Show that the knowledge of the vector function $b = b(s)$ (binormal vector) of a curve α, with nonzero torsion everywhere, determines the curvature $k(s)$ and the absolute value of the torsion $\tau(s)$ of α.

***16.** Show that the knowledge of the vector function $n = n(s)$ (normal vector) of a curve α, with nonzero torsion everywhere, determines the curvature $k(s)$ and the torsion $\tau(s)$ of α.

17. In general, a curve α is called a *helix* if the tangent lines of α make a constant angle with a fixed direction. Assume that $\tau(s) \neq 0$, $s \in I$, and prove that:

***a.** α is a helix if and only if $k/\tau = \text{const.}$

***b.** α is a helix if and only if the lines containing $n(s)$ and passing through $\alpha(s)$ are parallel to a fixed plane.

***c.** α is a helix if and only if the lines containing $b(s)$ and passing through $\alpha(s)$ make a constant angle with a fixed direction.

d. The curve

$$\alpha(s) = \left(\frac{a}{c}\int \sin\theta(s)\,ds, \frac{a}{c}\int \cos\theta(s)\,ds, \frac{b}{c}s\right),$$

where $a^2 = b^2 + c^2$, is a helix, and that $k/\tau = b/a$.

***18.** Let $\alpha\colon I \longrightarrow R^3$ be a parametrized regular curve (not necessarily by arc length) with $k(t) \neq 0$, $\tau(t) \neq 0$, $t \in I$. The curve α is called a *Bertrand curve* if there exists a curve $\bar{\alpha}\colon I \longrightarrow R^3$ such that the normal lines of α and $\bar{\alpha}$ at $t \in I$ are equal. In this case, $\bar{\alpha}$ is called a *Bertrand mate* of α, and we can write

$$\bar{\alpha}(t) = \alpha(t) + rn(t).$$

Prove that

a. r is constant.

b. α is a Bertrand curve if and only if there exists a linear relation

$$Ak(t) + B\tau(t) = 1, \qquad t \in I,$$

where A, B are nonzero constants and k and τ are the curvature and torsion of α, respectively.

c. If α has more than one Bertrand mate, it has infinitely many Bertrand mates. This case occurs if and only if α is a circular helix.

1-6. The Local Canonical Form†

One of the most effective methods of solving problems in geometry consists of finding a coordinate system which is adapted to the problem. In the study of local properties of a curve, in the neighborhood of the point s, we have a natural coordinate system, namely the Frenet trihedron at s. It is therefore convenient to refer the curve to this trihedron.

Let $\alpha: I \longrightarrow R^3$ be a curve parametrized by arc length without singular points of order 1. We shall write the equations of the curve, in a neighborhood of s_0, using the trihedron $t(s_0)$, $n(s_0)$, $b(s_0)$ as a basis for R^3. We may assume, without loss of generality, that $s_0 = 0$, and we shall consider the (finite) Taylor expansion

$$\alpha(s) = \alpha(0) + s\alpha'(0) + \frac{s^2}{2}\alpha''(0) + \frac{s^3}{6}\alpha'''(0) + R,$$

where $\lim_{s\to 0} R/s^3 = 0$. Since $\alpha'(0) = t$, $\alpha''(0) = kn$, and

$$\alpha'''(0) = (kn)' = k'n + kn' = k'n - k^2t - k\tau b,$$

we obtain

$$\alpha(s) - \alpha(0) = \left(s - \frac{k^2s^3}{3!}\right)t + \left(\frac{s^2k}{2} + \frac{s^3k'}{3!}\right)n - \frac{s^3}{3!}k\tau b + R,$$

where all terms are computed at $s = 0$.

Let us now take the system $Oxyz$ in such a way that the origin O agrees with $\alpha(0)$ and that $t = (1, 0, 0)$, $n = (0, 1, 0)$, $b = (0, 0, 1)$. Under these conditions, $\alpha(s) = (x(s), y(s), z(s))$ is given by

$$\begin{aligned} x(s) &= s - \frac{k^2s^3}{6} + R_x, \\ y(s) &= \frac{k}{2}s^2 + \frac{k's^3}{6} + R_y, \\ z(s) &= -\frac{k\tau}{6}s^3 + R_z, \end{aligned} \tag{1}$$

where $R = (R_x, R_y, R_z)$. The representation (1) is called the *local canonical form* of α, in a neighborhood of $s = 0$. In Fig. 1-18 is a rough sketch of the projections of the trace of α, for s small, in the tn, tb, and nb planes.

---

†This section may be omitted on a first reading.

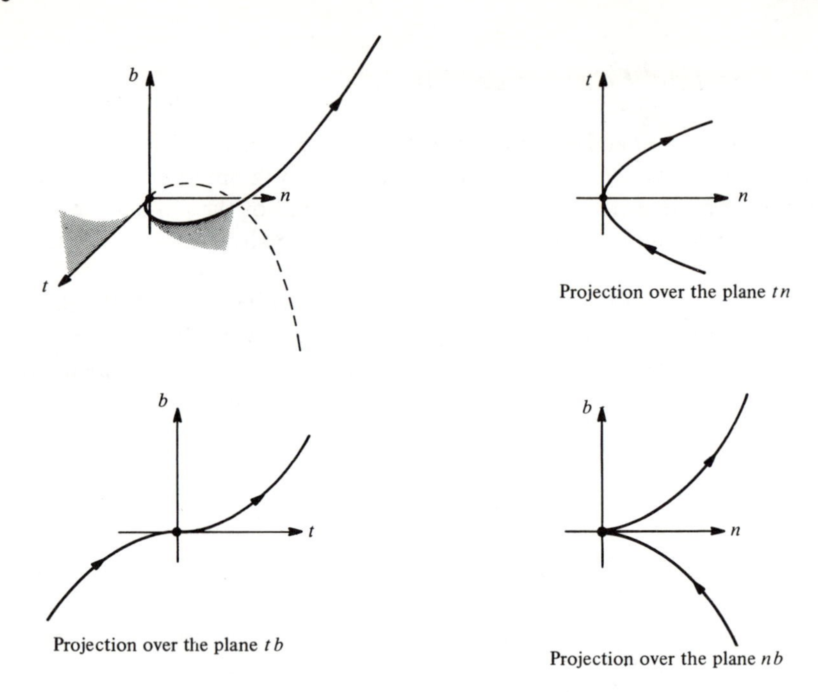

Figure 1-18

Below we shall describe some geometrical applications of the local canonical form. Further applications will be found in the Exercises.

A first application is the following interpretation of the sign of the torsion. From the third equation of (1) it follows that if $\tau < 0$ and s is sufficiently small, then $z(s)$ increases with s. Let us make the convention of calling the "positive side" of the osculating plane that side toward which b is pointing. Then, since $z(0) = 0$, when we describe the curve in the direction of increasing arc length, the curve will cross the osculating plane at $s = 0$, pointing toward the positive side (see Fig. 1-19). If, on the contrary, $\tau > 0$, the curve (described in the direction of increasing arc length) will cross the osculating plane pointing to the side opposite the positive side.

Figure 1-19

The helix of Exercise 1 of Sec. 1-5 has negative torsion. An example of a curve with positive torsion is the helix

$$\alpha(s) = \left(a \cos \frac{s}{c},\ a \sin \frac{s}{c},\ -b\frac{s}{c}\right)$$

obtained from the first one by a reflection in the xz plane (see Fig. 1-19).

Remark. It is also usual to define torsion by $b' = -\tau n$. With such a definition, the torsion of the helix of Exercise 1 becomes positive.

Another consequence of the canonical form is the existence of a neighborhood $J \subset I$ of $s = 0$ such that $\alpha(J)$ is entirely contained in the one side of the rectifying plane toward which the vector n is pointing (see Fig. 1-18). In fact, since $k > 0$, we obtain, for s sufficiently small, $y(s) \geq 0$, and $y(s) = 0$ if and only if $s = 0$. This proves our claim.

As a last application of the canonical form, we mention the following property of the osculating plane. The osculating plane at s is the limit position of the plane determined by the tangent line at s and the point $\alpha(s + h)$ when $h \to 0$. To prove this, let us assume that $s = 0$. Thus, every plane containing the tangent at $s = 0$ is of the form $z = cy$ or $y = 0$. The plane $y = 0$ is the rectifying plane that, as seen above, contains no points near $\alpha(0)$ (except $\alpha(0)$ itself) and that may therefore be discarded from our considerations. The condition for the plane $z = cy$ to pass through $s + h$ is ($s = 0$)

$$c = \frac{z(h)}{y(h)} = \frac{-\frac{k}{6}\tau h^3 + \cdots}{\frac{k}{2}h^2 + \frac{k^2}{6}h^3 + \cdots}.$$

Letting $h \to 0$, we see that $c \to 0$. Therefore, the limit position of the plane $z(s) = c(h)y(s)$ is the plane $z = 0$, that is, the osculating plane, as we wished.

EXERCISES

***1.** Let $\alpha: I \to R^3$ be a curve parametrized by arc length with curvature $k(s) \neq 0$, $s \in I$. Let P be a plane satisfying both of the following conditions:

1. P contains the tangent line at s.
2. Given any neighborhood $J \subset I$ of s, there exist points of $\alpha(J)$ in both sides of P.

Prove that P is the osculating plane of α at s.

2. Let $\alpha: I \to R^3$ be a curve parametrized by arc length, with curvature $k(s) \neq 0$, $s \in I$. Show that *a. The osculating plane at s is the limit position of the plane passing through $\alpha(s)$, $\alpha(s + h_1)$, $\alpha(s + h_2)$ when $h_1, h_2 \longrightarrow 0$.

b. The limit position of the circle passing through $\alpha(s)$, $\alpha(s + h_1)$, $\alpha(s + h_2)$ when $h_1, h_2 \longrightarrow 0$ is a circle in the osculating plane at s, the center of which is on the line that contains $n(s)$ and the radius of which is the radius of curvature $1/k(s)$; this circle is called the *osculating circle* at s.

3. Show that the curvature $k(t) \neq 0$ of a regular parametrized curve $\alpha: I \longrightarrow R^3$ is the curvature at t of the plane curve $\pi \circ \alpha$, where π is the normal projection of α over the osculating plane at t.

1-7. Global Properties of Plane Curves†

In this section we want to describe some results that belong to the global differential geometry of curves. Even in the simple case of plane curves, the subject already offers examples of nontrivial theorems and interesting questions. To develop this material here, we must assume some plausible facts without proofs; we shall try to be careful by stating these facts precisely. Although we want to come back later, in a more systematic way, to global differential geometry (Chap. 5), we believe that this early presentation of the subject is both stimulating and instructive.

This section contains three topics in order of increasing difficulty: (A) the isoperimetric inequality, (B) the four-vertex theorem, and (C) the Cauchy-Crofton formula. The topics are entirely independent, and some or all of them can be omitted on a first reading.

A differentiable function on a closed interval $[a, b]$ is the restriction of a differentiable function defined on an open interval containing $[a, b]$.

A *closed plane curve* is a regular parametrized curve $\alpha: [a, b] \longrightarrow R^2$ such that α and all its derivatives agree at a and b; that is,

$$\alpha(a) = \alpha(b), \qquad \alpha'(a) = \alpha'(b), \qquad \alpha''(a) = \alpha''(b), \ldots.$$

The curve α is *simple* if it has no further self-intersections; that is, if $t_1, t_2 \in [a, b)$, $t_1 \neq t_2$, then $\alpha(t_1) \neq \alpha(t_2)$ (Fig. 1-20).

We usually consider the curve $\alpha: [0, l] \longrightarrow R^2$ parametrized by arc length s; hence, l is the length of α. Sometimes we refer to a simple closed curve C, meaning the trace of such an object. The curvature of α will be taken with a sign, as in Remark 1 of Sec. 1-5 (see Fig. 1-20).

We assume that *a simple closed curve C in the plane bounds a region of this plane* that is called the *interior* of C. This is part of the so-called Jordan

†This section may be omitted on a first reading.

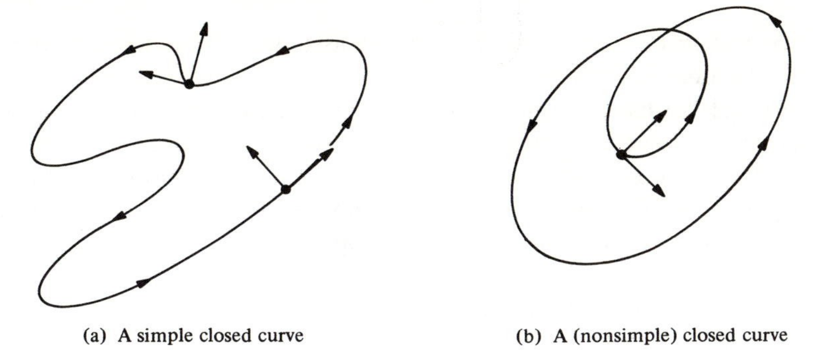

(a) A simple closed curve

(b) A (nonsimple) closed curve

Figure 1-20

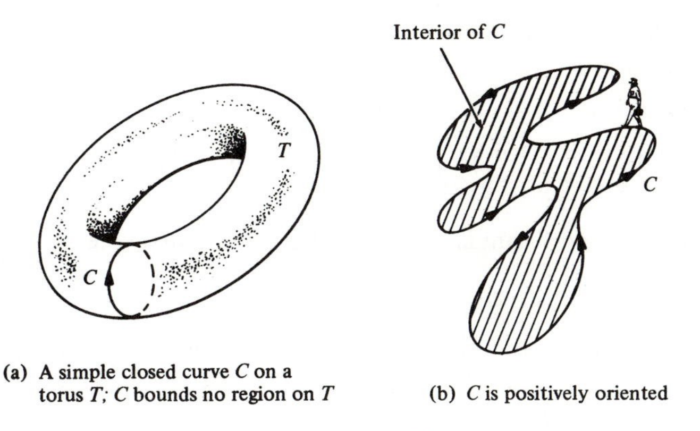

(a) A simple closed curve C on a torus T; C bounds no region on T

(b) C is positively oriented

Figure 1-21

curve theorem (a proof will be given in Sec. 5-6, Theorem 1), which does not hold, for instance, for simple curves on a torus (the surface of a doughnut; see Fig. 1-21(a)). Whenever we speak of the area bounded by a simple closed curve C, we mean the area of the interior of C. We assume further that the parameter of a simple closed curve can be so chosen that if one is going along the curve in the direction of increasing parameters, then the interior of the curve remains to the left (Fig. 1-21(b)). Such a curve will be called *positively oriented.*

A. The Isoperimetric Inequality

This is perhaps the oldest global theorem in differential geometry and is related to the following (isoperimetric) problem. *Of all simple closed curves in the plane with a given length l, which one bounds the largest area*? In this form, the problem was known to the Greeks, who also knew the solution, namely, the circle. A satisfactory proof of the fact that the circle is a solution to the isoperimetric problem took, however, a long time to appear. The main reason seems to be that the earliest proofs assumed that a solution should exist. It was only in 1870 that K. Weierstrass pointed out that many similar questions did not have solutions and gave a complete proof of the existence of a solution to the isoperimetric problem. Weierstrass' proof was somewhat hard, in the sense that it was a corollary of a theory developed by him to handle problems of maximizing (or minimizing) certain integrals (this theory is called calculus of variations and the isoperimetric problem is a typical example of the problems it deals with). Later, more direct proofs were found. The simple proof we shall present is due to E. Schmidt (1939). For another direct proof and further bibliography on the subject, one may consult Reference [10] in the Bibliography.

We shall make use of the following formula for the area A bounded by a positively oriented simple closed curve $\alpha(t) = (x(t), y(t))$, where $t \in [a, b]$ is an arbitrary parameter:

$$A = -\int_a^b y(t)x'(t)\,dt = \int_a^b x(t)y'(t)\,dt = \frac{1}{2}\int_a^b (xy' - yx')\,dt \tag{1}$$

Notice that the second formula is obtained from the first one by observing that

$$\int_a^b xy'\,dt = \int_a^b (xy)'\,dt - \int_a^b x'y\,dt = [xy(b) - xy(a)] - \int_a^b x'y\,dt$$
$$= -\int_a^b x'y\,dt,$$

since the curve is closed. The third formula is immediate from the first two.

To prove the first formula in Eq. (1), we consider initially the case of Fig. 1-22 where the curve is made up of two straight-line segments parallel

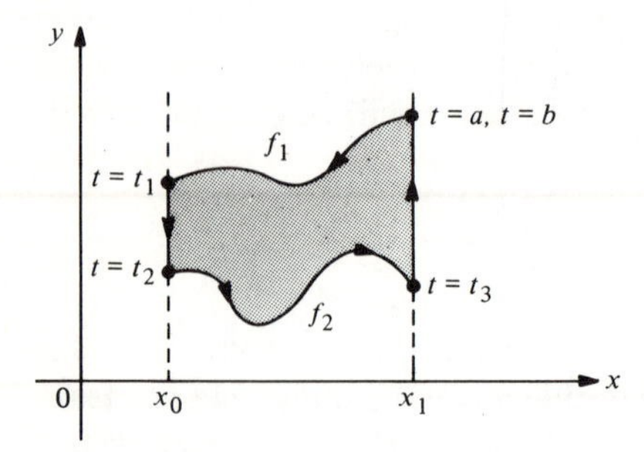

Figure 1-22

to the y axis and two arcs that can be written in the form

$$y = f_1(x) \quad \text{and} \quad y = f_2(x),\ x \in [x_0, x_1],\ f_1 > f_2.$$

Clearly, the area bounded by the curve is

$$A = \int_{x_0}^{x_1} f_1(x)\,dx - \int_{x_0}^{x_1} f_2(x)\,dx.$$

Since the curve is positively oriented, we obtain, with the notation of Fig. 1-22,

$$A = -\int_a^{t_1} y(t)x'(t)\,dt - \int_{t_2}^{t_3} y(t)x'(t)\,dt = -\int_a^b y(t)x'(t)\,dt,$$

since $x'(t) = 0$ along the segments parallel to the y axis. This proves Eq. (1) for this case.

To prove the general case, it must be shown that it is possible to divide the region bounded by the curve into a finite number of regions of the above type. This is clearly possible (Fig. 1-23) if *there exists a straight line E in the*

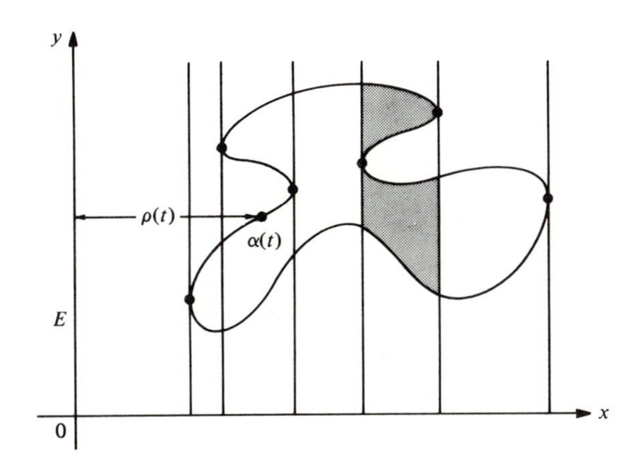

Figure 1-23

plane such that the distance $\rho(t)$ *of* $\alpha(t)$ *to this line is a function with finitely many critical points* (a critical point is a point where $\rho'(t) = 0$). The last assertion is true, but we shall not go into its proof. We shall mention, however, that Eq. (1) can also be obtained by using Stokes' (Green's) theorem in the plane (see Exercise 15).

THEOREM 1 (The Isoperimetric Inequality). *Let* C *be a simple closed plane curve with length* l, *and let* A *be the area of the region bounded by* C. *Then*

$$l^2 - 4\pi A \geq 0, \tag{2}$$

and equality holds if and only if C *is a circle.*

Proof. Let E and E' be two parallel lines which do not meet the closed curve C, and move them together until they first meet C. We thus obtain two parallel tangent lines to C, L and L', so that the curve is entirely contained in the strip bounded by L and L'. Consider a circle S^1 which is tangent to both L and L' and does not meet C. Let O be the center of S^1 and take a coordinate system with origin at O and the x axis perpendicular to L and L' (Fig. 1-24).

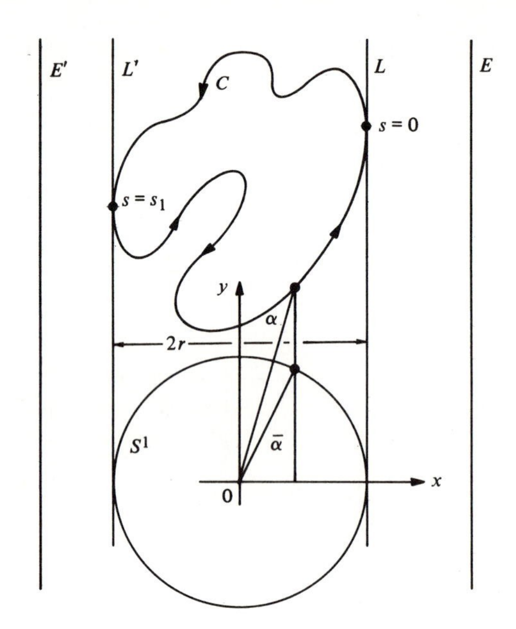

Figure 1-24

Parametrize C by arc length, $\alpha(s) = (x(s), y(s))$, so that it is positively oriented and the tangency points of L and L' are $s = 0$ and $s = s_1$, respectively.

We can assume that the equation of S^1 is

$$\bar{\alpha}(s) = (\bar{x}(s), \bar{y}(s)) = (x(s), \bar{y}(s)), \; s \in [0, l]$$

where $2r$ is the distance between L and L'. By using Eq. (1) and denoting by $\bar{A}$ the area bounded by S^1, we have

$$A = \int_0^l xy'\,ds, \qquad \bar{A} = \pi r^2 = -\int_0^l \bar{y}x'\,ds.$$

Thus,

$$\begin{aligned} A + \pi r^2 &= \int_0^l (xy' - \bar{y}x')\,ds \leq \int_0^l \sqrt{(xy' - \bar{y}x')^2}\,ds \\ &\leq \int_0^l \sqrt{(x^2 + \bar{y}^2)((x')^2 + (y')^2)}\,ds = \int_0^l \sqrt{\bar{x}^2 + \bar{y}^2}\,ds \qquad (3) \\ &= lr. \end{aligned}$$

We now notice the fact that the geometric mean of two positive numbers is smaller than or equal to their arithmetic mean, and equality holds if and only if they are equal. It follows that

$$\sqrt{A}\sqrt{\pi r^2} \leq \tfrac{1}{2}(A + \pi r^2) \leq \tfrac{1}{2}lr. \tag{4}$$

Therefore, $4\pi A r^2 \leq l^2 r^2$, and this gives Eq. (2).

Now, assume that equality holds in Eq. (2). Then equality must hold everywhere in Eqs. (3) and (4). From the equality in Eq. (4) it follows that $A = \pi r^2$. Thus, $l = 2\pi r$ and r does not depend on the choice of the direction of L. Furthermore, equality in Eq. (3) implies that

$$(xy' - \bar{y}x')^2 = (x^2 + \bar{y}^2)((x')^2 + (y')^2)$$

or

$$(xx' + \bar{y}y')^2 = 0;$$

that is,

$$\frac{x}{y'} = \frac{\bar{y}}{x'} = \frac{\sqrt{x^2 + \bar{y}^2}}{\sqrt{(y')^2 + (x')^2}} = \pm r.$$

Thus, $x = \pm ry'$. Since r does not depend on the choice of the direction of L, we can interchange x and y in the last relation and obtain $y = \pm rx'$. Thus,

$$x^2 + y^2 = r^2((x')^2 + (y')^2) = r^2$$

and C is a circle, as we wished. **Q.E.D.**

Remark 1. It is easily checked that the above proof can be applied to C^1 *curves*, that is, curves $\alpha(t) = (x(t), y(t))$, $t \in [a, b]$, for which we require only that the functions $x(t)$, $y(t)$ have continuous first derivatives (which, of course, agree at a and b if the curve is closed).

Remark 2. The isoperimetric inequality holds true for a wide class of curves. Direct proofs have been found that work as long as we can define arc length and area for the curves under consideration. For the applications, it is convenient to remark that the theorem holds for *piecewise C^1 curves*, that is, continuous curves that are made up by a finite number of C^1 arcs. These curves can have a finite number of corners, where the tangent is discontinuous (Fig. 1-25).

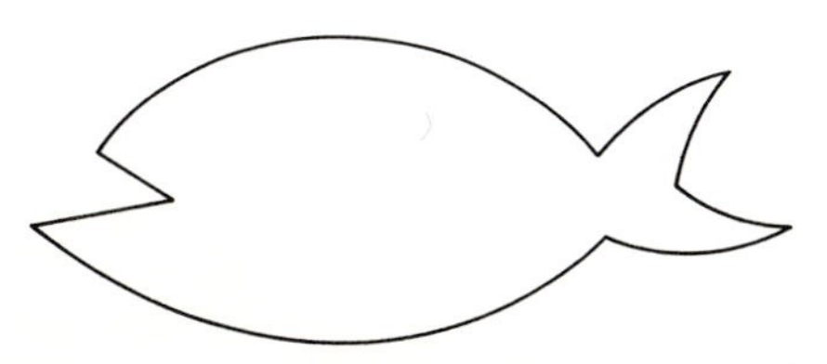

A piecewise C^1 curve **Figure 1-25**

B. The Four-Vertex Theorem

We shall need further general facts on plane closed curves.

Let $\alpha: [0, l] \longrightarrow R^2$ be a plane closed curve given by $\alpha(s) = (x(s), y(s))$. Since s is the arc length, the tangent vector $t(s) = (x'(s), y'(s))$ has unit length. It is convenient to introduce the *tangent indicatrix* $t: [0, l] \longrightarrow R^2$ that is given by $t(s) = (x'(s), y'(s))$; this is a differentiable curve, the trace of which is contained in a circle of radius 1 (Fig. 1-26). Observe that the velocity vector

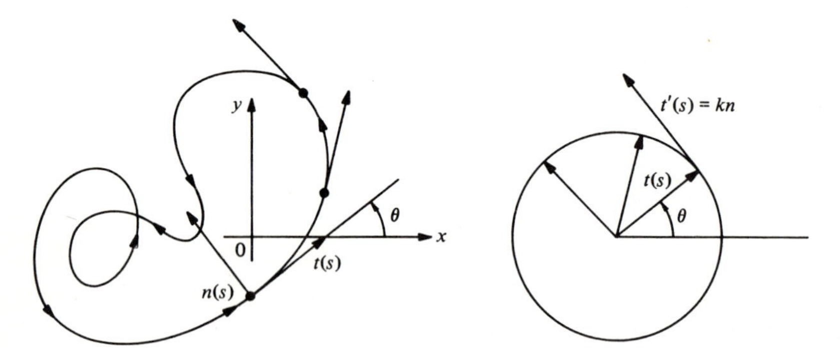

Figure 1-26

of the tangent indicatrix is

$$\frac{dt}{ds} = (x''(s), y''(s))$$
$$= \alpha''(s) = kn,$$

where n is the normal vector, oriented as in Remark 2 of Sec. 1-5, and k is the curvature of α.

Let $\theta(s)$, $0 < \theta(s) < 2\pi$, be the angle that $t(s)$ makes with the x axis; that is, $x'(s) = \cos\theta(s)$, $y'(s) = \sin\theta(s)$. Since

$$\theta(s) = \arctan\frac{y'(s)}{x'(s)},$$

$\theta = \theta(s)$ is locally well defined (that is, it is well defined in a small interval about each s) as a differentiable function and

$$\frac{dt}{ds} = \frac{d}{ds}(\cos\theta, \sin\theta)$$
$$= \theta'(-\sin\theta, \cos\theta) = \theta' n.$$

This means that $\theta'(s) = k(s)$ and suggests defining a global differentiable function $\theta\colon [0, l] \longrightarrow R$ by

$$\theta(s) = \int_0^s k(s)\, ds.$$

Since

$$\theta' = k = x'y'' - x''y' = \left(\text{arc tan}\, \frac{y'}{x'}\right)',$$

this global function agrees, up to constants, with the previous locally defined θ. Intuitively, $\theta(s)$ measures the total rotation of the tangent vector, that is, the total angle described by the point $t(s)$ on the tangent indicatrix, as we run the curve α from 0 to s. Since α is closed, this angle is an integer multiple I of 2π; that is,

$$\int_0^l k(s)\, ds = \theta(l) - \theta(0) = 2\pi I.$$

The integer I is called the *rotation index* of the curve α.

In Fig. 1-27 are some examples of curves with their rotation indices. Observe that the rotation index changes sign when we change the orientation of the curve. Furthermore, the definition is so set that the rotation index of a positively oriented simple closed curve is positive.

An important global fact about the rotation index is given in the following theorem, which will be proved later in the book (Sec. 5-6, Theorem 2).

THE THEOREM OF TURNING TANGENTS. *The rotation index of a simple closed curve is* ± 1, *where the sign depends on the orientation of the curve.*

A regular, plane (not necessarily closed) curve $\alpha\colon [a, b] \longrightarrow R^2$ is *convex* if, for all $t \in [a, b]$, the trace $\alpha([a, b])$ of α lies entirely on one side of the closed half-plane determined by the tangent line at t (Fig. 1-28).

A *vertex* of a regular plane curve $\alpha\colon [a, b] \longrightarrow R^2$ is a point $t \in [a, b]$ where $k'(t) = 0$. For instance, an ellipse with unequal axes has exactly four vertices, namely the points where the axes meet the ellipse (see Exercise 3). It is an interesting global fact that this is the least number of vertices for all closed convex curves.

THEOREM 2 (The Four-Vertex Theorem). *A simple closed convex curve has at least four vertices.*

Before starting the proof, we need a lemma.

LEMMA. *Let* $\alpha\colon [0, 1] \longrightarrow \mathrm{R}^2$ *be a plane closed curve parametrized by arc length and let* A, B, *and* C *be arbitrary real numbers. Then*

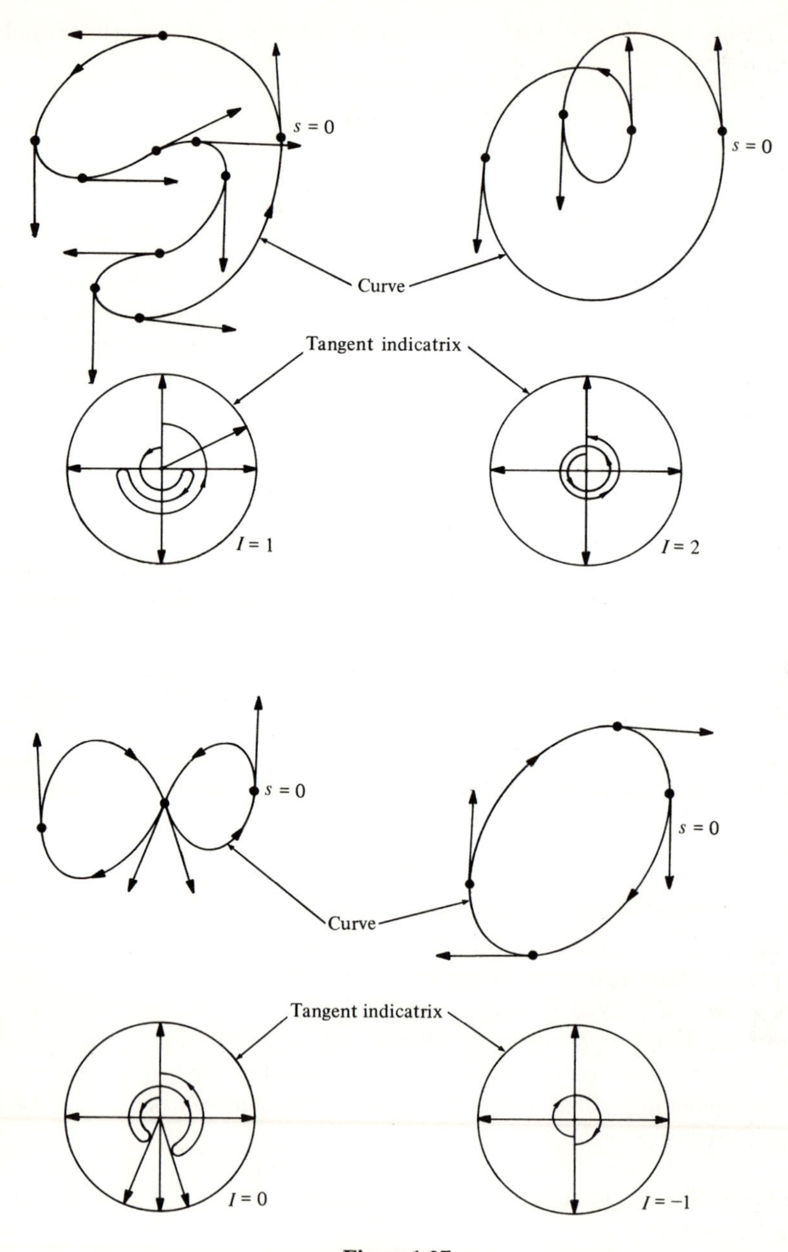

$s = 0$
$s = 0$
Curve
Tangent indicatrix
$I = 1$
$I = 2$
$s = 0$
$s = 0$
Curve
Tangent indicatrix
$I = 0$
$I = -1$

Figure 1-27

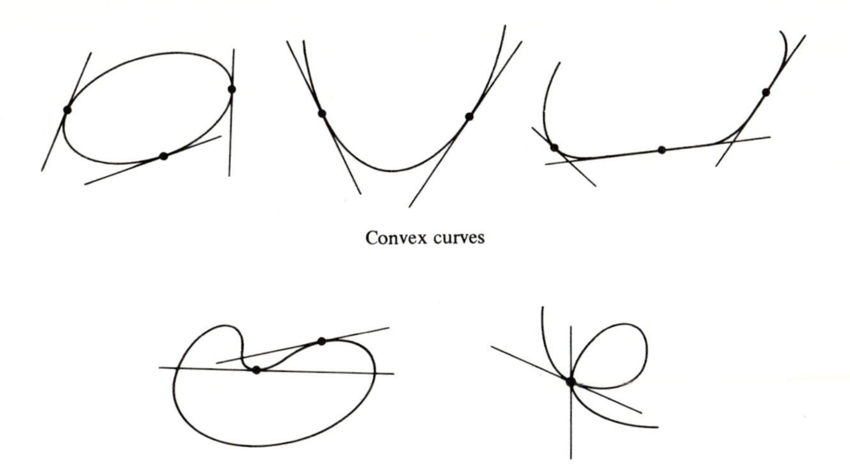

Convex curves

Nonconvex curves

Figure 1-28

$$\int_0^1 (\mathrm{Ax} + \mathrm{By} + \mathrm{C})\frac{\mathrm{dk}}{\mathrm{ds}}\,\mathrm{ds} = 0, \tag{5}$$

where the functions $\mathrm{x} = \mathrm{x(s)}$, $\mathrm{y} = \mathrm{y(s)}$ *are given by* $\alpha(\mathrm{s}) = (\mathrm{x(s)}, \mathrm{y(s)})$, *and* k *is the curvature of* α.

Proof of the Lemma. Recall that there exists a differentiable function $\theta\colon [0, l] \longrightarrow R$ such that $x'(s) = \cos\theta$, $y'(s) = \sin\theta$. Thus, $k(s) = \theta'(s)$ and

$$x'' = -ky', \qquad y'' = kx'.$$

Therefore, since the functions involved agree at 0 and l,

$$\int_0^l k'\,ds = 0,$$

$$\int_0^l xk'\,ds = -\int_0^l kx'\,dx = -\int_0^l y''\,ds = 0,$$

$$\int_0^l yk'\,ds = -\int_0^l ky'\,ds = \int_0^l x''\,ds = 0. \qquad \textbf{Q.E.D.}$$

Proof of the Theorem. Parametrize the curve by arc length, $\alpha\colon [0, l] \longrightarrow R^2$. Since $k = k(s)$ is *a continuous function on the closed interval* $[0, l]$, *it reaches a maximum and a minimum on* $[0, l]$ (this is a basic fact in real functions; a proof can be found, for instance, in the appendix to Chap. 5, Prop. 10). Thus, α has at least two vertices, $\alpha(s_1) = p$ and $\alpha(s_2) = q$. Let L be the straight line passing through p and q, and let β and γ be the two arcs of C which are determined by the points p and q.

We claim that each of these arcs lies on a definite side of L. Otherwise, it meets L in a point r distinct from p and q (Fig. 1-29(*a*)). By convexity, and since p, q, r are distinct points on C, the tangent line at the intermediate point, say p, has to agree with L. Again, by convexity, this implies that L is tangent to C at the three points p, q, and r. But then the tangent to a point

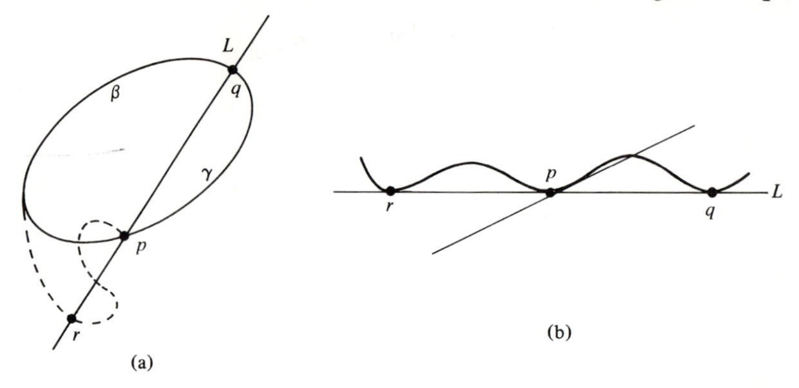

Figure 1-29

near p (the intermediate point) will have q and r on distinct sides, unless the whole segment rq of L belongs to C (Fig. 1-29(b)). This implies that $k = 0$ at p and q. Since these are points of maximum and minimum for k, $k \equiv 0$ on C, a contradiction.

Let $Ax + By + C = 0$ be the equation of L. If there are no further vertices, $k'(s)$ keeps a constant sign on each of the arcs β and γ. We can then arrange the sign of all the coefficients A, B, C so that the integral in Eq. (5) is positive. This contradiction shows that there is a third vertex and that $k'(s)$ changes sign on β or γ, say, on β. Since p and q are points of maximum and minimum, $k'(s)$ changes sign twice on β. Thus, there is a fourth vertex.

Q.E.D.

The four-vertex theorem has been the subject of many investigations. The theorem also holds for simple, closed (not necessarily convex) curves, but the proof is harder. For further literature on the subject, see Reference [10].

Later (Sec. 5-6, Prop. 1) we shall prove that *a plane closed curve is convex if and only if it is simple and can be oriented so that its curvature is positive or zero*. From that, and the proof given above, we see that we can reformulate the statement of the four-vertex theorem as follows. *The curvature function of a closed convex curve is* (*nonnegative and*) *either constant or else has at least two maxima and two minima*. It is then natural to ask whether such curvature functions do characterize the convex curves. More precisely, we can ask the following question. *Let* $k\colon [a, b] \longrightarrow R$ *be a differentiable nonnegative function such that k agrees, with all its derivatives, at a and b. Assume that k is either constant or else has at least two maxima and two minima. Is there a simple closed curve* $\alpha: [a, b] \longrightarrow R^2$ *such that the curvature of* α *at* t *is* $k(t)$*?*

For the case where $k(t)$ is strictly positive, H. Gluck answered the above question affirmatively (see H. Gluck, "The Converse to the Four Vertex Theorem," *L'Enseignement Mathématique* T. XVII, fasc. 3–4 (1971), 295–309). His methods, however, do not apply to the case $k \geq 0$.

C. The Cauchy-Crofton Formula

Our last topic in this section will be dedicated to finding a theorem which, roughly speaking, describes the following situation. Let C be a regular curve in the plane. We look at all straight lines in the plane that meet C and assign to each such line a *multiplicity* which is the number of its interesection points with C (Fig. 1-30).

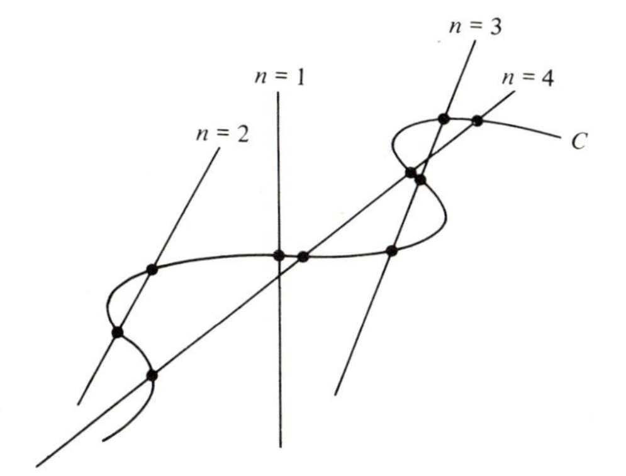

Figure 1-30. n is the multiplicity of the corresponding straight line.

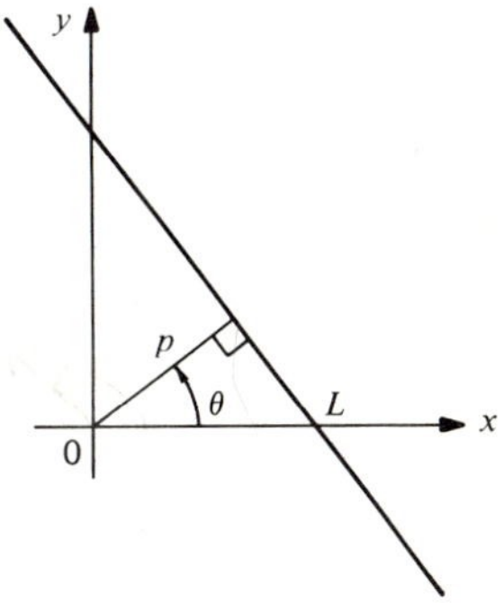

Figure 1-31 L is determined by p and θ.

We first want to find a way of assigning a measure to a given subset of straight lines in the plane. It should not be too surprising that this is possible. After all, we assign a measure (area) to point subsets of the plane. Once we realize that a straight line can be determined by two parameters (for instance, p and θ in Fig. 1-31), we can think of the straight lines in the plane as points in a region of a certain plane. Thus, what we want is to find a "reasonable" way of measuring "areas" in such a plane.

Having chosen this measure, we want to apply it and find the measure of the set of straight lines (counted with multiplicities) which meet C. The result is quite interesting and can be stated as follows.

THEOREM 3 (The Cauchy-Crofton Formula). *Let* C *be a regular plane curve with length* l. *The measure of the set of straight lines (counted with multiplicities) which meet* C *is equal to* $2l$.

Before going into the proof we must define what we mean by a reasonable measure in the set of straight lines in the plane. First, let us choose a convenient system of coordinates for such a set. A straight line L in the plane is determined by the distance $p \geq 0$ from L to the origin O of the coordinates and by the angle θ, $0 \leq \theta < 2\pi$, which a half-line starting at 0 and normal to L makes with the x axis (Fig. 1-31). The equation of L in terms of these parameters is easily seen to be

$$x \cos \theta + y \sin \theta = p.$$

Thus we can replace the set of all straight lines in the plane by the set

$$\mathfrak{L} = \{(p, \theta) \in R^2; p \geq 0, 0 \leq \theta < 2\pi\}.$$

We will show that, up to a choice of units, there is only one reasonable measure in this set.

To decide what we mean by reasonable, let us look more closely at the usual measure of areas in R^2. We need a definition.

A *rigid motion* in R^2 is a map $F: R^2 \to R^2$ given by $(\bar{x}, \bar{y}) \to (x, y)$, where (Fig. 1-32)

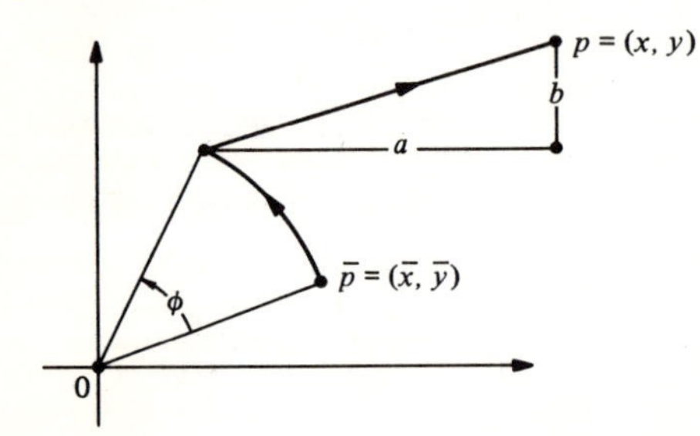

Figure 1-32

$$\begin{aligned} x &= a + \bar{x} \cos \varphi - \bar{y} \sin \varphi \\ y &= b + \bar{x} \sin \varphi + \bar{y} \cos \varphi. \end{aligned} \tag{6}$$

Now, to define the area of a set $S \subset R^2$ we consider the double integral

$$\iint_S dx\, dy;$$

that is, we integrate the "element of area" $dx\, dy$ over S. When this integral exists in some sense, we say that S is *measurable* and define the area of S as the value of the above integral. From now on, we shall assume that all the integrals involved in our discussions do exist.

Notice that we could have chosen some other element of area, say, $xy^2\, dx\, dy$. The reason for the choice of $dx\, dy$ is that, up to a factor, this is the only element of area that is invariant under rigid motions. More precisely, we have the following proposition.

PROPOSITION 1. *Let* f(x, y) *be a continuous function defined in* R^2. *For any set* $S \subset R^2$, *define the area* A *of* S *by*

$$A(S) = \iint_S f(x, y)\, dx\, dy$$

(*of course, we are considering only those sets for which the above integral exists*). *Assume that* A *is invariant under rigid motions; that is, if* S *is any set and* $\bar{S} = F^{-1}(S)$, *where* F *is the rigid motion* (6), *we have*

$$A(\bar{S}) = \iint_{\bar{S}} f(\bar{x}, \bar{y})\, d\bar{x}\, d\bar{y} = \iint_S f(x, y)\, dx\, dy = A(S).$$

Then f(x, y) = *const.*

Proof. We recall the formula for change of variables in multiple integrals (Buck, *Advanced Calculus*, p. 301, or Exercise 15 of this section):

$$\iint_S f(x, y)\, dx\, dy = \iint_{\bar{S}} f(x(\bar{x}, \bar{y}), y(\bar{x}, \bar{y})) \frac{\partial(x, y)}{\partial(\bar{x}, \bar{y})}\, d\bar{x}\, d\bar{y}. \tag{7}$$

Here, $x = x(\bar{x}, \bar{y})$, $y = y(\bar{x}, \bar{y})$ are functions with continuous partial derivatives which define the transformation of variables $T: R^2 \to R^2$, $\bar{S} = T^{-1}(S)$, and

$$\frac{\partial(x, y)}{\partial(\bar{x}, \bar{y})} = \begin{vmatrix} \dfrac{\partial x}{\partial \bar{x}} & \dfrac{\partial x}{\partial \bar{y}} \\ \dfrac{\partial y}{\partial \bar{x}} & \dfrac{\partial y}{\partial \bar{y}} \end{vmatrix}$$

is the Jacobian of the transformation T. In our particular case, the transformation is the rigid motion (6) and the Jacobian is

$$\frac{\partial(x, y)}{\partial(\bar{x}, \bar{y})} = \begin{vmatrix} \cos\varphi & -\sin\varphi \\ \sin\varphi & \cos\varphi \end{vmatrix} = 1.$$

By using this fact and Eq. (7), we obtain

$$\iint_{\bar{S}} f(x(\bar{x}, \bar{y}), y(\bar{x}, \bar{y}))\, d\bar{x}\, d\bar{y} = \iint_{\bar{S}} f(\bar{x}, \bar{y})\, d\bar{x}\, d\bar{y}.$$

Since this is true for all S, we have

$$f(x(\bar{x}, \bar{y}), y(\bar{x}, \bar{y})) = f(\bar{x}, \bar{y}).$$

We now use the fact that for any pair of points (x, y), $(\bar{x}, \bar{y})$ in R^2 there exists a rigid motion F such that $F(\bar{x}, \bar{y}) = (x, y)$. Thus,

$$f(x, y) = (f \circ F)(\bar{x}, \bar{y}) = f(\bar{x}, \bar{y}),$$

and $f(x, y) = \text{const.}$, as we wished. **Q.E.D.**

Remark 3. The above proof rests upon two facts: first, that the Jacobian of a rigid motion is 1, and, second, that the rigid motions are transitive on points of the plane; that is, given two points in the plane there exists a rigid motion taking one point into the other.

With these preparations, we can finally define a measure in the set $\mathcal{L}$. We first observe that the rigid motion (6) induces a transformation on $\mathcal{L}$. In fact, Eq. (6) maps the line $x \cos\theta + y \sin\theta = p$ into the line

$$\bar{x}\cos(\theta - \varphi) + \bar{y}\sin(\theta - \varphi) = p - a\cos\theta - b\sin\theta.$$

This means that the transformation induced by Eq. (6) on $\mathcal{L}$ is

$$\begin{aligned} \bar{p} &= p - a\cos\theta - b\sin\theta, \\ \bar{\theta} &= \theta - \varphi. \end{aligned}$$

It is easily checked that the Jacobian of the above transformation is 1 and that such transformations are also transitive on the set of lines in the plane. We then define the measure of a set $\mathcal{S} \subset \mathcal{L}$ as

$$\iint_{\mathcal{S}} dp\, d\theta.$$

In the same way as in Prop. 1, we can then prove that this is, up to a constant factor, the only measure on $\mathcal{L}$ that is invariant under rigid motions. This measure is, therefore, as reasonable as it can be.

We can now sketch a proof of Theorem 3.

Sketch of Proof of Theorem 3. First assume that the curve C is a segment of a straight line with length l. Since our measure is invariant under rigid motions, we can assume that the coordinate system has its origin 0 in the middle point of C and that the x axis is in the direction of C. Then the measure of the set of straight lines that meet C is (Fig. 1-33)

$$\iint dp\, d\theta = \int_0^{2\pi}\left(\int_0^{|\cos\theta|\,(l/2)} dp\right) d\theta = \int_0^{2\pi} \frac{l}{2}|\cos\theta|\, d\theta = 2l.$$

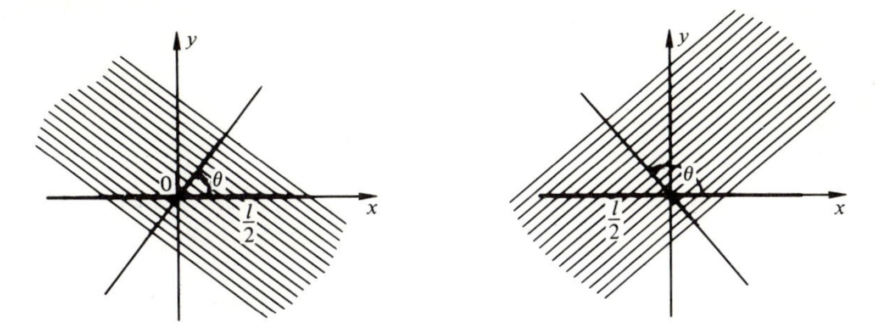

Figure 1-33

Next, let C be a polygonal line composed of a finite number of segments C_i with length l_i ($\sum l_i = l$). Let $n = n(p, \theta)$ be the number of intersection points of the straight line (p, θ) with C. Then, by summing up the results for each segment C_i, we obtain

$$\iint n\, dp\, d\theta = 2 \sum_i l_i = 2l, \tag{8}$$

which is the Cauchy-Crofton formula for a polygonal line.

Finally, by a limiting process, it is possible to extend the above formula to any regular curve, and this will prove Theorem 3. **Q.E.D.**

It should be remarked that the general ideas of this topic belong to a branch of geometry known under the name of integral geometry. A survey of the subject can be found in L. A. Santaló, "Integral Geometry," in *Studies in Global Geometry and Analysis*, edited by S. S. Chern, The Mathematical Association of America, 1967, 147–193.

The Cauchy-Crofton formula can be used in many ways. For instance, if a curve is not rectifiable (see Exercise 9, Sec. 1-3) but the left-hand side of Eq. (8) has a meaning, this can be used to define the "length" of such a curve. Equation (8) can also be used to obtain an efficient way of estimating lengths of curves. Indeed, a good approximation for the integral in Eq. (8) is given as follows.† Consider a family of parallel straight lines such that two consecutive lines are at a distance r. Rotate this family by angles of $\pi/4$, $2\pi/4$, $3\pi/4$ in order to obtain four families of straight lines. Let n be the number of intersection points of a curve C with all these lines. Then

$$\frac{1}{2} nr \frac{\pi}{4}$$

†I want to thank Robert Gardner for suggesting this application and the example that follows.

is an approximation to the integral

$$\frac{1}{2}\iint n\,dp\,d\theta = \text{length of } C$$

and therefore gives an estimate for the length of C. To have an idea of how good this estimate can be, let us work out an example.

Example. Figure 1-34 is a drawing of an electron micrograph of a circular DNA molecule and we want to estimate its length. The four families

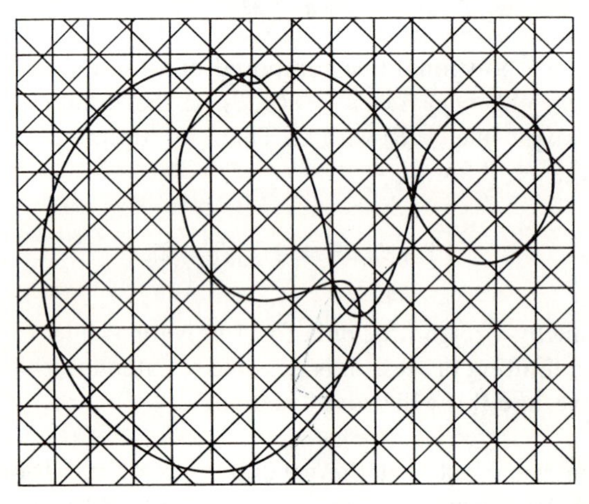

Figure 1-34. Reproduced from H. Ris and B. C. Chandler, *Cold Spring Harbor Symp. Quant. Biol.* 28, 2 (1963), with permission.

of straight lines at a distance of 7 millimeters and angles of $\pi/4$ are drawn over the picture (a more practical way would be to have this family drawn once and for all on transparent paper). The number of interesection points is found to be 153. Thus,

$$\frac{1}{2}n\frac{\pi}{4} = \frac{1}{2}153\frac{3.14}{4} \sim 60.$$

Since the reference line in the picture represents 1 micrometer ($= 10^{-6}$ meter) and measures, in our scale, 25 millimeters, $r = \frac{25}{7}$, and thus the length of this DNA molecule, from our values, is approximately

$$60\left(\frac{25}{7}\right) \sim 16.6 \text{ micrometers.}$$

The actual value is 16.3 micrometers.

EXERCISES

***1.** Is there a simple closed curve in the plane with length equal to 6 feet and bounding an area of 3 square feet?

***2.** Let $\overline{AB}$ be a segment of straight line and let $l >$ length of AB. Show that the curve C joining A and B, with length l, and such that together with $\overline{AB}$ bounds the largest possible area is an arc of a circle passing through A and B (Fig. 1-35).

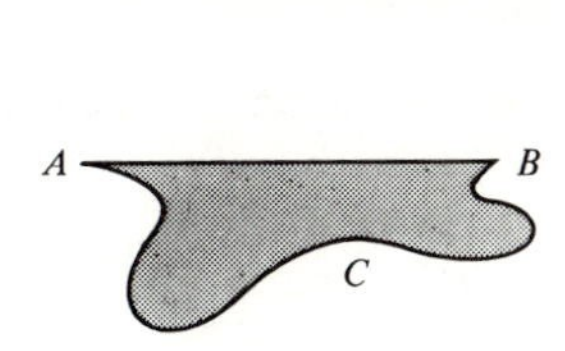

Figure 1-35

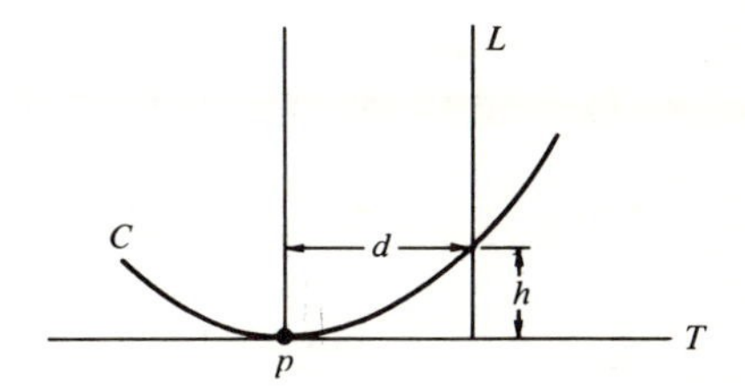

Figure 1-36

3. Compute the curvature of the ellipse

$$x = a\cos t, \qquad y = b\sin t, \qquad t \in [0, 2\pi],\ a \neq b,$$

and show that it has exactly four vertices, namely, the points $(a, 0)$, $(-a, 0)$, $(0, b)$, $(0, -b)$.

***4.** Let C be a plane curve and let T be the tangent line at a point $p \in C$. Draw a line L parallel to the normal line at p and at a distance d of p (Fig. 1-36). Let h be the length of the segment determined on L by C and T (thus, h is the "height" of C relative to T). Prove that

$$|k(p)| = \lim_{d \to 0} \frac{2h}{d^2},$$

where $k(p)$ is the curvature of C at p.

***5.** If a closed plane curve C is contained inside a disk of radius r, prove that there exists a point $p \in C$ such that the curvature k of C at p satisfies $|k| \geq 1/r$.

6. Let $\alpha(s)$, $s \in [0, l]$ be a closed convex plane curve positively oriented. The curve

$$\beta(s) = \alpha(s) - rn(s),$$

where r is a positive constant and n is the normal vector, is called a *parallel* curve to α (Fig. 1-37). Show that

a. Length of β = length of $\alpha + 2\pi r$.

b. $A(\beta) = A(\alpha) + rl + \pi r^2$.

c. $k_\beta(s) = k_\alpha(s)/(1 + r)$.

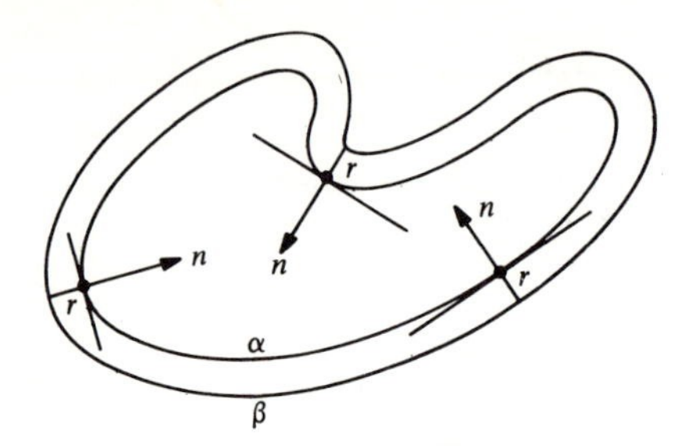

Figure 1-37

For (a)-(c), $A(\ )$ denotes the area bounded by the corresponding curve, and k_α, k_β are the curvatures of α and β, respectively.

7. Let $\alpha: R \longrightarrow R^2$ be a plane curve defined in the entire real line R. Assume that α does not pass through the origin $0 = (0, 0)$ and that both

$$\lim_{t\to+\infty} |\alpha(t)| = \infty \quad \text{and} \quad \lim_{t\to-\infty} |\alpha(t)| = \infty.$$

a. Prove that there exists a point $t_0 \in R$ such that $|\alpha(t_0)| \leq |\alpha(t)|$ for all $t \in R$.

b. Show, by an example, that the assertion in part a is false if one does not assume that both $\lim_{t\to+\infty} |\alpha(t)| = \infty$ and $\lim_{t\to-\infty} |\alpha(t)| = \infty$.

8. ***a.** Let $\alpha(s)$, $s \in [0, l]$, be a plane simple closed curve. Assume that the curvature $k(s)$ satisfies $0 < k(s) \leq c$, where c is a constant (thus, α is less curved than a circle of radius $1/c$). Prove that

$$\text{length of } \alpha \geq \frac{2\pi}{c}.$$

b. In part a replace the assumption of being simple by "α has rotation index N." Prove that

$$\text{length of } \alpha \geq \frac{2\pi N}{c}.$$

***9.** A set $K \subset R^2$ is *convex* if given any two points $p, q \in K$ the segment of straight line $\overline{pq}$ is contained in K (Fig. 1-38). Prove that a simple closed convex curve bounds a convex set.

10. Let C be a convex plane curve. Prove geometrically that C has no self-intersections.

***11.** Given a nonconvex simple closed plane curve C, we can consider its *convex hull* H (Fig. 1-39), that is, the boundary of the smallest convex set containing the interior of C. The curve H is formed by arcs of C and by the segments of the tangents to C that bridge "the nonconvex gaps" (Fig. 1-39). It can be proved that H is a C^1 closed convex curve. Use this to show that, in the isoperimetric problem, we can restrict ourselves to convex curves.

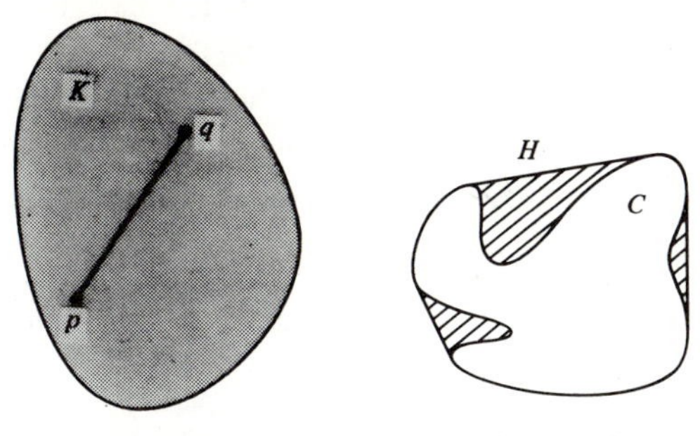

Figure 1-38 Figure 1-39

***12.** Consider a unit circle S^1 in the plane. Show that the ratio $M_1/M_2 = \frac{1}{3}$, where M_2 is the measure of the set of straight lines in the plane that meet S^1 and M_1 is the measure of all such lines that determine in S^1 a chord of length $> \sqrt{3}$. Intuitively, this ratio is the probability that a straight line that meets S^1 determines in S^1 a chord longer than the side of an equilateral triangle inscribed in S^1 (Fig. 1-40).

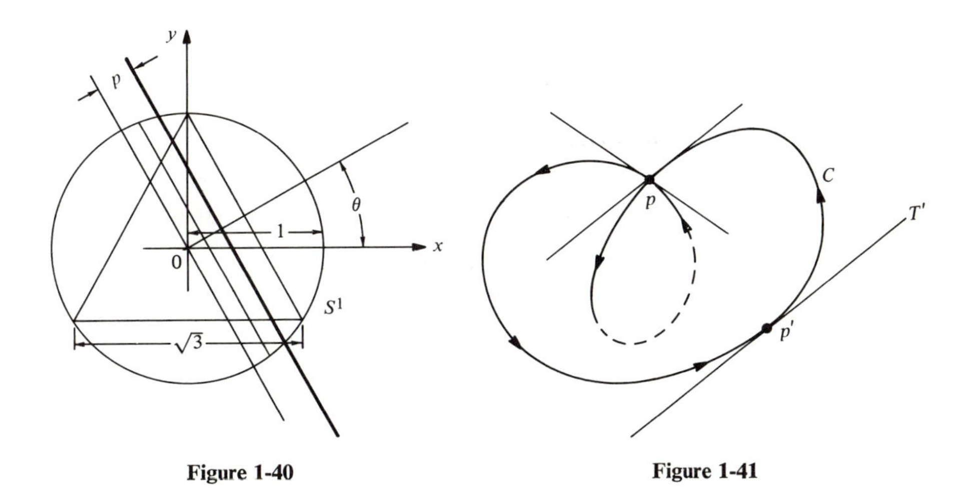

Figure 1-40 Figure 1-41

13. Let C be an oriented plane closed curve with curvature $k > 0$. Assume that C has at least one point p of self-intersection. Prove that

a. There is a point $p' \in C$ such that the tangent line T' at p' is parallel to some tangent at p.

b. The rotation angle of the tangent in the positive arc of C made up by $pp'p$ is $> \pi$ (Fig. 1-41).

c. The rotation index of C is ≥ 2.

14. a. Show that if a straight line L meets a closed convex curve C, then either L is tangent to C or L intersects C in exactly two points.

b. Use part a to show that the measure of the set of lines that meet C (without multiplicities) is equal to the length of C.

15. Green's theorem in the plane is a basic fact of calculus and can be stated as follows. Let a simple closed plane curve be given by $\alpha(t) = (x(t), y(t))$, $t \in [a, b]$. Assume that α is positively oriented, let C be its trace, and let R be the interior of C. Let $p = p(x, y)$, $q = q(x, y)$ be real functions with continuous partial derivatives p_x, p_y, q_x, q_y. Then

$$\int_R (q_x - p_y)dx\, dy = \int_C \left(p\frac{dx}{dt} + q\frac{dy}{dt}\right) dt, \tag{9}$$

where in the second integral it is understood that the functions p and q are restricted to α and the integral is taken between the limits $t = a$ and $t = b$. In parts a and b below we propose to derive, from Green's theorem, a formula for the area of R and the formula for the change of variables in double integrals (cf. Eqs. (1) and (7) in the text).

a. Set $q = x$ and $p = -y$ in Eq. (9) and conclude that

$$A(R) = \iint_R dx\, dy = \frac{1}{2}\int_a^b \left(x(t)\frac{dy}{dt} - y(t)\frac{dx}{dt}\right) dt.$$

b. Let $f(x, y)$ be a real function with continuous partial derivatives and T: $R^2 \longrightarrow R^2$ be a transformation of coordinates given by the functions $x = x(u, v)$, $y = y(u, v)$, which also admit continuous partial derivatives. Choose in Eq. (9) $p = 0$ and q so that $q_x = f$. Apply successively Green's theorem, the map T, and Green's theorem again to obtain

$$\begin{aligned}\iint_R f(x, y)\, dx\, dy &= \int_C q\, dy = \iint_{T^{-1}(C)} (q\circ T)(y_u u'(t) + y_v v'(t))\, dt \\ &= \iint_{T^{-1}(R)} \left\{\frac{\partial}{\partial u}((q\circ T)y_v) - \frac{\partial}{\partial v}((q\circ T)y_u)\right\} du\, dv.\end{aligned}$$

Show that

$$\begin{aligned}&\frac{\partial}{\partial u}(q(x(u, v), y(u, v))y_v) - \frac{\partial}{\partial v}(q(x(u, v), y(u, v))y_u) \\ &\qquad = f(x(u, v), y(u, v))(x_u y_v - x_v y_u) = f\frac{\partial(x, y)}{\partial(u, v)}.\end{aligned}$$

Put that together with the above and obtain the transformation formula for double integrals:

$$\iint_R f(x, y)\, dx\, dy = \int_{T^{-1}(R)} f(x(u, v), y(u, v))\frac{\partial(x, y)}{\partial(u, v)}\, du\, dv.$$

2 Regular Surfaces

2-1. Introduction

In this chapter, we shall begin the study of surfaces. Whereas in the first chapter we used mainly elementary calculus of one variable, we shall now need some knowledge of calculus of several variables. Specifically, we need to know some facts about continuity and differentiability of functions and maps in R^2 and R^3. What we need can be found in any standard text of advanced calculus, for instance, Buck *Advanced Calculus*; we have included a brief review of some of this material in an appendix to Chap. 2.

In Sec. 2-2 we shall introduce the basic concept of a regular surface in R^3. In contrast to the treatment of curves in Chap. 1, regular surfaces are defined as sets rather than maps. The goal of Sec. 2-2 is to describe some criteria that are helpful in trying to decide whether a given subset of R^3 is a regular surface.

In Sec. 2-3 we shall show that it is possible to define what it means for a function on a regular surface to be differentiable, and in Sec. 2-4 we shall show that the usual notion of differential in R^2 can be extended to such functions. Thus, regular surfaces in R^3 provide a natural setting for two-dimensional calculus.

Of course, curves can also be treated from the same point of view, that is, as subsets of R^3 which provide a natural setting for one-dimensional calculus. We shall mention them briefly in Sec. 2-3.

Sections 2-2 and 2-3 are crucial to the rest of the book. A beginner may find the proofs in these sections somewhat difficult. If so, the proofs can be omitted on a first reading.

In Sec. 2-5 we shall introduce the first fundamental form, a natural instrument to treat metric questions (lengths of curves, areas of regions, etc.) on a regular surface. This will become a very important issue when we reach Chap. 4.

Sections 2-6 through 2-8 are optional on a first reading. In Sec. 2-6, we shall treat the idea of orientation on regular surfaces. This will be needed in Chaps. 3 and 4. For the benefit of those who omit this section, we shall review the notion of orientation at the beginning of Chap. 3.

2-2. Regular Surfaces; Inverse Images of Regular Values†

In this section we shall introduce the notion of a regular surface in R^3. Roughly speaking, a regular surface in R^3 is obtained by taking pieces of a plane, deforming them, and arranging them in such a way that the resulting figure has no sharp points, edges, or self-intersections and so that it makes sense to speak of a tangent plane at points of the figure. The idea is to define a set that is, in a certain sense, two-dimensional and that also is smooth enough so that the usual notions of calculus can be extended to it. By the end of Sec. 2-4, it should be completely clear that the following definition is the right one.

DEFINITION 1. *A subset* $\mathrm{S} \subset \mathrm{R}^3$ *is a* regular surface *if, for each* $\mathrm{p} \in \mathrm{S}$, *there exists a neighborhood* V *in* R^3 *and a map* $\mathbf{x}\colon \mathrm{U} \to \mathrm{V} \cap \mathrm{S}$ *of an open set* $\mathrm{U} \subset \mathrm{R}^2$ *onto* $\mathrm{V} \cap \mathrm{S} \subset \mathrm{R}^3$ *such that (Fig. 2-1)*

1. $\mathbf{x}$ *is differentiable. This means that if we write*

$$\mathbf{x}(\mathrm{u}, \mathrm{v}) = (\mathrm{x}(\mathrm{u}, \mathrm{v}), \mathrm{y}(\mathrm{u}, \mathrm{v}), \mathrm{z}(\mathrm{u}, \mathrm{v})), \qquad (\mathrm{u}, \mathrm{v}) \in \mathrm{U},$$

the functions $\mathrm{x}(\mathrm{u}, \mathrm{v})$, $\mathrm{y}(\mathrm{u}, \mathrm{v})$, $\mathrm{z}(\mathrm{u}, \mathrm{v})$ *have continuous partial derivatives of all orders in* U.

2. $\mathbf{x}$ *is a homeomorphism. Since* $\mathbf{x}$ *is continuous by condition* 1, *this means that* $\mathbf{x}$ *has an inverse* $\mathbf{x}^{-1}\colon \mathrm{V} \cap \mathrm{S} \to \mathrm{U}$ *which is continuous; that is,* $\mathbf{x}^{-1}$ *is the restriction of a continuous map* $\mathrm{F}\colon \mathrm{W} \subset \mathrm{R}^3 \to \mathrm{R}^2$ *defined on an open set* W *containing* $\mathrm{V} \cap \mathrm{S}$.

3. *(The regularity condition.) For each* $\mathrm{q} \in \mathrm{U}$, *the differential* $d\mathbf{x}_{\mathrm{q}}\colon \mathrm{R}^2 \to \mathrm{R}^3$ *is one-to-one.*‡

We shall explain condition 3 in a short while.

†Proofs in this section may be omitted on a first reading.

‡In italic context, letter symbols are roman so they can be distinguished from the surrounding text.

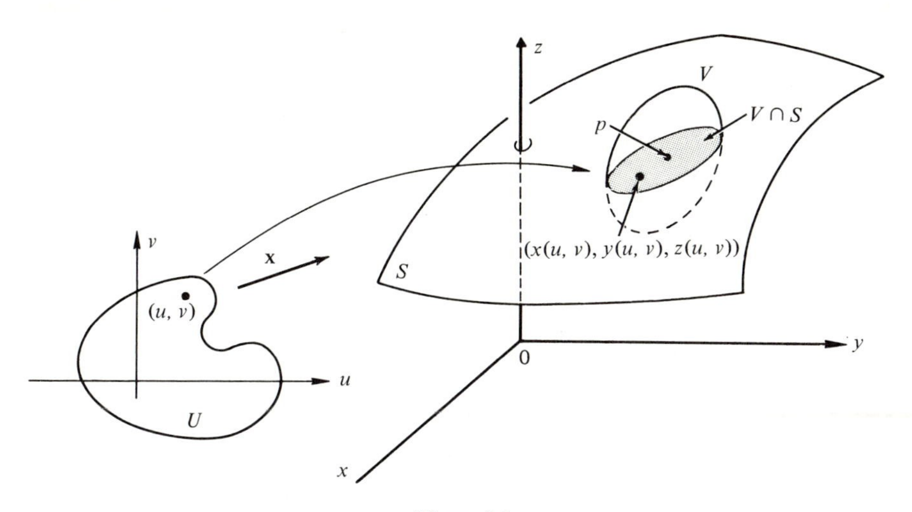

Figure 2-1

The mapping $\mathbf{x}$ is called a *parametrization* or a *system of* (local) *coordinates* in (a neighborhood of) p. The neighborhood $V \cap S$ of p in S is called a *coordinate neighborhood.*

To give condition 3 a more familiar form, let us compute the matrix of the linear map $d\mathbf{x}_q$ in the canonical bases $e_1 = (1, 0)$, $e_2 = (0, 1)$ of R^2 with coordinates (u, v) and $f_1 = (1, 0, 0), f_2 = (0, 1, 0), f_3 = (0, 0, 1)$ of R^3, with coordinates (x, y, z).

Let $q = (u_0, v_0)$. The vector e_1 is tangent to the curve $u \longrightarrow (u, v_0)$ whose image under $\mathbf{x}$ is the curve

$$u \longrightarrow (x(u, v_0), y(u, v_0), z(u, v_0)).$$

This image curve (called the *coordinate curve* $v = v_0$) lies on S and has at $\mathbf{x}(q)$ the tangent vector (Fig. 2-2)

$$\left(\frac{\partial x}{\partial u}, \frac{\partial y}{\partial u}, \frac{\partial z}{\partial u}\right) = \frac{\partial \mathbf{x}}{\partial u},$$

where the derivatives are computed at (u_0, v_0) and a vector is indicated by its components in the basis $\{f_1, f_2, f_3\}$. By the definition of differential (appendix to Chap. 2, Def. 1),

$$d\mathbf{x}_q(e_1) = \left(\frac{\partial x}{\partial u}, \frac{\partial y}{\partial u}, \frac{\partial z}{\partial u}\right) = \frac{\partial \mathbf{x}}{\partial u}.$$

Similarly, using the coordinate curve $u = u_0$ (image by $\mathbf{x}$ of the curve $v \longrightarrow (u_0, v)$), we obtain

$$d\mathbf{x}_q(e_2) = \left(\frac{\partial x}{\partial v}, \frac{\partial y}{\partial v}, \frac{\partial z}{\partial v}\right) = \frac{\partial \mathbf{x}}{\partial v}.$$

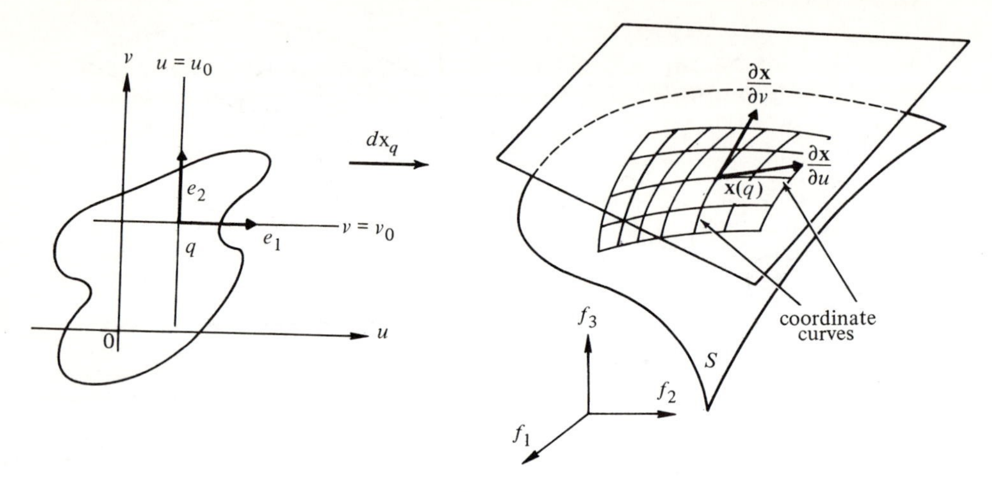

Figure 2-2

Thus, the matrix of the linear map $d\mathbf{x}_q$ in the referred basis is

$$d\mathbf{x}_q = \begin{pmatrix} \dfrac{\partial x}{\partial u} & \dfrac{\partial x}{\partial v} \\ \dfrac{\partial y}{\partial u} & \dfrac{\partial y}{\partial v} \\ \dfrac{\partial z}{\partial u} & \dfrac{\partial z}{\partial v} \end{pmatrix}.$$

Condition 3 of Def. 1 may now be expressed by requiring the two column vectors of this matrix to be linearly independent; or, equivalently, that the vector product $\partial\mathbf{x}/\partial u \wedge \partial\mathbf{x}/\partial v \neq 0$; or, in still another way, that one of the minors of order 2 of the matrix of $d\mathbf{x}_q$, that is, one of the Jacobian determinants

$$\frac{\partial(x, y)}{\partial(u, v)} = \begin{vmatrix} \dfrac{\partial x}{\partial u} & \dfrac{\partial x}{\partial v} \\ \dfrac{\partial y}{\partial u} & \dfrac{\partial y}{\partial v} \end{vmatrix}, \qquad \frac{\partial(y, z)}{\partial(u, v)}, \qquad \frac{\partial(x, z)}{\partial(u, v)},$$

be different from zero at q.

Remark 1. Definition 1 deserves a few comments. First, in contrast to our treatment of curves in Chap. 1, we have defined a surface as a subset S of R^3, and not as a map. This is achieved by covering S with the traces of parametrizations which satisfy conditions 1, 2, and 3.

Condition 1 is very natural if we expect to do some differential geometry on S. The one-to-oneness in condition 2 has the purpose of preventing self-intersections in regular surfaces. This is clearly necessary if we are to speak about, say, *the* tangent plane at a point $p \in S$ (see Fig. 2-3(a)). The continuity of the inverse in condition 2 has a more subtle purpose which can be fully understood only in the next section. For the time being, we shall mention that this condition is essential to proving that certain objects defined in terms of a parametrization do not depend on this parametrization but only on the set S itself. Finally, as we shall show in Sec. 2.4, condition 3 will guarantee the existence of a "tangent plane" at all points of S (see Fig. 2-3(b)).

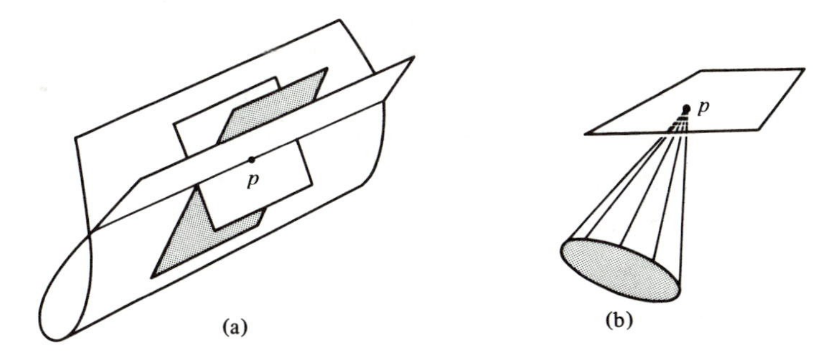

Figure 2-3. Some situations to be avoided in the definition of a regular surface.

Example 1. Let us show that the unit sphere

$$S^2 = \{(x, y, z) \in R^3; x^2 + y^2 + z^2 = 1\}$$

is a regular surface.

We first verify that the map $\mathbf{x}_1 : U \subset R^2 \longrightarrow R^3$ given by

$$\mathbf{x}_1(x, y) = (x, y, +\sqrt{1 - (x^2 + y^2)}), \qquad (x, y) \in U,$$

where $R^2 = \{(x, y, z) \in R^3; z = 0\}$ and $U = \{(x, y) \in R^2; x^2 + y^2 < 1\}$, is a parametrization of S^2. Observe that $\mathbf{x}_1(U)$ is the (open) part of S^2 above the xy plane.

Since $x^2 + y^2 < 1$, the function $+\sqrt{1 - (x^2 + y^2)}$ has continuous partial derivatives of all orders. Thus, $\mathbf{x}_1$ is differentiable and condition 1 holds.

Condition 3 is easily verified, since

$$\frac{\partial(x, y)}{\partial(x, y)} \equiv 1.$$

To check condition 2, we observe that $\mathbf{x}_1$ is one-to-one and that $\mathbf{x}_1^{-1}$ is the restriction of the (continuous) projection $\pi(x, y, z) = (x, y)$ to the set $\mathbf{x}_1(U)$. Thus, $\mathbf{x}_1^{-1}$ is continuous in $\mathbf{x}_1(U)$.

We shall now cover the whole sphere with similar parametrizations as follows. We define $\mathbf{x}_2 \colon U \subset R^2 \longrightarrow R^3$ by

$$\mathbf{x}_2(x, y) = (x, y, -\sqrt{1 - (x^2 + y^2)}),$$

check that $\mathbf{x}_2$ is a parametrization, and observe that $\mathbf{x}_1(U) \cup \mathbf{x}_2(U)$ covers S^2 minus the equator

$$\{(x, y, z) \in R^3; x^2 + y^2 = 1, z = 0\}.$$

Then, using the xz and zy planes, we define the parametrizations

$$\begin{aligned}
\mathbf{x}_3(x, z) &= (x, +\sqrt{1 - (x^2 + z^2)}, z), \\
\mathbf{x}_4(x, z) &= (x, -\sqrt{1 - (x^2 + z^2)}, z), \\
\mathbf{x}_5(y, z) &= (+\sqrt{1 - (y^2 + z^2)}, y, z), \\
\mathbf{x}_6(y, z) &= (-\sqrt{1 - (y^2 + z^2)}, y, z),
\end{aligned}$$

which, together with $\mathbf{x}_1$ and $\mathbf{x}_2$, cover S^2 completely (Fig. 2-4) and show that S^2 is a regular surface.

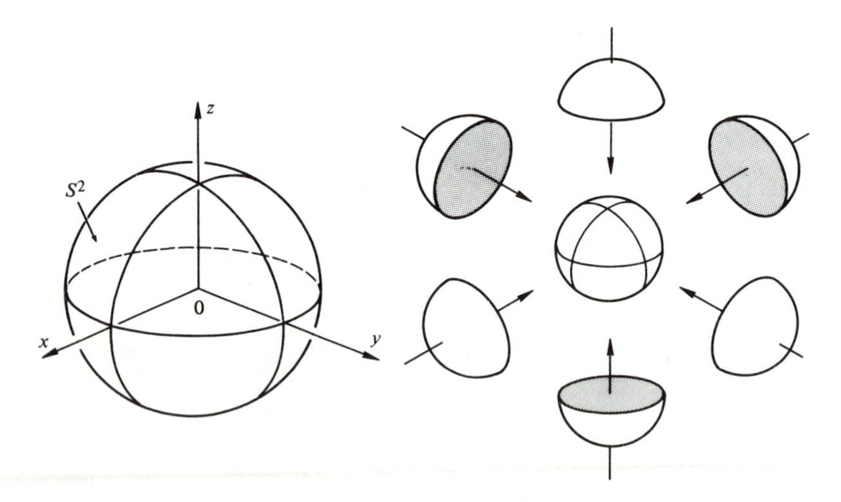

Figure 2-4

For most applications, it is convenient to relate parametrizations to the geographical coordinates on S^2. Let $V = \{(\theta, \varphi); 0 < \theta < \pi, 0 < \varphi < 2\pi\}$ and let $\mathbf{x} \colon V \longrightarrow R^3$ be given by

$$\mathbf{x}(\theta, \varphi) = (\sin\theta\cos\varphi, \sin\theta\sin\varphi, \cos\theta).$$

Clearly, $\mathbf{x}(V) \subset S^2$. We shall prove that $\mathbf{x}$ is a parametrization of S^2. θ is

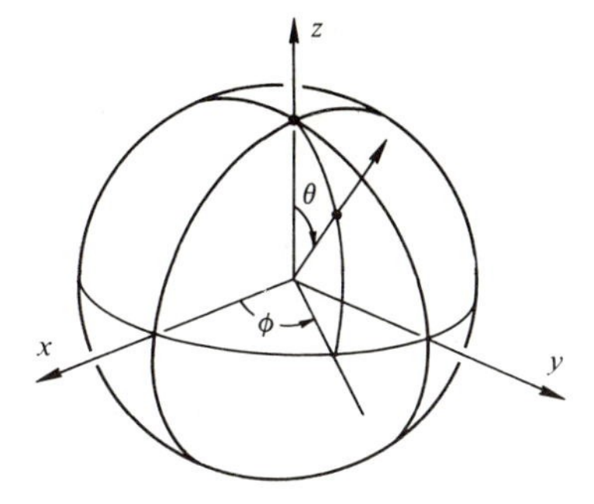

Figure 2-5

usually called the *colatitude* (the complement of the latitude) and φ the *longitude* (Fig. 2-5).

It is clear that the functions $\sin\theta\cos\varphi$, $\sin\theta\sin\varphi$, $\cos\theta$ have continuous partial derivatives of all orders; hence, $\mathbf{x}$ is differentiable. Moreover, in order that the Jacobian determinants

$$\frac{\partial(x, y)}{\partial(\theta, \varphi)} = \cos\theta\sin\theta,$$

$$\frac{\partial(y, z)}{\partial(\theta, \varphi)} = \sin^2\theta\cos\varphi,$$

$$\frac{\partial(x, z)}{\partial(\theta, \varphi)} = \sin^2\theta\sin\varphi$$

vanish simultaneously, it is necessary that

$$\cos^2\theta\sin^2\theta + \sin^4\theta\cos^2\varphi + \sin^4\theta\sin^2\varphi = \sin^2\theta = 0.$$

This does not happen in V, and so conditions 1 and 3 of Def. 1 are satisfied.

Next, we observe that given $(x, y, z) \in S^2 - C$, where C is the semicircle

$$C = \{(x, y, z) \in S^2; y = 0, x \geq 0\},$$

θ is uniquely determined by $\theta = \cos^{-1} z$, since $0 < \theta < \pi$. By knowing θ, we find $\sin\varphi$ and $\cos\varphi$ from $x = \sin\theta\cos\varphi$, $y = \sin\theta\sin\varphi$, and this determines φ uniquely ($0 < \varphi < 2\pi$). It follows that $\mathbf{x}$ has an inverse $\mathbf{x}^{-1}$. To complete the verification of condition 2, we should prove that $\mathbf{x}^{-1}$ is continuous. However, since we shall soon prove (Prop. 4) that this verification is not necessary provided we already know that the set S is a regular surface, we shall not do that here.

We remark that $\mathbf{x}(V)$ only omits a semicircle of S^2 (including the two poles) and that S^2 can be covered with the coordinate neighborhoods of two parametrizations of this type.

In Exercise 16 we shall indicate how to cover S^2 with another useful set of coordinate neighborhoods.

Example 1 shows that deciding whether a given subset of R^3 is a regular surface directly from the definition may be quite tiresome. Before going into further examples, we shall present two propositions which will simplify this task. Proposition 1 shows the relation which exists between the definition of a regular surface and the graph of a function $z = f(x, y)$. Proposition 2 uses the inverse function theorem and relates the definition of a regular surface with the subsets of the form $f(x, y, z) =$ constant.

PROPOSITION 1. *If* $f: U \to R$ *is a differentiable function in an open set* U *of* R^2, *then the graph of* f, *that is, the subset of* R^3 *given by* $(x, y, f(x, y))$ *for* $(x, y) \in U$, *is a regular surface.*

Proof: It suffices to show that the map $\mathbf{x}: U \to R^3$ given by

$$\mathbf{x}(u, v) = (u, v, f(u, v))$$

is a parametrization of the graph whose coordinate neighborhood covers every point of the graph. Condition 1 is clearly satisfied, and condition 3 also offers no difficulty since $\partial(x, y)/\partial(u, v) \equiv 1$. Finally, each point (x, y, z) of the graph is the image under $\mathbf{x}$ of the unique point $(u, v) = (x, y) \in U$. $\mathbf{x}$ is therefore one-to-one, and since $\mathbf{x}^{-1}$ is the restriction to the graph of f of the (continuous) projection of R^3 onto the xy plane, $\mathbf{x}^{-1}$ is continuous.
Q.E.D.

Before stating Prop. 2, we shall need a definition.

DEFINITION 2. *Given a differentiable map* $F: U \subset R^n \to R^m$ *defined in an open set* U *of* R^n *we say that* $p \in U$ *is a* critical point *of* F *if the differential* $dF_p: R^n \to R^m$ *is not a surjective (or onto) mapping. The image* $F(p) \in R^m$ *of a critical point is called a* critical value *of* F. *A point of* R^m *which is not a critical value is called a* regular value *of* F.

The terminology is evidently motivated by the particular case in which $f: U \subset R \to R$ is a real-valued function of a real variable. A point $x_0 \in U$ is critical if $f'(x_0) = 0$, that is, if the differential df_{x_0} carries all the vectors in R to the zero vector (Fig. 2-6). Notice that any point $a \notin f(U)$ is trivially a regular value of f.

If $f: U \subset R^3 \to R$ is a differentiable function, then df_p applied to the vector $(1, 0, 0)$ is obtained by calculating the tangent vector at $f(p)$ to the curve

$$x \longrightarrow f(x, y_0, z_0).$$

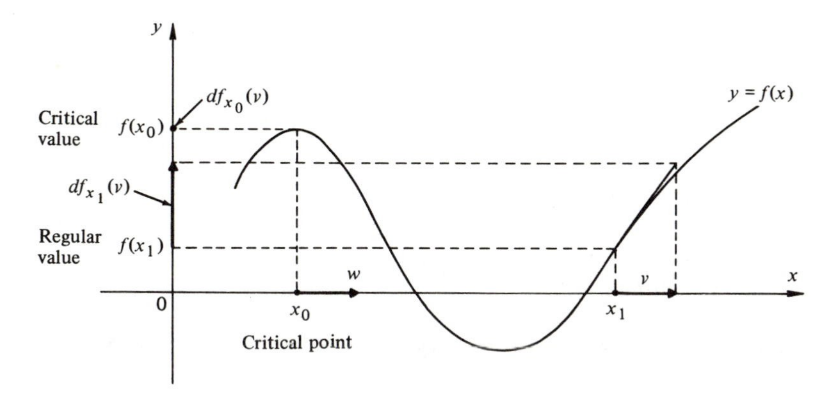

Figure 2-6

It follows that

$$df_p(1, 0, 0) = \frac{\partial f}{\partial x}(x_0, y_0, z_0) = f_x$$

and analogously that

$$df_p(0, 1, 0) = f_y, \qquad df_p(0, 0, 1) = f_z.$$

We conclude that the matrix of df_p in the basis (1, 0, 0), (0, 1, 0), (0, 0, 1) is given by

$$df_p = (f_x, f_y, f_z).$$

Note, in this case, that to say that df_p is not surjective is equivalent to saying that $f_x = f_y = f_z = 0$ at p. Hence, $a \in f(U)$ is a regular value of $f: U \subset R^3 \to R$ if and only if $f_x, f_y,$ and f_z do not vanish simultaneously at any point in the inverse image

$$f^{-1}(a) = \{(x, y, z) \in U: f(x, y, z) = a\}.$$

PROPOSITION 2. *If* $f: U \subset R^3 \to R$ *is a differentiable function and* $a \in f(U)$ *is a regular value of* f, *then* $f^{-1}(a)$ *is a regular surface in* R^3.

Proof. Let $p = (x_0, y_0, z_0)$ be a point of $f^{-1}(a)$. Since a is a regular value of f, it is possible to assume, by renaming the axis if necessary, that $f_z \neq 0$ at p. We define a mapping $F: U \subset R^3 \to R^3$ by

$$F(x, y, z) = (x, y, f(x, y, z)),$$

and we indicate by (u, v, t) the coordinates of a point in R^3 where F takes its values. The differential of F at p is given by

$$dF_p = \begin{pmatrix} 1 & 0 & 0 \\ 0 & 1 & 0 \\ f_x & f_y & f_z \end{pmatrix},$$

whence

$$\det(dF_p) = f_z \neq 0.$$

We can therefore apply the inverse function theorem (cf. the appendix to Chap. 2), which guarantees the existence of neighborhoods V of p and W of $F(p)$ such that $F: V \longrightarrow W$ is invertible and the inverse $F^{-1}: W \longrightarrow V$ is differentiable (Fig. 2-7). It follows that the coordinate functions of F^{-1}, i.e., the functions

$$x = u, \qquad y = v, \qquad z = g(u, v, t), \qquad (u, v, t) \in W,$$

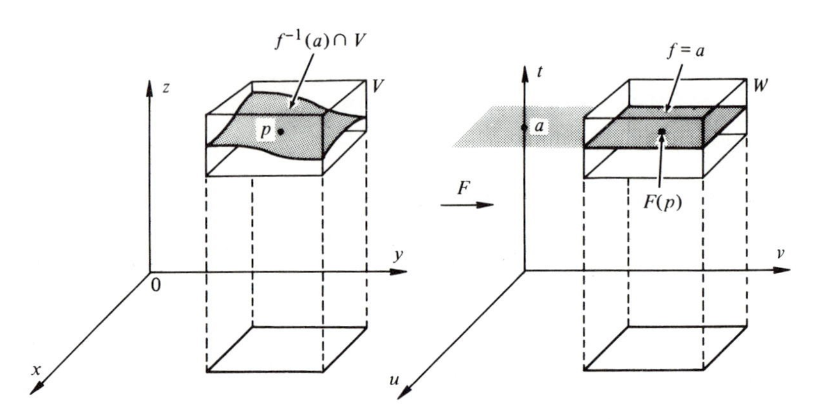

Figure 2-7

are differentiable. In particular, $z = g(u, v, a) = h(x, y)$ is a differentiable function defined in the projection of V onto the xy plane. Since

$$F(f^{-1}(a) \cap V) = W \cap \{(u, v, t); t = a\},$$

we conclude that the graph of h is $f^{-1}(a) \cap V$. By Prop. 1, $f^{-1}(a) \cap V$ is a coordinate neighborhood of p. Therefore, every $p \in f^{-1}(a)$ can be covered by a coordinate neighborhood, and so $f^{-1}(a)$ is a regular surface. **Q.E.D.**

Remark 2. The proof consists essentially of using the inverse function theorem "to solve for z" in the equation $f(x, y, z) = a$, which can be done in a neighborhood of p if $f_z(p) \neq 0$. This fact is a special case of the general implicit function theorem, which follows from the inverse function theorem and is, in fact, equivalent to it.

Example 2. The ellipsoid

$$\frac{x^2}{a^2} + \frac{y^2}{b^2} + \frac{z^2}{c^2} = 1$$

is a regular surface. In fact, it is the set $f^{-1}(0)$ where

$$f(x, y, z) = \frac{x^2}{a^2} + \frac{y^2}{b^2} + \frac{z^2}{c^2} - 1$$

is a differentiable function and 0 is a regular value of f. This follows from the fact that the partial derivatives $f_x = 2x/a^2$, $f_y = 2y/b^2$, $f_z = 2z/c^2$ vanish simultaneously only at the point $(0, 0, 0)$, which does not belong to $f^{-1}(0)$. This example includes the sphere as a particular case ($a = b = c = 1$).

The examples of regular surfaces presented so far have been connected subsets of R^3. A surface $S \subset R^3$ is said to be *connected* if any two of its points can be joined by a continuous curve in S. In the definition of a regular surface we made no restrictions on the connectedness of the surfaces, and the following example shows that the regular surfaces given by Prop. 2 may not be connected.

Example 3. The hyperboloid of two sheets $-x^2 - y^2 + z^2 = 1$ is a regular surface, since it is given by $S = f^{-1}(0)$, where 0 is a regular value of $f(x, y, z) = -x^2 - y^2 + z^2 - 1$ (Fig. 2-8). Note that the surface S is not connected; that is, given two points in two distinct sheets ($z > 0$ and $z < 0$) it is not possible to join them by a continuous curve $\alpha(t) = (x(t), y(t), z(t))$ contained in the surface; otherwise, z changes sign and, for some t_0, we have $z(t_0) = 0$, which means that $\alpha(t_0) \notin S$.

Incidentally, the argument of Example 3 may be used to prove a property of connected surfaces that we shall use repeatedly. *If* $f: S \subset R^3 \to R$ *is a nonzero continuous function defined on a connected surface* S, *then* f *does not change sign on* S.

To prove this, we use the intermediate value theorem (appendix to Chap. 2, Prop. 4). Assume, by contradiction, that $f(p) > 0$ and $f(q) < 0$ for some points $p, q \in S$. Since S is connected, there exists a continuous curve $\alpha: [a, b] \to S$ with $\alpha(a) = p$, $\alpha(b) = q$. By applying the intermediate value theorem to the continuous function $f \circ \alpha: [a, b] \to R$, we find that there exists $c \in (a, b)$ with $f \circ \alpha(c) = 0$; that is, f is zero at $\alpha(c)$, a contradiction.

Example 4. The torus T is a "surface" generated by rotating a circle S^1 of radius r about a straight line belonging to the plane of the circle and at a distance $a > r$ away from the center of the circle (Fig. 2-9).

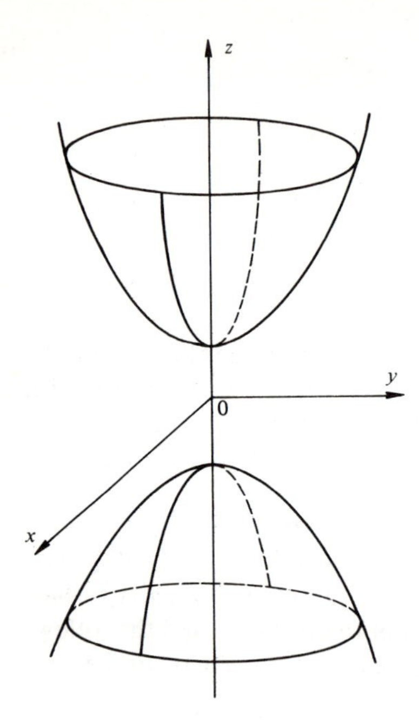

Figure 2-8. A nonconnected surface: $-y^2 - x^2 + z^2 = 1$.

Let S^1 be the circle in the yz plane with its center in the point $(0, a, 0)$. Then S^1 is given by $(y - a)^2 + z^2 = r^2$, and the points of the figure T obtained by rotating this circle about the z axis satisfy the equation

$$z^2 = r^2 - (\sqrt{x^2 + y^2} - a)^2.$$

Therefore, T is the inverse image of r^2 by the function

$$f(x, y, z) = z^2 + (\sqrt{x^2 + y^2} - a)^2.$$

This function is differentiable for $(x, y) \neq (0, 0)$, and since

$$\frac{\partial f}{\partial z} = 2z, \qquad \frac{\partial f}{\partial y} = \frac{2y(\sqrt{x^2 + y^2} - a)}{\sqrt{x^2 + y^2}},$$

$$\frac{\partial f}{\partial x} = \frac{2x(\sqrt{x^2 + y^2} - a)}{\sqrt{x^2 + y^2}},$$

r^2 is a regular value of f. It follows that the torus T is a regular surface.

Proposition 1 says that the graph of a differentiable function is a regular surface. The following proposition provides a local converse of this; that is, any regular surface is locally the graph of a differentiable function.

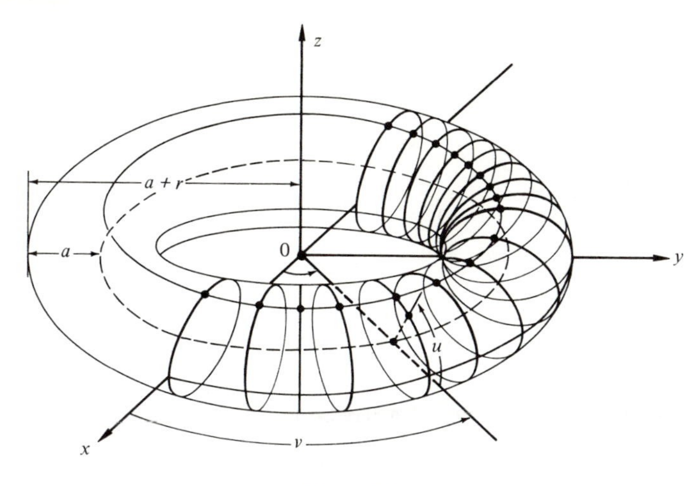

Figure 2-9

PROPOSITION 3. *Let* $S \subset R^3$ *be a regular surface and* $p \in S$. *Then there exists a neighborhood* V *of* p *in* S *such that* V *is the graph of a differentiable function which has one of the following three forms:* $z = f(x, y)$, $y = g(x, z)$, $x = h(y, z)$.

Proof. Let $\mathbf{x}: U \subset R^2 \to S$ be a parametrization of S in p, and write $\mathbf{x}(u, v) = (x(u, v), y(u, v), z(u, v))$, $(u, v) \in U$. By condition 3 of Def. 1, one of the Jacobian determinants

$$\frac{\partial(x, y)}{\partial(u, v)}, \qquad \frac{\partial(y, z)}{\partial(u, v)}, \qquad \frac{\partial(z, x)}{\partial(u, v)}$$

is not zero at $\mathbf{x}^{-1}(p) = q$.

Suppose first that $(\partial(x, y)/\partial(u, v))(q) \neq 0$, and consider the map $\pi \circ \mathbf{x}: U \to R^2$, where π is the projection $\pi(x, y, z) = (x, y)$. Then $\pi \circ \mathbf{x}(u, v) = (x(u, v), y(u, v))$, and, since $(\partial(x, y)/\partial(u, v))(q) \neq 0$, we can apply the inverse function theorem to guarantee the existence of neighborhoods V_1 of q, V_2 of $\pi \circ \mathbf{x}(q)$ such that $\pi \circ \mathbf{x}$ maps V_1 diffeomorphically onto V_2 (Fig. 2-10). It follows that π restricted to $\mathbf{x}(V_1) = V$ is one-to-one and that there is a differentiable inverse $(\pi \circ \mathbf{x})^{-1}: V_2 \to V_1$. Observe that, since $\mathbf{x}$ is a homeomorphism, V is a neighborhood of p in S. Now, if we compose the map $(\pi \circ \mathbf{x})^{-1}: (x, y) \to (u(x, y), v(x, y))$ with the function $(u, v) \to z(u, v)$, we find that V is the graph of the differentiable function $z = z(u(x, y), v(x, y)) = f(x, y)$, and this settles the first case.

The remaining cases can be treated in the same way, yielding $x = h(y, z)$ and $y = g(x, z)$. **Q.E.D.**

The next proposition says that if we already know that S is a regular surface and we have a candidate $\mathbf{x}$ for a parametrization, we do not have to

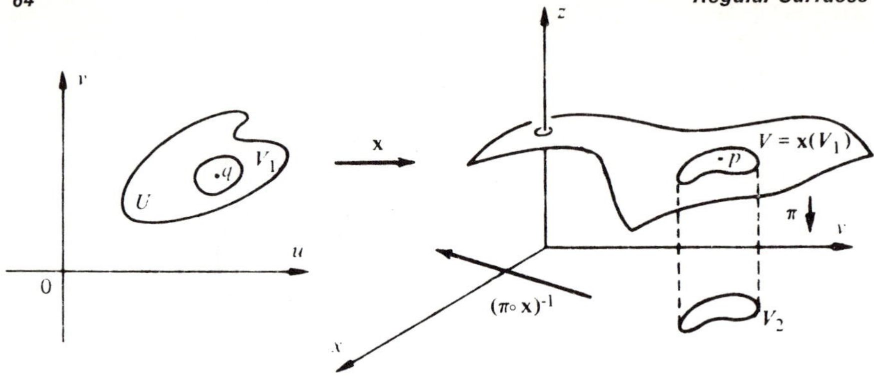

Figure 2-10

check that $\mathbf{x}^{-1}$ is continuous, provided that the other conditions hold. This remark was used in Example 1.

PROPOSITION 4. *Let* $p \in S$ *be a point of a regular surface* S *and let* $\mathbf{x}: U \subset R^2 \longrightarrow R^3$ *be a map with* $p \in \mathbf{x}(U)$ *such that conditions* 1 *and* 3 *of Def.* 1 *hold. Assume that* $\mathbf{x}$ *is one-to-one. Then* $\mathbf{x}^{-1}$ *is continuous.*

Proof. The first part of the proof is similar to the proof of Prop. 3. Write $\mathbf{x}(u, v) = (x(u, v), y(u, v), z(u, v))$, $(u, v) \in U$, and let $q \in U$. By conditions 1 and 3 we may assume, interchanging the coordinate axis of R^3 if necessary, that $(\partial(x, y)/\partial(u, v))(q) \neq 0$. Let $\pi: R^3 \longrightarrow R^2$ be the projection $\pi(x, y, z) = (x, y)$. From the inverse function theorem, we obtain neighborhoods V_1 of q in U and V_2 of $\pi \circ \mathbf{x}(q)$ in R^2 such that $\pi \circ \mathbf{x}$ maps V_1 diffeomorphically onto V_2.

Assume now that $\mathbf{x}$ is one-to-one. Then, restricted to $\mathbf{x}(V_1)$,

$$\mathbf{x}^{-1} = (\pi \circ \mathbf{x})^{-1} \circ \pi$$

(see Fig. 2-10). Thus, $\mathbf{x}^{-1}$ is continuous as a composition of continuous maps. Since q is arbitrary, $\mathbf{x}^{-1}$ is continuous in $\mathbf{x}(U)$. **Q.E.D.**

Example 5. The one-sheeted cone C, given by

$$z = +\sqrt{x^2 + y^2}, \qquad (x, y) \in R^2,$$

is not a regular surface. Observe that we cannot conclude this from the fact alone that the "natural" parametrization

$$(x, y) \longrightarrow (x, y, +\sqrt{x^2 + y^2})$$

is not differentiable; there could be other parametrizations satisfying Def. 1.

To show that this is not the case, we use Prop. 3. If C were a regular surface, it would be, in a neighborhood of $(0, 0, 0) \in C$, the graph of a differentiable function having one of three forms: $y = h(x, z)$, $x = g(y, z)$, $z = f(x, y)$. The two first forms can be discarded by the simple fact that the projections of C over the xz and yz planes are not one-to-one. The last form would have to agree, in a neighborhood of $(0, 0, 0)$, with $z = +\sqrt{x^2 + y^2}$. Since $z = +\sqrt{x^2 + y^2}$ is not differentiable at $(0, 0)$, this is impossible.

Example 6. A parametrization for the torus T of Example 4 can be given by (Fig. 2-9)

$$\mathbf{x}(u, v) = ((r \cos u + a) \cos v, (r \cos u + a) \sin v, r \sin u),$$

where $0 < u < 2\pi$, $0 < v < 2\pi$.

Condition 1 of Def. 1 is easily checked, and condition 3 reduces to a straightforward computation, which is left as an exercise. Since we know that T is a regular surface, condition 2 is equivalent, by Prop. 4, to the fact that $\mathbf{x}$ is one-to-one.

To prove that $\mathbf{x}$ is one-to-one, we first observe that $\sin u = z/r$; also, if $\sqrt{x^2 + y^2} \leq a$, then $\pi/2 \leq u \leq 3\pi/2$, and if $\sqrt{x^2 + y^2} \geq a$, then either $0 < u \leq \pi/2$ or $3\pi/2 \leq u < 2\pi$. Thus, given (x, y, z), this determines u, $0 < u < 2\pi$, uniquely. By knowing u, x, and y we find $\cos v$ and $\sin v$. This determines v uniquely, $0 < v < 2\pi$. Thus, $\mathbf{x}$ is one-to-one.

It is easy to see that the torus can be covered by three such coordinate neighborhoods.

EXERCISES†

1. Show that the cylinder $\{(x, y, z) \in R^3; x^2 + y^2 = 1\}$ is a regular surface, and find parametrizations whose coordinate neighborhoods cover it.

2. Is the set $\{(x, y, z) \in R^3; z = 0 \text{ and } x^2 + y^2 \leq 1\}$ a regular surface? Is the set $\{(x, y, z) \in R^3; z = 0, \text{ and } x^2 + y^2 < 1\}$ a regular surface?

3. Show that the two-sheeted cone, with its vertex at the origin, that is, the set $\{(x, y, z) \in R^3; x^2 + y^2 - z^2 = 0\}$, is not a regular surface.

4. Let $f(x, y, z) = z^2$. Prove that 0 is not a regular value of f and yet that $f^{-1}(0)$ is a regular surface.

***5.** Let $P = \{(x, y, z) \in R^3; x = y\}$ (a plane) and let $\mathbf{x}\colon U \subset R^2 \longrightarrow R^3$ be given by

$$\mathbf{x}(u, v) = (u + v, u + v, uv),$$

where $U = \{(u, v) \in R^2; u > v\}$. Clearly, $\mathbf{x}(U) \subset P$. Is $\mathbf{x}$ a parametrization of P?

†Those who have omitted the proofs in this section should also omit Exercises 17–19.

6. Give another proof of Prop. 1 by applying Prop. 2 to $h(x, y, z) = f(x, y) - z$.

7. Let $f(x, y, z) = (x + y + z - 1)^2$.

a. Locate the critical points and critical values of f.

b. For what values of c is the set $f(x, y, z) = c$ a regular surface?

c. Answer the questions of parts a and b for the function $f(x, y, z) = xyz^2$.

8. Let $\mathbf{x}(u, v)$ be as in Def. 1. Verify that $d\mathbf{x}_q: R^2 \longrightarrow R^3$ is one-to-one if and only if

$$\frac{\partial \mathbf{x}}{\partial u} \wedge \frac{\partial \mathbf{x}}{\partial v} \neq 0.$$

9. Let V be an open set in the xy plane. Show that the set

$$\{(x, y, z) \in R^3; z = 0 \text{ and } (x, y) \in V\}$$

is a regular surface.

10. Let C be a figure "8" in the xy plane and let S be the cylindrical surface over C (Fig. 2-11); that is,

$$S = \{(x, y, z) \in R^3; (x, y) \in C\}.$$

Is the set S a regular surface?

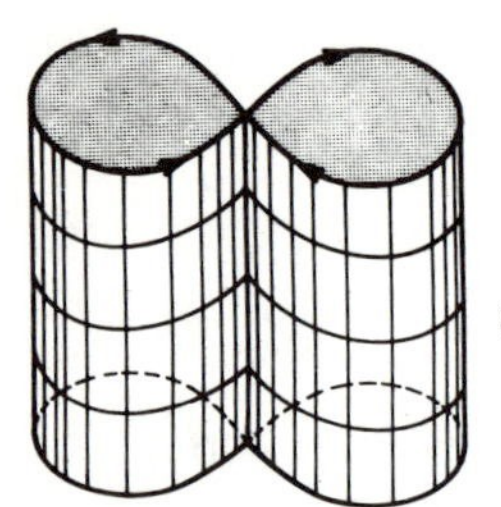

Figure 2-11

11. Show that the set $S = \{(x, y, z) \in R^3; z = x^2 - y^2\}$ is a regular surface and check that parts a and b are parametrizations for S:

a. $\mathbf{x}(u, v) = (u + v, u - v, 4uv)$, $(u, v) \in R^2$.

***b.** $\mathbf{x}(u, v) = (u \cosh v, u \sinh v, u^2)$, $(u, v) \in R^2$, $u \neq 0$.

Which parts of S do these parametrizations cover?

12. Show that $\mathbf{x}: U \subset R^2 \longrightarrow R^3$ given by

$$\mathbf{x}(u, v) = (a \sin u \cos v, b \sin u \sin v, c \cos u), \qquad a, b, c \neq 0,$$

where $0 < u < \pi$, $0 < v < 2\pi$, is a parametrization for the ellipsoid

$$\frac{x^2}{a^2} + \frac{y^2}{b^2} + \frac{z^2}{c^2} = 1.$$

Describe geometrically the curves $u = \text{const.}$ on the ellipsoid.

***13.** Find a parametrization for the hyperboloid of two sheets $\{(x, y, z) \in R^3; -x^2 - y^2 + z^2 = 1\}$.

14. A half-line $[0, \infty)$ is perpendicular to a line E and rotates about E from a given initial position while its origin 0 moves along E. The movement is such that when $[0, \infty)$ has rotated through an angle θ, the origin is at a distance $d = \sin^2(\theta/2)$ from its initial position on E. Verify that by removing the line E from the image of the rotating line we obtain a regular surface. If the movement were such that $d = \sin(\theta/2)$, what else would need to be excluded to have a regular surface?

***15.** Let two points $p(t)$ and $q(t)$ move with the same speed, p starting from $(0, 0, 0)$ and moving along the z axis and q starting at $(a, 0, 0)$, $a \neq 0$, and moving parallel to the y axis. Show that the line joining $p(t)$ to $q(t)$ describes a set in R^3 given by $y(x - a) + zx = 0$. Is this a regular surface?

16. One way to define a system of coordinates for the sphere S^2, given by $x^2 + y^2 + (z - 1)^2 = 1$, is to consider the so-called *stereographic projection* $\pi: S^2 \sim \{N\} \longrightarrow R^2$ which carries a point $p = (x, y, z)$ of the sphere S^2 minus the north pole $N = (0, 0, 2)$ onto the intersection of the xy plane with the straight line which connects N to p (Fig. 2-12). Let $(u, v) = \pi(x, y, z)$, where $(x, y, z) \in S^2 \sim \{N\}$ and $(u, v) \in xy$ plane.

a. Show that $\pi^{-1}: R^2 \longrightarrow S^2$ is given by

$$\pi^{-1}\begin{cases} x = \dfrac{4u}{u^2 + v^2 + 4}, \\ y = \dfrac{4v}{u^2 + v^2 + 4}, \\ z = \dfrac{2(u^2 + v^2)}{u^2 + v^2 + 4}. \end{cases}$$

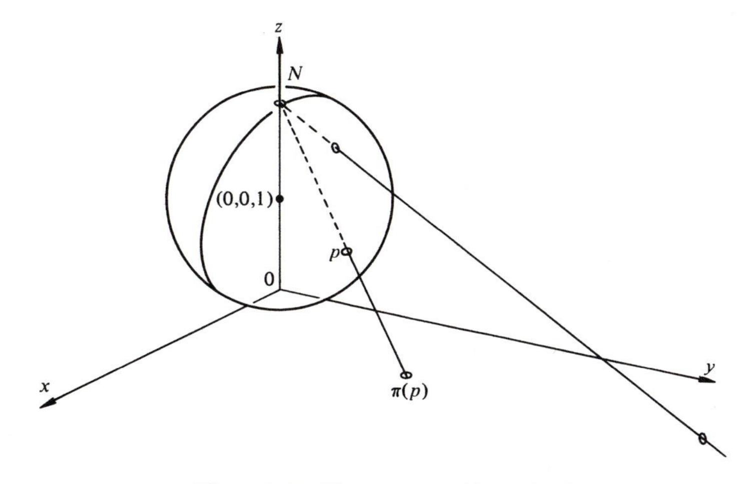

Figure 2-12. The stereographic projection.

b. Show that it is possible, using stereographic projection, to cover the sphere with two coordinate neighborhoods.

17. Define a regular curve in analogy with a regular surface. Prove that

a. The inverse image of a regular value of a differentiable function

$$f\colon U \subset R^2 \longrightarrow R$$

is a regular plane curve. Give an example of such a curve which is not connected.

b. The inverse image of a regular value of a differentiable map

$$F\colon U \subset R^3 \longrightarrow R^2$$

is a regular curve in R^3. Show the relationship between this proposition and the classical way of defining a curve in R^3 as the intersection of two surfaces.

***c.** The set $C = \{(x, y) \in R^2; x^2 = y^3\}$ is not a regular curve.

***18.** Suppose that $f(x, y, z) = u = \text{const.}$, $g(x, y, z) = v = \text{const.}$,

$$h(x, y, z) = w = \text{const.},$$

describe three families of regular surfaces and assume that at (x_0, y_0, z_0) the Jacobian

$$\frac{\partial(f, g, h)}{\partial(x, y, z)} \neq 0.$$

Prove that in a neighborhood of (x_0, y_0, z_0) the three families will be described by a mapping $F(u, v, w) = (x, y, z)$ of an open set of R^3 into R^3, where a local parametrization for the surface of the family $f(x, y, z) = u$, for example, is obtained by setting $u = \text{const.}$ in this mapping. Determine F for the case where the three families of surfaces are

$f(x, y, z) = x^2 + y^2 + z^2 = u = \text{const.}$; (spheres with center (0, 0, 0));

$g(x, y, z) = \dfrac{y}{x} = v = \text{const.}$, (planes through the z axis);

$h(x, y, z) = \dfrac{x^2 + y^2}{z^2} = w = \text{const.}$, (cones with vertex at (0, 0, 0)).

***19.** Let $\alpha\colon (-3, 0) \longrightarrow R^2$ be defined by (Fig. 2-13)

$$\alpha(t)\begin{cases} = (0, -(t+2)), & \text{if } t \in (-3, -1), \\ = \text{regular parametrized curve joining } p = (0, -1) \text{ to } q = \left(\frac{1}{\pi}, 0\right), & \text{if } t \in \left(-1, -\frac{1}{\pi}\right), \\ = \left(-t, -\sin\frac{1}{t}\right), & \text{if } t \in \left(-\frac{1}{\pi}, 0\right). \end{cases}$$

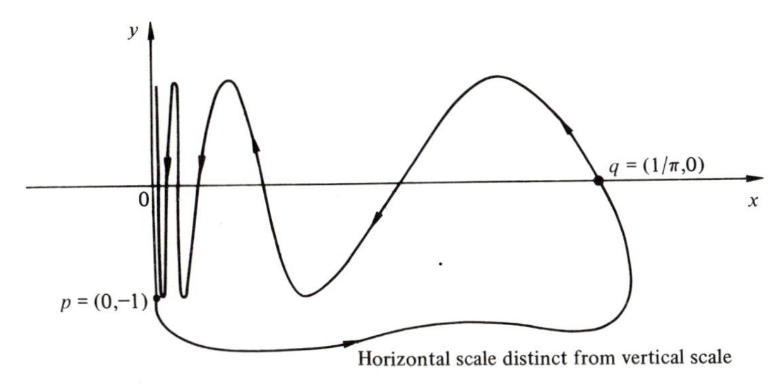

Figure 2-13

It is possible to define the curve joining p to q so that all the derivatives of α are continuous at the corresponding points and α has no self-intersections. Let C be the trace of α.

a. Is C a regular curve?

b. Let a normal line to the plane R^2 run through C so that it describes a "cylinder" S. Is S a regular surface?

2-3. Change of Parameters; Differentiable Functions on Surfaces†

Differential geometry is concerned with those properties of surfaces which depend on their behavior in a neighborhood of a point. The definition of a regular surface given in Sec. 2-2 is adequate for this purpose. According to this definition, each point p of a regular surface belongs to a coordinate neighborhood. The points of such a neighborhood are characterized by their coordinates, and we should be able, therefore, to define the local properties which interest us in terms of these coordinates.

For example, it is important that we be able to define what it means for a function $f: S \longrightarrow R$ to be differentiable at a point p of a regular surface S. A natural way to proceed is to choose a coordinate neighborhood of p, with coordinates u, v, and say that f is differentiable at p if its expression in the coordinates u and v admits continuous partial derivatives of all orders.

The same point of S can, however, belong to various coordinate neighborhoods (in the sphere of Example 1 of Sec. 2-2 any point of the interior of the first octant belongs to three of the given coordinate neighborhoods). Moreover, other coordinate systems could be chosen in a neighborhood of p (the points referred to on the sphere could also be parametrized by geo-

†Proofs in this section may be omitted on a first reading.

graphical coordinates or by stereographic projection; cf. Exercise 16, Sec. 2-2). For the above definition to make sense, it is necessary that it does not depend on the chosen system of coordinates. In other words, it must be shown that when p belongs to two coordinate neighborhoods, with parameters (u, v) and (ξ, η), it is possible to pass from one of these pairs of coordinates to the other by means of a differentiable transformation.

The following proposition shows that this is true.

PROPOSITION 1 (Change of Parameters). *Let* p *be a point of a regular surface* S, *and let* $\mathbf{x}\colon U \subset R^2 \to S$, $\mathbf{y}\colon V \subset R^2 \to S$ *be two parametrizations of* S *such that* $p \in \mathbf{x}(U) \cap \mathbf{y}(V) = W$. *Then the "change of coordinates"* $h = \mathbf{x}^{-1} \circ \mathbf{y}\colon \mathbf{y}^{-1}(W) \to \mathbf{x}^{-1}(W)$ (Fig. 2-14) *is a diffeomorphism; that is,* h *is differentiable and has a differentiable inverse* h^{-1}.

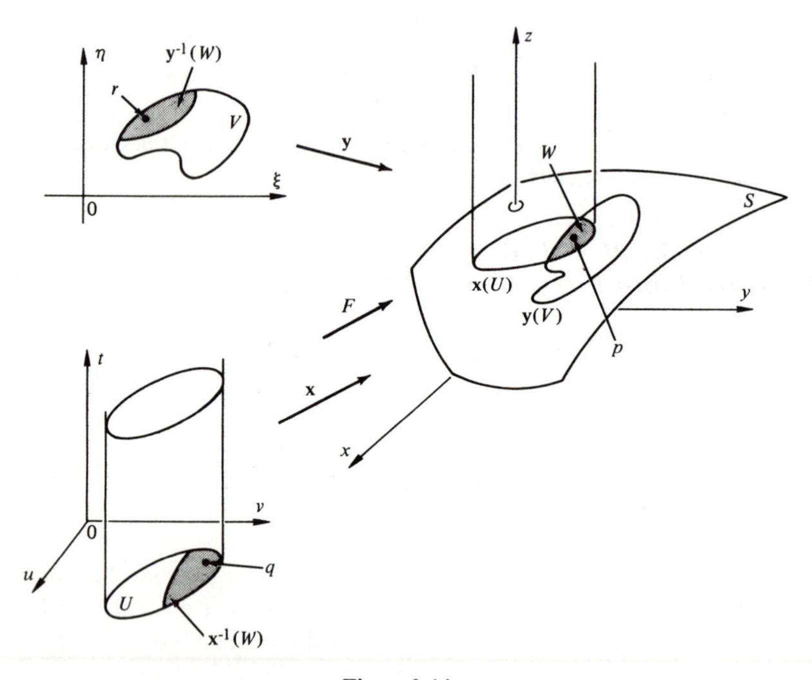

Figure 2-14

In other words, if $\mathbf{x}$ and $\mathbf{y}$ are given by

$$\mathbf{x}(u, v) = (x(u, v), y(u, v), z(u, v)), \qquad (u, v) \in U,$$
$$\mathbf{y}(\xi, \eta) = (x(\xi, \eta), y(\xi, \eta), z(\xi, \eta)), \qquad (\xi, \eta) \in V,$$

then the change of coordinates h, given by

$$u = u(\xi, \eta), \qquad v = v(\xi, \eta), \qquad (\xi, \eta) \in \mathbf{y}^{-1}(W),$$

has the property that the functions u and v have continuous partial derivatives of all orders, and the map h can be inverted, yielding

$$\xi = \xi(u, v), \qquad \eta = \eta(u, v), \qquad (u, v) \in \mathbf{x}^{-1}(W),$$

where the functions ξ and η also have partial derivatives of all orders. Since

$$\frac{\partial(u, v)}{\partial(\xi, \eta)} \cdot \frac{\partial(\xi, \eta)}{\partial(u, v)} = 1,$$

this implies that the Jacobian determinants of both h and h^{-1} are nonzero everywhere.

Proof of Prop. 1. $h = \mathbf{x}^{-1} \circ \mathbf{y}$ is a homeomorphism, since it is composed of homeomorphisms (cf. the appendix to Chap. 2, Prop. 3). It is not possible to conclude, by an analogous argument, that h is differentiable, since $\mathbf{x}^{-1}$ is defined in an open subset of S, and we do not yet know what is meant by a differentiable function on S.

We proceed in the following way. Let $r \in \mathbf{y}^{-1}(W)$ and set $q = h(r)$. Since $\mathbf{x}(u, v) = (x(u, v), y(u, v), z(u, v))$ is a parametrization, we can assume, by renaming the axis if necessary, that

$$\frac{\partial(x, y)}{\partial(u, v)}(q) \neq 0.$$

We extend $\mathbf{x}$ to a map $F\colon U \times R \longrightarrow R^3$ defined by

$$F(u, v, t) = (x(u, v), y(u, v), z(u, v) + t), \qquad (u, v) \in U, t \in R.$$

Geometrically, F maps a vertical cylinder C over U into a "vertical cylinder" over $\mathbf{x}(U)$ by mapping each section of C with height t into the surface $\mathbf{x}(u, v) + te_3$, where e_3 is the unit vector of the z axis (Fig. 2-14).

It is clear that F is differentiable and that the restriction $F|U \times \{0\} = \mathbf{x}$. Calculating the determinant of the differential dF_q, we obtain

$$\begin{vmatrix} \dfrac{\partial x}{\partial u} & \dfrac{\partial x}{\partial v} & 0 \\ \dfrac{\partial y}{\partial u} & \dfrac{\partial y}{\partial v} & 0 \\ \dfrac{\partial z}{\partial u} & \dfrac{\partial z}{\partial v} & 1 \end{vmatrix} = \frac{\partial(x, y)}{\partial(u, v)}(q) \neq 0.$$

It is possible therefore to apply the inverse function theorem, which guarantees the existence of a neighborhood M of $\mathbf{x}(q)$ in R^3 such that F^{-1} exists and is differentiable in M.

By the continuity of $\mathbf{y}$, there exists a neighborhood N of r in V such that $\mathbf{y}(N) \subset M$ (appendix to Chap. 2, Prop. 2). Notice that, restricted to N, $h|N = F^{-1} \circ \mathbf{y}|N$ is a composition of differentiable maps. Thus, we can apply the chain rule for maps (appendix to Chap. 2, Prop. 8) and conclude that h is differentiable at r. Since r is arbitrary, h is differentiable on $\mathbf{y}^{-1}(W)$.

Exactly the same argument can be applied to show that the map h^{-1} is differentiable, and so h is a diffeomorphism. **Q.E.D.**

We shall now give an explicit definition of what is meant by a differentiable function on a regular surface.

DEFINITION 1. *Let* $f\colon V \subset S \to R$ *be a function defined in an open subset* V *of a regular surface* S. *Then* f *is said to be* differentiable at $p \in V$ *if, for some parametrization* $\mathbf{x}\colon U \subset R^2 \to S$ *with* $p \in \mathbf{x}(U) \subset V$, *the composition* $f \circ \mathbf{x}\colon U \subset R^2 \to R$ *is differentiable at* $\mathbf{x}^{-1}(p)$. f *is* differentiable *in* V *if it is differentiable at all points of* V.

It follows immediately from the last proposition that the definition given does not depend on the choice of the parametrization $\mathbf{x}$. In fact, if $\mathbf{y}\colon V \subset R^2 \to S$ is another parametrization with $p \in \mathbf{x}(V)$, and if $h = \mathbf{x}^{-1} \circ \mathbf{y}$, then $f \circ \mathbf{y} = f \circ \mathbf{x} \circ h$ is also differentiable, whence the asserted independence.

Remark 1. We shall frequently make the notational abuse of indicating f and $f \circ \mathbf{x}$ by the same symbol $f(u, v)$, and say that $f(u, v)$ is *the expression* of f in the system of coordinates $\mathbf{x}$. This is equivalent to identifying $\mathbf{x}(U)$ with U and thinking of (u, v), indifferently, as a point of U and as a point of $\mathbf{x}(U)$ with coordinates (u, v). From now on, abuses of language of this type will be used without further comment.

Example 1. Let S be a regular surface and $V \subset R^3$ be an open set such that $S \subset V$. Let $f\colon V \subset R^3 \to R$ be a differentiable function. Then the restriction of f to S is a differentiable function on S. In fact, for any $p \in S$ and any parametrization $\mathbf{x}\colon U \subset R^2 \to S$ in p, the function $f \circ \mathbf{x}\colon U \to R$ is differentiable. In particular, the following are differentiable functions:

1. The *height function* relative to a unit vector $v \in R^3$, $h\colon S \to R$, given by $h(p) = p \cdot v$, $p \in S$, where the dot denotes the usual inner product in R^3. $h(p)$ is the height of $p \in S$ relative to a plane normal to v and passing through the origin of R^3 (Fig. 2-15).

2. The square of the distance from a fixed point $p_0 \in R^3$, $f(p) = |p - p_0|^2$, $p \in S$. The need for taking the square comes from the fact that the distance $|p - p_0|$ is not differentiable at $p = p_0$.

Remark 2. The proof of Prop. 1 makes essential use of the fact that the inverse of a parametrization is continuous. Since we need Prop. 1 to be able

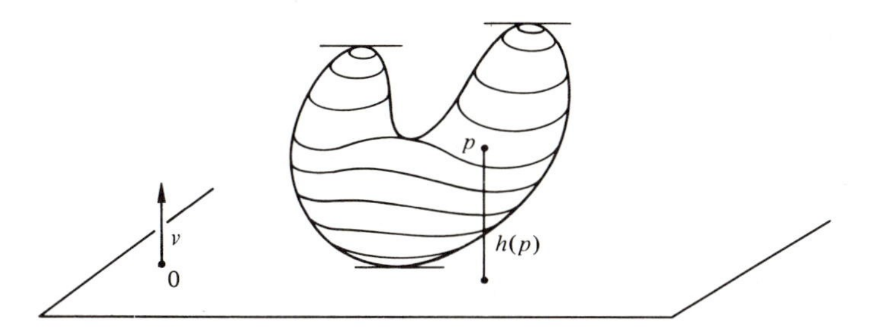

Figure 2-15

to define differentiable functions on surfaces (a vital concept), we cannot dispose of this condition in the definition of a regular surface (cf. Remark 1 of Sec. 2-2).

The definition of differentiability can be easily extended to mappings between surfaces. A continuous map $\varphi\colon V_1 \subset S_1 \to S_2$ of an open set V_1 of a regular surface S_1 to a regular surface S_2 is said to be *differentiable at* $p \in V_1$ if, given parametrizations

$$\mathbf{x}_1\colon U_1 \subset R^2 \longrightarrow S_1 \qquad \mathbf{x}_2\colon U_2 \subset R^2 \longrightarrow S_2,$$

with $p \in \mathbf{x}_1(U)$ and $\varphi(\mathbf{x}_1(U_1)) \subset \mathbf{x}_2(U_2)$, the map

$$\mathbf{x}_2^{-1} \circ \varphi \circ \mathbf{x}_1\colon U_1 \longrightarrow U_2$$

is differentiable at $q = \mathbf{x}_1^{-1}(p)$ (Fig. 2-16).

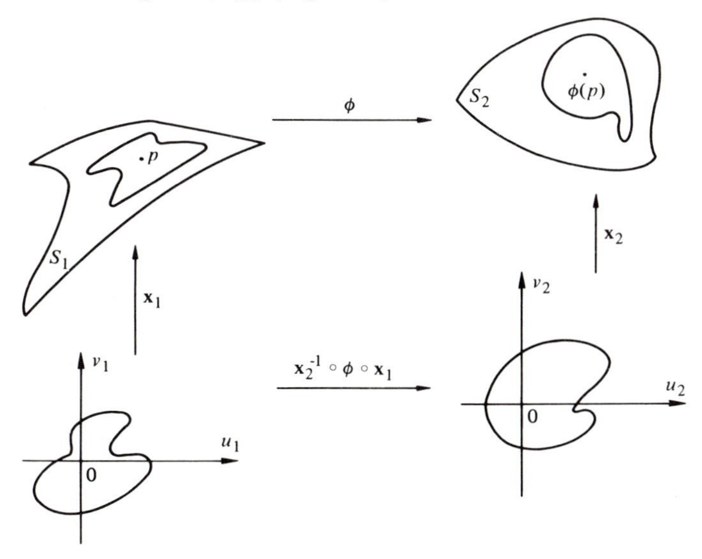

Figure 2-16

In other words, φ is differentiable if when expressed in local coordinates as $\varphi(u_1, v_1) = (\varphi_1(u_1, v_1), \varphi_2(u_1, v_1))$ the functions φ_1 and φ_2 have continuous partial derivatives of all orders.

The proof that this definitions does not depend on the choice of parametrizations is left as an exercise.

We should mention that the natural notion of equivalence associated with differentiability is the notion of diffeomorphism. Two regular surfaces S_1 and S_2 are *diffeomorphic* if there exists a differentiable map $\varphi: S_1 \longrightarrow S_2$ with a differentiable inverse $\varphi^{-1}: S_2 \longrightarrow S_1$. Such a φ is called a *diffeomorphism* from S_1 to S_2. The notion of diffeomorphism plays the same role in the study of regular surfaces that the notion of isomorphism plays in the study of vector spaces or the notion of congruence plays in Euclidean geometry. In other words, from the point of view of differentiability, two diffeomorphic surfaces are indistinguishable.

Example 2. If $\mathbf{x}: U \subset R^2 \longrightarrow S$ is a parametrization, $\mathbf{x}^{-1}: \mathbf{x}(U) \longrightarrow R^2$ is differentiable. In fact, for any $p \in \mathbf{x}(U)$ and any parametrization $\mathbf{y}: V \subset R^2 \longrightarrow S$ in p, we have that $\mathbf{x}^{-1} \circ \mathbf{y}: \mathbf{y}^{-1}(W) \longrightarrow \mathbf{x}^{-1}(W)$, where

$$W = \mathbf{x}(U) \cap \mathbf{y}(V),$$

is differentiable. This shows that U and $\mathbf{x}(U)$ are diffeomorphic (i.e., every regular surface is locally diffeomorphic to a plane) and justifies the identification made in Remark 1.

Example 3. Let S_1 and S_2 be regular surfaces. Assume that $S_1 \subset V \subset R^3$, where V is an open set of R^3, and that $\varphi: V \longrightarrow R^3$ is a differentiable map such that $\varphi(S_1) \subset S_2$. Then the restriction $\varphi \mid S_1: S_1 \longrightarrow S_2$ is a differentiable map. In fact, given $p \in S_1$ and parametrizations $\mathbf{x}_1: U_1 \longrightarrow S_1$, $\mathbf{x}_2: U_2 \longrightarrow S_2$, with $p \in \mathbf{x}_1(U_1)$ and $\varphi(\mathbf{x}_1(U_1)) \subset \mathbf{x}_2(U_2)$, we have that the map

$$\mathbf{x}_2^{-1} \circ \varphi \circ \mathbf{x}_1: U_1 \longrightarrow U_2$$

is differentiable. The following are particular cases of this general example:

1. Let S be symmetric relative to the xy plane; that is, if $(x, y, z) \in S$, then also $(x, y, -z) \in S$. Then the map $\sigma: S \longrightarrow S$, which takes $p \in S$ into its symmetrical point, is differentiable, since it is the restriction to S of $\sigma: R^3 \longrightarrow R^3$, $\sigma(x, y, z) = (x, y, -z)$. This, of course, generalizes to surfaces symmetric relative to any plane of R^3.

2. Let $R_{z,\theta}: R^3 \longrightarrow R^3$ be the rotation of angle θ about the z axis, and let $S \subset R^3$ be a regular surface invariant by this rotation; i.e., if $p \in S$, $R_{z,\theta}(p) \in S$. Then the restriction $R_{z,\theta}: S \longrightarrow S$ is a differentiable map.

3. Let $\varphi\colon R^3 \to R^3$ be given by $\varphi(x, y, z) = (xa, yb, zc)$, where a, b, and c are nonzero real numbers. φ is clearly differentiable, and the restriction $\varphi \mid S^2$ is a differentiable map from the sphere

$$S^2 = \{(x, y, z) \in R^3;\, x^2 + y^2 + z^2 = 1\}$$

into the ellipsoid

$$\left\{(x, y, z) \in R^3;\, \frac{x^2}{a^2} + \frac{y^2}{b^2} + \frac{z^2}{c^2} = 1\right\}$$

(cf. Example 6 of the appendix to Chap 2).

Remark 3. Proposition 1 implies (cf. Example 2) that a parametrization $\mathbf{x}\colon U \subset R^2 \to S$ is a diffeomorphism of U onto $\mathbf{x}(U)$. Actually, we can now characterize the regular surfaces as those subsets $S \subset R^3$ which are locally diffeomorphic to R^2; that is, for each point $p \in S$, there exists a neighborhood V of p in S, an open set $U \subset R^2$, and a map $\mathbf{x}\colon U \to V$, which is a diffeomorphism. This pretty characterization could be taken as the starting point of a treatment of surfaces (see Exercise 13).

At this stage we could return to the theory of curves and treat them from the point of view of this chapter, i.e., as subsets of R^3. We shall mention only certain fundamental points and leave the details to the reader.

The symbol I will denote an open interval of the line R. A *regular curve* in R^3 is a subset $C \subset R^3$ with the following property: For each point $p \in C$ there is a neighborhood V of p in R^3 and a differentiable homeomorphism $\alpha\colon I \subset R \to V \cap C$ such that the differential $d\alpha_t$ is one-to-one for each $t \in I$ (Fig. 2-17).

It is possible to prove (Exercise 15) that the change of parameters is given (as with surfaces) by a diffeomorphism. From this fundamental result, it is possible to decide when a given property obtained by means of a parametrization is independent of that parametrization. Such a property will then be a local property of the set C.

For example, it is shown that the arc length, defined in Chap. 1, is independent of the parametrization chosen (Exercise 15) and is, therefore, a property of the set C. Since it is always possible to locally parametrize a regular curve C by arc length, the properties (curvature, torsion, etc.) determined by means of this parametrization are local properties of S. This shows that the local theory of curves developed in Chap. 1 is valid for regular curves.

Sometimes a surface is defined by displacing a certain regular curve in a specified way. This occurs in the following example.

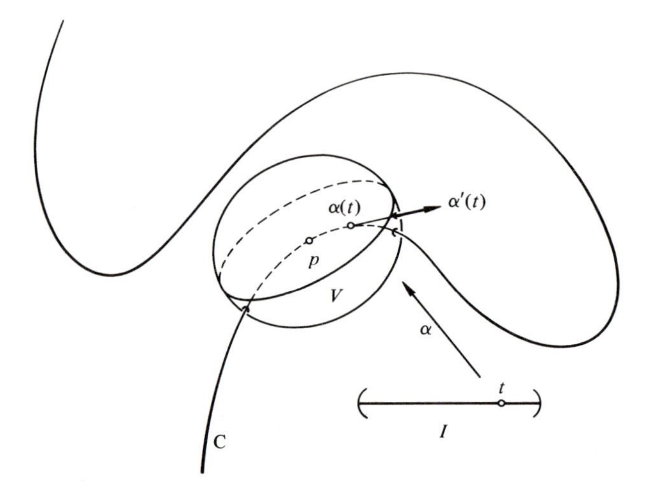

Figure 2-17. A regular curve.

Example 4 (*Surfaces of Revolution*). Let $S \subset R^3$ be the set obtained by rotating a regular plane curve C about an axis in the plane which does not meet the curve; we shall take the xz plane as the plane of the curve and the z axis as the rotation axis. Let

$$x = f(v), \qquad z = g(v), \qquad a < v < b, \qquad f(v) > 0,$$

be a parametrization for C and denote by u the rotation angle about the z axis. Thus, we obtain a map

$$\mathbf{x}(u, v) = (f(v) \cos u, \ f(v) \sin u, \ g(v))$$

from the open set $U = \{(u, v) \in R^2; 0 < u < 2\pi, a < v < b\}$ into S (Fig. 2-18).

We shall soon see that $\mathbf{x}$ satisfies the conditions for a parametrization in the definition of a regular surface. Since S can be entirely covered by similar parametrizations, it follows that S is a regular surface which is called a *surface of revolution*. The curve C is called the *generating curve* of S, and the z axis is the *rotation axis* of S. The circles described by the points of C are called the *parallels* of S, and the various positions of C on S are called the *meridians* of S.

To show that $\mathbf{x}$ is a parametrization of S we must check conditions 1, 2, and 3 of Def. 1, Sec. 2-2. Conditions 1 and 3 are straightforward, and we leave them to the reader. To show that $\mathbf{x}$ is a homeomorphism, we first show that $\mathbf{x}$ is one-to-one. In fact, since $(f(v), g(v))$ is a parametrization of C, given z and $x^2 + y^2 = (f(v))^2$, we can determine v uniquely. Thus, $\mathbf{x}$ is one-to-one.

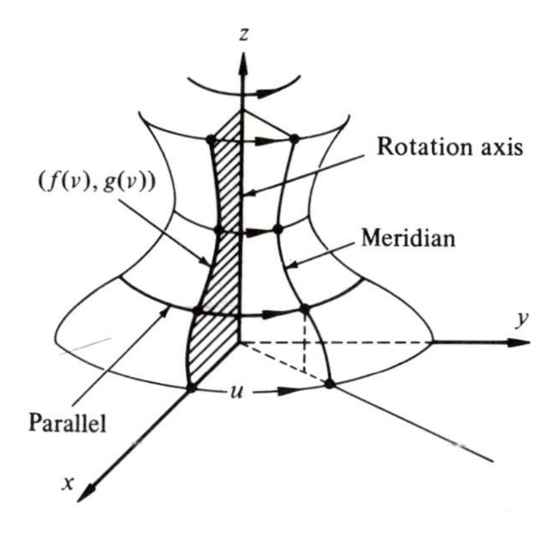

Figure 2-18. A surface of revolution.

We remark that, again because $(f(v), g(v))$ is a parametrization of C, v is a continuous function of z and of $\sqrt{x^2 + y^2}$ and thus a continuous function of (x, y, z).

To prove that $\mathbf{x}^{-1}$ is continuous, it remains to show that u is a continuous function of (x, y, z). To see this, we first observe that if $u \neq \pi$, we obtain, since $f(v) \neq 0$,

$$\tan\frac{u}{2} = \frac{\sin\dfrac{u}{2}}{\cos\dfrac{u}{2}} = \frac{2\sin\dfrac{u}{2}\cos\dfrac{u}{2}}{2\cos^2\dfrac{u}{2}} = \frac{\sin u}{1 + \cos u}$$

$$= \frac{\dfrac{y}{f(v)}}{1 + \dfrac{x}{f(v)}} = \frac{y}{x + \sqrt{x^2 + y^2}};$$

hence,

$$u = 2\tan^{-1}\frac{y}{x + \sqrt{x^2 + y^2}}.$$

Thus, if $u \neq \pi$, u is a continuous function of (x, y, z). By the same token, if u is in a small interval about π, we obtain

$$u = 2\,\mathrm{cotan}^{-1}\frac{y}{-x + \sqrt{x^2 + y^2}}.$$

Thus, u is a continuous function of (x, y, z). This shows that $\mathbf{x}^{-1}$ is continuous and completes the verification.

Remark 4. There is a slight problem with our definition of surface of revolution. If $C \subset R^2$ is a closed regular plane curve which is symmetric relative to an axis r of R^3, then, by rotating C about r, we obtain a surface which can be proved to be regular and should also be called a surface of revolution (when C is a circle and r contains a diameter of C, the surface is a sphere). To fit it in our definition, we would have to exclude two of its points, namely, the points where r meets C. For technical reasons, we want to maintain the previous terminology and shall call the latter surfaces *extended surfaces of revolution.*

A final comment should now be made on our definition of surface. We have chosen to define a (regular) surface as a subset of R^3. If we want to consider global, as well as local, properties of surfaces, this is the correct setting. The reader might have wondered, however, why we have not defined surface simply as a parametrized surface, as in the case of curves. This can be done, and in fact a certain amount of the classical literature in differential geometry was presented that way. No serious harm is done as long as only local properties are considered. However, basic global concepts, like orientation (to be treated in Secs. 2-6 and 3-1), have to be omitted, or treated inadequately, with such an approach.

In any case, the notion of parametrized surface is sometimes useful and should be included here.

DEFINITION 2. *A* parametrized surface $\mathbf{x}: U \subset R^2 \to R^3$ *is a differentiable map* $\mathbf{x}$ *from an open set* $U \subset R^2$ *into* R^3. *The set* $\mathbf{x}(U) \subset R^3$ *is called the* trace *of* $\mathbf{x}$. $\mathbf{x}$ *is* regular *if the differential* $d\mathbf{x}_q: R^2 \to R^3$ *is one-to-one for all* $q \in U$ (*i.e., the vectors* $\partial\mathbf{x}/\partial u$, $\partial\mathbf{x}/\partial v$ *are linearly independent for all* $q \in U$). *A point* $p \in U$ *where* $d\mathbf{x}_q$ *is not one-to-one is called a* singular point *of* $\mathbf{x}$.

Observe that a parametrized surface, even when regular, may have self-intersections in its trace.

Example 5. Let $\alpha: I \to R^3$ be a regular parametrized curve. Define

$$\mathbf{x}(t, v) = \alpha(t) + v\alpha'(t), \qquad (t, v) \in I \times R.$$

$\mathbf{x}$ is a parametrized surface called the *tangent surface* of α (Fig. 2-19).

Assume now that the curvature $k(t)$, $t \in I$, of α is nonzero for all $t \in I$, and restrict the domain of $\mathbf{x}$ to $U = \{(t, v) \in I \times R;\ v \neq 0\}$. Then

$$\frac{\partial \mathbf{x}}{\partial t} = \alpha'(t) + v\alpha''(t), \qquad \frac{\partial \mathbf{x}}{\partial v} = \alpha'(t)$$

and

$$\frac{\partial \mathbf{x}}{\partial t} \wedge \frac{\partial \mathbf{x}}{\partial v} = v\alpha''(t) \wedge \alpha'(t) \neq 0, \qquad (t, v) \in U,$$

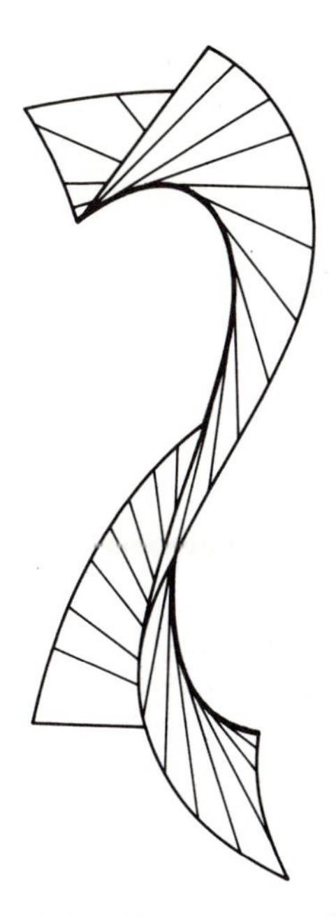

Figure 2-19. The tangent surface.

since, for all t, the curvature (cf. Exercise 12 of Sec. 1-5)

$$k(t) = \frac{|\alpha''(t) \wedge \alpha'(t)|}{|\alpha'(t)|^3}$$

is nonzero. It follows that the restriction $\mathbf{x}\colon U \longrightarrow R^3$ is a regular parametrized surface, the trace of which consists of two connected pieces whose common boundary is the set $\alpha(I)$.

The following proposition shows that we can extend the local concepts and properties of differential geometry to regular parametrized surfaces.

PROPOSITION 2. *Let* $\mathbf{x}\colon \mathrm{U} \subset \mathrm{R}^2 \longrightarrow \mathrm{R}^3$ *be a regular parametrized surface and let* $\mathrm{q} \in \mathrm{U}$. *Then there exists a neighborhood* V *of* q *in* R^2 *such that* $\mathbf{x}(\mathrm{V}) \subset \mathrm{R}^3$ *is a regular surface.*

Proof. This is again a consequence of the inverse function theorem. Write

$$\mathbf{x}(u, v) = (x(u, v), y(u, v), z(u, v)).$$

By regularity, we can assume that $(\partial(x, y)/\partial(u, v))(q) \neq 0$. Define a map $F\colon U \times R \longrightarrow R^3$ by

$$F(u, v, t) = (x(u, v), y(u, v), z(u, v) + t), \qquad (u, v) \in U, t \in R.$$

Then

$$\det(dF_q) = \frac{\partial(x, y)}{\partial(u, v)}(q) \neq 0.$$

By the inverse function theorem, there exist neighborhoods W_1 of q and W_2 of $F(q)$ such that $F: W_1 \longrightarrow W_2$ is a diffeomorphism. Set $V = W_1 \cap U$ and observe that the restriction $F|V = \mathbf{x}|V$. Thus, $\mathbf{x}(V)$ is diffeomorphic to V, and hence a regular surface. **Q.E.D.**

EXERCISES†

*1. Let $S^2 = \{(x, y, z) \in R^3; x^2 + y^2 + z^2 = 1\}$ be the unit sphere and let $A: S^2 \longrightarrow S^2$ be the (*antipodal*) map $A(x, y, z) = (-x, -y, -z)$. Prove that A is a diffeomorphism.

2. Let $S \subset R^3$ be a regular surface and $\pi: S \longrightarrow R^2$ be the map which takes each $p \in S$ into its orthogonal projection over $R^2 = \{(x, y, z) \in R^3; z = 0\}$. Is π differentiable?

3. Show that the paraboloid $z = x^2 + y^2$ is diffeomorphic to a plane.

4. Construct a diffeomorphism between the ellipsoid

$$\frac{x^2}{a^2} + \frac{y^2}{b^2} + \frac{z^2}{c^2} = 1$$

and the sphere $x^2 + y^2 + z^2 = 1$.

*5. Let $S \subset R^3$ be a regular surface, and let $d: S \longrightarrow R$ be given by $d(p) = |p - p_0|$, where $p \in S$, $p_0 \in R^3$, $p_0 \notin S$; that is, d is the distance from p to a fixed point p_0 not in S. Prove that d is differentiable.

6. Prove that the definition of a differentiable map between surfaces does not depend on the parametrizations chosen.

7. Prove that the relation "S_1 is diffeomorphic to S_2" is an equivalence relation in the set of regular surfaces.

*8. Let $S^2 = \{(x, y, z) \in R^3; x^2 + y^2 + z^2 = 1\}$ and $H = \{(x, y, z) \in R^3; x^2 + y^2 - z^2 = 1\}$. Denote by $N = (0, 0, 1)$ and $S = (0, 0, -1)$ the north and south poles of S^2, respectively, and let $F: S^2 - \{N\} \cup \{S\} \longrightarrow H$ be defined as follows: For each $p \in S^2 - \{N\} \cup \{S\}$ let the perpendicular from p to the z axis meet $0z$ at q. Consider the half-line l starting at q and containing p. Then $F(p) = l \cap H$ (Fig. 2-20). Prove that F is differentiable.

9. **a.** Define the notion of differentiable function on a regular curve. What does one need to prove for the definition to make sense? Do not prove it now. If

†Those who have omitted the proofs of this section should also omit Exercises 13–16.

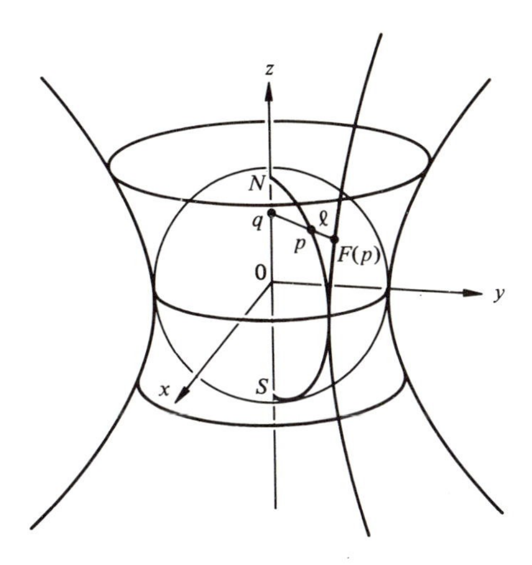

Figure 2-20

you have not omitted the proofs in this section, you will be asked to do it in Exercise 15.

b. Show that the map $E: R \longrightarrow S^1 = \{(x, y) \in R^2; x^2 + y^2 = 1\}$ given by

$$E(t) = (\cos t, \sin t), \qquad t \in R,$$

is differentiable (geometrically, E "wraps" R around S^1).

10. Let C be a plane regular curve which lies in one side of a straight line r of the plane and meets r at the points p, q (Fig. 2-21). What conditions should C satisfy to ensure that the rotation of C about r generates an extended (regular) surface of revolution?

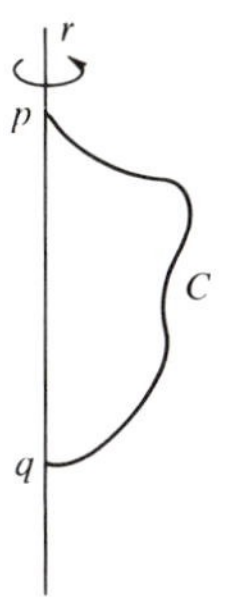

Figure 2-21

11. Prove that the rotations of a surface of revolution S about its axis are diffeomorphisms of S.

12. Parametrized surfaces are often useful to describe sets Σ which are regular surfaces except for a finite number of points and a finite number of lines. For instance, let C be the trace of a regular parametrized curve $\alpha: (a, b) \longrightarrow R^3$ which does not pass through the origin $O = (0, 0, 0)$. Let Σ be the set generated by the

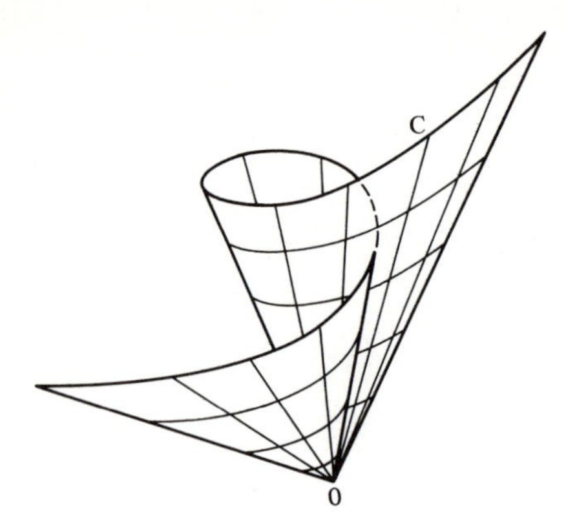

Figure 2-22

displacement of a straight line l passing through a moving point $p \in C$ and the fixed point 0 (a cone with vertex 0; see Fig. 2-22).

a. Find a parametrized surface $\mathbf{x}$ whose trace is Σ.

b. Find the points where $\mathbf{x}$ is not regular.

c. What should be removed from Σ so that the remaining set is a regular surface?

***13.** Show that the definition of differentiability of a function $f\colon V \subset S \to R$ given in the text (Def. 1) is equivalent to the following: f is differentiable in $p \in V$ if it is the restriction to V of a differentiable function defined in an open set of R^3 containing p. (Had we started with this definition of differentiability, we could have defined a surface as a set which is locally diffeomorphic to R^2; see Remark 3.)

14. Let $A \subset S$ be a subset of a regular surface S. Prove that A is itself a regular surface if and only if A is open in S; that is, $A = U \cap S$, where U is an open set in R^3.

15. Let C be a regular curve and let $\alpha\colon I \subset R \to C$, $\beta\colon J \subset R \to C$ be two parametrizations of C in a neighborhood of $p \in \alpha(I) \cap \beta(I) = W$. Let

$$h = \alpha^{-1} \circ \beta\colon \beta^{-1}(W) \longrightarrow \alpha^{-1}(W)$$

be the change of parameters. Prove that

a. h is a diffeomorphism.

b. The absolute value of the arc length of C in W does not depend on which parametrization is chosen to define it, that is,

$$\left|\int_{t_0}^{t} |\alpha'(t)|\,dt\right| = \left|\int_{\tau_0}^{\tau} |\beta'(\tau)|\,d\tau\right|, \qquad t = h(\tau),\ t \in I,\ \tau \in J.$$

***16.** Let $R^2 = \{(x, y, z) \in R^3; z = -1\}$ be identified with the complex plane $\mathbb{C}$ by setting $(x, y, -1) = x + iy = \zeta \in \mathbb{C}$. Let $P\colon \mathbb{C} \to \mathbb{C}$ be the complex polynomial

$$P(\zeta) = a_0\zeta^n + a_1\zeta^{n-1} + \cdots + a_n, \qquad a_0 \neq 0, a_i \in \mathbb{C}, i = 1, \ldots, n.$$

Denote by π_N the stereographic projection of $S^2 = \{(x, y, z) \in R^3; x^2 + y^2 + z^2 = 1\}$ from the north pole $N = (0, 0, 1)$ onto R^2. Prove that the map $F: S^2 \to S^2$ given by

$$F(p) = \pi_N^{-1} \circ P \circ \pi_N(p), \qquad \text{if } p \in S^2 - \{N\},$$
$$F(N) = N$$

is differentiable.

2-4. The Tangent Plane; The Differential of a Map

In this section we shall show that condition 3 in the definition of a regular surface S guarantees that for every $p \in S$ the set of tangent vectors to the parametrized curves of S, passing through p, constitutes a plane.

By a *tangent vector* to S, at a point $p \in S$, we mean the tangent vector $\alpha'(0)$ of a differentiable parametrized curve $\alpha: (-\epsilon, \epsilon) \to S$ with $\alpha(0) = p$.

PROPOSITION 1. *Let* $\mathbf{x}$: $U \subset R^2 \to S$ *be a parametrization of a regular surface* S *and let* q $\in$ U. *The vector subspace of dimension* 2,

$$d\mathbf{x}_q(R^2) \subset R^3,$$

coincides with the set of tangent vectors to S *at* $\mathbf{x}$(q).

Proof. Let w be a tangent vector at $\mathbf{x}(q)$, that is, let $w = \alpha'(0)$, where $\alpha: (-\epsilon, \epsilon) \to \mathbf{x}(U) \subset S$ is differentiable and $\alpha(0) = \mathbf{x}(q)$. By Example 2 of Sec. 2-3, the curve $\beta = \mathbf{x}^{-1} \circ \alpha: (-\epsilon, \epsilon) \to U$ is differentiable. By definition of the differential (appendix to Chap. 2, Def. 1), we have $d\mathbf{x}_q(\beta'(0)) = w$. Hence, $w \in d\mathbf{x}_q(R^2)$ (Fig. 2-23).

On the other hand, let $w = d\mathbf{x}_q(v)$, where $v \in R^2$. It is clear that v is the velocity vector of the curve $\gamma: (-\epsilon, \epsilon) \to U$ given by

$$\gamma(t) = tv + q, \qquad t \in (-\epsilon, \epsilon).$$

By the definition of the differential, $w = \alpha'(0)$, where $\alpha = \mathbf{x} \circ \gamma$. This shows that w is a tangent vector. **Q.E.D.**

By the above proposition, the plane $d\mathbf{x}_q(R^2)$, which passes through $\mathbf{x}(q) = p$, does not depend on the parametrization $\mathbf{x}$. This plane will be called the

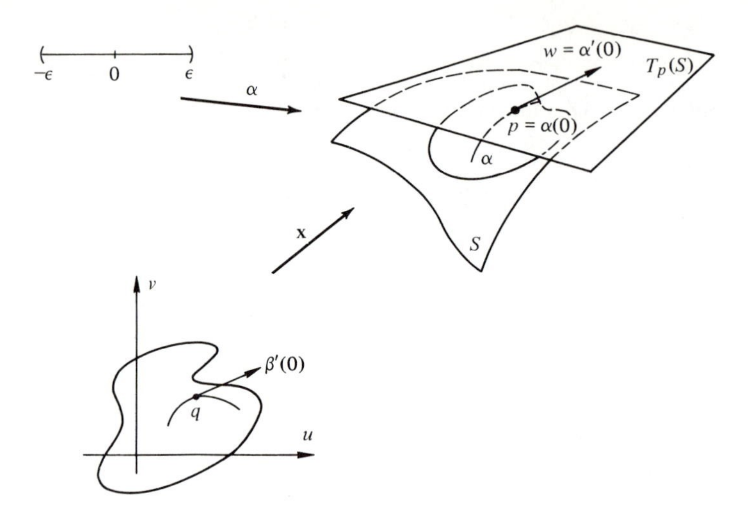

Figure 2-23

tangent plane to S at p and will be denoted by $T_p(S)$. The choice of the parametrization $\mathbf{x}$ determines a basis $\{(\partial\mathbf{x}/\partial u(q), (\partial\mathbf{x}/\partial v)(q)\}$ of $T_p(S)$, called the basis associated to $\mathbf{x}$. Sometimes it is convenient to write $\partial\mathbf{x}/\partial u = \mathbf{x}_u$ and $\partial\mathbf{x}/\partial v = \mathbf{x}_v$.

The coordinates of a vector $w \in T_p(S)$ in the basis associated to a parametrization $\mathbf{x}$ are determined as follows. w is the velocity vector $\alpha'(0)$ of a curve $\alpha = \mathbf{x} \circ \beta$, where $\beta: (-\epsilon, \epsilon) \to U$ is given by $\beta(t) = (u(t), v(t))$, with $\beta(0) = q = \mathbf{x}^{-1}(p)$. Thus,

$$\alpha'(0) = \frac{d}{dt}(\mathbf{x} \circ \beta)(0) = \frac{d}{dt}\mathbf{x}(u(t), v(t))(0)$$
$$= \mathbf{x}_u(q)u'(0) + \mathbf{x}_v(q)v'(0) = w.$$

Thus, in the basis $\{\mathbf{x}_u(q), \mathbf{x}_v(q)\}$, w has coordinates $(u'(0), v'(0))$, where $(u(t), v(t))$ is the expression, in the parametrization $\mathbf{x}$, of a curve whose velocity vector at $t = 0$ is w.

With the notion of a tangent plane, we can talk about the differential of a (differentiable) map between surfaces. Let S_1 and S_2 be two regular surfaces and let $\varphi: V \subset S_1 \to S_2$ be a differentiable mapping of an open set V of S_1 into S_2. If $p \in V$, we know that every tangent vector $w \in T_p(S_1)$ is the velocity vector $\alpha'(0)$ of a differentiable parametrized curve $\alpha: (-\epsilon, \epsilon) \to V$ with $\alpha(0) = p$. The curve $\beta = \varphi \circ \alpha$ is such that $\beta(0) = \varphi(p)$, and therefore $\beta'(0)$ is a vector of $T_{\varphi(p)}(S_2)$ (Fig. 2-24).

PROPOSITION 2. *In the discussion above, given* w, *the vector* $\beta'(0)$ *does not depend on the choice of* α. *The map* $d\varphi_p: T_p(S_1) \to T_{\varphi(p)}(S_2)$ *defined by* $d\varphi_p(w) = \beta'(0)$ *is linear.*

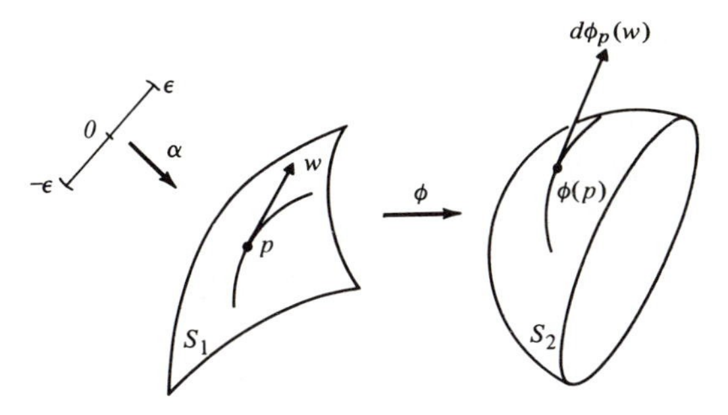

Figure 2-24

Proof. The proof is similar to the one given in Euclidean spaces (cf. Prop. 4, appendix to Chap. 2). Let $\mathbf{x}(u, v)$, $\bar{\mathbf{x}}(\bar{u}, \bar{v})$ be parametrizations in neighborhoods of p and $\varphi(p)$, respectively. Suppose that φ is expressed in these coordinates by

$$\varphi(u, v) = (\varphi_1(u, v), \varphi_2(u, v))$$

and that α is expressed by

$$\alpha(t) = (u(t), v(t)), \qquad t \in (-\epsilon, \epsilon).$$

Then $\beta(t) = (\varphi_1(u(t), v(t)), \varphi_2(u(t), v(t)))$, and the expression of $\beta'(0)$ in the basis $\{\bar{\mathbf{x}}_{\bar{u}}, \bar{\mathbf{x}}_{\bar{v}}\}$ is

$$\beta'(0) = \left(\frac{\partial \varphi_1}{\partial u}u'(0) + \frac{\partial \varphi_1}{\partial v}v'(0), \frac{\partial \varphi_2}{\partial u}u'(0) + \frac{\partial \varphi_2}{\partial v}v'(0)\right).$$

The relation above shows that $\beta'(0)$ depends only on the map φ and the coordinates $(u'(0), v'(0))$ of w in the basis $\{\mathbf{x}_u, \mathbf{x}_v\}$. $\beta'(0)$ is therefore independent of α. Moreover, the same relation shows that

$$\beta'(0) = d\varphi_p(w) = \begin{pmatrix} \dfrac{\partial \varphi_1}{du} & \dfrac{\partial \varphi_1}{\partial v} \\ \dfrac{\partial \varphi_2}{\partial u} & \dfrac{\partial \varphi_2}{\partial v} \end{pmatrix} \begin{pmatrix} u'(0) \\ v'(0) \end{pmatrix};$$

that is, $d\varphi_p$ is a linear mapping of $T_p(S_1)$ into $T_{\varphi(p)}(S_2)$ whose matrix in the bases $\{\mathbf{x}_u, \mathbf{x}_v\}$ of $T_p(S_1)$ and $\{\bar{\mathbf{x}}_{\bar{u}}, \bar{\mathbf{x}}_{\bar{v}}\}$ of $T_{\varphi(p)}(S_2)$ is just the matrix given above.

Q.E.D.

The linear map $d\varphi_p$ defined by Prop. 2 is called the *differential* of φ at $p \in S_1$. In a similar way we define the differential of a (differentiable) function $f: U \subset S \to R$ at $p \in U$ as a linear map $df_p: T_p(S) \to R$. We leave the details to the reader.

Example 1. Let $v \in R^3$ be a unit vector and let $h: S \to R$, $h(p) = v \cdot p$, $p \in S$, be the height function defined in Example 1 of Sec. 2-3. To compute $dh_p(w)$, $w \in T_p(S)$, choose a differentiable curve $\alpha: (-\epsilon, \epsilon) \to S$ with $\alpha(0) = p$, $\alpha'(0) = w$. Since $h(\alpha(t)) = \alpha(t) \cdot v$, we obtain

$$dh_p(w) = \frac{d}{dt} h(\alpha(t))|_{t=0} = \alpha'(0) \cdot v = w \cdot v.$$

Example 2. Let $S^2 \subset R^2$ be the unit sphere

$$S^2 = \{(x, y, z) \in R^3; x^2 + y^2 + z^2 = 1\}$$

and let $R_{z,\theta}: R^3 \to R^3$ be the rotation of angle θ about the z axis. Then $R_{z,\theta}$ restricted to S^2 is a differentiable map of S^2 (cf. Example 3 of Sec. 2-3). We shall compute $(dR_{z,\theta})_p(w)$, $p \in S^2$, $w \in T_p(S^2)$. Let $\alpha: (-\epsilon, \epsilon) \to S^2$ be a differentiable curve with $\alpha(0) = p$, $\alpha'(0) = w$. Then, since $R_{z,\theta}$ is linear,

$$(dR_{z,\theta})_p(w) = \frac{d}{dt}(R_{z,\theta} \circ \alpha(t)) = R_{z,\theta}(\alpha'(0)) = R_{z.\theta}(w).$$

Observe that $R_{z,\theta}$ leaves the north pole $N = (0, 0, 1)$ fixed, and that $(dR_{z,\theta})_N: T_N(S) \to T_N(S)$ is just a rotation of angle θ in the plane $T_N(S)$.

In retrospect, what we have been doing up to now is extending the notions of differential calculus in R^2 to regular surfaces. Since calculus is essentially a local theory, we defined an entity (the regular surface) which locally was a plane, up to diffeomorphisms, and this extension then became natural. It might be expected therefore that the basic inverse function theorem extends to differentiable mappings between surfaces.

We shall say that a mapping $\varphi: U \subset S_1 \to S_2$ is a *local diffeomorphism* at $p \in U$ if there exists a neighborhood $V \subset U$ of p such that φ restricted to V is a diffeomorphism onto an open set $\varphi(V) \subset S_2$. In these terms, the version of the inverse of function theorem for surfaces is expressed as follows.

PROPOSITION 3. *If* S_1 *and* S_2 *are regular surfaces and* $\varphi: U \subset S_1 \to S_2$ *is a differentiable mapping of an open set* $U \subset S_1$ *such that the differential* $d\varphi_p$ *of* φ *at* $p \in U$ *is an isomorphism, then* φ *is a local diffeomorphism at* p.

The proof is an immediate application of the inverse function theorem in R^2 and will be left as an exercise.

Of course, all other concepts of calculus, like critical points, regular values, etc., do extend naturally to functions and maps defined on regular surfaces.

The tangent plane also allows us to speak of the angle of two intersecting surfaces at a point of intersection.

Given a point p on a regular surface S, there are two unit vectors of R^3 that are normal to the tangent plane $T_p(S)$; each of them is called *a unit normal vector* at p. The straight line that passes through p and contains a unit normal vector at p is called the *normal line* at p. The *angle* of two intersecting surfaces at an intersection point p is the angle of their tangent planes (or their normal lines) at p (Fig. 2-25).

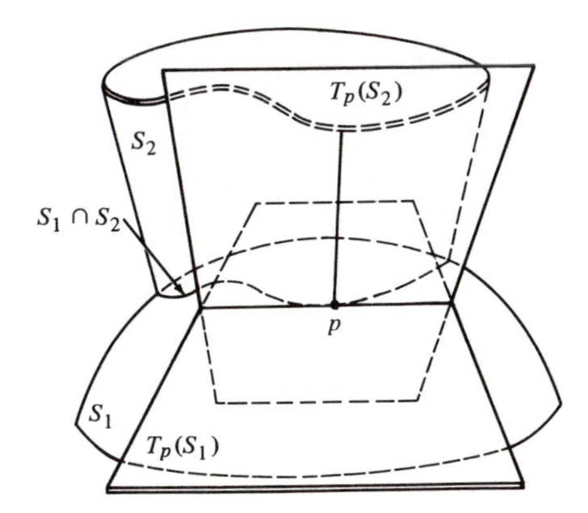

Figure 2-25

By fixing a parametrization $\mathbf{x}: U \subset R^2 \to S$ at $p \in S$, we can make a definite choice of a unit normal vector at each point $q \in \mathbf{x}(U)$ by the rule

$$N(q) = \frac{\mathbf{x}_u \wedge \mathbf{x}_v}{|\mathbf{x}_u \wedge \mathbf{x}_v|}(q).$$

Thus, we obtain a differentiable map $N: \mathbf{x}(U) \to R^3$. We shall see later (Secs. 2-6 and 3-1) that it is not always possible to extend this map differentiably to the whole surface S.

Before leaving this section, we shall make some observations on questions of differentiability.

The definition given for a regular surface requires that the parametrizations be of class C^∞, that is, that they possess continuous partial derivatives of *all* orders. For questions in differential geometry we need in general the existence and continuity of the partial derivatives only up to a certain order, which varies with the nature of the problem (very rarely do we need more than four derivatives).

For example, the existence and continuity of the tangent plane depends only on the existence and continuity of the first partial derivatives. It could happen, therefore, that the graph of a function $z = f(x, y)$ admits a tangent plane at every point but is not sufficiently differentiable to satisfy the definition of a regular surface. This occurs in the following example.

Example 3. Consider the graph of the function $z = \sqrt[3]{(x^2 + y^2)^2}$, generated by rotating the curve $z = x^{4/3}$ about the z axis. Since the curve is symmetric relative to the z axis and has a continuous derivative which vanishes at the origin, it is clear that the graph of $z = \sqrt[3]{(x^2 + y^2)^2}$ admits the xy plane as a tangent plane at the origin. However, the partial derivative z_{xx} does not exist at the origin, and the graph considered is not a regular surface as defined above (see Prop. 3 of Sec. 2-2).

We do not intend to get involved with this type of question. The hypothesis C^∞ in the definition was adopted precisely to avoid the study of the minimal conditions of differentiability required in each particular case. These nuances have their place, but they can eventually obscure the geometric nature of the problems treated here.

EXERCISES

***1.** Show that the equation of the tangent plane at (x_0, y_0, z_0) of a regular surface given by $f(x, y, z) = 0$, where 0 is a regular value of f, is

$$f_x(x_0, y_0, z_0)(x - x_0) + f_y(x_0, y_0, z_0)(y - y_0) + f_z(x_0, y_0, z_0)(z - z_0) = 0.$$

2. Determine the tangent planes of $x^2 + y^2 - z^2 = 1$ at the points $(x, y, 0)$ and show that they are all parallel to the z axis.

3. Show that the equation of the tangent plane of a surface which is the graph of a differentiable function $z = f(x, y)$, at the point $p_0 = (x_0, y_0)$, is given by

$$z = f(x_0, y_0) + f_x(x_0, y_0)(x - x_0) + f_y(x_0, y_0)(y - y_0).$$

Recall the definition of the differential df of a function $f\colon R^2 \longrightarrow R$ and show that the tangent plane is the graph of the differential df_p.

***4.** Show that the tangent planes of a surface given by $z = xf(y/x)$, $x \neq 0$, where f is a differentiable function, all pass through the origin $(0, 0, 0)$.

5. If a coordinate neighborhood of a regular surface can be parametrized in the form

$$\mathbf{x}(u, v) = \alpha_1(u) + \alpha_2(v),$$

where α_1 and α_2 are regular parametrized curves, show that the tangent planes along a fixed coordinate curve of this neighborhood are all parallel to a line.

6. Let $\alpha\colon I \longrightarrow R^3$ be a regular parametrized curve with everywhere nonzero curvature. Consider the tangent surface of α (Example 5 of Sec. 2-3)

$$\mathbf{x}(t, v) = \alpha(t) + v\alpha'(t), \qquad t \in I, v \neq 0.$$

Show that the tangent planes along the curve $\mathbf{x}(t, \text{const.})$ are all equal.

7. Let $f: S \longrightarrow R$ be given by $f(p) = |p - p_0|^2$, where $p \in S$ and p_0 is a fixed point of R^3 (see Example 1 of Sec. 2-3). Show that $df_p(w) = 2w \cdot (p - p_0)$, $w \in T_p(S)$.

8. Prove that if $L: R^3 \longrightarrow R^3$ is a linear map and $S \subset R^3$ is a regular surface invariant under L, i.e., $L(S) \subset S$, then the restriction $L|S$ is a differentiable map and

$$dL_p(w) = L(w), \qquad p \in S, w \in T_p(S).$$

9. Show that the parametrized surface

$$\mathbf{x}(u, v) = (v \cos u, v \sin u, au), \qquad a \neq 0,$$

is regular. Compute its normal vector $N(u, v)$ and show that along the coordinate line $u = u_0$ the tangent plane of $\mathbf{x}$ rotates about this line in such a way that the tangent of its angle with the z axis is proportional to the distance $v\ (= \sqrt{x^2 + y^2})$ of the point $\mathbf{x}(u_0, v)$ to the z axis.

10. (*Tubular Surfaces.*) Let $\alpha: I \longrightarrow R^3$ be a regular parametrized curve with nonzero curvature everywhere and arc length as parameter. Let

$$\mathbf{x}(s, v) = \alpha(s) + r(n(s) \cos v + b(s) \sin v), \qquad r = \text{const.} \neq 0, s \in I,$$

be a parametrized surface (the *tube* of radius r around α), where n is the normal vector and b is the binormal vector of α. Show that, when $\mathbf{x}$ is regular, its unit normal vector is

$$N(s, v) = -(n(s) \cos v + b(s) \sin v).$$

11. Show that the normals to a parametrized surface given by

$$\mathbf{x}(u, v) = (f(u) \cos v, f(u) \sin v, g(u)), \qquad f(u) \neq 0, g' \neq 0,$$

all pass through the z axis.

***12.** Show that each of the equations $(a, b, c \neq 0)$

$$x^2 + y^2 + z^2 = ax,$$
$$x^2 + y^2 + z^2 = by,$$
$$x^2 + y^2 + z^2 = cz$$

define a regular surface and that they all intersect orthogonally.

13. A *critical point* of a differentiable function $f: S \longrightarrow R$ defined on a regular surface S is a point $p \in S$ such that $df_p = 0$.

***a.** Let $f: S \longrightarrow R$ be given by $f(p) = |p - p_0|$, $p \in S$, $p_0 \notin S$ (cf. Exercise 5,

Sec. 2-3). Show that $p \in S$ is a critical point of f if and only if the line joining p to p_0 is normal to S at p.

b. Let $h: S \longrightarrow R$ be given by $h(p) = p \cdot v$, where $v \in R^3$ is a unit vector (cf. Example 1, Sec. 2-3). Show that $p \in S$ is a critical point of f if and only if v is a normal vector of S at p.

***14.** Let Q be the union of the three coordinate planes $x = 0$, $y = 0$, $z = 0$. Let $p = (x, y, z) \in R^3 - Q$.

a. Show that the equation in t,

$$\frac{x^2}{a-t} + \frac{y^2}{b-t} + \frac{z^2}{c-t} \equiv f(t) = 1, \qquad a > b > c > 0,$$

has three distinct real roots: t_1, t_2, t_3.

b. Show that for each $p \in R^3 - Q$, the sets given by $f(t_1) - 1 = 0$, $f(t_2) - 1 = 0$, $f(t_3) - 1 = 0$ are regular surfaces passing through p which are pairwise orthogonal.

15. Show that if all normals to a connected surface pass through a fixed point, the surface is contained in a sphere.

16. Let w be a tangent vector to a regular surface S at a point $p \in S$ and let $\mathbf{x}(u, v)$ and $\bar{\mathbf{x}}(\bar{u}, \bar{v})$ be two parametrizations at p. Suppose that the expressions of w in the bases associated to $\mathbf{x}(u, v)$ and $\bar{\mathbf{x}}(\bar{u}, \bar{v})$ are

$$w = \alpha_1 \mathbf{x}_u + \alpha_2 \mathbf{x}_v$$

and

$$w = \beta_1 \bar{\mathbf{x}}_{\bar{u}} + \beta_2 \bar{\mathbf{x}}_{\bar{v}}.$$

Show that the coordinates of w are related by

$$\beta_1 = \alpha_1 \frac{\partial \bar{u}}{\partial u} + \alpha_2 \frac{\partial \bar{u}}{\partial v}$$

$$\beta_2 = \alpha_1 \frac{\partial \bar{v}}{\partial u} + \alpha_2 \frac{\partial \bar{v}}{\partial v},$$

where $\bar{u} = \bar{u}(u, v)$ and $\bar{v} = \bar{v}(u, v)$ are the expressions of the change of coordinates.

***17.** Two regular surfaces S_1 and S_2 intersect *transversally* if whenever $p \in S_1 \cap S_2$ then $T_p(S_1) \neq T_p(S_2)$. Prove that if S_1 intersects S_2 transversally, then $S_1 \cap S_2$ is a regular curve.

18. Prove that if a regular surface S meets a plane P in a single point p, then this plane coincides with the tangent plane of S at p.

19. Let $S \subset R^3$ be a regular surface and $P \subset R^3$ be a plane. If all points of S are on the same side of P, prove that P is tangent to S at all points of $P \cap S$.

***20.** Show that the perpendicular projections of the center (0, 0, 0) of the ellipsoid

$$\frac{x^2}{a^2} + \frac{y^2}{b^2} + \frac{z^2}{c^2} = 1$$

onto its tangent planes constitute a regular surface given by

$$\{(x, y, z) \in R^3; (x^2 + y^2 + z^2)^2 = a^2x^2 + b^2y^2 + c^2z^2\} - \{(0, 0, 0)\}.$$

***21.** Let $f\colon S \longrightarrow R$ be a differentiable function on a connected regular surface S. Assume that $df_p = 0$ for all $p \in S$. Prove that f is constant on S.

***22.** Prove that if all normal lines to a connected regular surface S meet a fixed straight line, then S is a surface of revolution.

23. Prove that the map $F\colon S^2 \longrightarrow S^2$ defined in Exercise 16 of Sec. 2-3 has only a finite number of critical points (see Exercise 13).

24. (*Chain Rule.*) Show that if $\varphi\colon S_1 \longrightarrow S_2$ and $\psi\colon S_2 \longrightarrow S_3$ are differentiable maps and $p \in S_1$, then

$$d(\psi \circ \varphi)_p = d\psi_{\varphi(p)} \circ d\varphi_p.$$

25. Prove that if two regular curves C_1 and C_2 of a regular surface S are tangent at a point $p \in S$, and if $\varphi\colon S \longrightarrow S$ is a diffeomorphism, then $\varphi(C_1)$ and $\varphi(C_2)$ are regular curves which are tangent at $\varphi(p)$.

26. Show that if p is a point of a regular surface S, it is possible, by a convenient choice of the (x, y, z) coordinates, to represent a neighborhood of p in S in the form $z = f(x, y)$ so that $f(0, 0) = 0$, $f_x(0, 0) = 0$, $f_y(0, 0) = 0$. (This is equivalent to taking the tangent plane to S at p as the xy plane.)

27. (*Theory of Contact.*) Two regular surfaces, S and $\bar{S}$, in R^3, which have a point p in common, are said to have *contact of order* ≥ 1 at p if there exist parametrizations with the same domain $\mathbf{x}(u, v)$, $\bar{\mathbf{x}}(u, v)$ at p of S and $\bar{S}$, respectively, such that $\mathbf{x}_u = \bar{\mathbf{x}}_u$ and $\mathbf{x}_v = \bar{\mathbf{x}}_v$ at p. If, moreover, some of the second partial derivatives are different at p, the *contact* is said to be *of order exactly equal to* 1. Prove that

a. The tangent plane $T_p(S)$ of a regular surface S at the point p has contact of order ≥ 1 with the surface at p.

b. If a plane has contact of order ≥ 1 with a surface S at p, then this plane coincides with the tangent plane to S at p.

c. Two regular surfaces have contact of order ≥ 1 if and only if they have a common tangent plane at p, i.e., they are tangent at p.

d. If two regular surfaces S and $\bar{S}$ of R^3 have contact of order ≥ 1 at p and if $F\colon R^3 \longrightarrow R^3$ is a diffeomorphism of R^3, then the images $F(S)$ and $F(\bar{S})$ are regular surfaces which have contact of order ≥ 1 at $f(p)$ (that is, the notion of contact of order ≥ 1 is invariant under diffeomorphisms).

e. If two surfaces have contact of order ≥ 1 at p, then $\lim_{r\to 0} (d/r) = 0$, where d is the length of the segment which is determined by the intersections with the surfaces of some parallel to the common normal, at a distance r from this normal.

28. **a.** Define regular value for a differentiable function $f\colon S \longrightarrow R$ on a regular surface S.

b. Show that the inverse image of a regular value of a differentiable function on a regular surface S is a regular curve on S.

2-5. The First Fundamental Form; Area

So far we have looked at surfaces from the point of view of differentiability. In this section we shall begin the study of further geometric structures carried by the surface. The most important of these is perhaps the first fundamental form, which we shall now describe.

The natural inner product of $R^3 \supset S$ induces on each tangent plane $T_p(S)$ of a regular surface S an inner product, to be denoted by $\langle\ ,\ \rangle_p$: If $w_1, w_2 \in T_p(S) \subset R^3$, then $\langle w_1, w_2\rangle_p$ is equal to the inner product of w_1 and w_2 as vectors in R^3. To this inner product, which is a symmetric bilinear form (i.e., $\langle w_1, w_2\rangle = \langle w_2, w_1\rangle$ and $\langle w_1, w_2\rangle$ is linear in both w_1 and w_2), there corresponds a quadratic form $I_p\colon T_p(S) \longrightarrow R$ given by

$$I_p(w) = \langle w, w\rangle_p = |w|^2 \geq 0. \tag{1}$$

DEFINITION 1. *The quadratic form* I_p *on* $T_q(S)$, *defined by Eq. (1), is called the* first fundamental form *of the regular surface* $S \subset R^3$ *at* $p \in S$.

Therefore, the first fundamental form is merely the expression of how the surface S inherits the natural inner product of R^3. Geometrically, as we shall see in a while, the first fundamental form allows us to make measurements on the surface (lengths of curves, angles of tangent vectors, areas of regions) without referring back to the ambient space R^3 where the surface lies.

We shall now express the first fundamental form in the basis $\{\mathbf{x}_u, \mathbf{x}_v\}$ associated to a parametrization $\mathbf{x}(u, v)$ at p. Since a tangent vector $w \in T_p(S)$ is the tangent vector to a parametrized curve $\alpha(t) = \mathbf{x}(u(t), v(t))$, $t \in (-\epsilon, \epsilon)$, with $p = \alpha(0) = \mathbf{x}(u_0, v_0)$, we obtain

$$\begin{aligned}
I_p(\alpha'(0)) &= \langle \alpha'(0), \alpha'(0)\rangle_p \\
&= \langle \mathbf{x}_u u' + \mathbf{x}_v v', \mathbf{x}_u u' + \mathbf{x}_v v'\rangle_p \\
&= \langle \mathbf{x}_u, \mathbf{x}_u\rangle_p (u')^2 + 2\langle \mathbf{x}_u, \mathbf{x}_v\rangle_p u'v' + \langle \mathbf{x}_v, \mathbf{x}_v\rangle_p (v')^2 \\
&= E(u')^2 + 2Fu'v' + G(v')^2,
\end{aligned}$$

where the values of the functions involved are computed for $t = 0$, and

$$\begin{aligned}
E(u_0, v_0) &= \langle \mathbf{x}_u, \mathbf{x}_u\rangle_p, \\
F(u_0, v_0) &= \langle \mathbf{x}_u, \mathbf{x}_v\rangle_p, \\
G(u_0, v_0) &= \langle \mathbf{x}_v, \mathbf{x}_v\rangle_p
\end{aligned}$$

are the coefficients of the first fundamental form in the basis $\{\mathbf{x}_u, \mathbf{x}_v\}$ of $T_p(S)$. By letting p run in the coordinate neighborhood corresponding to $\mathbf{x}(u, v)$ we obtain functions $E(u, v)$, $F(u, v)$, $G(u, v)$ which are differentiable in that neighborhood.

From now on we shall drop the subscript p in the indication of the inner product $\langle \ , \ \rangle_p$ or the quadratic form I_p when it is clear from the context which point we are referring to. It will also be convenient to denote the natural inner product of R^3 by the same symbol $\langle \ , \ \rangle$ rather than the previous dot.

Example 1. A coordinate system for a plane $P \subset R^3$ passing through $p_0 = (x_0, y_0, z_0)$ and containing the orthonormal vectors $w_1 = (a_1, a_2, a_3)$, $w_2 = (b_1, b_2, b_3)$ is given as follows:

$$\mathbf{x}(u, v) = p_0 + uw_1 + vw_2, \qquad (u, v) \in R^2.$$

To compute the first fundamental form for an arbitrary point of P we observe that $\mathbf{x}_u = w_1$, $\mathbf{x}_v = w_2$; since w_1 and w_2 are unit orthogonal vectors, the functions E, F, G are constant and given by

$$E = 1, \qquad F = 0, \qquad G = 1.$$

In this trivial case, the first fundamental form is essentially the Pythagorean theorem in P; i.e., the square of the length of a vector w which has coordinates a, b in the basis $\{\mathbf{x}_u, \mathbf{x}_v\}$ is equal to $a^2 + b^2$.

Example 2. The right cylinder over the circle $x^2 + y^2 = 1$ admits the parametrization $\mathbf{x}\colon U \longrightarrow R^3$, where (Fig. 2-26)

$$\mathbf{x}(u, v) = (\cos u, \sin u, v),$$
$$U = \{(u, v) \in R^2; \quad 0 < u < 2\pi, \quad -\infty < v < \infty\}.$$

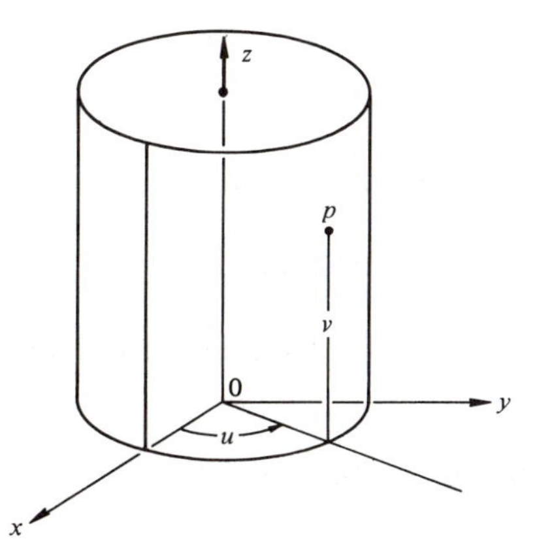

Figure 2-26

To compute the first fundamental form, we notice that

$$\mathbf{x}_u = (-\sin u, \cos u, 0), \qquad \mathbf{x}_v = (0, 0, 1),$$

and therefore

$$E = \sin^2 u + \cos^2 u = 1, \qquad F = 0, \qquad G = 1.$$

We remark that, although the cylinder and the plane are distinct surfaces, we obtain the same result in both cases. We shall return to this subject later (Sec. 4-2).

Example 3. Consider a helix that is given by (see Example 1, Sec. 1-2) $(\cos u, \sin u, au)$. Through each point of the helix, draw a line parallel to the xy plane and intersecting the z axis. The surface generated by these lines is called a *helicoid* and admits the following parametrization:

$$\mathbf{x}(u, v) = (v \cos u, v \sin u, au), \qquad 0 < u < 2\pi, \qquad -\infty < v < \infty.$$

$\mathbf{x}$ applies an open strip with width 2π of the uv plane onto that part of the helicoid which corresponds to a rotation of 2π along the helix (Fig. 2-27).

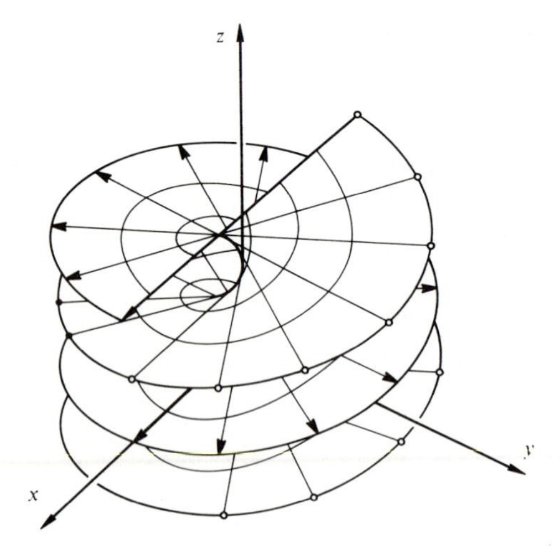

Figure 2-27. The helicoid.

The verification that the helicoid is a regular surface is straightforward and left to the reader.

The computation of the coefficients of the first fundamental form in the above parametrization gives

$$E(u, v) = v^2 + a^2, \qquad F(u, v) = 0, \qquad G(u, v) = 1.$$

As we mentioned before, the importance of the first fundamental form I comes from the fact that by knowing I we can treat metric questions on a regular surface without further references to the ambient space R^3. Thus, the arc length s of a parametrized curve $\alpha: I \longrightarrow S$ is given by

$$s(t) = \int_0^t |\alpha'(t)|\, dt = \int_0^t \sqrt{I(\alpha'(t))}\, dt.$$

In particular, if $\alpha(t) = \mathbf{x}(u(t), v(t))$ is contained in a coordinate neighborhood corresponding to the parametrization $\mathbf{x}(u, v)$, we can compute the arc length of α between, say, 0 and t by

$$s(t) = \int_0^t \sqrt{E(u')^2 + 2Fu'v' + G(v')^2}\, dt. \tag{2}$$

Also, the angle θ under which two parametrized regular curves $\alpha: I \longrightarrow S$, $\beta: I \longrightarrow S$ intersect at $t = t_0$ is given by

$$\cos\theta = \frac{\langle \alpha'(t_0), \beta'(t_0)\rangle}{|\alpha'(t_0)||\beta'(t_0)|}.$$

In particular, the angle φ of the coordinate curves of a parametrization $\mathbf{x}(u, v)$ is

$$\cos\varphi = \frac{\langle \mathbf{x}_u, \mathbf{x}_v\rangle}{|\mathbf{x}_u||\mathbf{x}_v|} = \frac{F}{\sqrt{EG}};$$

it follows that *the coordinate curves of a parametrization are orthogonal if and only if* F(u, v) = 0 *for all* (u, v). Such a parametrization is called an *orthogonal parametrization.*

Remark. Because of Eq. (2), many mathematicians talk about the "element" of arc length, ds of S, and write

$$ds^2 = E\,du^2 + 2F\,du\,dv + G\,dv^2,$$

meaning that if $\alpha(t) = \mathbf{x}(u(t), v(t))$ is a curve on S and $s = s(t)$ is its arc length, then

$$\left(\frac{ds}{dt}\right)^2 = E\left(\frac{du}{dt}\right)^2 + 2F\frac{du}{dt}\frac{dv}{dt} + G\left(\frac{dv}{dt}\right)^2.$$

Example 4. We shall compute the first fundamental form of a sphere at a point of the coordinate neighborhood given by the parametrization (cf. Example 1, Sec. 2-2)

$$\mathbf{x}(\theta, \varphi) = (\sin\theta\cos\varphi, \sin\theta\sin\varphi, \cos\theta).$$

First, observe that

$$\mathbf{x}_\theta(\theta, \varphi) = (\cos\theta\cos\varphi, \cos\theta\sin\varphi, -\sin\theta),$$
$$\mathbf{x}_\varphi(\theta, \varphi) = (-\sin\theta\sin\varphi, \sin\theta\cos\varphi, 0).$$

Hence,

$$E(\theta, \varphi) = \langle \mathbf{x}_\theta, \mathbf{x}_\theta \rangle = 1,$$
$$F(\theta, \varphi) = \langle \mathbf{x}_\theta, \mathbf{x}_\varphi \rangle = 0,$$
$$G(\theta, \varphi) = \langle \mathbf{x}_\varphi, \mathbf{x}_\varphi \rangle = \sin^2\theta.$$

Thus, if w is a tangent vector to the sphere at the point $\mathbf{x}(\theta, \varphi)$, given in the basis associated to $\mathbf{x}(\theta, \varphi)$ by

$$w = a\mathbf{x}_\theta + b\mathbf{x}_\varphi,$$

then the square of the length of w is given by

$$|w|^2 = I(w) = Ea^2 + 2Fab + Gb^2 = a^2 + b^2\sin^2\theta.$$

As an application, let us determine the curves in this coordinate neighborhood of the sphere which make a constant angle β with the meridians $\varphi = \text{const.}$ These curves are called *loxodromes* (rhumb lines) of the sphere.

We may assume that the required curve $\alpha(t)$ is the image by $\mathbf{x}$ of a curve $(\theta(t), \varphi(t))$ of the $\theta\varphi$ plane. At the point $\mathbf{x}(\theta, \varphi)$ where the curve meets the meridian $\varphi = \text{const.}$, we have

$$\cos\beta = \frac{\langle \mathbf{x}_\theta, \alpha'(t) \rangle}{|\mathbf{x}_\theta||\alpha'(t)|} = \frac{\theta'}{\sqrt{(\theta')^2 + (\varphi')^2\sin^2\theta}},$$

since in the basis $\{\mathbf{x}_\theta, \mathbf{x}_\varphi\}$ the vector $\alpha'(t)$ has coordinates (θ', φ') and the vector $\mathbf{x}_\theta$ has coordinates $(1, 0)$. It follows that

$$(\theta')^2\tan^2\beta - (\varphi')^2\sin^2\theta = 0$$

or

$$\frac{\theta'}{\sin\theta} = \pm\frac{\varphi'}{\tan\beta},$$

whence, by integration, we obtain the equation of the loxodromes

$$\log\tan\left(\frac{\theta}{2}\right) = \pm(\varphi + c)\operatorname{cotan}\beta,$$

where the constant of integration c is to be determined by giving one point $\mathbf{x}(\theta_0, \varphi_0)$ through which the curve passes.

Another metric question that can be treated by the first fundamental form is the computation (or definition) of the area of a bounded region of a regular surface S. A (regular) *domain* of S is an open and connected subset of S such that its boundary is the image of a circle by a differentiable homeomorphism which is regular (that is, its differential is nonzero) except at a finite number of points. A *region* of S is the union of a domain with its boundary (Fig. 2-28). A region of $S \subset R^3$ is *bounded* if it is contained in some ball of R^3.

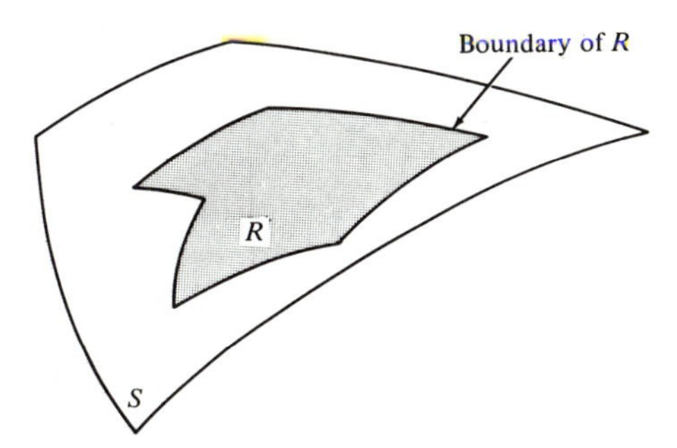

Figure 2-28

We shall consider bounded regions R which are contained in a coordinate neighborhood $\mathbf{x}(U)$ of a parametrization $\mathbf{x}: U \subset R^2 \to S$. In other words, R is the image by $\mathbf{x}$ of a bounded region $Q \subset U$.

The function $|\mathbf{x}_u \wedge \mathbf{x}_v|$, defined in U, measures the area of the parallelogram generated by the vectors $\mathbf{x}_u$ and $\mathbf{x}_v$. We first show that the integral

$$\int_Q |\mathbf{x}_u \wedge \mathbf{x}_v \ du\, dv$$

does not depend on the parametrization $\mathbf{x}$.

In fact, let $\bar{\mathbf{x}}: \bar{U} \subset R^2 \to S$ be another parametrization with $R \subset \bar{\mathbf{x}}(\bar{U})$ and set $\bar{Q} = \bar{\mathbf{x}}^{-1}(R)$. Let $\partial(u, v)/\partial(\bar{u}, \bar{v})$ be the Jacobian of the change of parameters $h = \mathbf{x}^{-1} \circ \bar{\mathbf{x}}$. Then

$$\begin{aligned}\iint_{\bar{Q}} |\bar{\mathbf{x}}_{\bar{u}} \wedge \bar{\mathbf{x}}_{\bar{v}}|\, d\bar{u}\, d\bar{v} &= \iint_{\bar{Q}} |\mathbf{x}_u \wedge \mathbf{x}_v| \left|\frac{\partial(u, v)}{\partial(\bar{u}, \bar{v})}\right| d\bar{u}\, d\bar{v} \\ &= \iint_Q |\mathbf{x}_u \wedge \mathbf{x}_v|\, du\, dv,\end{aligned}$$

where the last equality comes from the theorem of change of variables in multiple integrals (cf. Buck *Advanced Calculus*, p. 304). The asserted independence is therefore proved and we can make the following definition.

DEFINITION 2. *Let* R $\subset$ S *be a bounded region of a regular surface contained in the coordinate neighborhood of the parametrization* $\mathbf{x}\colon U \subset R^2 \to S$. *The positive number*

$$\iint_Q |\mathbf{x}_u \wedge \mathbf{x}_v|\, du\, dv = A(R), \qquad Q = \mathbf{x}^{-1}(R),$$

is called the area *of* R.

There are several geometric justifications for such a definition, and one of them will be presented in Sec. 2-8.

It is convenient to observe that

$$|\mathbf{x}_u \wedge \mathbf{x}_v|^2 + \langle \mathbf{x}_u, \mathbf{x}_v \rangle^2 = |\mathbf{x}_u|^2 \cdot |\mathbf{x}_v|^2,$$

which shows that the integrand of $A(R)$ can be written as

$$|\mathbf{x}_u \wedge \mathbf{x}_v| = \sqrt{EG - F^2}.$$

We should also remark that, in most examples, the restriction that the region R is contained in some coordinate neighborhood is not very serious, because there exist coordinate neighborhoods which cover the entire surface except for some curves, which do not contribute to the area.

Example 5. Let us compute the area of the torus of Example 6, Sec. 2-2. For that, we consider the coordinate neighborhood corresponding to the parametrization

$$\mathbf{x}(u, v) = ((a + r\cos u)\cos v, (a + r\cos u)\sin v, r\sin u),$$
$$0 < u < 2\pi, \qquad 0 < v < 2\pi,$$

which covers the torus, except for a meridian and a parallel. The coefficients of the first fundamental form are

$$E = r^2, \qquad F = 0, \qquad G = (r\cos u + a)^2;$$

hence,

$$\sqrt{EG - F^2} = r(r\cos u + a).$$

Now, consider the region R_ϵ obtained as the image by $\mathbf{x}$ of the region Q_ϵ (Fig. 2-29) given by ($\epsilon > 0$ and small),

$$Q_\epsilon = \{(u, v) \in R^2;\ 0 + \epsilon \le u \le 2\pi - \epsilon,\ 0 + \epsilon \le v \le 2\pi - \epsilon\}.$$

Using Def. 2, we obtain

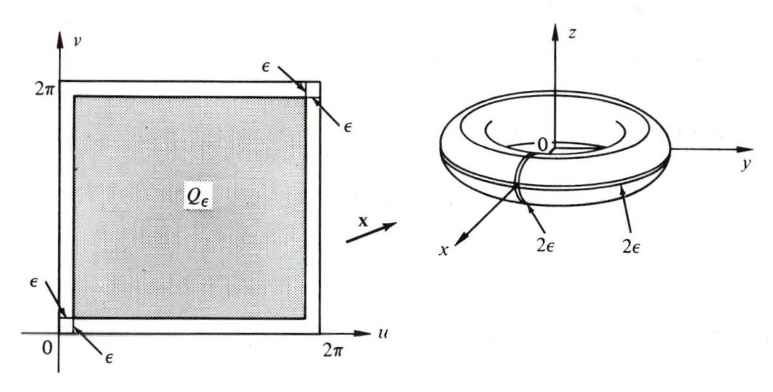

Figure 2-29

$$
\begin{aligned}
A(R_\epsilon) &= \iint_{Q_\varepsilon} r(r\cos u + a)\,du\,dv \\
&= \int_{0+\epsilon}^{2\pi-\epsilon} (r^2\cos u + ra)\,du \int_{0+\epsilon}^{2\pi-\epsilon} dv \\
&= r^2(2\pi - 2\epsilon)(\sin(2\pi-\epsilon) - \sin\epsilon) + ra(2\pi - 2\epsilon)^2.
\end{aligned}
$$

Letting $\epsilon \longrightarrow 0$ in the above expression, we obtain

$$A(T) = \lim_{\epsilon\to 0} A(R_\epsilon) = 4\pi^2 ra.$$

This agrees with the value found by elementary calculus, say, by using the theorem of Pappus for the area of surfaces of revolution (cf. Exercise 11).

EXERCISES

1. Compute the first fundamental forms of the following parametrized surfaces where they are regular:

a. $\mathbf{x}(u, v) = (a \sin u \cos v, b \sin u \sin v, c \cos u)$; ellipsoid.

b. $\mathbf{x}(u, v) = (au \cos v, bu \sin v, u^2)$; elliptic paraboloid.

c. $\mathbf{x}(u, v) = (au \cosh v, bu \sinh v, u^2)$; hyperbolic paraboloid.

d. $\mathbf{x}(u, v) = (a \sinh u \cos v, b \sinh u \sin v, c \cosh u)$; hyperboloid of two sheets.

2. Let $\mathbf{x}(u, v) = (\sin\theta\cos\varphi, \sin\theta\sin\varphi, \cos\theta)$ be a parametrization of the unit sphere S^2. Let P be the plane $x = z \operatorname{cotan} \alpha$, $0 < \alpha < \pi$, and β be the acute angle which the curve $P \cap S^2$ makes with the semimeridian $\varphi = \varphi_0$. Compute $\cos\beta$.

3. Obtain the first fundamental form of the sphere in the parametrization given by stereographic projection (cf. Exercise 16, Sec. 2-2).

4. Given the parametrized surface

$$\mathbf{x}(u, v) = (u \cos v, u \sin v, \log \cos v + u), \qquad -\frac{\pi}{2} < v < \frac{\pi}{2},$$

show that the two curves $\mathbf{x}(u_1, v)$, $\mathbf{x}(u_2, v)$ determine segments of equal lengths on all curves $\mathbf{x}(u, \text{const.})$.

5. Show that the area A of a bounded region R of the surface $z = f(x, y)$ is

$$A = \iint_Q \sqrt{1 + f_x^2 + f_y^2}\, dx\, dy,$$

where Q is the normal projection of R onto the xy plane.

6. Show that

$$\mathbf{x}(u, v) = (u \sin \alpha \cos v, u \sin \alpha \sin v, u \cos \alpha)$$
$$0 < u < \infty, \qquad 0 < v < 2\pi, \qquad \alpha = \text{const.},$$

is a parametrization of the cone with 2α as the angle of the vertex. In the corresponding coordinate neighborhood, prove that the curve

$$\mathbf{x}(c \exp(v \sin \alpha \cot \beta), v), \qquad c = \text{const.}, \beta = \text{const.},$$

intersects the generators of the cone ($v = \text{const.}$) under the constant angle β.

7. The coordinate curves of a parametrization $\mathbf{x}(u, v)$ constitute a *Tchebyshef net* if the lengths of the opposite sides of any quadrilateral formed by them are equal. Show that a necessary and sufficient condition for this is

$$\frac{\partial E}{\partial v} = \frac{\partial G}{\partial u} = 0.$$

***8.** Prove that whenever the coordinate curves constitute a Tchebyshef net (see Exercise 7) it is possible to reparametrize the coordinate neighborhood in such a way that the new coefficients of the first quadratic form are

$$E = 1, \qquad F = \cos \theta, \qquad G = 1,$$

where θ is the angle of the coordinate curves.

***9.** Show that a surface of revolution can always be parametrized so that

$$E = E(v), \qquad F = 0, \qquad G = 1.$$

10. Let $P = \{(x, y, z) \in R^3; z = 0\}$ be the xy plane and let $\mathbf{x}: U \longrightarrow P$ be a parametrization of P given by

$$\mathbf{x}(\rho, \theta) = (\rho \cos \theta, \rho \sin \theta),$$

where

$$U = \{(\rho, \theta) \in R^2; \rho > 0, 0 < \theta < 2\pi\}.$$

Compute the coefficients of the first fundamental form of P in this parametrization.

11. Let S be a surface of revolution and C its generating curve (cf. Example 4, Sec. 2-3). Let s be the arc length of C and denote by $\rho = \rho(s)$ the distance to the rotation axis of the point of C corresponding to s.

a. (*Pappus' Theorem.*) Show that the area of S is

$$2\pi \int_0^l \rho(s)\, ds,$$

where l is the length of C.

b. Apply part a to compute the area of a torus of revolution.

12. Show that the area of a regular tube of radius r around a curve α (cf. Exercise 10, Sec. 2-4) is $2\pi r$ times the length of α.

13. (*Generalized Helicoids.*) A natural generalization of both surfaces of revolution and helicoids is obtained as follows. Let a regular plane curve C, which does not meet an axis E in the plane, be displaced in a rigid screw motion about E, that is, so that each point of C describes a helix (or circle) with E as axis. The set S generated by the displacement of C is called a *generalized helicoid* with *axis E* and *generator C*. If the screw motion is a pure rotation about E, S is a surface of revolution; if C is a straight line perpendicular to E, S is (a piece of) the standard helicoid (cf. Example 3).

Choose the coordinate axes so that E is the z axis and C lies in the yz plane. Prove that

a. If $(f(s), g(s))$ is a parametrization of C by arc length s, $a < s < b$, $f(s) > 0$, then $\mathbf{x}: U \longrightarrow S$, where

$$U = \{(s, u) \in R^2; a < s < b, 0 < u < 2\pi\}$$

and

$$\mathbf{x}(s, u) = (f(s) \cos u, f(s) \sin u, g(s) + cu), \qquad c = \text{const.},$$

is a parametrization of S. Conclude that S is a regular surface.

b. The coordinate lines of the above parametrization are orthogonal (i.e., $F = 0$) if and only if $\mathbf{x}(U)$ is either a surface of revolution or (a piece of) the standard helicoid.

14. (*Gradient on Surfaces.*) The *gradient* of a differentiable function $f: S \longrightarrow R$ is a differentiable map $\text{grad} f: S \longrightarrow R^3$ which assigns to each point $p \in S$ a vector $\text{grad} f(p) \in T_p(S) \subset R^3$ such that

$$\langle \text{grad } f(p), v \rangle_p = df_p(v) \qquad \text{for all } v \in T_p(S).$$

Show that

a. If E, F, G are the coefficients of the first fundamental form in a parametrization $\mathbf{x}: U \subset R^2 \longrightarrow S$, then grad f on $\mathbf{x}(U)$ is given by

$$\operatorname{grad} f = \frac{f_u G - f_v F}{EG - F^2}\mathbf{x}_u + \frac{f_v E - f_u F}{EG - F^2}\mathbf{x}_v.$$

In particular, if $S = R^2$ with coordinates x, y,

$$\operatorname{grad} f = f_x e_1 + f_y e_2,$$

where $\{e_1, e_2\}$ is the canonical basis of R_2 (*thus, the definition agrees with the usual definition of gradient in the plane*).

b. If you let $p \in S$ be fixed and v vary in the unit circle $|v| = 1$ in $T_p(s)$, then $df_p(v)$ is maximum if and only if $v = \operatorname{grad} f/|\operatorname{grad} f|$ (*thus,* grad $f(p)$ *gives the direction of maximum variation of* f *at* p).

c. If grad $f \neq 0$ at all points of the *level curve* $C = \{q \in S;\ f(q) = \text{const.}\}$, then C is a regular curve on S and grad f is normal to C at all points of C.

15. (*Orthogonal Families of Curves.*)

a. Let E, F, G be the coefficients of the first fundamental form of a regular surface S in the parametrization $\mathbf{x}: U \subset R^2 \longrightarrow S$. Let $\varphi(u, v) = \text{const.}$ and $\psi(u, v) = \text{const.}$ be two families of regular curves on $\mathbf{x}(U) \subset S$ (cf. Exercise 28, Sec. 2-4). Prove that these two families are orthogonal (i.e., whenever two curves of distinct families meet, their tangent lines are orthogonal) if and only if

$$E\varphi_v\psi_v - F(\varphi_u\psi_v + \varphi_v\psi_u) + G\varphi_u\psi_u = 0.$$

b. Apply part a to show that on the coordinate neighborhood $\mathbf{x}(U)$ of the helicoid of Example 3 the two families of regular curves

$$v\cos u = \text{const.}, \qquad v \neq 0,$$
$$(v^2 + a^2)\sin^2 u = \text{const.}, \qquad v \neq 0, \qquad u \neq \pi,$$

are orthogonal.

2-6. Orientation of Surfaces†

In this section we shall discuss in what sense, and when, it is possible to orient a surface. Intuitively, since every point p of a regular surface S has a tangent plane $T_p(S)$, the choice of an orientation of $T_p(S)$ induces an orientation in a neighborhood of p, that is, a notion of positive movement along sufficiently

†This section may be omitted on a first reading.

small closed curves about each point of the neighborhood (Fig. 2-30). If it is possible to make this choice for each $p \in S$ so that in the intersection of any two neighborhoods the orientations coincide, then S is said to be orientable. If this is not possible, S is called nonorientable.

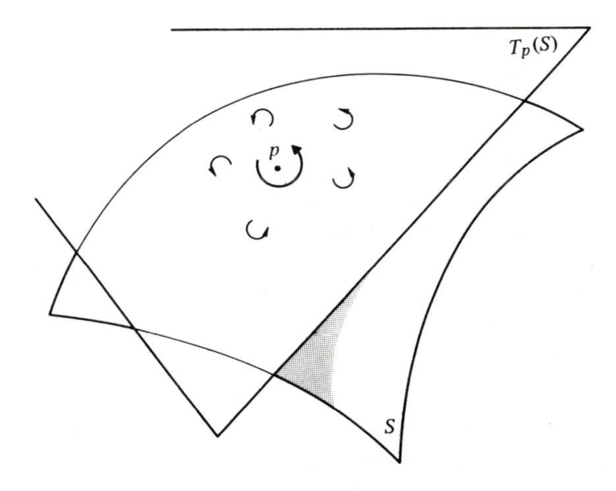

Figure 2-30

We shall now make these ideas precise. By fixing a parametrization $\mathbf{x}(u, v)$ of a neighborhood of a point p of a regular surface S, we determine an orientation of the tangent plane $T_p(S)$, namely, the orientation of the associated ordered basis $\{\mathbf{x}_u, \mathbf{x}_v\}$. If p belongs to the coordinate neighborhood of another parametrization $\bar{\mathbf{x}}(\bar{u}, \bar{v})$, the new basis $\{\bar{\mathbf{x}}_{\bar{u}}, \bar{\mathbf{x}}_{\bar{v}}\}$ is expressed in terms of the first one by

$$\bar{\mathbf{x}}_{\bar{u}} = \mathbf{x}_u \frac{\partial u}{\partial \bar{u}} + \mathbf{x}_v \frac{\partial v}{\partial \bar{u}},$$

$$\mathbf{x}_{\bar{v}} = \mathbf{x}_u \frac{\partial u}{\partial \bar{v}} + \mathbf{x}_v \frac{\partial v}{\partial \bar{v}},$$

where $u = u(\bar{u}, \bar{v})$ and $v = v(\bar{u}, \bar{v})$ are the expressions of the change of coordinates. The bases $\{\mathbf{x}_u, \mathbf{x}_v\}$ and $\{\bar{\mathbf{x}}_{\bar{u}}, \bar{\mathbf{x}}_{\bar{v}}\}$ determine, therefore, the same orientation of $T_p(S)$ if and only if the Jacobian

$$\frac{\partial(u, v)}{\partial(\bar{u}, \bar{v})}$$

of the coordinate change is positive.

DEFINITION 1. *A regular surface* S *is called* orientable *if it is possible to cover it with a family of coordinate neighborhoods in such a way that if a* *point* p ∈ S *belongs to two neighborhoods of this family, then the change of coordinates has positive Jacobian at* p. *The choice of such a family is called an* orientation *of* S, *and* S, *in this case, is called* oriented. *If such a choice is not possible, the surface is called* nonorientable.

Example 1. A surface which is the graph of a differentiable function (cf. Sec. 2-2, Prop. 1) is an orientable surface. In fact, all surfaces which can be covered by one coordinate neighborhood are trivially orientable.

Example 2. The sphere is an orientable surface. Instead of proceeding to a direct calculation, let us resort to a general argument. The sphere can be covered by two coordinate neighborhoods (using stereographic projection; see Exercise 16 of Sec. 2-2), with parameters (u, v) and $(\bar{u}, \bar{v})$, in such a way that the intersection W of these neighborhoods (the sphere minus two points) is a connected set. Fix a point p in W. If the Jacobian of the coordinate change at p is negative, we interchange u and v in the first system, and the Jacobian becomes positive. Since the Jacobian is different from zero in W and positive at $p \in W$, it follows from the connectedness of W that the Jacobian is everywhere positive. There exists, therefore, a family of coordinate neighborhoods satisfying Def. 1, and so the sphere is orientable.

By the argument just used, it is clear that *if a regular surface can be covered by two coordinate neighborhoods whose intersection is connected, then the surface is orientable.*

Before presenting an example of a nonorientable surface, we shall give a geometric interpretation of the idea of orientability of a regular surface in R^3.

As we have seen in Sec. 2-4, given a system of coordinates $\mathbf{x}(u, v)$ at p, we have a definite choice of a unit normal vector N at p by the rule

$$N = \frac{\mathbf{x}_u \wedge \mathbf{x}_v}{|\mathbf{x}_u \wedge \mathbf{x}_v|}(p). \tag{1}$$

Taking another system of local coordinates $\bar{\mathbf{x}}(\bar{u}, \bar{v})$ at p, we see that

$$\bar{\mathbf{x}}_{\bar{u}} \wedge \bar{\mathbf{x}}_{\bar{v}} = (\mathbf{x}_u \wedge \mathbf{x}_v)\frac{\partial(u, v)}{\partial(\bar{u}, \bar{v})}, \tag{2}$$

where $\partial(u, v)/\partial(\bar{u}, \bar{v})$ is the Jacobian of the coordinate change. Hence, N will preserve its sign or change it, depending on whether $\partial(u, v)/\partial(\bar{u}, \bar{v})$ is positive or negative, respectively.

By a differentiable *field of unit normal vectors* on an open set $U \subset S$, we shall mean a differentiable map $N\colon U \longrightarrow R^3$ which associates to each $q \in U$ a unit normal vector $N(q) \in R^3$ to S at q.

PROPOSITION 1. *A regular surface* $S \subset R^3$ *is orientable if and only if there exists a differentiable field of unit normal vectors* $N: S \longrightarrow R^3$ *on* S.

Proof. If S is orientable, it is possible to cover it with a family of coordinate neighborhoods so that, in the intersection of any two of them, the change of coordinates has a positive Jacobian. At the points $p = \mathbf{x}(u, v)$ of each neighborhood, we define $N(p) = N(u, v)$ by Eq. (1). $N(p)$ is well defined, since if p belongs to two coordinate neighborhoods, with parameters (u, v) and $(\bar{u}, \bar{v})$, the normal vector $N(u, v)$ and $N(\bar{u}, \bar{v})$ coincide by Eq. (2). Moreover, by Eq. (1), the coordinates of $N(u, v)$ in R^3 are differentiable functions of (u, v), and thus the mapping $N: S \longrightarrow R^3$ is differentiable, as desired.

On the other hand, let $N: S \longrightarrow R^3$ be a differentiable field of unit normal vectors, and consider a family of *connected* coordinate neighborhoods covering S. For the points $p = \mathbf{x}(u, v)$ of each coordinate neighborhood $\mathbf{x}(U)$, $U \subset R^2$, it is possible, by the continuity of N and, if necessary, by interchanging u and v, to arrange that

$$N(p) = \frac{\mathbf{x}_u \wedge \mathbf{x}_v}{|\mathbf{x}_u \wedge \mathbf{x}_v|}.$$

In fact, the inner product

$$\left\langle N(p), \frac{\mathbf{x}_u \wedge \mathbf{x}_v}{|\mathbf{x}_u \wedge \mathbf{x}_v|} \right\rangle = f(p) = \pm 1$$

is a continuous function on $\mathbf{x}(U)$. Since $\mathbf{x}(U)$ is connected, the sign of f is constant. If $f = -1$, we interchange u and v in the parametrization, and the assertion follows.

Proceeding in this manner with all the coordinate neighborhoods, we have that in the intersection of any two of them, say, $\mathbf{x}(u, v)$ and $\bar{\mathbf{x}}(\bar{u}, \bar{v})$, the Jacobian

$$\frac{\partial(u, v)}{\partial(\bar{u}, \bar{v})}$$

is certainly positive; otherwise, we would have

$$\frac{\mathbf{x}_u \wedge \mathbf{x}_v}{|\mathbf{x}_u \wedge \mathbf{x}_v|} = N(p) = -\frac{\bar{\mathbf{x}}_{\bar{u}} \wedge \bar{\mathbf{x}}_{\bar{v}}}{|\bar{\mathbf{x}}_{\bar{u}} \wedge \bar{\mathbf{x}}_{\bar{v}}|} = -N(p),$$

which is a contradiction. Hence, the given family of coordinate neighborhoods after undergoing certain interchanges of u and v satisfies the conditions of Def. 1, and S is, therefore, orientable. **Q.E.D.**

Remark. As the proof shows, we need only to require the existence of a *continuous* unit vector field on S for S to be orientable. Such a vector field will be automatically differentiable.

Example 3. We shall now describe an example of a nonorientable surface, the so-called *Möbius strip*. This surface is obtained (see Fig. 2-31) by considering the circle S^1 given by $x^2 + y^2 = 4$ and the open segment AB given in the yz plane by $y = 2$, $|z| < 1$. We move the center c of AB along S^1 and turn AB about c in the cz plane in such a manner that when c has passed through an angle u, AB has rotated by an angle $u/2$. When c completes one trip around the circle, AB returns to its initial position, with its end points inverted.

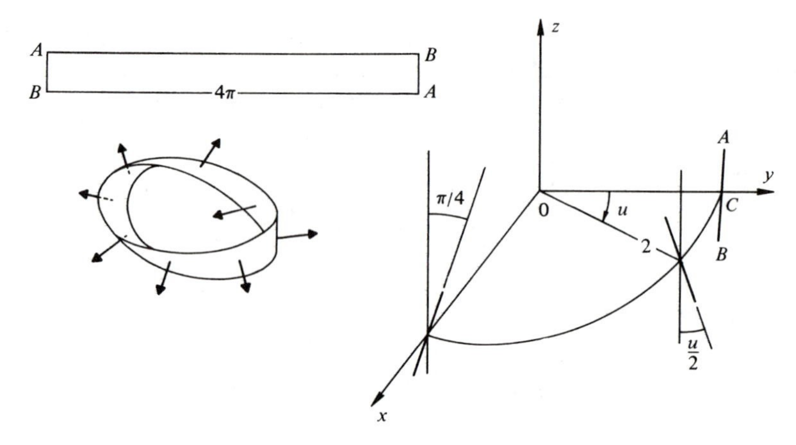

Figure 2-31

From the point of view of differentiability, it is as if we had identified the opposite (vertical) sides of a rectangle giving a twist to the rectangle so that each point of the side AB was identified with its symmetric point (Fig. 2-31).

It is geometrically evident that the Möbius strip M is a regular, nonorientable surface. In fact, if M were orientable, there would exist a differentiable field $N: M \longrightarrow R^3$ of unit normal vectors. Taking these vectors along the circle $x^2 + y^2 = 4$ we see that after making one trip the vector N returns to its original position as $-N$, which is a contradiction.

We shall now give an analytic proof of the facts mentioned above.

A system of coordinates $\mathbf{x}: U \longrightarrow M$ for the Möbius strip is given by

$$\mathbf{x}(u, v) = \left(\left(2 - v \sin \frac{u}{2}\right) \sin u, \left(2 - v \sin \frac{u}{2}\right) \cos u, v \cos \frac{u}{2}\right),$$

where $0 < u < 2\pi$ and $-1 < v < 1$. The corresponding coordinate neighborhood omits the points of the open interval $u = 0$. Then by taking the origin of the u's at the x axis, we obtain another parametrization $\bar{\mathbf{x}}(\bar{u}, \bar{v})$ given by

$$x = \left\{2 - \bar{v}\sin\left(\frac{\pi}{4} + \frac{\bar{u}}{2}\right)\right\}\cos\bar{u},$$

$$y = -\left\{2 - \bar{v}\sin\left(\frac{\pi}{4} + \frac{\bar{u}}{2}\right)\right\}\sin\bar{u},$$

$$z = \bar{v}\cos\left(\frac{\pi}{4} + \frac{\bar{u}}{2}\right),$$

whose coordinate neighborhood omits the interval $u = \pi/2$. These two coordinate neighborhoods cover the Möbius strip and can be used to show that it is a regular surface.

Observe that the intersection of the two coordinate neighborhoods is not connected but consists of two connected components:

$$W_1 = \left\{x(u, v): \frac{\pi}{2} < u < 2\pi\right\},$$

$$W_2 = \left\{x(u, v): 0 < u < \frac{\pi}{2}\right\}.$$

The change of coordinates is given by

$$\left.\begin{aligned} \bar{u} &= u - \frac{\pi}{2} \\ \bar{v} &= v \end{aligned}\right\} \quad \text{in } W_1,$$

and

$$\left.\begin{aligned} \bar{u} &= \frac{3\pi}{2} + u \\ \bar{v} &= -v \end{aligned}\right\} \quad \text{in } W_2.$$

It follows that

$$\frac{\partial(\bar{u}, \bar{v})}{\partial(u, v)} = 1 > 0 \qquad \text{in } W_1$$

and that

$$\frac{\partial(\bar{u}, \bar{v})}{\partial(u, v)} = -1 < 0 \qquad \text{in } W_2.$$

To show that the Möbius strip is nonorientable, we suppose that it is possible to define a differentiable field of unit normal vectors $N: M \to R^3$. Interchanging u and v if necessary, we can assume that

$$N(p) = \frac{\mathbf{x}_u \wedge \mathbf{x}_v}{|\mathbf{x}_u \wedge \mathbf{x}_v|}$$

for any p in the coordinate neighborhood of $\mathbf{x}(u, v)$. Analogously, we may assume that

$$N(p) = \frac{\bar{\mathbf{x}}_{\bar{u}} \wedge \bar{\mathbf{x}}_{\bar{v}}}{|\bar{\mathbf{x}}_{\bar{u}} \wedge \bar{\mathbf{x}}_{\bar{v}}|}$$

at all points of the coordinate neighborhood of $\bar{\mathbf{x}}(\bar{u}, \bar{v})$. However, the Jacobian of the change of coordinates must be -1 in either W_1 or W_2 (depending on what changes of the type $u \longrightarrow v$, $\bar{u} \longrightarrow \bar{v}$ has to be made). If p is a point of that component of the intersection, then $N(p) = -N(p)$, which is a contradiction.

We have already seen that a surface which is the graph of a differentiable function is orientable. We shall now show that a surface which is the inverse image of a regular value of a differentiable function is also orientable. This is one of the reasons it is relatively difficult to construct examples of non-orientable, regular surfaces in R^3.

PROPOSITION 2. *If a regular surface is given by* $S = \{(x, y, z) \in R^3; f(x, y, z) = a\}$, *where* $f\colon U \subset R^3 \longrightarrow R$ *is differentiable and* a *is a regular value of* f, *then* S *is orientable.*

Proof. Given a point $(x_0, y_0, z_0) = p \in S$, consider the parametrized curve $(x(t), y(t), z(t))$, $t \in I$, on S passing through p for $t = t_0$. Since the curve is on S, we have

$$f(x(t), y(t), z(t)) = a$$

for all $t \in I$. By differentiating both sides of this expression with respect to t, we see that at $t = t_0$

$$f_x(p)\left(\frac{dx}{dt}\right)_{t_0} + f_y(p)\left(\frac{dy}{dt}\right)_{t_0} + f_z(p)\left(\frac{dz}{dt}\right)_{t_0} = 0.$$

This shows that the tangent vector to the curve at $t = t_0$ is perpendicular to the vector (f_x, f_y, f_z) at p. Since the curve and the point are arbitrary, we conclude that

$$N(x, y, z) = \left(\frac{f_x}{\sqrt{f_x^2 + f_y^2 + f_z^2}}, \frac{f_y}{\sqrt{f_x^2 + f_y^2 + f_z^2}}, \frac{f_z}{\sqrt{f_x^2 + f_y^2 + f_z^2}}\right)$$

is a differentiable field of unit normal vectors on S. Together with Prop. 1, this implies that S is orientable as desired. **Q.E.D.**

A final remark. Orientation is definitely not a local property of a regular surface. Locally, every regular surface is diffeomorphic to an open set in the plane, and hence orientable. Orientation is a global property, in the sense that it involves the whole surface. We shall have more to say about global properties later in this book (Chap. 5).

EXERCISES

1. Let S be a regular surface covered by coordinate neighborhoods V_1 and V_2. Assume that $V_1 \cap V_2$ has two connected components, W_1, W_2, and that the Jacobian of the change of coordinates is positive in W_1 and negative in W_2. Prove that S is nonorientable.

2. Let S_2 be an orientable regular surface and $\varphi\colon S_1 \longrightarrow S_2$ be a differentiable map which is a local diffeomorphism at every $p \in S_1$. Prove that S_1 is orientable.

3. Is it possible to give a meaning to the notion of area for a Möbius strip? If so, set up an integral to compute it.

4. Let S be an orientable surface and let $\{U_\alpha\}$ and $\{V_\beta\}$ be two families of coordinate neighborhoods which cover S (that is, $\bigcup U_\alpha = S = \bigcup V_\beta$) and satisfy the conditions of Def. 1 (that is, in each of the families, the coordinate changes have positive Jacobian). We say that $\{U_\alpha\}$ and $\{V_\beta\}$ determine the *same orientation* of S if the union of the two families again satisfies the conditions of Def. 1.

Prove that a regular, connected, orientable surface can have only two distinct orientations.

5. Let $\varphi\colon S_1 \longrightarrow S_2$ be a diffeomorphism.

a. Show that S_1 is orientable if and only if S_2 is orientable (*thus, orientability is preserved by diffeomorphisms*).

b. Let S_1 and S_2 be orientable and oriented. Prove that the diffeomorphism φ induces an orientation in S_2. Use the antipodal map of the sphere (Exercise 1, Sec. 2-3) to show that this orientation may be distinct (cf. Exercise 4) from the initial one (*thus, orientation itself may not be preserved by diffeomorphisms; note, however, that if* S_1 *and* S_2 *are connected, a diffeomorphism either preserves or "reverses" the orientation*).

6. Define the notion of orientation of a regular curve $C \subset R^3$, and show that if C is connected, there exist at most two distinct orientations in the sense of Exercise 4 (actually there exist exactly two, but this is harder to prove).

7. Show that if a regular surface S contains an open set diffeomorphic to a Möbius strip, then S is nonorientable.

2-7. A Characterization of Compact Orientable Surfaces†

The converse of Prop. 2 of Sec. 2-6, namely, that *an orientable surface in* R^3 *is the inverse image of a regular value of some differentiable function*, is true

†This section may be omitted on a first reading.

and nontrivial to prove. Even in the particular case of compact surfaces (defined in this section), the proof is instructive and offers an interesting example of a global theorem in differential geometry. This section will be dedicated entirely to the proof of this converse statement.

Let $S \subset R^3$ be an orientable surface. The crucial point of the proof consists of showing that one may choose, on the normal line through $p \in S$, an open interval I_p around p of length, say, $2\epsilon_p$ (ϵ_p varies with p) in such a way that if $p \neq q \in S$, then $I_p \cap I_q = \phi$. Thus, the union $\bigcup I_p$, $p \in S$, constitutes an open set V of R^3, which contains S and has the property that through each point of V there passes a unique normal line to S; V is then called a *tubular neighborhood* of S (Fig. 2-32).

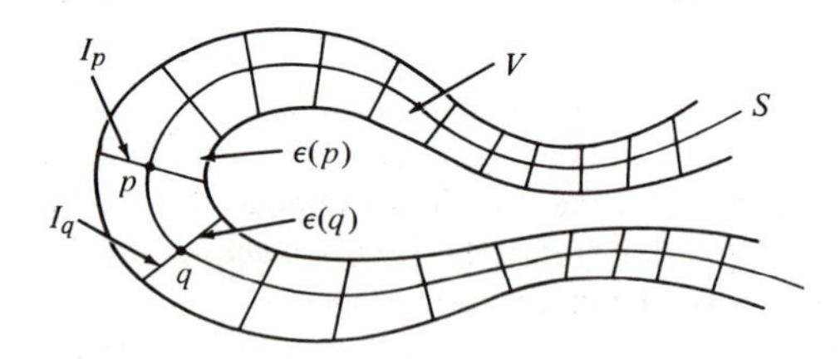

Figure 2-32. A tubular neighborhood.

Let us assume, for the moment, the existence of a tubular neighborhood V of an orientable surface S. We can then define a function $g: V \to R$ as follows: Fix an orientation for S. Observe that no two segments I_p and I_q, $p \neq q$, of the tubular neighborhood V intersect. Thus, through each point $P \in V$ there passes a unique normal line to S which meets S at a point p; by definition, $g(P)$ is the distance from p to P, with a sign given by the direction of the unit normal vector at p. If we can prove that g is a differentiable function and that 0 is a regular value of g, we shall have that $S = g^{-1}(0)$, as we wished to prove.

We shall now start the proof of the existence of a tubular neighborhood of an orientable surface. We shall first prove a local version of this fact; that is, we shall show that for each point p of a regular surface there exists a neighborhood of p which has a tubular neighborhood.

PROPOSITION 1. *Let* S *be a regular surface and* $\mathbf{x}: U \to S$ *be a parametrization of a neighborhood of a point* $p = \mathbf{x}(u_0, v_0) \in S$. *Then there exists a neighborhood* $W \subset \mathbf{x}(U)$ *of* p *in* S *and a number* $\epsilon > 0$ *such that the segments of the normal lines passing through points* $q \in W$, *with center at* q *and length* 2ϵ, *are disjoint* (*that is,* W *has a tubular neighborhood*).

Proof. Consider the map $F: U \times R \to R^3$ given by

$$F(u, v; t) = \mathbf{x}(u, v) + tN(u, v), \qquad (u, v) \in U, \qquad t \in R,$$

where $N(u, v) = (N_x, N_y, N_z)$ is the unit normal vector at

$$\mathbf{x}(u, v) = (x(u, v), y(u, v), z(u, v)).$$

Geometrically, F maps the point $(u, v; t)$ of the "cylinder" $U \times R$ in the point of the normal line to S at a distance t from $\mathbf{x}(u, v)$. F is clearly differentiable and its Jacobian at $t = 0$ is given by

$$\begin{vmatrix} \dfrac{\partial x}{\partial u} & \dfrac{\partial y}{\partial u} & \dfrac{\partial z}{\partial u} \\ \dfrac{\partial x}{\partial v} & \dfrac{\partial y}{\partial v} & \dfrac{\partial z}{\partial v} \\ N_x & N_y & N_z \end{vmatrix} = |\mathbf{x}_u \wedge \mathbf{x}_v| \neq 0.$$

By the inverse function theorem, there exists a parallelepiped in $U \times R$, say,

$$u_0 - \delta < u < u_0 + \delta, \qquad v_0 - \delta < v < v_0 + \delta, \qquad -\epsilon < t < \epsilon,$$

restricted to which F is one-to-one. But this means that in the image W by F of the rectangle

$$u_0 - \delta < u < u_0 + \delta, \qquad v_0 - \delta < v < v_0 + \delta$$

the segments of the normal lines with centers $q \in W$ and of length $< 2\epsilon$ do not meet. **Q.E.D.**

At this point, it is convenient to observe the following. The fact that the function $g: V \to R$, defined above by assuming the existence of a tubular neighborhood V, is differentiable and has 0 as a regular value is a local fact and can be proved at once.

PROPOSITION 2. *Assume the existence of a tubular neighborhood* $V \subset R^3$ *of an orientable surface* $S \subset R^3$, *and choose an orientation for* S. *Then the function* $g: V \to R$, *defined as the oriented distance from a point of* V *to the foot of the unique normal line passing through this point, is differentiable and has zero as a regular value.*

Proof. Let us look again at the map $F: U \times R \to R^3$ defined in Prop. 1, where we now assume that the parametrization $\mathbf{x}$ is compatible with the given orientation. Denoting by x, y, z the coordinates of $F(u, v, t) = \mathbf{x}(u, v) + tN(u, v)$ we can write

$$F(u, v, t) = (x(u, v, t), y(u, v, t), z(u, v, t)).$$

Since the Jacobian $\partial(x, y, z)/\partial(u, v, t)$ is different from zero at $t = 0$, we can invert F in some parallelepiped Q,

$$u_0 - \delta < u < u_0 + \delta, \qquad v_0 - \delta < v < v_0 + \delta, \qquad -\epsilon < t < \epsilon,$$

to obtain a differentiable map

$$F^{-1}(x, y, z) = (u(x, y, z), v(x, y, z), t(x, y, z)),$$

where $(x, y, z) \in F(Q) = V$. But the function $g\colon V \longrightarrow R$ in the statement of Prop. 2 is precisely $t = t(x, y, z)$. Thus, g is differentiable. Furthermore, 0 is a regular value of g; otherwise

$$\frac{\partial t}{\partial x} = \frac{\partial t}{\partial y} = \frac{\partial t}{\partial z} = 0$$

for some point where $t = 0$; hence, the differential dF^{-1} would be singular for $t = 0$, which is a contradiction. **Q.E.D.**

To pass from the local to the global, that is, to prove the existence of a tubular neighborhood of an entire orientable surface, we need some topological arguments. We shall restrict ourselves to compact surfaces, which we shall now define.

Let A be a subset of R^3. We say that $p \in R^3$ is a *limit point* of A if every neighborhood of p in R^3 contains a point of A distinct from p. A is said to be *closed* if it contains all its limit points. A is *bounded* if it is contained in some ball of R^3. If A is closed and bounded, it is called a *compact set.*

The sphere and the torus are compact surfaces. The paraboloid of revolution $z = x^2 + y^2$, $(x, y) \in R^2$, is a closed surface, but, being unbounded, it is not a compact surface. The disk $x^2 + y^2 < 1$ in the plane and the Möbius strip are bounded but not closed and therefore are noncompact.

We shall need some properties of compact subsets of R^3, which we shall now state. The distance between two points $p, q \in R^3$ will be denoted by $d(p, q)$.

PROPERTY 1 (Bolzano-Weierstrass). *Let* $A \subset R^3$ *be a compact set. Then every infinite subset of* A *has at least one limit point in* A.

PROPERTY 2 (Heine-Borel). *Let* $A \subset R^3$ *be a compact set and* $\{U_\alpha\}$ *be a family of open sets of* A *such that* $\bigcup_\alpha U_\alpha = A$. *Then it is possible to choose a finite number* $U_{k_1}, U_{k_2}, \ldots, U_{k_n}$ *of* U_α *such that* $\bigcup U_{k_i} = A$, $i = 1, \ldots, n$.

PROPERTY 3 (Lebesgue). *Let* $A \subset R^3$ *be a compact set and* $\{U_\alpha\}$ *a family of open sets of* A *such that* $\bigcup_\alpha U_\alpha = A$. *Then there exists a number* $\delta > 0$

(the Lebesgue number of the family $\{U_\alpha\}$*) such that whenever two points* $p, q \in A$ *are at a distance* $d(p, q) < \delta$ *then* p *and* q *belong to some* U_α.

Properties 1 and 2 are usually proved in courses of advanced calculus. For completeness, we shall now prove Property 3. Later in this book (appendix to Chap. 5), we shall treat compact sets in R^n in a more systematic way and shall present proofs of Properties 1 and 2.

Proof of Property 3. Let us assume that there is no $\delta > 0$ satisfying the conditions in the statement; that is, given $1/n$ there exist points p_n and q_n such that $d(p_n, q_n) < 1/n$ but p_n and q_n do not belong to the same open set of the family $\{U_\alpha\}$. Setting $n = 1, 2, \ldots$, we obtain two infinite sets of points $\{p_n\}$ and $\{q_n\}$ which, by Property 1, have limit points p and q, respectively. Since $d(p_n, q_n) < 1/n$, we may choose these limit points in such a way that $p = q$. But $p \in U_\alpha$ for some α, because $p \in A = \bigcup_\alpha U_\alpha$, and since U_α is an open set, there is an open ball $B_\epsilon(p)$, with center in p, such that $B_\epsilon(p) \subset U_\alpha$. Since p is a limit point of $\{p_n\}$ and $\{q_n\}$, there exist, for n sufficiently large, points p_n and q_n in $B_\epsilon(p) \subset U_\alpha$; that is, p_n and q_n belong to the same U_α, which is a contradiction. **Q.E.D.**

Using Properties 2 and 3, we shall now prove the existence of a tubular neighborhood of an orientable compact surface.

PROPOSITION 3. *Let* $S \subset R^3$ *be a regular, compact, orientable surface. Then there exists a number* $\epsilon > 0$ *such that whenever* $p, q \in S$ *the segments of the normal lines of length* 2ϵ*, centered in* p *and* q*, are disjoint (that is,* S *has a tubular neighborhood).*

Proof. By Prop. 1, for each $p \in S$ there exists a neighborhood W_p and a number $\epsilon_p > 0$ such that the proposition holds for points of W_p with $\epsilon = \epsilon_p$. Letting p run through S, we obtain a family $\{W_p\}$ with $\bigcup_{p \in S} W_p = S$. By compactness (Property 2), it is possible to choose a finite number of the W_p's, say, $W_1, \ldots, W_k$ (corresponding to $\epsilon_1, \ldots, \epsilon_k$) such that $\bigcup W_i = S$, $i = 1, \ldots, k$. We shall show that the required ϵ is given by

$$\epsilon < \min\left(\epsilon_1, \ldots, \epsilon_k, \frac{\delta}{2}\right),$$

where δ is the Lebesgue number of the family $\{W_i\}$ (Property 3).

In fact, let two points $p, q \in S$. If both belong to some W_i, $i = 1, \ldots, k$, the segments of the normal lines with centers in p and q and length 2ϵ do not meet, since $\epsilon < \epsilon_i$. If p and q do not belong to the same W_i, then $d(p, q) \geq \delta$; were the segments of the normal lines, centered in p and q and of length 2ϵ, to meet at point $Q \in R^3$, we would have

$$2\epsilon \geq d(p, Q) + d(Q, q) \geq d(p, q) \geq \delta,$$

which contradicts the definition of ϵ. **Q.E.D.**

Putting together Props. 1, 2, and 3, we obtain the following theorem, which is the main goal of this section.

THEOREM. *Let* $S \subset R^3$ *be a regular compact orientable surface. Then there exists a differentiable function* $g: V \longrightarrow R$, *defined in an open set* $V \subset R^3$, *with* $V \supset S$ *(precisely a tubular neighborhood of* S*), which has zero as a regular value and is such that* $S = g^{-1}(0)$.

Remark 1. It is possible to prove the existence of a tubular neighborhood of an orientable surface, even if the surface is not compact; the theorem is true, therefore, without the restriction of compactness. The proof is, however, more technical. In this general case, the $\epsilon(p) > 0$ is not constant as in the compact case but may vary with p.

Remark 2. It is possible to prove that a regular compact surface in R^3 is orientable; the hypothesis of orientability in the theorem (the compact case) is therefore unnecessary. A proof of this fact can be found in H. Samelson, "Orientability of Hypersurfaces in R^n," *Proc. A.M.S.* 22 (1969), 301–302.

2-8. A Geometric Definition of Area†

In this section we shall present a geometric justification for the definition of area given in Sec. 2-5. More precisely, we shall give a geometric definition of area and shall prove that in the case of a bounded region of a regular surface such a definition leads to the formula given for the area in Sec. 2-5.

To define the area of a region $R \subset S$ we shall start with a *partition* $\mathcal{P}$ of R into a finite number of regions R_i, that is, we write $R = \bigcup_i R_i$, where the intersection of two such regions R_i is either empty or made up of boundary points of both regions (Fig. 2-33). The *diameter* of R_i is the supremum of the distances (in R^3) of any two points in R_i; the largest diameter of the R_i's of a given partition $\mathcal{P}$ is called the *norm* μ of $\mathcal{P}$. If we now take a partition of each R_i, we obtain a second partition of R, which is said to *refine* $\mathcal{P}$.

Given a partition

$$R = \bigcup_i R_i$$

of R, we choose arbitrarily points $p_i \in R_i$ and project R_i onto the tangent

†This section may be omitted on a first reading.

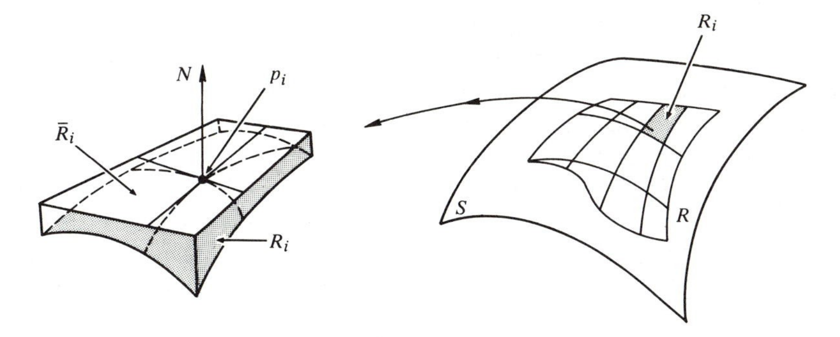

Figure 2-33

plane at p_i in the direction of the normal line at p_i; this projection is denoted by $\bar{R}_i$ and its area by $A(\bar{R}_i)$. The sum $\sum_i A(\bar{R}_i)$ is an approximation of what we understand intuitively by the area of R.

If, by choosing partitions $\mathcal{P}_1, \ldots, \mathcal{P}_n, \ldots$ more and more refined and such that the norm μ_n of $\mathcal{P}_n$ converges to zero, there exists a limit of $\sum_i A(\bar{R}_i)$ and this limit is independent of all choices, then we say that R has an *area* $A(R)$ defined by

$$A(R) = \lim_{\mu_n \to 0} \sum_i A(\bar{R}_i).$$

An instructive discussion of this definition can be found in R. Courant, *Differential and Integral Calculus*, Vol. II, Wiley-Interscience, New York, 1936, p. 311.

We shall show that a bounded region of a regular surface does have an area. We shall restrict ourselves to bounded regions contained in a coordinate neighborhood and shall obtain an expression for the area in terms of the coefficients of the first fundamental form in the corresponding coordinate system.

PROPOSITION. *Let* $\mathbf{x}: U \to S$ *be a coordinate system in a regular surface* S *and let* $R = \mathbf{x}(Q)$ *be a bounded region of* S *contained in* $\mathbf{x}(U)$. *Then* R *has an area given by*

$$A(R) = \iint_Q |\mathbf{x}_u \wedge \mathbf{x}_v| \, du \, dv.$$

Proof. Consider a partition, $R = \bigcup_i R_i$, of R. Since R is bounded and closed (hence, compact), we can assume that the partition is sufficiently refined so that any two normal lines of R_i are never orthogonal. In fact, because the normal lines vary continuously in S, there exists for each $p \in R$ a neighborhood of p in S where any two normals are never orthogonal; these neighborhoods constitute a family of open sets covering R, and considering a partition of R the norm of which is smaller than the Lebesgue number of the covering (Sec. 2-7, Property 3 of compact sets), we shall satisfy the required condition.

Fix a region R_i of the partition and choose a point $p_i \in R_i = \mathbf{x}(Q_i)$. We want to compute the area of the normal projection $\bar{R}_i$ of R_i onto the tangent plane at p_i. To do this, consider a new system of axes $p_i\bar{x}\bar{y}\bar{z}$ in R^3, obtained from $Oxyz$ by a translation Op_i, followed by a rotation which takes the z axis into the normal line at p_i in such a way that both systems have the same orientation (Fig. 2-34). In the new axes, the parametrization can be written

$$\bar{\mathbf{x}}(u, v) = (\bar{x}(u, v), \bar{y}(u, v), \bar{z}(u, v)),$$

where the explicit form of $\bar{\mathbf{x}}(u, v)$ does not interest us; it is enough to know that the vector $\bar{\mathbf{x}}(u, v)$ is obtained from the vector $\mathbf{x}(u, v)$ by a translation followed by an orthogonal linear map.

We observe that $\partial(\bar{x}, \bar{y})/\partial(u, v) \neq 0$ in Q_i; otherwise, the $\bar{z}$ component of some normal vector in R_i is zero and there are two orthogonal normal lines in R_i, a contradiction of our assumptions.

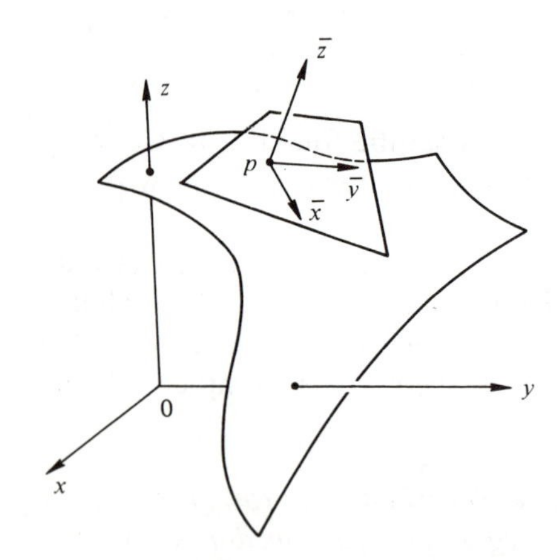

Figure 2-34

The expression of $A(\bar{R}_i)$ is given by

$$A(\bar{R}_i) = \iint_{\bar{R}_i} d\bar{x}\, d\bar{y}.$$

Since $\partial(\bar{x}, \bar{y})/\partial(u, v) \neq 0$, we can consider the change of coordinates $\bar{x} = \bar{x}(u, v)$, $\bar{y} = \bar{y}(u, v)$ and transform the above expression into

$$A(\bar{R}_i) = \iint_{Q_i} \frac{\partial(\bar{x}, \bar{y})}{\partial(u, v)}\, du\, dv.$$

We remark now that, at p_i, the vectors $\bar{\mathbf{x}}_u$ and $\bar{\mathbf{x}}_v$ belong to the $\bar{x}\bar{y}$ plane; therefore,

$$\frac{\partial \bar{z}}{\partial u} = \frac{\partial \bar{z}}{\partial v} = 0 \qquad \text{at } p_i;$$

hence,

$$\left|\frac{\partial(\bar{x}, \bar{y})}{\partial(u, v)}\right| = \left|\frac{\partial \bar{\mathbf{x}}}{\partial u} \wedge \frac{\partial \bar{\mathbf{x}}}{\partial v}\right| \qquad \text{at } p_i.$$

It follows that

$$\left|\frac{\partial(\bar{x}, \bar{y})}{\partial(u, v)}\right| - \left|\frac{\partial \bar{\mathbf{x}}}{\partial u} \wedge \frac{\partial \bar{\mathbf{x}}}{\partial v}\right| = \epsilon_i(u, v), \qquad (u, v) \in Q_i,$$

where $\epsilon_i(u, v)$ is a continuous function in Q_i with $\epsilon_i(\mathbf{x}^{-1}(p_i)) = 0$. Since the length of a vector is preserved by translations and orthogonal linear maps, we obtain

$$\left|\frac{\partial \mathbf{x}}{\partial u} \wedge \frac{\partial \mathbf{x}}{\partial v}\right| = \left|\frac{\partial \bar{\mathbf{x}}}{\partial u} \wedge \frac{\partial \bar{\mathbf{x}}}{\partial v}\right| = \left|\frac{\partial(\bar{x}, \bar{y})}{\partial(u, v)}\right| - \epsilon_i(x, y).$$

Now let M_i and m_i be the maximum and the minimum of the continuous function $\epsilon_i(u, v)$ in the compact region Q_i; thus,

$$m_i \leq \left|\frac{\partial(\bar{x}, \bar{y})}{\partial(u, v)}\right| - \left|\frac{\partial \mathbf{x}}{\partial u} \wedge \frac{\partial \mathbf{x}}{\partial v}\right| \leq M_i;$$

hence,

$$m_i \iint_{Q_i} du\, dv \leq A(\bar{R}_i) - \iint_{Q_i} \left|\frac{\partial \mathbf{x}}{\partial u} \wedge \frac{\partial \mathbf{x}}{\partial v}\right| du\, dv \leq M_i \iint_{Q_i} du\, dv.$$

Doing the same for all R_i, we obtain

$$\sum_i m_i A(Q_i) \leq \sum_i A(\bar{R}_i) - \iint_Q |\mathbf{x}_u \wedge \mathbf{x}_v|\, du\, dv \leq \sum_i M_i A(Q_i).$$

Now, refine more and more the given partition in such a way that the norm $\mu \to 0$. Then $M_i \to m_i$. Therefore, there exists the limit of $\sum_i A(\bar{R}_i)$, given by

$$A(R) = \iint_Q \left|\frac{\partial \mathbf{x}}{\partial u} \wedge \frac{\partial \mathbf{x}}{\partial v}\right| du\, dv,$$

which is clearly independent of the choice of the partitions and of the point p_i in each partition. **Q.E.D.**

Appendix A Brief Review of Continuity and Differentiability

R^n will denote the set of n-tuples $(x_1, \ldots, x_n)$ of real numbers. Although we use only the cases $R^1 = R$, R^2, and R^3, the more general notion of R^n unifies the definitions and brings in no additional difficulties; the reader may think in R^2 or R^3, if he wishes so. In these particular cases, we shall use the following more traditional notation: x or t for R, (x, y) or (u, v) for R^2, and (x, y, z) for R^3.

A. Continuity in R^n

We start by making precise the notion of a point being ϵ-close to a given point $p_0 \in R^n$.

A *ball* (or *open ball*) in R^n with center $p_0 = (x_1^0, \ldots, x_n^0)$ and radius $\epsilon > 0$ is the set

$$B_\epsilon(p_0) = \{(x_1, \ldots, x_n) \in R^n; (x_1 - x_1^0)^2 + \cdots + (x_n - x_n^0) < \epsilon^2\}.$$

Thus, in R, $B_\epsilon(p_0)$ is an open interval with center p_0 and length 2ϵ; in R^2, $B_\epsilon(p_0)$ is the interior of a disk with center p_0 and radius ϵ; in R^3, $B_\epsilon(p_0)$ is the interior of a region bounded by a sphere of center p_0 and radius ϵ (see Fig. A2-1).

A set $U \subset R^n$ is an *open set* if for each $p \in U$ there is a ball $B_\epsilon(p) \subset U$; intuitively this means that points in U are entirely surrounded by points of U, or that points sufficiently close to points of U still belong to U.

For instance, the set

$$\{(x, y) \in R^2; a < x < b, c < y < d\}$$

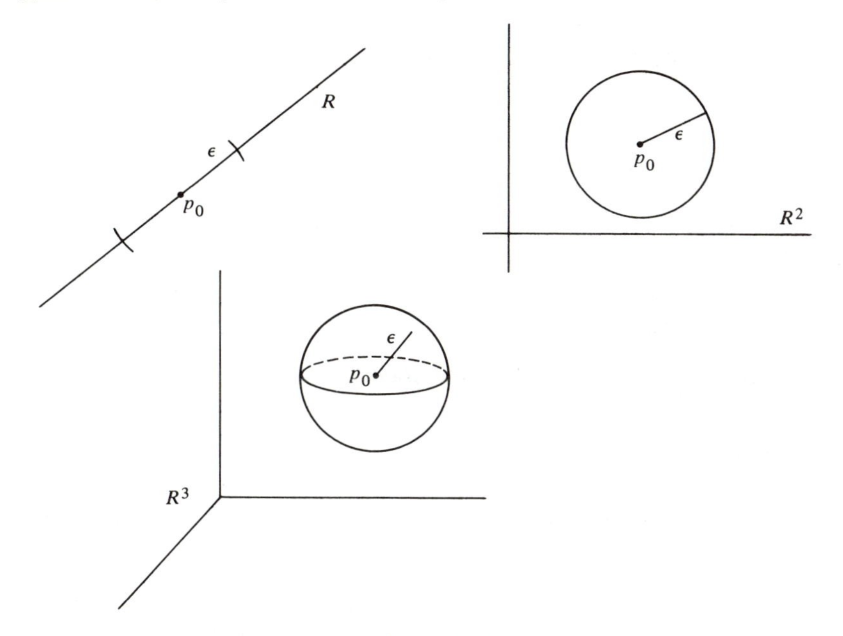

Figure A2-1

is easily seen to be open in R^2. However, if one of the strict inequalities, say $x < b$, is replaced by $x \leq b$, the set is no longer open; no ball with center at the point $(b, (d + c)/2)$, which belongs to the set, can be contained in the set (Fig. A2-2).

It is convenient to say that an open set in R^n containing a point $p \in R^n$ is a *neighborhood* of p.

From now on, $U \subset R^n$ will denote an open set in R^n.

We recall that a real function $f: U \subset R \to R$ of a real variable is continuous at $x_0 \in U$ if given an $\epsilon > 0$ there exists a $\delta > 0$ such that if $|x - x_0| < \delta$, then

$$|f(x) - f(x_0)| < \epsilon.$$

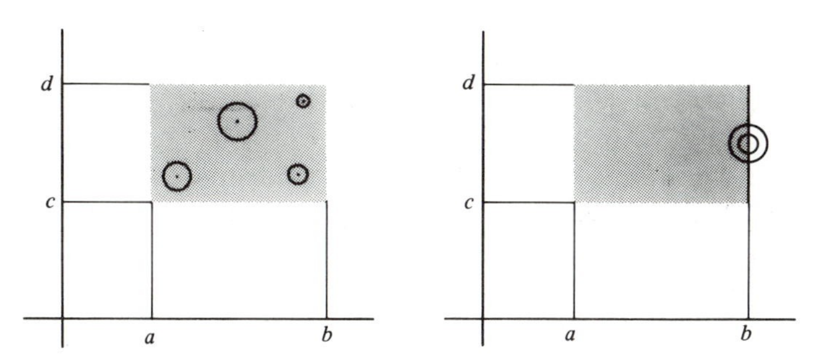

Figure A2-2

Similarly, a real function $f: U \subset R^2 \longrightarrow R$ of two real variables is continuous at $(x_0, y_0) \in U$ if given an $\epsilon > 0$ there exists a $\delta > 0$ such that if $(x - x_0)^2 + (y - y_0)^2 < \delta^2$, then

$$|f(x, y) - f(x_0, y_0)| < \epsilon.$$

The notion of ball unifies these definitions as particular cases of the following general concept:

A map $F: U \subset R^n \longrightarrow R^m$ is *continuous* at $p \in U$ if given $\epsilon > 0$, there exists a $\delta > 0$ such that

$$F(B_\delta(p)) \subset B_\epsilon(F(p)).$$

In other words, F is continuous at p if points arbitrarily close to $F(p)$ are images of points sufficiently close to p. It is easily seen that in the particular cases of $n = 1, 2$ and $m = 1$, this agrees with the previous definitions. We say that F is *continuous* in U if F is *continuous* for all $p \in U$ (Fig. A2-3).

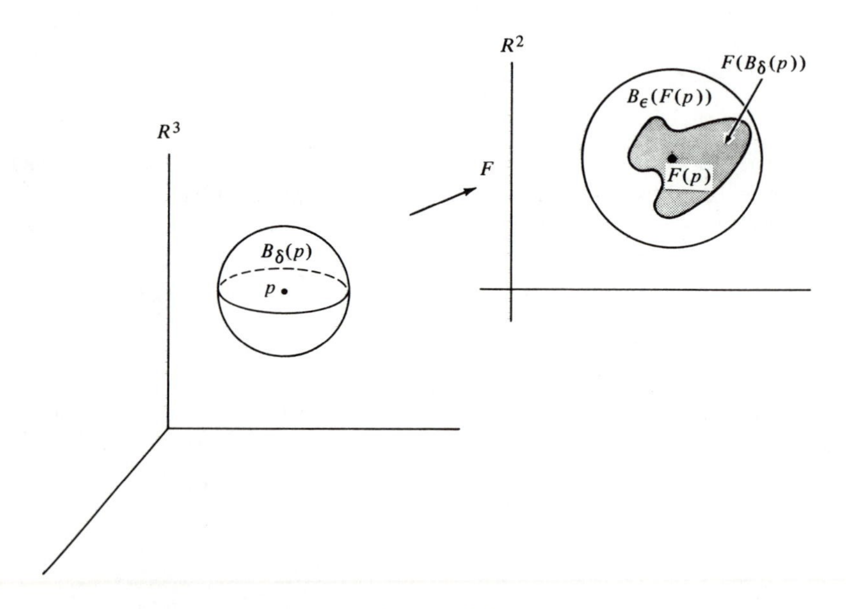

Figure A2-3

Given a map $F: U \subset R^n \longrightarrow R^m$, we can determine m functions of n variables as follows. Let $p = (x_1, \ldots, x_n) \in U$ and $f(p) = (y_1, \ldots, y_m)$. Then we can write

$$y_1 = f_1(x_1, \ldots, x_n), \ldots, y_m = f_m(x_1, \ldots, x_n).$$

The functions $f_i: U \longrightarrow R$, $i = 1, \ldots, m$, are the *component functions* of F.

Example 1 (*Symmetry*). Let $F: R^3 \longrightarrow R^3$ be the map which assigns to each $p \in R^3$ the point which is symmetric to p with respect to the origin $O \in R^3$. Then $F(p) = -p$, or

$$F(x, y, z) = (-x, -y, -z),$$

and the component functions of F are

$$f_1(x, y, z) = -x, \qquad f_2(x, y, z) = -y, \qquad f_3(x, y, z) = -z.$$

Example 2 (*Inversion*). Let $F: R^2 - \{(0, 0)\} \longrightarrow R^2$ be defined as follows. Denote by $|p|$ the distance to the origin $(0, 0) = O$ of a point $p \in R^2$. By definition, $F(p), p \neq 0$, belongs to the half-line Op and is such that $|F(p)|\cdot|p| = 1$. Thus, $F(p) = p/|p|^2$, or

$$F(x, y) = \left(\frac{x}{x^2 + y^2}, \frac{y}{x^2 + y^2}\right), \qquad (x, y) \neq (0, 0),$$

and the component functions of F are

$$f_1(x, y) = \frac{x}{x^2 + y^2}, \qquad f_2(x, y) = \frac{y}{x^2 + y^2}.$$

Example 3 (*Projection*). Let $\pi: R^3 \longrightarrow R^2$ be the projection $\pi(x, y, z) = (x, y)$. Then $f_1(x, y, z) = x, f_2(x, y, z) = y$.

The following proposition shows that the continuity of the map F is equivalent to the continuity of its component functions.

PROPOSITION 1. $\mathrm{F}: \mathrm{U} \subset \mathrm{R}^n \longrightarrow \mathrm{R}^m$ *is continuous if and only if each component function* $\mathrm{f}_i: \mathrm{U} \subset \mathrm{R}^n \longrightarrow \mathrm{R}$, $\mathrm{i} = 1, \ldots, \mathrm{m}$, *is continuous.*

Proof. Assume that F is continuous at $p \in U$. Then given $\epsilon > 0$, there exists $\delta > 0$ such that $F(B_\delta(p)) \subset B_\epsilon(F(p))$. Thus, if $q \in B_\delta(p)$, then

$$F(q) \in B_\epsilon(F(p)),$$

that is,

$$(f_1(q) - f_1(p))^2 + \cdots + (f_m(q) - f_m(p))^2 < \epsilon^2,$$

which implies that, for each $i = 1, \ldots, m$, $|f_i(q) - f_i(p)| < \epsilon$. Therefore, given $\epsilon > 0$ there exists $\delta > 0$ such that if $q \in S_\delta(p)$, then $|f_i(q) - f_i(p)| < \epsilon$. Hence, each f_i is continuous at p.

Conversely, let f_i, $i = 1, \ldots, m$, be continuous at p. Then given $\epsilon > 0$ there exists $\delta_i > 0$ such that if $q \in S_{\delta_i}(p)$, then $|f_i(q) - f_i(p)| < \epsilon/\sqrt{m}$. Set

$\delta < \min \delta_i$ and let $q \in S_\delta(p)$. Then

$$(f_1(q) - f_1(p))^2 + \cdots + (f_m(q) - f_m(p))^2 < \epsilon^2,$$

and hence, the continuity of F at p. **Q.E.D.**

It follows that the maps in Examples 1, 2, and 3 are continuous.

Example 4. Let $F\colon U \subset R \longrightarrow R^m$. Then

$$F(t) = (x_1(t), \ldots, x_m(t)), \qquad t \in U.$$

This is usually called a *vector-valued function*, and the component functions of F are the components of the vector $F(t) \in R^m$. When F is continuous, or, equivalently, the functions $x_i(t)$, $i = 1, \ldots, m$, are continuous, we say that F is a *continuous curve* in R^n.

In most applications, it is convenient to express the continuity in terms of neighborhoods instead of balls.

PROPOSITION 2. *A map* $F\colon U \subset R^n \longrightarrow R^m$ *is continuous at* $p \in U$ *if and only if, given a neighborhood* V *of* $F(p)$ *in* R^m *there exists a neighborhood* W *of* p *in* R^n *such that* $F(W) \subset V$.

Proof. Assume that F is continuous at p. Since V is an open set containing $F(p)$, it contains a ball $B_\epsilon(F(p))$ for some $\epsilon > 0$. By continuity, there exists a ball $B_\delta(p) = W$ such that

$$F(W) = F(B_\delta(p)) \subset B_\epsilon(F(p)) \subset V,$$

and this proves that the condition is necessary.

Conversely, assume that the condition holds. Let $\varepsilon > 0$ be given and set $V = B_\epsilon(F(p))$. By hypothesis, there exists a neighborhood W of p in R^n such that $F(W) \subset V$. Since W is open, there exists a ball $B_\delta(p) \subset W$. Thus,

$$F(B_\delta(p)) \subset F(W) \subset V = B_\epsilon(F(p)),$$

and hence the continuity of F at p. **Q.E.D.**

The composition of continuous maps yields a continuous map. More precisely, we have the following proposition.

PROPOSITION 3. *Let* $F\colon U \subset R^n \longrightarrow R^m$ *and* $G\colon V \subset R^m \longrightarrow R^k$ *be continuous maps, where* U *and* V *are open sets such that* $F(U) \subset V$. *Then* $G \circ F\colon U \subset R^n \longrightarrow R^k$ *is a continuous map.*

Proof. Let $p \in U$ and let V be a neighborhood of $G \circ F(p)$ in R^k. By continuity of G, there is a neighborhood Q of $F(p)$ in R^m with $G(Q) \subset V$. By continuity of F, there is a neighborhood W of p in R^n with $F(W) \subset Q$. Thus,

$$G \circ F(W) \subset G(Q) \subset V,$$

and hence the continuity of $G \circ F$. **Q.E.D.**

It is often necessary to deal with maps defined on arbitrary (not necessarily open) sets of R^n. To extend the previous ideas to this situation, we shall proceed as follows.

Let $F: A \subset R^n \longrightarrow R^m$ be a map, where A is an arbitrary set in R^n. We say that F is *continuous in* A if there exists an open set $U \subset R^n$, $U \supset A$, and a continuous map $\bar{F}: U \longrightarrow R^m$ such that the restriction $\bar{F}|A = F$. In other words, F is continuous in A if it is the restriction of a continuous map defined in an open set containing A.

It is clear that if $F: A \subset R^n \longrightarrow R^m$ is continuous, given a neighborhood V of $F(p)$ in R^m, $p \in A$, there exists a neighborhood W of p in R^n such that $F(W \cap A) \subset V$. For this reason, it is convenient to call the set $W \cap A$ a *neighborhood of* p *in* A (Fig. A2-4).

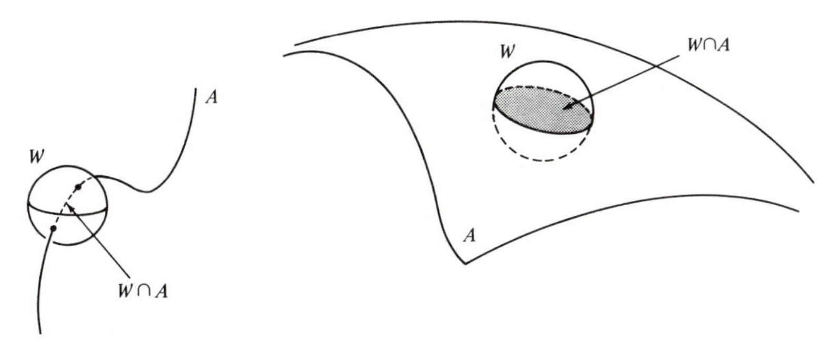

Figure A2-4

Example 5. Let

$$E = \left\{(x, y, z) \in R^3; \frac{x^2}{a^2} + \frac{y^2}{b^2} + \frac{z^2}{c^2} = 1\right\}$$

be an ellipsoid, and let $\pi: R^3 \longrightarrow R^2$ be the projection of Example 3. Then the restriction of π to E is a continuous map from E to R^2.

We say that a continuous map $F: A \subset R^n \longrightarrow R^n$ is a *homeomorphism* onto $F(A)$ if F is one-to-one and the inverse $F^{-1}: F(A) \subset R^n \longrightarrow R^n$ is continuous. In this case A and $F(A)$ are *homeomorphic sets.*

Example 6. Let $F: R^3 \longrightarrow R^3$ be given by

$$F(x, y, z) = (xa, yb, zc).$$

F is clearly continuous, and the restriction of F to the sphere

$$S^2 = \{(x, y, z) \in R^3;\ x^2 + y^2 + z^2 = 1\}$$

is a continuous map $\tilde{F}: S^2 \longrightarrow R^3$. Observe that $\tilde{F}(S^2) = E$, where E is the ellipsoid of Example 5. It is also clear that F is one-to-one and that

$$F^{-1}(x, y, z) = \left(\frac{x}{a}, \frac{y}{b}, \frac{z}{c}\right).$$

Thus, $\tilde{F}^{-1} = F^{-1} \mid E$ is continuous. Therefore, $\tilde{F}$ is a homeomorphism of the sphere S^2 onto the ellipsoid E.

Finally, we want to describe two properties of real continuous functions defined on a closed interval $[a, b]$,

$$[a, b] = \{x \in R;\ a \leq x \leq b\}$$

(Props. 4 and 5 below), and an important property of the closed interval $[a, b]$ itself. They will be used repeatedly in this book.

PROPOSITION 4 (The Intermediate Value Theorem). *Let* $f: [a, b] \longrightarrow R$ *be a continuous function defined on the closed interval* $[a, b]$. *Assume that* $f(a)$ *and* $f(b)$ *have opposite signs; that is,* $f(a)f(b) < 0$. *Then there exists a point* $c \in (a, b)$ *such that* $f(c) = 0$.

PROPOSITION 5. *Let* $f: [a, b]$ *be a continuous function defined in the closed interval* $[a, b]$. *Then* f *reaches its maximum and its minimum in* $[a, b]$; *that is, there exist points* $x_1, x_2 \in [a, b]$ *such that* $f(x_1) \leq f(x) \leq f(x_2)$ *for all* $x \in [a, b]$.

PROPOSITION 6 (Heine-Borel). *Let* $[a, b]$ *be a closed interval and let* I_α, $\alpha \in A$, *be a collection of open intervals in* $[a, b]$ *such that* $\bigcup_\alpha I_\alpha = [a, b]$. *Then it is possible to choose a finite number* $I_{k_1}, I_{k_2}, \ldots, I_{k_n}$ *of* I_α *such that* $\bigcup I_{k_i} = I$, $i = 1, \ldots, n$.

These propositions are standard theorems in courses on advanced calculus, and we shall not prove them here. However, proofs are provided in the appendix to Chap. 5 (Props. 6, 13, and 11, respectively).

B. *Differentiability in* R^n

Let $f\colon U \subset R \longrightarrow R$. The *derivative* $f'(x_0)$ of f at $x_0 \in U$ is the limit (when it exists)

$$f'(x_0) = \lim_{h \to 0} \frac{f(x_0 + h) - f(x_0)}{h}, \qquad x_0 + h \in U.$$

When f has derivatives at all points of a neighborhood V of x_0, we can consider the derivative of $f'\colon V \longrightarrow R$ at x_0, which is called the *second derivative* $f''(x_0)$ of f at x_0, and so forth. f is *differentiable* at x_0 if it has continuous derivatives of all orders at x_0. f is *differentiable* in U if it is differentiable at all points in U.

Remark. We use the word differentiable for what is sometimes called infinitely differentiable (or of class C^∞). Our usage should not be confused with the usage of elementary calculus, where a function is called differentiable if its first derivative exists.

Let $F\colon U \subset R^2 \longrightarrow R$. The *partial derivative of* f *with respect to* x *at* $(x_0, y_0) \in U$, denoted by $(\partial f/\partial x)(x_0, y_0)$, is (when it exists) the derivative at x_0 of the function of one variable: $x \longrightarrow f(x, y_0)$. Similarly, the partial derivative with respect to y at (x_0, y_0), $(\partial f/\partial y)(x_0, y_0)$, is defined as the derivative at y_0 of $y \longrightarrow f(x_0, y)$. When f has partial derivatives at all points of a neighborhood V of (x_0, y_0), we can consider the *second partial derivatives at* (x_0, y_0):

$$\frac{\partial}{\partial x}\left(\frac{\partial f}{\partial x}\right) = \frac{\partial^2 f}{\partial x^2}, \qquad \frac{\partial}{\partial x}\left(\frac{\partial f}{\partial y}\right) = \frac{\partial^2 f}{\partial x\, \partial y},$$
$$\frac{\partial}{\partial y}\left(\frac{\partial f}{\partial x}\right) = \frac{\partial^2 f}{\partial y\, \partial x}, \qquad \frac{\partial}{\partial y}\left(\frac{\partial f}{\partial y}\right) = \frac{\partial^2 f}{\partial y^2},$$

and so forth. f is *differentiable* at (x_0, y_0) if it has continuous partial derivatives of all orders at (x_0, y_0). f is *differentiable* in U if it is differentiable at all points of U. We sometimes denote partial derivatives by

$$\frac{\partial f}{\partial x} = f_x, \qquad \frac{\partial f}{\partial y} = f_y, \qquad \frac{\partial^2 f}{\partial x^2} = f_{xx}, \qquad \frac{\partial^2 f}{\partial x\, \partial y} = f_{xy}, \qquad \frac{\partial^2 f}{\partial y^2} = f_{yy}.$$

It is an important fact that when f is differentiable the partial derivatives of f are independent of the order in which they are performed; that is,

$$\frac{\partial^2 f}{\partial x\, \partial y} = \frac{\partial^2 f}{\partial y\, \partial x}, \qquad \frac{\partial^3 f}{\partial^2 x\, \partial y} = \frac{\partial^3 f}{\partial x\, \partial y\, \partial x}, \qquad \text{etc.}$$

The definitions of partial derivatives and differentiability are easily extended to functions $f: U \subset R^n \to R$. For instance, $(\partial f/\partial x_3)(x_1^0, x_2^0, \ldots, x_n^0)$ is the derivative of the function of one variable

$$x_3 \longrightarrow f(x_1^0, x_2^0, x_3, x_4^0, \ldots, x_n^0).$$

A further important fact is that partial derivatives obey the so-called *chain rule*. For instance, if $x = x(u, v)$, $y = y(u, v)$, $z = z(u, v)$ are real differentiable functions in $U \subset R^2$ and $f(x, y, z)$ is a real differentiable function in R^3, then the composition $f(x(u, v), y(u, v), z(u, v))$ is a differentiable function in U, and the partial derivative of f with respect to, say, u is given by

$$\frac{\partial f}{\partial u} = \frac{\partial f}{\partial x}\frac{\partial x}{\partial u} + \frac{\partial f}{\partial y}\frac{\partial y}{\partial u} + \frac{\partial f}{\partial z}\frac{\partial z}{\partial u}.$$

We are now interested in extending the notion of differentiability to maps $F: U \subset R^n \to R^m$. We say that F is *differentiable* at $p \in U$ if its component functions are differentiable at p; that is, by writing

$$F(x_1, \ldots, x_n) = (f_1(x_1, \ldots, x_n), \ldots, f_m(x_1, \ldots, x_n)),$$

the functions f_i, $i = 1, \ldots, m$, have continuous partial derivatives of all orders at p. F is *differentiable* in U if it is differentiable at all points in U.

For the case $m = 1$, this repeats the previous definition. For the case $n = 1$, we obtain the notion of a (parametrized) *differentiable curve* in R^m. In Chap. 1, we have already seen such an object in R^3. For our purposes, we need to extend the definition of tangent vector of Chap. 1 to the present situation. A *tangent vector* to a map $\alpha: U \subset R \to R^m$ at $t_0 \in U$ is the vector in R^m

$$\alpha'(t_0) = (x_1'(t_0), \ldots, x_m'(t_0)).$$

Example 7. Let $F: U \subset R^2 \to R^3$ be given by

$$F(u, v) = (\cos u \cos v, \cos u \sin v, \cos^2 v), \qquad (u, v) \in U.$$

The component functions of F, namely,

$$f_1(u, v) = \cos u \cos v, \qquad f_2(u, v) = \cos u \sin v, \qquad f_3(u, v) = \cos^2 v$$

have continuous partial derivatives of all orders in U. Thus, F is differentiable in U.

Example 8. Let $\alpha: U \subset R \to R^4$ be given by

$$\alpha(t) = (t^4, t^3, t^2, t), \qquad t \in U.$$

Then α is a differentiable curve in R^4, and the tangent vector to α at t is $\alpha'(t) = (4t^3, 3t^2, 2t, 1)$.

Example 9. Given a vector $w \in R^m$ and a point $p_0 \in U \subset R^m$, we can always find a differentiable curve $\alpha: (-\epsilon, \epsilon) \longrightarrow U$ with $\alpha(0) = p_0$ and $\alpha'(0) = w$. Simply define $\alpha(t) = p_0 + tw$, $t \in (-\epsilon, \epsilon)$. By writing $p_0 = (x_1^0, \ldots, x_m^0)$ and $w = (w_1, \ldots, w_m)$, the component functions of α are $x_i(t) = x_i^0 + tw_i$, $i = 1, \ldots, m$. Thus, α is differentiable, $\alpha(0) = p_0$ and

$$\alpha'(0) = (x_1'(0), \ldots, x_m'(0)) = (w_1, \ldots, w_m) = w.$$

We shall now introduce the concept of differential of a differentiable map. It will play an important role in this book.

DEFINITION 1. *Let* $F: U \subset R^n \longrightarrow R^m$ *be a differentiable map. To each* $p \in U$ *we associate a linear map* $dF_p: R^n \longrightarrow R^m$ *which is called the* differential *of* F *at* p *and is defined as follows. Let* $w \in R^n$ *and let* $\alpha: (-\epsilon, \epsilon) \longrightarrow U$ *be a differentiable curve such that* $\alpha(0) = p$, $\alpha'(0) = w$. *By the chain rule, the curve* $\beta = F \circ \alpha: (-\epsilon, \epsilon) \longrightarrow R^m$ *is also differentiable.* Then (Fig. A2-5)

$$dF_p(w) = \beta'(0).$$

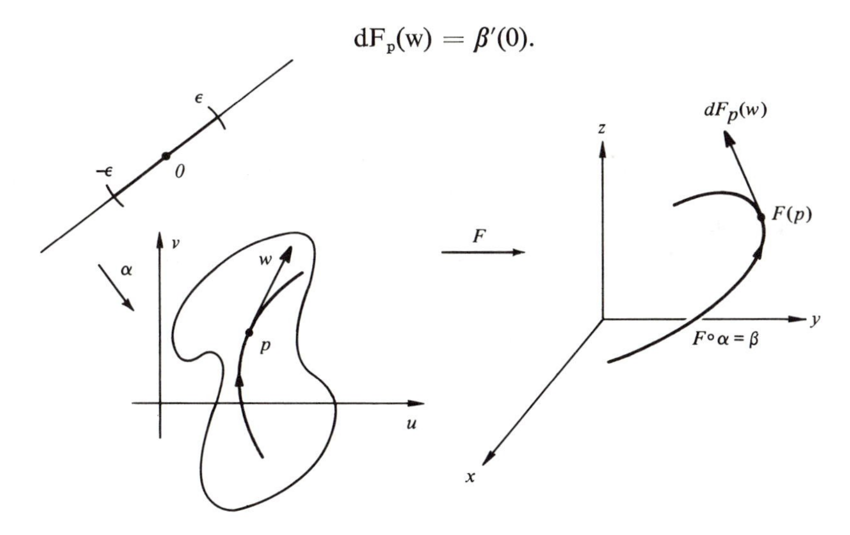

Figure A2-5

PROPOSITION 7. *The above definition of* dF_p *does not depend on the choice of the curve which passes through* p *with tangent vector* w, *and* dF_p *is, in fact, a linear map.*

Proof. To simplify notation, we work with the case $F: U \subset R^2 \longrightarrow R^3$. Let (u, v) be coordinates in R^2 and (x, y, z) be coordinates in R^3. Let

$e_1 = (1, 0)$, $e_2 = (0, 1)$ be the canonical basis in R^2 and $f_1 = (1, 0, 0)$, $f_2 = (0, 1, 0)$, $f_3 = (0, 0, 1)$ be the canonical basis in R^3. Then we can write $\alpha(t) = (u(t), v(t))$, $t \in (-\epsilon, \epsilon)$,

$$\alpha'(0) = w = u'(0)e_1 + v'(0)e_2,$$

$F(u, v) = (x(u, v), y(u, v), z(u, v))$, and

$$\beta(t) = F \circ \alpha(t) = (x(u(t), v(t)), y(u(t), v(t)), z(u(t), v(t))).$$

Thus, using the chain rule and taking the derivatives at $t = 0$, we obtain

$$\begin{aligned}\beta'(0) &= \left(\frac{\partial x}{\partial u}\frac{du}{dt} + \frac{\partial x}{\partial v}\frac{dv}{dt}\right)f_1 + \left(\frac{\partial y}{\partial u}\frac{du}{dt} + \frac{\partial y}{\partial v}\frac{dv}{dt}\right)f_2 \\ &\quad + \left(\frac{\partial z}{\partial u}\frac{du}{dt} + \frac{\partial z}{\partial v}\frac{dv}{dt}\right)f_3 \\ &= \begin{pmatrix} \frac{\partial x}{\partial u} & \frac{\partial x}{\partial v} \\ \frac{\partial y}{\partial u} & \frac{\partial y}{dv} \\ \frac{\partial z}{\partial u} & \frac{\partial z}{\partial v} \end{pmatrix} \begin{pmatrix} \frac{du}{dt} \\ \\ \frac{dv}{dt} \end{pmatrix} = dF_p(w).\end{aligned}$$

This shows that dF_p is represented, in the canonical bases of R^2 and R^3, by a matrix which depends only on the partial derivatives at p of the component functions x, y, z of F. Thus, dF_p is a linear map, and clearly $dF_p(w)$ does not depend on the choice of α.

The reader will have no trouble in extending this argument to the more general situation. **Q.E.D.**

The matrix of $dF_p\colon R^n \longrightarrow R^m$ in the canonical bases of R^n and R^m, that is, the matrix $(\partial f_i/\partial x_j)$, $i = 1, \ldots, m$, $j = 1, \ldots, n$, is called the *Jacobian matrix* of F at p. When $n = m$, this is a square matrix and its determinant is called the *Jacobian determinant;* it is usual to denote it by

$$\det\left(\frac{\partial f_i}{\partial x_j}\right) = \frac{\partial(f_1, \ldots, f_n)}{\partial(x_1, \ldots, x_n)}.$$

Remark. There is no agreement in the literature regarding the notation for the differential. It is also of common usage to call dF_p the derivative of F at p and to denote it by $F'(p)$.

Example 10. Let $F\colon R^2 \longrightarrow R^2$ be given by

$$F(x, y) = (x^2 - y^2, 2xy), \qquad (x, y) \in R^2.$$

F is easily seen to be differentiable, and its differential dF_p at $p = (x, y)$ is

$$dF_p = \begin{pmatrix} 2x & -2y \\ 2y & 2x \end{pmatrix}.$$

For instance, $dF_{(1,1)}(2, 3) = (-2, 10)$.

One of the advantages of the notion of differential of a map is that it allows us to express many facts of calculus in a geometric language. Consider, for instance, the following situation: Let $F: U \subset R^2 \to R^3$, $G: V \subset R^3 \to R^2$ be differentiable maps, where U and V are open sets such that $F(U) \subset V$. Let us agree on the following set of coordinates,

$$\begin{array}{ccccc} U \subset R^2 & \xrightarrow{F} & V \subset R^3 & \xrightarrow{G} & R^2 \\ (u, v) & & (x, y, z) & & (\xi, \eta) \end{array}$$

and let us write

$$F(u, v) = (x(u, v), y(u, v), z(u, v)),$$
$$G(x, y, z) = (\xi(x, y, z), \eta(x, y, z)).$$

Then

$$G \circ F(u, v) = (\xi(x(u, v), y(u, v), z(u, v)), \eta(x(u, v), y(u, v), z(u, v))),$$

and, by the chain rule, we can say that $G \circ F$ is differentiable and compute the partial derivatives of its component functions. For instance,

$$\frac{\partial \xi}{\partial u} = \frac{\partial \xi}{\partial x}\frac{\partial x}{\partial u} + \frac{\partial \xi}{\partial y}\frac{\partial y}{\partial u} + \frac{\partial \xi}{\partial z}\frac{\partial z}{\partial u}.$$

Now, a simple way of expressing the above situation is by using the following general fact.

PROPOSITION 8 (The Chain Rule for Maps). *Let* $\mathrm{F}: \mathrm{U} \subset \mathrm{R}^n \to \mathrm{R}^m$ *and* $\mathrm{G}: \mathrm{V} \subset \mathrm{R}^m \to \mathrm{R}^k$ *be differentiable maps, where* U *and* V *are open sets such that* $\mathrm{F(U)} \subset \mathrm{V}$. *Then* $\mathrm{G} \circ \mathrm{F}: \mathrm{U} \to \mathrm{R}^k$ *is a differentiable map, and*

$$\mathrm{d(G \circ F)_p = dG_{F(p)} \circ dF_p}, \qquad \mathrm{p} \in \mathrm{U}.$$

Proof. The fact that $G \circ F$ is differentiable is a consequence of the chain rule for functions. Now, let $w_1 \in R^n$ be given and let us consider a curve $\alpha: (-\epsilon_2, \epsilon_2) \to U$, with $\alpha(0) = p$, $\alpha'(0) = w_1$. Set $dF_p(w_1) = w_2$ and observe that $dG_{F(p)}(w_2) = (d/dt)(G \circ F \circ \alpha)|_{t=0}$. Then

$$d(G \circ F)_p(w_1) = \frac{d}{dt}(G \circ F \circ \alpha)_{t=0} = dG_{F(p)}(w_2) = dG_{F(p)} \circ dF_p(w_1).$$

Q.E.D.

Notice that, for the particular situation we were considering before, the relation $d(G \circ F)_p = dG_{F(p)} \circ dF_p$ is equivalent to the following product of Jacobian matrices,

$$\begin{pmatrix} \dfrac{\partial\xi}{\partial u} & \dfrac{\partial\xi}{\partial v} \\ \\ \dfrac{\partial\eta}{\partial u} & \dfrac{\partial\eta}{\partial v} \end{pmatrix} = \begin{pmatrix} \dfrac{\partial\xi}{\partial x} & \dfrac{\partial\xi}{\partial y} & \dfrac{\partial\xi}{\partial z} \\ \\ \dfrac{\partial\eta}{\partial x} & \dfrac{\partial\eta}{\partial y} & \dfrac{\partial\eta}{\partial z} \end{pmatrix} \begin{pmatrix} \dfrac{\partial x}{\partial u} & \dfrac{\partial x}{\partial v} \\ \dfrac{\partial y}{\partial u} & \dfrac{\partial y}{\partial v} \\ \dfrac{\partial z}{\partial u} & \dfrac{\partial z}{\partial v} \end{pmatrix},$$

which contains the expressions of all partial derivatives $\partial\xi/\partial u$, $\partial\xi/\partial v$, $\partial\eta/\partial u$, $\partial\eta/\partial v$. Thus, the simple expression of the chain rule for maps embodies a great deal of information on the partial derivatives of their component functions.

An important property of a differentiable function $f\colon (a, b) \subset R \to R$ defined in an open interval (a, b) is that if $f'(x) \equiv 0$ on (a, b), then f is constant on (a, b). This generalizes for differentiable functions of several variables as follows.

We say that an open set $U \subset R^n$ is *connected* if given two points $p, q \in U$ there exists a continuous map $\alpha\colon [a, b] \to U$ such that $\alpha(a) = p$ and $\alpha(b) = q$. This means that two points of U can be joined by a continuous curve in U or that U is made up of one single "piece."

PROPOSITION 9. *Let* $f\colon U \subset R^n \to R$ *be a differentiable function defined on a connected open subset* U *of* R^n. *Assume that* $df_p\colon R^n \to R$ *is zero at every point* $p \in U$. *Then* f *is constant on* U.

Proof. Let $p \in U$ and let $B_\delta(p) \subset U$ be an open ball around p and contained in U. Any point $q \in B_\epsilon(p)$ can be joined to p by the "radial" segment $\beta\colon [0, 1] \to U$, where $\beta(t) = tq + (1 - t)p$, $t \in [0, 1]$ (Fig. A2-6). Since U is open, we can extend β to $(0 - \epsilon, 1 + \epsilon)$. Now, $f \circ \beta\colon (0 - \epsilon, 1 + \epsilon) \to R$ is a function defined in an open interval, and

$$d(f \circ \beta)_t = (df \circ d\beta)_t = 0,$$

since $df \equiv 0$. Thus,

$$\frac{d}{dt}(f \circ \beta) = 0$$

for all $t \in (0 - \epsilon, 1 + \epsilon)$, and hence $(f \circ \beta) = \text{const}$. This means that $f(\beta(0)) = f(p) = f(\beta(1)) = f(q)$; that is, f is constant on $B_\delta(p)$.

Thus, the proposition is proved locally; that is, each point of U has a

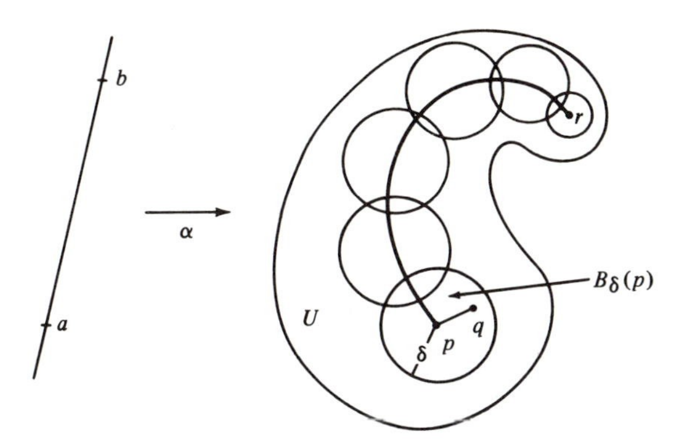

Figure A2-6

neighborhood such that f is constant on that neighborhood. Notice that so far we have not used the connectedness of U. We shall need it now to show that these constants are all the same.

Let r be an arbitrary point of U. Since U is connected, there exists a continuous curve $\alpha: [a, b] \to U$, with $\alpha(a) = p, \alpha(b) = r$. The function $f \circ \alpha: [a, b] \to R$ is continuous in $[a, b]$. By the first part of the proof, for each $t \in [a, b]$, there exists an interval I_t, open in $[a, b]$, such that $f \circ \alpha$ is constant on I_t. Since $\bigcup_t I_t = [a, b]$, we can apply the Heine-Borel theorem (Prop. 6). Thus, we can choose a finite number $I_1, \ldots, I_k$ of the intervals I_t so that $\bigcup_i I_i = [a, b]$, $i = 1, \ldots, k$. We can assume, by renumbering the intervals, if necessary, that two consecutive intervals overlap. Thus, $f \circ \alpha$ is constant in the union of two consecutive intervals. It follows that f is constant on $[a, b]$; that is,

$$f(\alpha(a)) = f(p) = f(\alpha(b)) = f(r).$$

Since r is arbitrary, f is constant on U. **Q.E.D.**

One of the most important theorems of differential calculus is the so-called inverse function theorem, which, in the present notation, says the following. (Recall that a linear map A is an isomorphism if the matrix of A is invertible.)

INVERSE FUNCTION THEOREM. *Let* $\mathrm{F}: \mathrm{U} \subset \mathrm{R}^n \to \mathrm{R}^n$ *be a differentiable mapping and suppose that at* $\mathrm{p} \in \mathrm{U}$ *the differential* $d\mathrm{F}_p: \mathrm{R}^n \to \mathrm{R}^n$ *is an isomorphism. Then there exists a neighborhood* V *of* p *in* U *and a neighborhood* W *of* F(p) *in* R^n *such that* $\mathrm{F}: \mathrm{V} \to \mathrm{W}$ *has a differentiable inverse* $\mathrm{F}^{-1}: \mathrm{W} \to \mathrm{V}$.

A differentiable mapping $F: V \subset R^n \to W \subset R^n$, where V and W are open sets, is called a *diffeomorphism* of V with W if F has a differentiable inverse.

The inverse function theorem asserts that if at a point $p \in U$ the differential dF_p is an isomorphism, then F is a diffeomorphism in a neighborhood of p. In other words, an assertion about the differential of F at a point implies a similar assertion about the behavior of F in a neighborhood of the point.

This theorem will be used repeatedly in this book. A proof can be found, for instance, in Buck, *Advanced Calculus*, p. 285.

Example 11. Let $F: R^2 \longrightarrow R^2$ be given by

$$F(x, y) = (e^x \cos y, e^x \sin y), \qquad (x, y) \in R^2.$$

The component functions of F, namely, $u(x, y) = e^x \cos y$, $v(x, y) = e^x \sin y$, have continuous partial derivatives of all orders. Thus, F is differentiable.

It is instructive to see, geometrically, how F transforms curves of the xy plane. For instance, the vertical line $x = x_0$ is mapped into the circle $u = e^{x_0} \cos y$, $v = e^{x_0} \sin y$ of radius e^{x_0}, and the horizontal line $y = y_0$ is mapped into the half-line $u = e^x \cos y_0$, $v = e^x \sin y_0$ with slope $\tan y_0$. It follows that (Fig. A2-7)

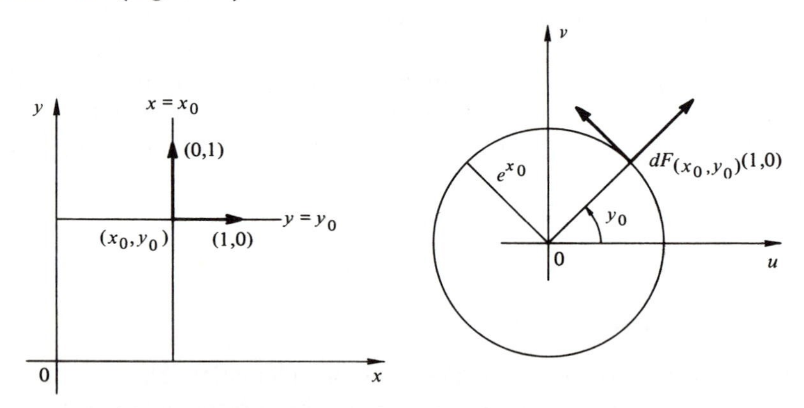

Figure A2-7

$$\begin{aligned}
dF_{(x_0, y_0)}(1, 0) &= \frac{d}{dx}(e^x \cos y_0, e^x \sin y_0)\big|_{x=x_0} \\
&= (e^{x_0} \cos y_0, e^{x_0} \sin y_0), \\
dF_{(x_0, y_0)}(1, 0) &= \frac{d}{dy}(e^{x_0} \cos y, e^{x_0} \sin y)\big|_{y=y_0} \\
&= (-e^{x_0} \sin y_0, e^{x_0} \cos y_0).
\end{aligned}$$

This can be most easily checked by computing the Jacobian matrix of F,

$$dF_{(x,y)} = \begin{pmatrix} \dfrac{\partial u}{\partial x} & \dfrac{\partial u}{\partial y} \\ \dfrac{\partial v}{\partial x} & \dfrac{\partial v}{\partial y} \end{pmatrix} = \begin{pmatrix} e^x \cos y & -e^x \sin y \\ e^x \sin y & e^x \cos y \end{pmatrix},$$

and applying it to the vectors (1, 0) and (0, 1) at (x_0, y_0).

We notice that the Jacobian determinant $\det(dF_{(x,y)}) = e^x \neq 0$, and thus dF_p is nonsingular for all $p = (x, y) \in R^2$ (this is also clear from the previous geometric considerations). Therefore, we can apply the inverse function theorem to conclude that F is locally a diffeomorphism.

Observe that $F(x, y) = F(x, y + 2\pi)$. Thus, F is not one-to-one and has no global inverse. For each $p \in R^2$, the inverse function theorem gives neighborhoods V of p and W of $F(p)$ so that the restriction $F\colon V \to W$ is a diffeomorphism. In our case, V may be taken as the strip $\{-\infty < x < \infty, 0 < y < 2\pi\}$ and W as $R^2 - \{(0, 0)\}$. However, as the example shows, even if the conditions of the theorem are satisfied everywhere and the domain of definition of F is very simple, a global inverse of F may fail to exist.

3 The Geometry of the Gauss Map

3-1. Introduction

As we have seen in Chap. 1, the consideration of the rate of change of the tangent line to a curve C led us to an important geometric entity, namely, the curvature of C. In this chapter we shall extend this idea to regular surfaces; that is, we shall try to measure how rapidly a surface S pulls away from the tangent plane $T_p(S)$ in a neighborhood of a point $p \in S$. This is equivalent to measuring the rate of change at p of a unit normal vector field N on a neighborhood of p. As we shall see shortly, this rate of change is given by a linear map on $T_p(S)$ which happens to be self-adjoint (see the appendix to Chap. 3). A surprisingly large number of local properties of S at p can be derived from the study of this linear map.

In Sec. 3-2, we shall introduce the relevant definitions (the Gauss map, principal curvatures and principal directions, Gaussian and mean curvatures, etc.) without using local coordinates. In this way, the geometric content of the definitions is clearly brought up. However, for computational as well as for theoretical purposes, it is important to express all concepts in local coordinates. This is taken up in Sec. 3-3.

Sections 3-2 and 3-3 contain most of the material of Chap. 3 that will be used in the remaining parts of this book. The few exceptions will be explicitly pointed out. For completeness, we have proved the main properties of self-adjoint linear maps in the appendix to Chap. 3. Furthermore, for those who have omitted Sec. 2-6, we have included a brief review of orientation for surfaces at the beginning of Sec. 3-2.

Section 3-4 contains a proof of the fact that at each point of a regular surface there exists an orthogonal parametrization, that is, a parametrization such that its coordinate curves meet orthogonally. The techniques used here are interesting in their own right and yield further results. However, for a short course it might be convenient to assume these results and omit the section.

In Sec. 3-5 we shall take up two interesting special cases of surfaces, namely, the ruled surfaces and the minimal surfaces. They are treated independently so that one (or both) of them can be omitted on a first reading.

3-2. The Definition of the Gauss Map and Its Fundamental Properties

We shall begin by briefly reviewing the notion of orientation for surfaces.

As we have seen in Sec. 2-4, given a parametrization $\mathbf{x}: U \subset R^2 \to S$ of a regular surface S at a point $p \in S$, we can choose a unit normal vector at each point of $\mathbf{x}(U)$ by the rule

$$N(q) = \frac{\mathbf{x}_u \wedge \mathbf{x}_v}{|\mathbf{x}_u \wedge \mathbf{x}_v|}(q), \qquad q \in \mathbf{x}(U).$$

Thus, we have a differentiable map $N: \mathbf{x}(U) \to R^3$ that associates to each $q \in \mathbf{x}(U)$ a unit normal vector $N(q)$.

More generally, if $V \subset S$ is an open set in S and $N: V \to R^3$ is a differentiable map which associates to each $q \in V$ a unit normal vector at q, we say that *N is a differentiable field of unit normal vectors on V*.

It is a striking fact that not all surfaces admit a differentiable field of unit normal vectors *defined on the whole surface*. For instance, on the Möbius strip of Fig. 3-1 one cannot define such a field. This can be seen intuitively by

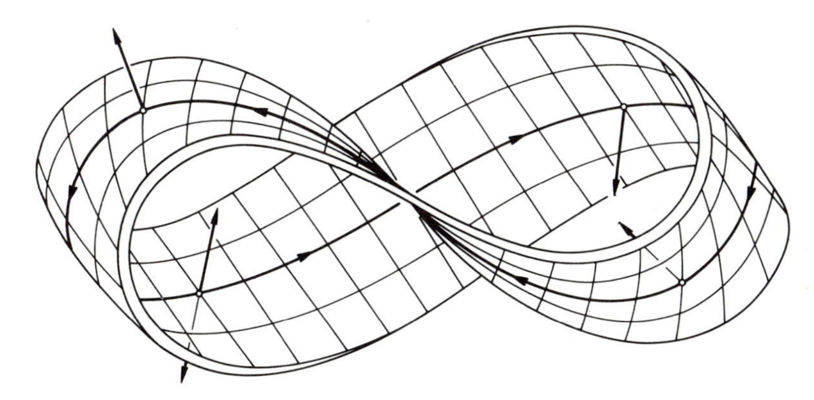

Figure 3-1. The Möbius strip.

going around once along the middle circle of the figure: After one turn, the vector field N would come back as $-N$, a contradiction to the continuity of N. Intuitively, one cannot, on the Möbius strip, make a consistent choice of a definite "side"; moving around the surface, we can go continuously to the "other side" without leaving the surface.

We shall say that a regular surface is *orientable* if it admits a differentiable field of unit normal vectors defined on the whole surface; the choice of such a field N is called an *orientation* of S.

For instance, the Möbius strip referred to above is not an orientable surface. Of course, every surface covered by a single coordinate system (for instance, surfaces represented by graphs of differentiable functions) is trivially orientable. Thus, every surface is locally orientable, and orientation is definitely a global property in the sense that it involves the whole surface.

An orientation N on S induces an orientation on each tangent space $T_p(S)$, $p \in S$, as follows. Define a basis $\{v, w\} \in T_p(S)$ to be *positive* if $\langle v \wedge w, N \rangle$ is positive. It is easily seen that the set of all positive bases of $T_p(S)$ is an orientation for $T_p(S)$ (cf. Sec. 1-4).

Further details on the notion of orientation are given in Sec. 2-6. However, for the purpose of Chaps. 3 and 4, the present description will suffice.

Throughout this chapter, S will denote a regular orientable surface in which an orientation (i.e., a differentiable field of unit normal vectors N) has been chosen; this will be simply called a surface S with an orientation N.

DEFINITION 1. *Let* $\mathrm{S} \subset \mathrm{R}^3$ *be a surface with an orientation* N. *The map* $\mathrm{N}: \mathrm{S} \longrightarrow \mathrm{R}^3$ *takes its values in the unit sphere*

$$\mathrm{S}^2 = \{(\mathrm{x}, \mathrm{y}, \mathrm{z}) \in \mathrm{R}^3;$$

The map $\mathrm{N}: \mathrm{S} \longrightarrow \mathrm{S}^2$, *thus defined, is called the* Gauss map *of* S (*Fig. 3-2*).†

It is straightforward to verify that the Gauss map is differentiable. The differential dN_p of N at $p \in S$ is a linear map from $T_p(S)$ to $T_{N(p)}(S^2)$. Since $T_p(S)$ and $T_{N(p)}(S^2)$ are parallel planes, dN_p can be looked upon as a linear map on $T_p(S)$.

The linear map $dN_p\colon T_p(S) \longrightarrow T_p(S)$ operates as follows. For each parametrized curve $\alpha(t)$ in S with $\alpha(0) = p$, we consider the parametrized curve $N \circ \alpha(t) = N(t)$ in the sphere S^2; this amounts to restricting the normal vector N to the curve $\alpha(t)$. The tangent vector $N'(0) = dN_p(\alpha'(0))$ is a vector in $T_p(S)$ (Fig. 3-3). It measures the rate of change of the normal vector N, restricted to the curve $\alpha(t)$, at $t = 0$. Thus, dN_p measures how N pulls away from $N(p)$ in

†In italic context, letter symbols set in roman rather than italics.

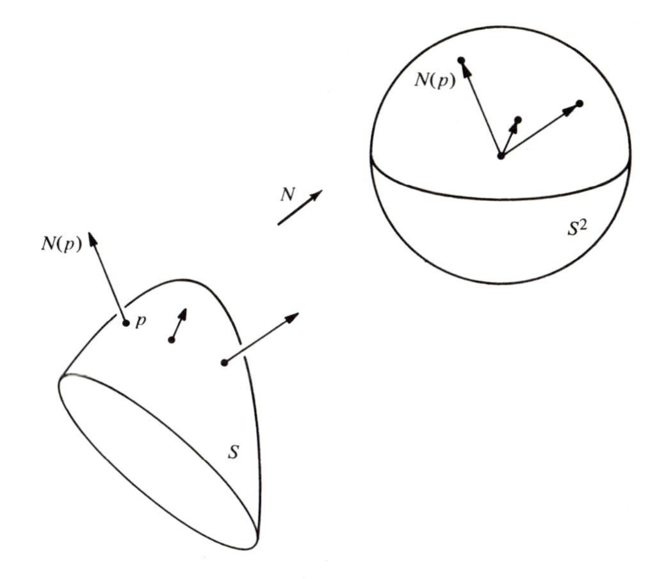

Figure 3-2. The Gauss map.

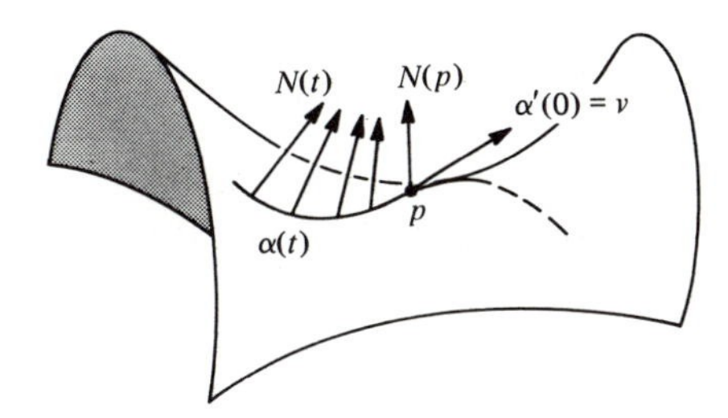

Figure 3-3

a neighborhood of p. In the case of curves, this measure is given by a number, the curvature. In the case of surfaces, this measure is characterized by a linear map.

Example 1. For a plane P given by $ax + by + cz + d = 0$, the unit normal vector $N = (a, b, c)/\sqrt{a^2 + b^2 + c^2}$ is constant, and therefore $dN \equiv 0$ (Fig. 3-4).

Example 2. Consider the unit sphere

$$S^2 = \{(x, y, z) \in R^3; x^2 + y^2 + z^2 = 1\}.$$

If $\alpha(t) = (x(t), y(t), z(t))$ is a parametrized curve in S^2, then

$$2xx' + 2yy' + 2zz' = 0,$$

which shows that the vector (x, y, z) is normal to the sphere at the point

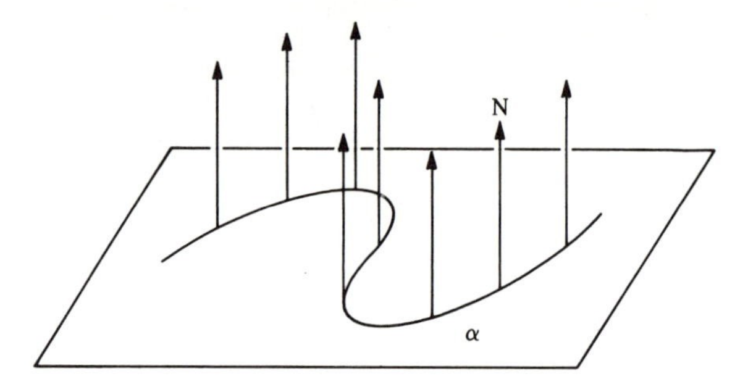

Figure 3-4. Plane: $dN_p = 0$.

(x, y, z). Thus, $\bar{N} = (x, y, z)$ and $N = (-x, -y, -z)$ are fields of unit normal vectors in S^2. We fix an orientation in S^2 by choosing $N = (-x, -y, -z)$ as a normal field. Notice that N points toward the center of the sphere.

Restricted to the curve $\alpha(t)$, the normal vector

$$N(t) = (-x(t), -y(t), -z(t))$$

is a vector function of t, and therefore

$$dN(x'(t), y'(t), z'(t)) = N'(t) = (-x'(t), -y'(t), -z'(t));$$

that is, $dN_p(v) = -v$ for all $p \in S^2$ and all $v \in T_p(S^2)$. Notice that with the choice of $\bar{N}$ as a normal field (that is, with the opposite orientation) we would have obtained $d\bar{N}_p(v) = v$ (Fig. 3-5).

Example 3. Consider the cylinder $\{(x, y, z) \in R^3; x^2 + y^2 = 1\}$. By an argument similar to that of the previous example, we see that $\bar{N} = (x, y, 0)$

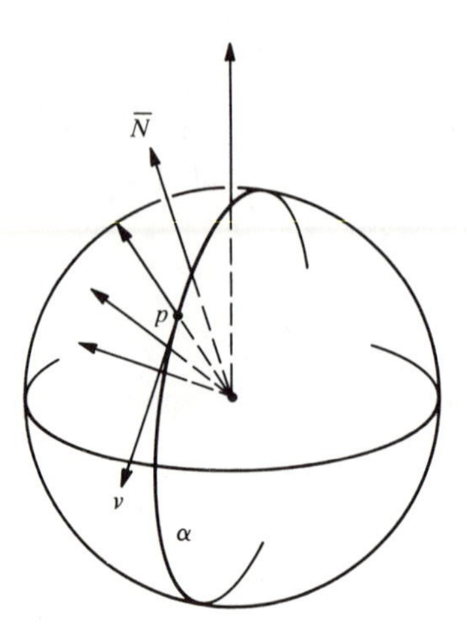

Figure 3-5. Unit sphere: $d\bar{N}_p(v) = v$.

and $N = (-x, -y, 0)$ are unit normal vectors at (x, y, z). We fix an orientation by choosing $N = (-x, -y, 0)$ as the normal vector field.

By considering a curve $(x(t), y(t), z(t))$ contained in the cylinder, that is, with $(x(t))^2 + (y(t))^2 = 1$, we are able to see that, along this curve, $N(t) = (-x(t), -y(t), 0)$ and therefore

$$dN(x'(t), y'(t), z'(t)) = N'(t) = (-x'(t), -y'(t), 0).$$

We conclude the following: If v is a vector tangent to the cylinder and parallel to the z axis, then

$$dN(v) = 0 = 0v;$$

if w is a vector tangent to the cylinder and parallel to the xy plane, then $dN(w) = -w$ (Fig. 3-6). It follows that the vectors v and w are eigenvectors of dN with eigenvalues 0 and -1, respectively (see the appendix to Chap. 3).

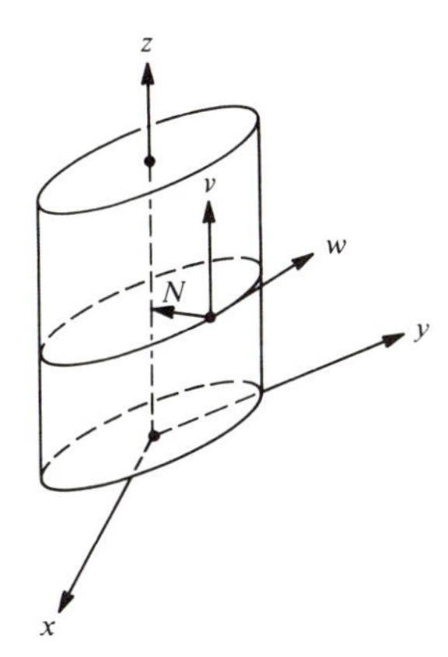

Figure 3-6

Example 4. Let us analyze the point $p = (0, 0, 0)$ of the hyperbolic paraboloid $z = y^2 - x^2$. For this, we consider a parametrization $\mathbf{x}(u, v)$ given by

$$\mathbf{x}(u, v) = (u, v, v^2 - u^2),$$

and compute the normal vector $N(u, v)$. We obtain successively

$$\mathbf{x}_u = (1, 0, -2u),$$
$$\mathbf{x}_v = (0, 1, 2v),$$
$$N = \left(\frac{u}{\sqrt{u^2 + v^2 + \frac{1}{4}}}, \frac{-v}{\sqrt{u^2 + v^2 + \frac{1}{4}}}, \frac{1}{2\sqrt{u^2 + v^2 + \frac{1}{4}}}\right).$$

Notice that at $p = (0, 0, 0)$ $\mathbf{x}_u$ and $\mathbf{x}_v$ agree with the unit vectors along the x and y axes, respectively. Therefore, the tangent vector at p to the curve

$\alpha(t) = \mathbf{x}(u(t), v(t))$, with $\alpha(0) = p$, has, in R^3, coordinates $(u'(0), v'(0), 0)$ (Fig. 3-7). Restricting $N(u, v)$ to this curve and computing $N'(0)$, we obtain

$$N'(0) = (2u'(0), -2v'(0), 0),$$

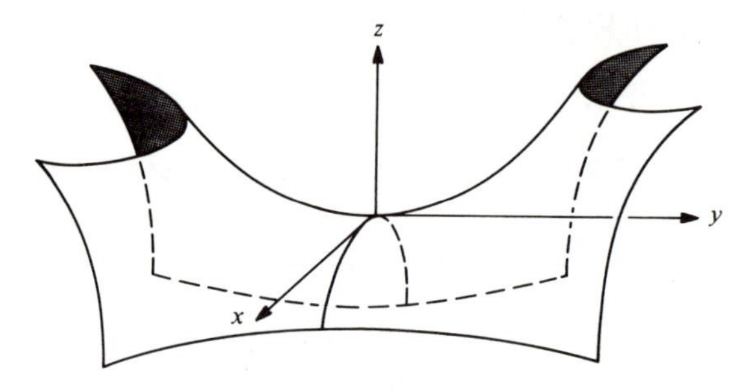

Figure 3-7

and therefore, at p,

$$dN_p(u'(0), v'(0), 0) = (2u'(0), -2v'(0), 0).$$

It follows that the vectors $(1, 0, 0)$ and $(0, 1, 0)$ are eigenvectors of dN_p with eigenvalues 2 and -2, respectively.

Example 5. The method of the previous example, applied to the point $p = (0, 0, 0)$ of the paraboloid $z = x^2 + ky^2$, $k > 0$, shows that the unit vectors of the x axis and the y axis are eigenvectors of dN_p, with eigenvalues 2 and $2k$, respectively (assuming that N is pointing outwards from the region bounded by the paraboloid).

An important fact about dN_p is contained in the following proposition.

PROPOSITION 1. *The differential* dN_p: $T_p(S) \to T_p(S)$ *of the Gauss map is a self-adjoint linear map* (cf. the appendix to Chap. 3).

Proof. Since dN_p is linear, it suffices to verify that $\langle dN_p(w_1), w_2 \rangle = \langle w_1, dN_p(w_2) \rangle$ for a basis $\{w_1, w_2\}$ of $T_p(S)$. Let $\mathbf{x}(u, v)$ be a parametrization of S at p and $\{\mathbf{x}_u, \mathbf{x}_v\}$ the associated basis of $T_p(S)$. If $\alpha(t) = \mathbf{x}(u(t), v(t))$ is a parametrized curve in S, with $\alpha(0) = p$, we have

$$\begin{aligned} dN_p(\alpha'(0)) &= dN_p(\mathbf{x}_u u'(0) + \mathbf{x}_v v'(0)) \\ &= \frac{d}{dt} N(u(t), v(t)) \Big|_{t=0} \\ &= N_u u'(0) + N_v v'(0); \end{aligned}$$

in particular, $dN_p(\mathbf{x}_u) = N_u$ and $dN_p(\mathbf{x}_v) = N_v$. Therefore, to prove that dN_p

is self-adjoint, it suffices to show that

$$\langle N_u, \mathbf{x}_v \rangle = \langle \mathbf{x}_u, N_v \rangle.$$

To see this, take the derivatives of $\langle N, \mathbf{x}_u \rangle = 0$ and $\langle N, \mathbf{x}_v \rangle = 0$, relative to v and u, respectively, and obtain

$$\langle N_v, \mathbf{x}_u \rangle + \langle N, \mathbf{x}_{uv} \rangle = 0,$$
$$\langle N_u, \mathbf{x}_v \rangle + \langle N, \mathbf{x}_{vu} \rangle = 0.$$

Thus,

$$\langle N_u, \mathbf{x}_v \rangle = -\langle N, \mathbf{x}_{uv} \rangle = \langle N_v, \mathbf{x}_u \rangle. \qquad \textbf{Q.E.D.}$$

The fact that $dN_p \colon T_p(S) \longrightarrow T_p(S)$ is a self-adjoint linear map allows us to associate to dN_p a quadratic form Q in $T_p(S)$, given by $Q(v) = \langle dN_p(v), v \rangle$, $v \in T_p(S)$ (cf. the appendix to Chap. 3). To obtain a geometric interpretation of this quadratic form, we need a few definitions. For reasons that will be clear shortly, we shall use the quadratic form $-Q$.

DEFINITION 2. *The quadratic form* II_p, *defined in* $\mathrm{T_p(S)}$ *by* $\mathrm{II_p(v)} = -\langle \mathrm{dN_p(v), v} \rangle$ *is called the* second fundamental form *of* S *at* p.

DEFINITION 3. *Let* C *be a regular curve in* S *passing through* $\mathrm{p} \in \mathrm{S}$, k *the curvature of* C *at* p, *and* $\cos\theta = \langle \mathrm{n, N} \rangle$, *where* n *is the normal vector to* C *and* N *is the normal vector to* S *at* p. *The number* $\mathrm{k_n} = \mathrm{k}\cos\theta$ *is then called the* normal curvature *of* $\mathrm{C} \subset \mathrm{S}$ *at* p.

In other words, k_n is the length of the projection of the vector kn over the normal to the surface at p, with a sign given by the orientation N of S at p (Fig. 3-8).

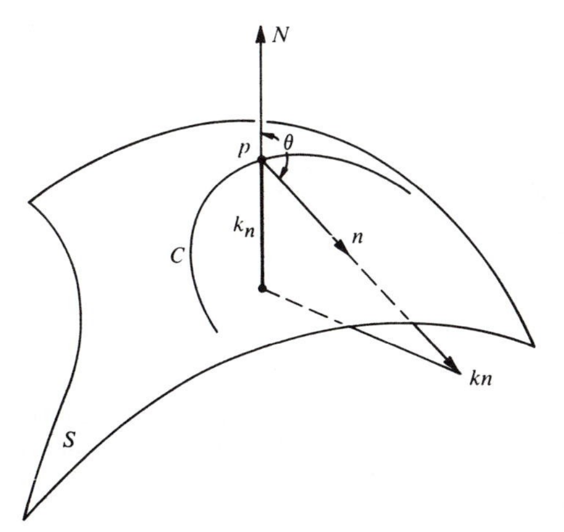

Figure 3-8

Remark. The normal curvature of C does not depend on the orientation of C but changes sign with a change of orientation for the surface.

To give an interpretation of the second fundamental form II_p, consider a regular curve $C \subset S$ parametrized by $\alpha(s)$, where s is the arc length of C, and with $\alpha(0) = p$. If we denote by $N(s)$ the restriction of the normal vector N to the curve $\alpha(s)$, we have $\langle N(s), \alpha'(s)\rangle = 0$. Hence,

$$\langle N(s), \alpha''(s)\rangle = -\langle N'(s), \alpha'(s)\rangle.$$

Therefore,

$$\begin{aligned} II_p(\alpha'(0)) &= -\langle dN_p(\alpha'(0)), \alpha'(0)\rangle \\ &= -\langle N'(0), \alpha'(0)\rangle = \langle N(0), \alpha''(0)\rangle \\ &= \langle N, kn\rangle(p) = k_n(p). \end{aligned}$$

In other words, the value of the second fundamental form II_p for a unit vector $v \in T_p(S)$ is equal to the normal curvature of a regular curve passing through p and tangent to v. In particular, we obtained the following result.

PROPOSITION 2 (Meusnier). *All curves lying on a surface* S *and having at a given point* p $\in$ S *the same tangent line have at this point the same normal curvatures.*

The above proposition allows us to speak of the *normal curvature along a given direction at* p. It is convenient to use the following terminology. Given a unit vector $v \in T_p(S)$, the intersection of S with the plane containing v and $N(p)$ is called the *normal section* of S at p along v (Fig. 3-9). In a neighborhood of p, a normal section of S at p is a regular plane curve on S whose normal vector n at p is $\pm N(p)$ or zero; its curvature is therefore equal to the absolute value of the normal curvature along v at p. With this terminology, the above

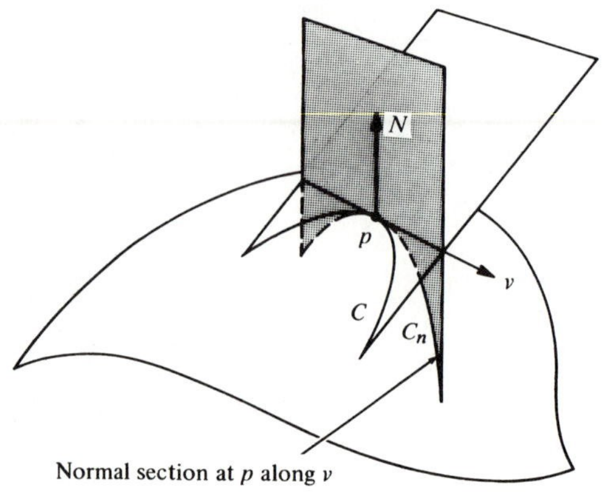

Figure 3-9. Meusnier theorem: C and C_n have the same normal curvature at p along v.

proposition says that the absolute value of the normal curvature at p of a curve $\alpha(s)$ is equal to the curvature of the normal section of S at p along $\alpha'(0)$.

Example 6. Consider the surface of revolution obtained by rotating the curve $z = y^4$ about the z axis (Fig. 3-10). We shall show that at $p = (0, 0, 0)$ the differential $dN_p = 0$. To see this, we observe that the curvature of the curve $z = y^4$ at p is equal to zero. Moreover, since the xy plane is a tangent plane to the surface at p, the normal vector $N(p)$ is parallel to the z axis. Therefore, any normal section at p is obtained from the curve $z = y^4$ by rotation; hence, it has curvature zero. It follows that all normal curvatures are zero at p, and thus $dN_p = 0$.

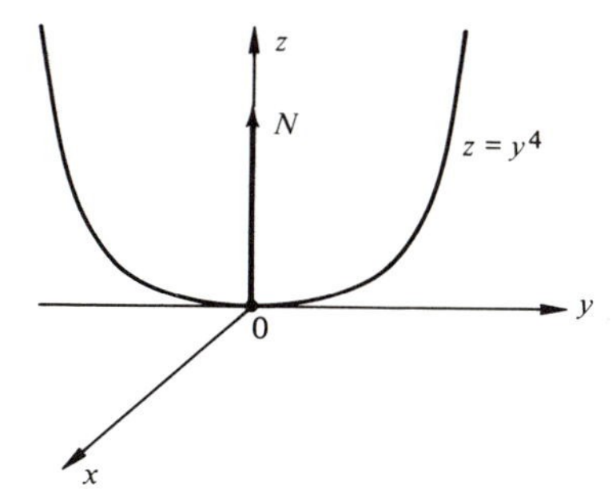

Figure 3-10

Example 7. In the plane of Example 1, all normal sections are straight lines; hence, all normal curvatures are zero. Thus, the second fundamental form is identically zero at all points. This agrees with the fact that $dN \equiv 0$.

In the sphere S^2 of Example 2, with $\bar{N}$ as orientation, the normal sections through a point $p \in S^2$ are circles with radius 1 (Fig. 3-11). Thus, all normal curvatures are equal to 1, and the second fundamental form is $II_p(v) = 1$ for all $p \in S^2$ and all $v \in T_p(S)$.

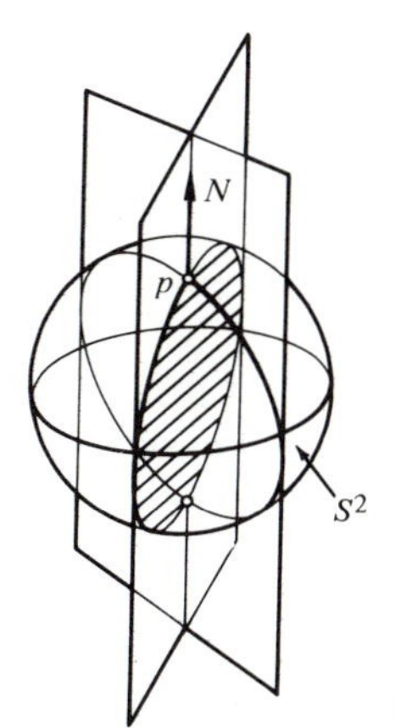

Figure 3-11. Normal sections on a sphere.

In the cylinder of Example 3, the normal sections at a point p vary from a circle perpendicular to the axis of the cylinder to a straight line parallel to the axis of the cylinder, passing through a family of ellipses (Fig. 3-12). Thus, the normal curvatures varies from 1 to 0. It is not hard to see geometrically that 1 is the maximum and 0 is the minimum of the normal curvature at p.

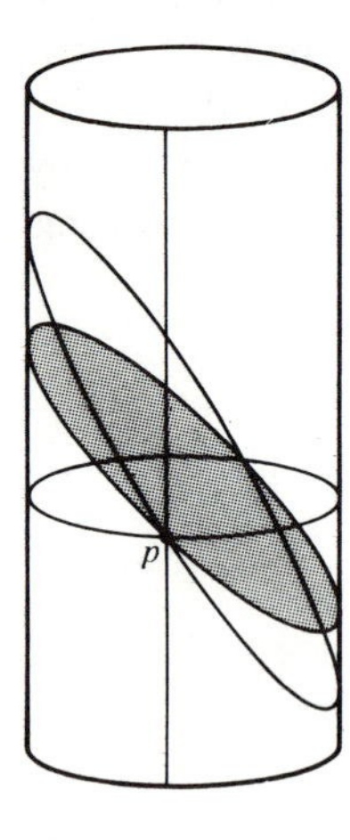

Figure 3-12. Normal sections on a cylinder.

However, an application of the theorem on quadratic forms of the appendix to Chap. 3 gives a simple proof of that. In fact, as we have seen in Example 3, the vectors w and v (corresponding to the directions of the normal curvatures 1 and 0, respectively) are eigenvectors of dN_p with eigenvalues -1 and 0, respectively. Thus, the second fundamental form takes up its extreme values in these vectors, as we claimed. Notice that this procedure allows us to check that such extreme values are 1 and 0.

We leave it to the reader to analyze the normal sections at the point $p = (0, 0, 0)$ of the hyperbolic paraboloid of Example 4.

Let us come back to the linear map dN_p. The theorem of the appendix to Chap. 3 shows that for each $p \in S$ there exists an orthonormal basis $\{e_1, e_2\}$ of $T_p(S)$ such that $dN_p(e_1) = -k_1 e_1$, $dN_p(e_2) = -k_2 e_2$. Moreover, k_1 and k_2 ($k_1 \geq k_2$) are the maximum and minimum of the second fundamental form II_p restricted to the unit circle of $T_p(S)$; that is, they are the extreme values of the normal curvature at p.

DEFINITION 4. *The maximum normal curvature* k_1 *and the minimum normal curvature* k_2 *are called the* principal curvatures *at* p*; the corresponding directions, that is, the directions given by the eigenvectors* e_1, e_2*, are called* principal directions *at* p.

For instance, in the plane all directions at all points are principal directions. The same happens with a sphere. In both cases, this comes from the fact that the second fundamental form at each point is constant (cf. Example 7); thus, all directions are extremals for the normal curvature.

In the cylinder of Example 3, the vectors v and w give the principal directions at p, corresponding to the principal curvatures 0 and 1, respectively. In the hyperbolic paraboloid of Example 4, the x and y axes are along the principal directions with principal curvatures -2 and 2, respectively.

DEFINITION 5. *If a regular connected curve* C *on* S *is such that for all* $p \in C$ *the tangent line of* C *is a principal direction at* p, *then* C *is said to be a* line of curvature *of* C.

PROPOSITION 3 (Olinde Rodrigues). *A necessary and sufficient condition for a connected regular curve* C *on* S *to be a line of curvature of* S *is that*

$$N'(t) = \lambda(t)\alpha'(t),$$

for any parametrization $\alpha(t)$ *of* C, *where* $N(t) = N \circ \alpha(t)$ *and* $\lambda(t)$ *is a differentiable function of* t. *In this case,* $-\lambda(t)$ *is the (principal) curvature along* $\alpha'(t)$.

Proof. It suffices to observe that if $\alpha'(t)$ is contained in a principal direction, then $\alpha'(t)$ is an eigenvector of dN and

$$dN(\alpha'(t)) = N'(t) = \lambda(t)\alpha'(t).$$

The converse is immediate. **Q.E.D.**

The knowledge of the principal curvatures at p allows us to compute easily the normal curvature along a given direction of $T_p(S)$. In fact, let $v \in T_p(S)$ with $|v| = 1$. Since e_1 and e_2 form an orthonormal basis of $T_p(S)$, we have

$$v = e_1 \cos\theta + e_2 \sin\theta,$$

where θ is the angle from e_1 to v in the orientation of $T_p(S)$. The normal curvature k_n along v is given by

$$\begin{aligned} k_n = II_p(v) &= -\langle dN_p(v), v\rangle \\ &= -\langle dN_p(e_1\cos\theta + e_2\sin\theta), e_1\cos\theta + e_2\sin\theta\rangle \\ &= \langle e_1k_1\cos\theta + e_2k_2\sin\theta, e_1\cos\theta + e_2\sin\theta\rangle \\ &= k_1\cos^2\theta + k_2\sin^2\theta. \end{aligned}$$

The last expression is known classically as the *Euler formula*; actually, it is just the expression of the second fundamental form in the basis $\{e_1, e_2\}$.

Given a linear map $A: V \longrightarrow V$ of a vector space of dimension 2 and given a basis $\{v_1, v_2\}$ of V, we recall that

$$\text{determinant of } A = a_{11}a_{22} - a_{12}a_{21}, \qquad \text{trace of } A = a_{11} + a_{22},$$

where (a_{ij}) is the matrix of A in the basis $\{v_1, v_2\}$. It is known that these numbers do not depend on the choice of the basis $\{v_1, v_2\}$ and are, therefore, attached to the linear map A.

In our case, the determinant of dN is the product $(-k_1)(-k_2) = k_1k_2$ of the principal curvatures, and the trace of dN is the negative $-(k_1 + k_2)$ of the sum of principal curvatures. If we change the orientation of the surface, the determinant does not change (the fact that the dimension is even is essential here); the trace, however, changes sign.

DEFINITION 6. *Let* $p \in S$ *and let* $dN_p: T_p(S) \longrightarrow T_p(S)$ *be the differential of the Gauss map. The determinant of* dN_p *is the* Gaussian curvature K *of* S *at* p. *The negative of half of the trace of* dN_p *is called the* mean curvature H *of* S *at* p.

In terms of the principal curvatures we can write

$$K = k_1k_2, \qquad H = \frac{k_1 + k_2}{2}.$$

DEFINITION 7. *A point of a surface* S *is called*

1. Elliptic *if* $det(dN_p) > 0$.
2. Hyperbolic *if* $det(dN_p) < 0$.
3. Parabolic *if* $det(dN_p) = 0$, *with* $dN_p \neq 0$.
4. Planar *if* $dN_p = 0$.

It is clear that this classification does not depend on the choice of the orientation.

At an elliptic point the Gaussian curvature is positive. Both principal curvatures have the same sign, and therefore all curves passing through this point have their normal vectors pointing toward the same side of the tangent plane. The points of a sphere are elliptic points. The point (0, 0, 0) of the paraboloid $z = x^2 + ky^2$, $k > 0$ (cf. Example 5), is also an elliptic point.

At a hyperbolic point, the Gaussian curvature is negative. The principal curvatures have opposite signs, and therefore there are curves through p whose normal vectors at p point toward any of the sides of the tangent plane at p. The point (0, 0, 0) of the hyperbolic paraboloid $z = y^2 - x^2$ (cf. Example 4) is a hyperbolic point.

At a parabolic point, the Gaussian curvature is zero, but one of the principal curvatures is not zero. The points of a cylinder (cf. Example 3) are parabolic points.

Finally, at a planar point, all principal curvatures are zero. The points of a plane trivially satisfy this condition. A nontrivial example of a planar point was given in Example 6.

DEFINITION 8. *If at* p $\in$ S, $k_1 = k_2$, *then* p *is called an* umbilical point *of* S*; in particular, the planar points* ($k_1 = k_2 = 0$) *are umbilical points.*

All the points of a sphere and a plane are umbilical points. Using the method of Example 6, we can verify that the point (0, 0, 0) of the paraboloid $z = x^2 + y^2$ is a (nonplanar) umbilical point.

We shall now prove the interesting fact that the only surfaces made up entirely of umbilical points are essentially spheres and planes.

PROPOSITION 4. *If all points of a connected surface* S *are umbilical points, then* S *is either contained in a sphere or in a plane.*

Proof. Let $p \in S$ and let $\mathbf{x}(u, v)$ be a parametrization of S at p such that the coordinate neighborhood V is connected.

Since each $q \in V$ is an umbilical point, we have, for any vector $w = a_1\mathbf{x}_u + a_2\mathbf{x}_v$ in $T_q(S)$,

$$dN(w) = \lambda(q)w,$$

where $\lambda = \lambda(q)$ is a real differentiable function in V.

We first show that $\lambda(q)$ is constant in V. For that, we write the above equation as

$$N_u a_1 + N_v a_2 = \lambda(\mathbf{x}_u a_1 + \mathbf{x}_v a_2);$$

hence, since w is arbitrary,

$$N_u = \lambda\mathbf{x}_u,$$
$$N_v = \lambda\mathbf{x}_v.$$

Differentiating the first equation in u and the second one in v and subtracting the resulting equations, we obtain

$$\lambda_u\mathbf{x}_v - \lambda_v\mathbf{x}_u = 0.$$

Since $\mathbf{x}_u$ and $\mathbf{x}_v$ are linearly independent, we conclude that

$$\lambda_u = \lambda_v = 0$$

for all $q \in V$. Since V is connected, λ is constant in V, as we claimed.

If $\lambda \equiv 0$, $N_u = N_v = 0$ and therefore $N = N_0 =$ constant in V. Thus, $\langle \mathbf{x}(u, v), N_0 \rangle_u = \langle \mathbf{x}(u, v), N_0 \rangle_v = 0$; hence,

$$\langle \mathbf{x}(u, v), N_0 \rangle = \text{const.},$$

and all points $\mathbf{x}(u, v)$ of V belong to a plane.

If $\lambda \neq 0$, then the point $\mathbf{x}(u, v) - (1/\lambda)N(u, v) = \mathbf{y}(u, v)$ is fixed, because

$$(\mathbf{x}(u, v) - \frac{1}{\lambda}N(u, v))_u = (\mathbf{x}(u, v) - \frac{1}{\lambda}N(u, v))_v = 0.$$

Since

$$|\mathbf{x}(u, v) - \mathbf{y}|^2 = \frac{1}{\lambda^2},$$

all points of V are contained in a sphere of center $\mathbf{y}$ and radius $1/|\lambda|$.

This proves the proposition locally, that is, for a neighborhood of a point $p \in S$. To complete the proof we observe that, since S is connected, given any other point $r \in S$, there exists a continuous curve $\alpha: [0, 1] \rightarrow S$ with $\alpha(0) = p$, $\alpha(1) = r$. For each point $\alpha(t) \in S$ of this curve there exists a neighborhood V_t in S contained in a sphere or in a plane and such that $\alpha^{-1}(V_t)$ is an open interval of $[0, 1]$. The union $\bigcup \alpha^{-1}(V_t)$, $t \in [0, 1]$, covers $[0, 1]$ and since $[0, 1]$ is a closed interval, it is covered by finitely many elements of the family $\{\alpha^{-1}(V_t)\}$ (cf. the Heine-Borel theorem, Prop. 6 of the appendix to Chap. 2). Thus, $c([0, 1])$ is covered by a finite number of the neighborhoods V_t.

If the points of one of these neighborhoods are on a plane, all the others will be on the same plane. Since r is arbitrary, all the points of S belong to this plane.

If the points of one of these neighborhoods are on a sphere, the same argument shows that all points on S belong to a sphere, and this completes the proof. **Q.E.D.**

DEFINITION 9. *Let* p *be a point in* S. *An* asymptotic direction *of* S *at* p *is a direction of* $T_p(S)$ *for which the normal curvature is zero. An* asymptotic curve *of* S *is a regular connected curve* $C \subset S$ *such that for each* $p \in C$ *the tangent line of* C *at* p *is an asymptotic direction.*

It follows at once from the definition that at an elliptic point there are no asymptotic directions.

A useful geometric interpretation of the asymptotic directions is given by means of the Dupin indicatrix, which we shall now describe.

Let p be a point in S. The *Dupin indicatrix* at p is the set of vectors w of $T_p(S)$ such that $II_p(w) = \pm 1$.

To write the equations of the Dupin indicatrix in a more convenient form, let (ξ, η) be the cartesian coordinates of $T_p(S)$ in the orthonormal basis $\{e_1, e_2\}$, where e_1 and e_2 are eigenvectors of dN_p. Given $w \in T_p(S)$, let ρ and θ be "polar coordinates" defined by $w = \rho v$, with $|v| = 1$ and $v = e_1 \cos\theta + e_2 \sin\theta$, if $\rho \neq 0$. By Euler's formula,

$$\begin{aligned} \pm 1 = II_p(w) &= \rho^2 II_p(v) \\ &= k_1 \rho^2 \cos^2\theta + k_2 \rho^2 \sin^2\theta \\ &= k_1 \xi^2 + k_2 \eta^2, \end{aligned}$$

where $w = \xi e_1 + \eta e_2$. Thus, the coordinates (ξ, η) of a point of the Dupin indicatrix satisfy the equation

$$k_1 \xi^2 + k_2 \eta^2 = \pm 1; \tag{1}$$

hence, the Dupin indicatrix is a union of conics in $T_p(S)$. We notice that the normal curvature along the direction determined by w is $k_n(v) = II_p(v) = \pm(1/\rho^2)$.

For an elliptic point, the Dupin indicatrix is an ellipse (k_1 and k_2 have the same sign); this ellipse degenerates into a circle if the point is an umbilical nonplanar point ($k_1 = k_2 \neq 0$).

For a hyperbolic point, k_1 and k_2 have opposite signs. The Dupin indicatrix is therefore made up of two hyperbolas with a common pair of asymptotic lines (Fig. 3-13). Along the directions of the asymptotes, the normal curvature is zero; they are therefore asymptotic directions. This justifies the terminology and shows that a hyperbolic point has *exactly two* asymptotic directions.

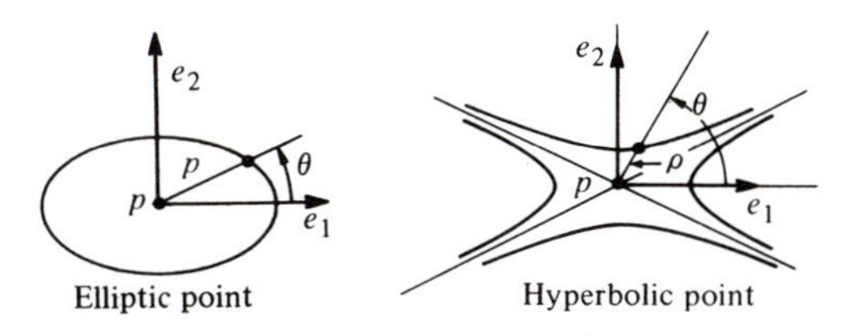

Figure 3-13. The Dupin indicatrix.

For a parabolic point, one of the principal curvatures is zero, and the Dupin indicatrix degenerates into a pair of parallel lines. The common direction of these lines is the only asymptotic direction at the given point.

In Example 5 of Sec. 3-4 we shall show an interesting property of the Dupin indicatrix.

Closely related with the concept of asymptotic direction is the concept of conjugate directions, which we shall now define.

DEFINITION 10. *Let* p *be a point on a surface* S. *Two nonzero vectors* $w_1, w_2 \in T_p(S)$ *are* conjugate *if* $\langle dN_p(w_1), w_2\rangle = \langle w_1, dN_p(w_2)\rangle = 0$. *Two directions* r_1, r_2 *at* p *are* conjugate *if a pair of nonzero vectors* w_1, w_2 *parallel to* r_1 *and* r_2, *respectively, are conjugate.*

It is immediate to check that the definition of conjugate directions does not depend on the choice of the vectors w_1 and w_2 on r_1 and r_2.

It follows from the definition that the principal directions are conjugate and that an asymptotic direction is conjugate to itself. Furthermore, at a nonplanar umbilic, every orthogonal pair of directions is a pair of conjugate directions, and at a planar umbilic each direction is conjugate to any other direction.

Let us assume that $p \in S$ is not an umbilical point, and let $\{e_1, e_2\}$ be the orthonormal basis of $T_p(S)$ determined by $dN_p(e_1) = -k_1 e_1$, $dN_p(e_2) = -k_2 e_2$. Let θ and φ be the angles that a pair of directions r_1 and r_2 make with e_1. We claim that r_1 and r_2 are conjugate if and only if

$$k_1 \cos\theta \cos\varphi = -k_2 \sin\theta \sin\varphi. \tag{2}$$

In fact, r_1 and r_2 are conjugate if and only if the vectors

$$w_1 = e_1 \cos\theta + e_2 \sin\theta, \qquad w_2 = e_1 \cos\varphi + e_2 \sin\varphi$$

are conjugate. Thus,

$$0 = \langle dN_p(w_1), w_2\rangle = -k_1 \cos\theta\cos\varphi - k_2 \sin\theta \sin\varphi.$$

Hence, condition (2) follows.

When both k_1 and k_2 are nonzero (i.e., p is either an elliptic or a hyperbolic point), condition (2) leads to a geometric construction of conjugate directions

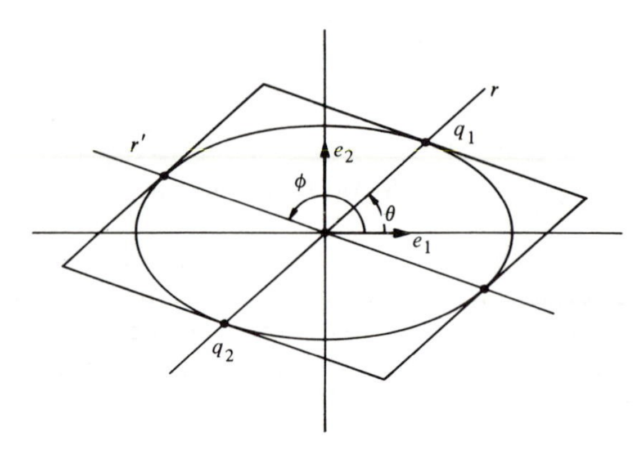

Figure 3-14. Construction of conjugate directions.

in terms of the Dupin indicatrix at p. We shall describe the construction at an elliptic point, the situation at a hyperbolic point being similar. Let r be a straight line through the origin of $T_p(S)$ and consider the intersection points q_1, q_2 of r with the Dupin indicatrix (Fig. 3-14). The tangent lines of the Dupin indicatrix at q_1 and q_2 are parallel, and their common direction r' is conjugate to r. We shall leave the proofs of these assertions to the Exercises (Exercise 12).

EXERCISES

1. Show that at a hyperbolic point, the principal directions bissect the asymptotic directions.

2. Show that if a surface is tangent to a plane along a curve, then the points of this curve are either parabolic or planar.

3. Let $C \subset S$ be a regular curve on a surface S with Gaussian curvature $K > 0$. Show that the curvature k of C at p satisfies

$$k \geq \min(|k_1|, |k_2|),$$

where k_1 and k_2 are the principal curvatures of S at p.

4. Assume that a surface S has the property that $|k_1| \leq 1$, $|k_2| \leq 1$ everywhere. Is it true that the curvature k of a curve on S also satisfies $|k| \leq 1$?

5. Show that the mean curvature H at $p \in S$ is given by

$$H = \frac{1}{\pi} \int_0^\pi k_n(\theta)\, d\theta,$$

where $k_n(\theta)$ is the normal curvature at p along a direction making an angle θ with a fixed direction.

6. Show that the sum of the normal curvatures for any pair of orthogonal directions, at a point $p \in S$, is constant.

7. Show that if the mean curvature is zero at a nonplanar point, then this point has two orthogonal asymptotic directions.

8. Describe the region of the unit sphere covered by the image of the Gauss map of the following surfaces:

a. Paraboloid of revolution $z = x^2 + y^2$.

b. Hyperboloid of revolution $x^2 + y^2 - z^2 = 1$.

c. Catenoid $x^2 + y^2 = \cosh^2 z$.

9. Prove that

a. The image $N \circ \alpha$ by the Gauss map $N\colon S \longrightarrow S^2$ of a parametrized regular curve $\alpha: I \longrightarrow S$ which contains no planar or parabolic points is a parametrized regular curve on the sphere S^2 (called the *spherical image* of α).

b. If $C = \alpha(I)$ is a line of curvature, and k is its curvature at p, then

$$k = |k_n k_N|,$$

where k_n is the normal curvature at p along the tangent line of C and k_N is the curvature of the spherical image $N(C) \subset S^2$ at $N(p)$.

10. Assume that the osculating plane of a line of curvature $C \subset S$, which is nowhere tangent to an asymptotic direction, makes a constant angle with the tangent plane of S along C. Prove that C is a plane curve.

11. Let p be an elliptic point of a surface S, and let r and r' be conjugate directions at p. Let r vary in $T_p(S)$ and show that the minimum of the angle of r with r' is reached at a unique pair of directions in $T_p(S)$ that are symmetric with respect to the principal directions.

12. Let p be a hyperbolic point of a surface S, and let r be a direction in $T_p(S)$. Describe and justify a geometric construction to find the conjugate direction r' of r in terms of the Dupin indicatrix (cf. the construction at the end of Sec. 3-2).

***13.** (*Theorem of Beltrami-Enneper.*) Prove that the absolute value of the torsion τ at a point of an asymptotic curve, whose curvature is nowhere zero, is given by

$$|\tau| = \sqrt{-K},$$

where K is the Gaussian curvature of the surface at the given point.

***14.** If the surface S_1 intersects the surface S_2 along the regular curve C, then the curvature k of C at $p \in C$ is given by

$$k^2 \sin^2 \theta = \lambda_1^2 + \lambda_2^2 - 2\lambda_1\lambda_2 \cos \theta,$$

where λ_1 and λ_2 are the normal curvatures at p, along the tangent line to C, of S_1 and S_2, respectively, and θ is the angle made up by the normal vectors of S_1 and S_2 at p.

15. (*Theorem of Joachimstahl.*) Suppose that S_1 and S_2 intersect along a regular curve C and make an angle $\theta(p)$, $p \in C$. Assume that C is a line of curvature of S_1. Prove that $\theta(p)$ is constant if and only if C is a line of curvature of S_2.

***16.** Show that the meridians of a torus are lines of curvature.

17. Show that if $H \equiv 0$ on S and S has no planar points, then the Gauss map $N: S \longrightarrow S^2$ has the following property:

$$\langle dN_p(w_1), dN_p(w_2) \rangle = -K(p)\langle w_1, w_2 \rangle$$

for all $p \in S$ and all $w_1, w_2 \in T_p(S)$. Show that the above condition implies that the angle of two intersecting curves on S^2 and the angle of their spherical images (cf. Exercise 9) are equal up to a sign.

***18.** Let $\lambda_1, \ldots, \lambda_m$ be the normal curvatures at $p \in S$ along directions making angles $0, 2\pi/m, \ldots, (m-1)2\pi/m$ with a principal direction. Prove that

$$\lambda_1 + \cdots + \lambda_m = mH,$$

where H is the mean curvature at p.

***19.** Let $C \subset S$ be a regular curve in S. Let $p \in C$ and $\alpha(s)$ be a parametrization of C in p by arc length so that $\alpha(0) = p$. Choose in $T_p(S)$ an orthonormal positive basis $\{t, h\}$, where $t = \alpha'(0)$. The *geodesic torsion* τ_g of $C \subset S$ at p is defined by

$$\tau_g = \left\langle \frac{dN}{ds}(0), h \right\rangle.$$

Prove that

a. $\tau_g = (k_1 - k_2) \cos \varphi \sin \varphi$, where φ is the angle from e_1 to t.

b. If τ is the torsion of C, n is the (principal) normal vector of C and $\cos \theta = \langle N, n \rangle$, then

$$\frac{d\theta}{ds} = \tau - \tau_g.$$

c. The lines of curvature of S are characterized by having geodesic torsion identically zero.

***20.** (*Dupin's Theorem.*) Three families of surfaces are said to form a *triply orthogonal system* in an open set $U \subset R^3$ if a unique surface of each family passes through each point $p \in U$ and if the three surfaces that pass through p are pairwise orthogonal. Use part c of Exercise 19 to prove Dupin's theorem: *The surfaces of a triply orthogonal system intersect each other in lines of curvature.*

3-3. The Gauss Map in Local Coordinates

In the preceding section, we introduced some concepts related to the local behavior of the Gauss map. To emphasize the geometry of the situation, the definitions were given without the use of a coordinate system. Some simple examples were then computed directly from the definitions; this procedure, however, is inefficient in handling general situations. In this section, we shall obtain the expressions of the second fundamental form and of the differential of the Gauss map in a coordinate system. This will give us a systematic method for computing specific examples. Moreover, the general expressions thus obtained are essential for a more detailed investigation of the concepts introduced above.

All parametrization $\mathbf{x}: U \subset R^2 \to S$ considered in this section are assumed to be compatible with the orientation N of S; that is, in $\mathbf{x}(U)$,

$$N = \frac{\mathbf{x}_u \wedge \mathbf{x}_v}{|\mathbf{x}_u \wedge \mathbf{x}_v|}.$$

Let $\mathbf{x}(u, v)$ be a parametrization at a point $p \in S$ of a surface S, and let $\alpha(t) = \mathbf{x}(u(t), v(t))$ be a parametrized curve on S, with $\alpha(0) = p$. To simplify the notation, we shall make the convention that all functions to appear below denote their values at the point p.

The tangent vector to $\alpha(t)$ at p is $\alpha' = \mathbf{x}_u u' + \mathbf{x}_v v'$ and

$$dN(\alpha') = N'(u(t), v(t)) = N_u u' + N_v v'.$$

Since N_u and N_v belong to $T_p(S)$, we may write

$$\begin{aligned} N_u &= a_{11}\mathbf{x}_u + a_{21}\mathbf{x}_v, \\ N_v &= a_{12}\mathbf{x}_u + a_{22}\mathbf{x}_v, \end{aligned} \tag{1}$$

and therefore,

$$dN(\alpha') = (a_{11}u' + a_{12}v')\mathbf{x}_u + (a_{21}u' + a_{22}v')\mathbf{x}_v;$$

hence,

$$dN\begin{pmatrix} u' \\ v' \end{pmatrix} = \begin{pmatrix} a_{11} & a_{12} \\ a_{21} & a_{22} \end{pmatrix}\begin{pmatrix} u' \\ v' \end{pmatrix}.$$

This shows that in the basis $\{\mathbf{x}_u, \mathbf{x}_v\}$, dN is given by the matrix (a_{ij}), $i, j = 1, 2$. Notice that this matrix is not necessarily symmetric, unless $\{\mathbf{x}_u, \mathbf{x}_v\}$ is an orthonormal basis.

On the other hand, the expression of the second fundamental form in the basis $\{\mathbf{x}_u, \mathbf{x}_v\}$ is given by

$$\begin{aligned} II_p(\alpha') &= -\langle dN(\alpha'), \alpha' \rangle = -\langle N_u u' + N_v v', \mathbf{x}_u u' + \mathbf{x}_v v' \rangle \\ &= e(u')^2 + 2fu'v' + g(v')^2, \end{aligned}$$

where, since $\langle N, \mathbf{x}_u \rangle = \langle N, \mathbf{x}_v \rangle = 0$,

$$\begin{aligned} e &= -\langle N_u, \mathbf{x}_u \rangle = \langle N, \mathbf{x}_{uu} \rangle, \\ f &= -\langle N_v, \mathbf{x}_u \rangle = \langle N, \mathbf{x}_{uv} \rangle = \langle N, \mathbf{x}_{vu} \rangle = -\langle N_u, \mathbf{x}_v \rangle, \\ g &= -\langle N_v, \mathbf{x}_v \rangle = \langle N, \mathbf{x}_{vv} \rangle. \end{aligned}$$

We shall now obtain the values of a_{ij} in terms of the coefficientes e, f, g. From Eq. (1), we have

$$\begin{aligned} -f &= \langle N_u, \mathbf{x}_v \rangle = a_{11}F + a_{21}G, \\ -f &= \langle N_v, \mathbf{x}_u \rangle = a_{12}E + a_{22}F, \\ -e &= \langle N_u, \mathbf{x}_u \rangle = a_{11}E + a_{21}F, \\ -g &= \langle N_v, \mathbf{x}_v \rangle = a_{12}F + a_{22}G, \end{aligned} \tag{2}$$

where E, F, and G are the coefficients of the first fundamental form in the basis $\{\mathbf{x}_u, \mathbf{x}_v\}$ (cf. Sec. 2-5). Relations (2) may be expressed in matrix form by

$$-\begin{pmatrix} e & f \\ f & g \end{pmatrix} = \begin{pmatrix} a_{11} & a_{21} \\ a_{12} & a_{22} \end{pmatrix} \begin{pmatrix} E & F \\ F & G \end{pmatrix}; \tag{3}$$

hence,

$$\begin{pmatrix} a_{11} & a_{21} \\ a_{12} & a_{22} \end{pmatrix} = -\begin{pmatrix} e & f \\ f & g \end{pmatrix} \begin{pmatrix} E & F \\ F & G \end{pmatrix}^{-1},$$

where $(\quad)^{-1}$ means the inverse matrix of $(\quad)$. It is easily checked that

$$\begin{pmatrix} E & F \\ F & G \end{pmatrix}^{-1} = \frac{1}{EG - F^2} \begin{pmatrix} G & -F \\ -F & E \end{pmatrix},$$

whence the following expressions for the coefficients (a_{ij}) of the matrix of dN in the basis $\{x_u, x_v\}$:

$$a_{11} = \frac{fF - eG}{EG - F^2},$$

$$a_{12} = \frac{gF - fG}{EG - F^2},$$

$$a_{21} = \frac{eF - fE}{EG - F^2},$$

$$a_{22} = \frac{fF - gE}{EG - F^2}.$$

For completeness, it should be mentioned that relations (1), with the above values, are known as the *equations of Weingarten.*

From Eq. (3) we immediately obtain

$$K = \det(a_{ij}) = \frac{eg - f^2}{EG - F^2}. \tag{4}$$

To compute the mean curvature, we recall that $-k_1$, $-k_2$ are the eigenvalues of dN. Therefore, k_1 and k_2 satisfy the equation

$$dN(v) = -kv = -kIv \qquad \text{for some } v \in T_p(S),\ v \neq 0,$$

where I is the identity map. It follows that the linear map $dN + kI$ is not invertible; hence, it has zero determinant. Thus,

$$\det\begin{pmatrix} a_{11} + k & a_{12} \\ a_{21} & a_{22} + k \end{pmatrix} = 0$$

or

$$k^2 + k(a_{11} + a_{22}) + a_{11}a_{22} - a_{21}a_{12} = 0.$$

Since k_1 and k_2 are the roots of the above quadratic equation, we conclude that

$$H = \frac{1}{2}(k_1 + k_2) = -\frac{1}{2}(a_{11} + a_{22}) = \frac{1}{2}\frac{eG - 2fF + gE}{EG - F^2}; \qquad (5)$$

hence,

$$k^2 - 2Hk + K = 0,$$

and therefore,

$$k = H \pm \sqrt{H^2 - K}. \qquad (6)$$

From this relation, it follows that if we choose $k_1(q) \geq k_2(q)$, $q \in S$, then the functions k_1 and k_2 are continuous in S. Moreover, k_1 and k_2 are differentiable in S, except perhaps at the umbilical points ($H^2 = K$) of S.

In the computations of this chapter, it will be convenient to write for short

$$\langle u \wedge v, w \rangle = (u, v, w) \qquad \text{for any } u, v, w \in R^3.$$

We recall that this is merely the determinant of the 3×3 matrix whose columns (or lines) are the components of the vectors u, v, w in the canonical basis of R^3.

Example 1. We shall compute the Gaussian curvature of the points of the torus covered by the parametrization (cf. Example 6 of Sec. 2-2)

$$\mathbf{x}(u, v) = ((a + r\cos u)\cos v, (a + r\cos u)\sin v, r\sin u),$$
$$0 < u < 2\pi, \qquad 0 < v < 2\pi.$$

For the computation of the coefficients e, f, g, we need to know N (and thus $\mathbf{x}_u$ and $\mathbf{x}_v$), $\mathbf{x}_{uu}$, $\mathbf{x}_{uv}$, and $\mathbf{x}_{vv}$:

$$\begin{aligned}
\mathbf{x}_u &= (-r\sin u\cos v, -r\sin u\sin v, r\cos u),\\
\mathbf{x}_v &= (-(a + r\cos u)\sin v, (a + r\cos u)\cos v, 0),\\
\mathbf{x}_{uu} &= (-r\cos u\cos v, -r\cos u\sin v, -r\sin u),\\
\mathbf{x}_{uv} &= (r\sin u\sin v, -r\sin u\cos v, 0),\\
\mathbf{x}_{vv} &= (-(a + r\cos u)\cos v, -(a + r\cos u)\sin v, 0).
\end{aligned}$$

From these, we obtain

$$E = \langle \mathbf{x}_u, \mathbf{x}_u \rangle = r^2, \qquad F = \langle \mathbf{x}_u, \mathbf{x}_v \rangle = 0,$$
$$G = \langle \mathbf{x}_v, \mathbf{x}_v \rangle = (a + r\cos u)^2.$$

Introducing the values just obtained in $e = \langle N, \mathbf{x}_{uu} \rangle$, we have, since $|\mathbf{x}_u \wedge \mathbf{x}_v| = \sqrt{EG - F^2}$,

$$e = \left\langle \frac{\mathbf{x}_u \wedge \mathbf{x}_v}{|\mathbf{x}_u \wedge \mathbf{x}_v|}, \mathbf{x}_{uu} \right\rangle = \frac{(\mathbf{x}_u, \mathbf{x}_v, \mathbf{x}_{uu})}{\sqrt{EG - F^2}} = \frac{r^2(a + r\cos u)}{r(a + r\cos u)} = r.$$

Similarly, we obtain

$$f = \frac{(\mathbf{x}_u, \mathbf{x}_v, \mathbf{x}_{uv})}{r(a + r\cos u)} = 0,$$

$$g = \frac{(\mathbf{x}_u, \mathbf{x}_v, \mathbf{x}_{vv})}{r(a + r\cos u)} = \cos u(a + r\cos u).$$

Finally, since $K = (eg - f^2)/(EG - F^2)$, we have that

$$K = \frac{\cos u}{r(a + r\cos u)}.$$

From this expression, it follows that $K = 0$ along the parallels $u = \pi/2$ and $u = 3\pi/2$; the points of such parallels are therefore parabolic points. In the region of the torus given by $\pi/2 < u < 3\pi/2$, K is negative (notice that $r > 0$ and $a > r$); the points in this region are therefore hyperbolic points. In the region given by $0 < u < \pi/2$ or $3\pi/2 < u < 2\pi$, the curvature is positive and the points are elliptic points (Fig. 3-15).

As an application of the expression for the second fundamental form in coordinates, we shall prove a proposition which gives information about the position of a surface in the neighborhood of an elliptic or a hyperbolic point, relative to the tangent plane at this point. For instance, if we look at an elliptic point of the torus of Example 1, we find that the surface lies on one side of the tangent plane at such a point (see Fig. 3-15). On the other hand, if p is a hyperbolic point of the torus T and $V \subset T$ is any neighborhood of p, we can find points of V on both sides of $T_p(S)$, however small V may be.

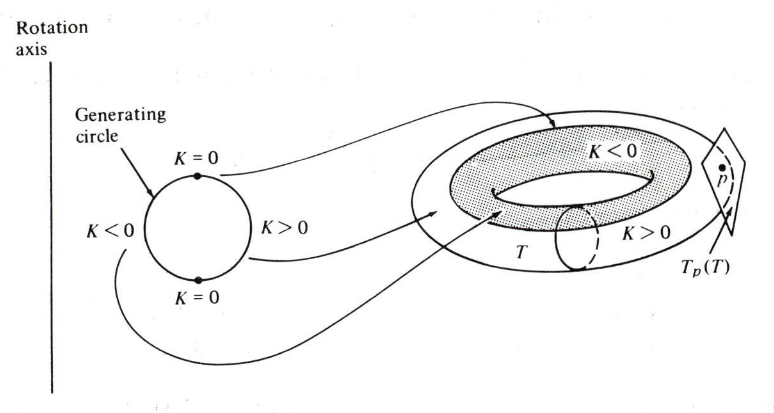

Figure 3-15

This example reflects a general local fact that is described in the following proposition.

PROPOSITION 1. *Let* p $\in$ S *be an elliptic point of a surface* S. *Then there exists a neighborhood* V *of* p *in* S *such that all points in* V *belong to the same side of the tangent plane* $T_p(S)$. *Let* p $\in$ S *be a hyperbolic point. Then in each neighborhood of* p *there exist points of* S *in both sides of* $T_p(S)$.

Proof. Let $\mathbf{x}(u, v)$ be a parametrization in p, with $\mathbf{x}(0, 0) = p$. The distance d from a point $q = \mathbf{x}(u, v)$ to the tangent plane $T_p(S)$ is given by (Fig. 3-16)

$$d = \langle \mathbf{x}(u, v) - \mathbf{x}(0, 0), N(p) \rangle.$$

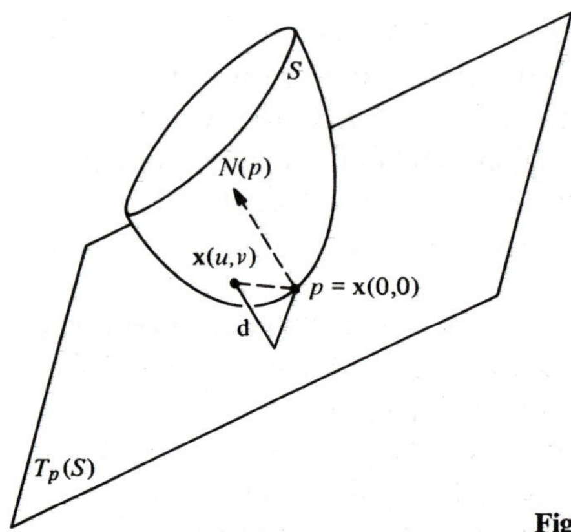

Figure 3-16

Since $\mathbf{x}(u, v)$ is differentiable, we have Taylor's formula:

$$\mathbf{x}(u, v) = \mathbf{x}(0, 0) + \mathbf{x}_u u + \mathbf{x}_v v + \tfrac{1}{2}(\mathbf{x}_{uu}u^2 + 2\mathbf{x}_{uv}uv + \mathbf{x}_{vv}v^2) + \bar{R},$$

where the derivatives are taken at (0, 0) and the remainder $\bar{R}$ satisfies the condition

$$\lim_{(u,v)\to(0,0)} \frac{\bar{R}}{u^2 + v^2} = 0.$$

It follows that

$$\begin{aligned} d &= \langle \mathbf{x}(u, v) - \mathbf{x}(0, 0), N(p) \rangle \\ &= \tfrac{1}{2}\{\langle \mathbf{x}_{uu}, N(p)\rangle u^2 + 2\langle \mathbf{x}_{uv}, N(p)\rangle uv + \langle \mathbf{x}_{vv}, N(p)\rangle v^2\} + R \\ &= \tfrac{1}{2}(eu^2 + 2fuv + gv^2) + R = \tfrac{1}{2}II_p(w) + R, \end{aligned}$$

where $w = \mathbf{x}_u u + \mathbf{x}_v v$, $R = \langle \bar{R}, N(p) \rangle$, and $\lim_{w\to 0} (R/|w|^2) = 0$.

For an elliptic point p, $II_p(w)$ has a fixed sign. Therefore, for all (u, v) sufficiently near p, d has the same sign as $II_p(w)$; that is, all such (u, v) belong to the same side of $T_p(S)$.

For a hyperbolic point p, in each neighborhood of p there exist points (u, v) and $(\bar{u}, \bar{v})$ such that $II_p(w/|w|)$ and $II_p(\bar{w}/|\bar{w}|)$ have opposite signs (here $\bar{w} = \mathbf{x}_u\bar{u} + \mathbf{x}_v\bar{v}$); such points belong therefore to distinct sides of $T_p(S)$.

Q.E.D.

No such statement as Prop. 1 can be made in a neighborhood of a parabolic or a planar point. In the above examples of parabolic and planar points (cf. Examples 3 and 6 of Sec. 3-1) the surface lies on one side of the tangent plane and may have a line in common with this plane. In the following examples we shall show that an entirely different situation may occur.

Example 2. The "monkey saddle" (see Fig. 3-17) is given by

$$x = u, \qquad y = v, \qquad z = u^3 - 3v^2u.$$

A direct computation shows that at $(0, 0)$ the coefficients of the second fundamental form are $e = f = g = 0$; the point $(0, 0)$ is therefore a planar point. In any neighborhood of this point, however, there are points in both sides of its tangent plane.

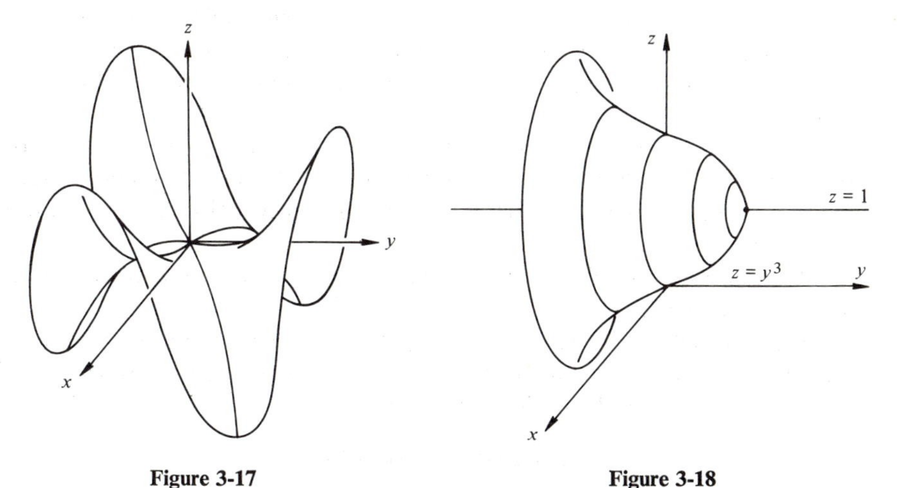

Figure 3-17 **Figure 3-18**

Example 3. Consider the surface obtained by rotating the curve $z = y^3$, $-1 < z < 1$, about the line $z = 1$ (see Fig. 3-18). A simple computation shows that the points generated by the rotation of the origin O are parabolic points. We shall omit this computation, because we shall prove shortly (Example 4) that the parallels and the meridians of a surface of revolution are lines of curvature; this, together with the fact that, for the points in question, the meridians (curves of the form $y = x^3$) have zero curvature and the parallel is a normal section, will imply the above statement.

Notice that in any neighborhood of such a parabolic point there exist points in both sides of the tangent plane.

The expression of the second fundamental form in local coordinates is particularly useful for the study of the asymptotic and principal directions. We first look at the asymptotic directions.

Let $\mathbf{x}(u, v)$ be a parametrization at $p \in S$, with $\mathbf{x}(0, 0) = p$, and let $e(u, v) = e$, $f(u, v) = f$, and $g(u, v) = g$ be the coefficients of the second fundamental form in this parametrization.

We recall that (see Def. 9 of Sec. 3-2) a connected regular curve C in the coordinate neighborhood of $\mathbf{x}$ is an asymptotic curve if and only if for any parametrization $\alpha(t) = \mathbf{x}(u(t), v(t))$, $t \in I$, of C we have $II(\alpha'(t)) = 0$, for all $t \in I$, that is, if and only if

$$e(u')^2 + 2fu'v' + g(v')^2 = 0, \qquad t \in I. \tag{7}$$

Because of that, Eq. (7) is called the *differential equation of the asymptotic curves*. In the next section we shall give a more precise meaning to this expression. For the time being, we want to draw from Eq. (7) only the following useful conclusion: *A necessary and sufficient condition for a parametrization in a neighborhood of a hyperbolic point* ($\mathrm{eg} - \mathrm{f}^2 < 0$) *to be such that the coordinate curves of the parametrization are asymptotic curves is that* $\mathrm{e} = \mathrm{g} = 0$.

In fact, if both curves $u = \text{const.}$, $v = v(t)$ and $u = u(t)$, $v = \text{const.}$ satisfy Eq. (7), we obtain $e = g = 0$. Conversely, if this last condition holds and $f \neq 0$, Eq. (7) becomes $fu'v' = 0$, which is clearly satisfied by the coordinate lines.

We shall now consider the principal directions, maintaining the notations already established.

A connected regular curve C in the coordinate neighborhood of $\mathbf{x}$ is a line of curvature if and only if for any parametrization $\alpha(t) = \mathbf{x}(u(t), v(t))$ of C, $t \in I$, we have (cf. Prop. 3 of Sec. 3-2)

$$dN(\alpha'(t)) = \lambda(t)\alpha'(t).$$

It follows that the functions $u'(t)$, $v'(t)$ satisfy the system of equations

$$\frac{fF - eG}{EG - F^2}u' + \frac{gF - fG}{EG - F^2}v' = \lambda u',$$

$$\frac{eF - fE}{EG - F^2}u' + \frac{fF - gE}{EG - F^2}v' = \lambda v'.$$

By eliminating λ in the above system, we obtain the *differential equation of the lines of curvature*,

$$(fE - eF)(u')^2 + (gE - eG)u'v' + (gF - fG)(v')^2 = 0,$$

which may be written, in a more symmetric way, as

$$\begin{vmatrix} (v')^2 & -u'v' & (u')^2 \\ E & F & G \\ e & f & g \end{vmatrix} = 0. \tag{8}$$

Using the fact that the principal directions are orthogonal to each other, it follows easily from Eq. (8) that *a necessary and sufficient condition for the coordinate curves of a parametrization to be lines of curvature in a neighborhood of a nonumbilical point is that* $F = f = 0$.

Example 4 (*Surfaces of Revolution*). Consider a surface of revolution parametrized by (cf. Example 4 of Sec. 2-3; we have changed f and g by φ and ψ, respectively)

$$\mathbf{x}(u, v) = (\varphi(v)\cos u, \varphi(v)\sin u, \psi(v)),$$
$$0 < u < 2\pi, \qquad a < v < b, \qquad \varphi(v) \neq 0.$$

The coefficients of the first fundamental form are given by

$$E = \varphi^2, \qquad F = 0, \qquad G = (\varphi')^2 + (\psi')^2.$$

It is convenient to assume that the rotating curve is parametrized by arc length, that is, that

$$(\varphi')^2 + (\psi')^2 = G = 1.$$

The computation of the coefficients of the second fundamental form is straightforward and yields

$$e = \frac{(\mathbf{x}_u, \mathbf{x}_v, \mathbf{x}_{uu})}{\sqrt{EG - F^2}} = \frac{1}{\sqrt{EG - F^2}}\begin{vmatrix} -\varphi\sin u & \varphi'\cos u & -\varphi\cos u \\ \varphi\cos u & \varphi'\sin u & -\varphi\sin u \\ 0 & \psi' & 0 \end{vmatrix}$$
$$= -\varphi\psi'$$
$$f = 0,$$
$$g = \psi'\varphi'' - \psi''\varphi'.$$

Since $F = f = 0$, we conclude that the parallels ($v = $ const.) and the meridians ($u =$ const.) of a surface of revolution are lines of curvature of such a surface (this fact was used in Example 3).

Because

$$K = \frac{eg - f^2}{EG - F^2} = -\frac{\psi'(\psi'\varphi'' - \psi''\varphi')}{\varphi}$$

and φ is always positive, it follows that the parabolic points are given by either $\psi' = 0$ (the tangent line to the generator curve is perpendicular to the axis of rotation) or $\varphi'\psi'' - \psi'\varphi'' = 0$ (the curvature of the generator curve is zero). A point which satisfies both conditions is a planar point, since these conditions imply that $e = f = g = 0$.

It is convenient to put the Gaussian curvature in still another form. By differentiating $(\varphi')^2 + (\psi')^2 = 1$ we obtain $\varphi'\varphi'' = -\psi'\psi''$. Thus,

$$K = -\frac{\psi'(\psi'\varphi'' - \psi''\varphi')}{\varphi} = -\frac{(\psi')^2\varphi'' + (\varphi')^2\varphi''}{\varphi} = -\frac{\varphi''}{\varphi}. \tag{9}$$

Equation (9) is a convenient expression for the Gaussian curvature of a surface of revolution. It can be used, for instance, to determine the surfaces of revolution of constant Gaussian curvature (cf. Exercise 7).

To compute the principal curvatures, we first make the following general observation: *If a parametrization of a regular surface is such that $F = f = 0$, then the principal curvatures are given by e/E and g/G.* In fact, in this case, the Gaussian and the mean curvatures are given by (cf. Eqs. (4) and (5))

$$K = \frac{eg}{EG}, \qquad H = \frac{1}{2}\frac{eG - gE}{EG}.$$

Since K is the product and $2H$ is the sum of the principal curvatures, our assertion follows at once.

Thus, the principal curvatures of a surface of revolution are given by

$$\frac{e}{E} = -\frac{\psi'\varphi}{\varphi^2} = -\frac{\psi'}{\varphi}, \qquad \frac{g}{G} = \psi'\varphi'' - \psi''\varphi'; \tag{10}$$

hence, the mean curvature of such a surface is

$$H = \frac{1}{2}\frac{-\psi' + \varphi(\psi'\varphi'' - \psi''\varphi')}{\varphi}. \tag{11}$$

Example 5. Very often a surface is given as the graph of a differentiable function (cf. Prop. 1, Sec. 2-2) $z = h(x, y)$, where (x, y) belong to an open set $U \subset R^2$. It is, therefore, convenient to have at hand formulas for the relevant concepts in this case. To obtain such formulas let us parametrize the surface by

$$\mathbf{x}(u, v) = (u, v, h(u, v)), \qquad (u, v) \in U,$$

where $u = x$, $v = y$. A simple computation shows that

$$\mathbf{x}_u = (1, 0, h_u), \qquad \mathbf{x}_v = (0, 1, h_v), \qquad \mathbf{x}_{uu} = (0, 0, h_{uu}),$$
$$\mathbf{x}_{uv} = (0, 0, h_{uv}), \qquad \mathbf{x}_{vv} = (0, 0, h_{vv}).$$

Thus

$$N(x, y) = \frac{(-h_x, -h_y, 1)}{(1 + h_x^2 + h_y^2)^{1/2}}$$

is a unit normal field on the surface, and the coefficients of the second fundamental form in this orientation are given by

$$e = \frac{h_{xx}}{(1 + h_x^2 + h_y^2)^{1/2}},$$
$$f = \frac{h_{xy}}{(1 + h_x^2 + h_y^2)^{1/2}},$$
$$g = \frac{h_{yy}}{(1 + h_x^2 + h_y^2)^{1/2}}.$$

From the above expressions, any needed formula can be easily computed. For instance, from Eqs. (4) and (5) we obtain the Gaussian and mean curvatures:

$$K = \frac{h_{xx}h_{yy} - h_{xy}^2}{(1 + h_x^2 + h_y^2)^2},$$
$$H = \frac{(1 + h_x^2)h_{yy} - 2h_xh_yh_{xy} + (1 + h_y^2)h_{xx}}{(1 + h_x^2 + h_y^2)^{3/2}}.$$

There is still another, perhaps more important, reason to study surfaces given by $z = h(x, y)$. It comes from the fact that locally any surface is the graph of a differentiable function (cf. Prop. 3, Sec. 2-2). Given a point p of a surface S, we can choose the coordinate axis of R^3 so that the origin O of the coordinates is at p and the z axis is directed along the positive normal of S at p (thus, the xy plane agrees with $T_p(S)$). It follows that a neighborhood of p in S can be represented in the form $z = h(x, y)$, $(x, y) \in U \subset R^2$, where U is an open set and h is a differentiable function (cf. Prop. 3, Sec. 2-2), with $h(0, 0) = p$, $h_x(0, 0) = 0$, $h_y(0, 0) = 0$ (Fig. 3-19).

The second fundamental form of S at p applied to the vector $(x, y) \in R^2$ becomes, in this case,

$$h_{xx}(0, 0)x^2 + 2h_{xy}(0, 0)xy + h_{yy}(0, 0)y^2.$$

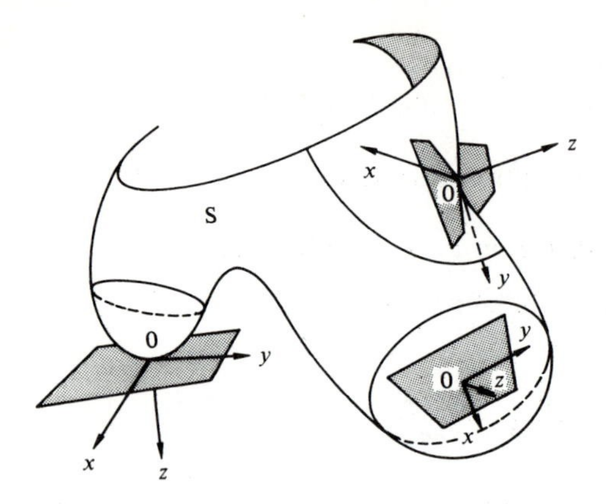

Figure 3-19. Each point of S has a neighborhood that can be written as $z = h(x, y)$.

In elementary calculus of two variables, the above quadratic form is known as the *Hessian* of h at $(0, 0)$. Thus, the Hessian of h at $(0, 0)$ is the second fundamental form of S at p.

Let us apply the above considerations to give a geometric interpretation of the Dupin indicatrix. With the notation as above, let $\epsilon > 0$ be a small number such that

$$C = \{(x, y) \in T_p(S); h(x, y) = \epsilon\}$$

is a regular curve (we may have to change the orientation of the surface to achieve $\epsilon > 0$). We want to show that if p is not a planar point, the curve C is "approximately" similar to the Dupin indicatrix of S at p (Fig. 3-20).

To see this, let us assume further that the x and y axes are directed along the principal directions, with the x axis along the direction of maximum principal curvature. Thus, $f = h_{xy}(0, 0) = 0$ and

$$k_1(p) = \frac{e}{E} = h_{xx}(0, 0), \qquad k_2(p) = \frac{g}{G} = h_{yy}(0, 0).$$

By developing $h(x, y)$ into a Taylor's expansion about $(0, 0)$, and taking into account that $h_x(0, 0) = 0 = h_y(0, 0)$, we obtain

$$\begin{aligned} h(x, y) &= \tfrac{1}{2}(h_{xx}(0, 0)x^2 + h_{xy}(0, 0)xy + h_{yy}(0, 0)y^2) + R \\ &= \tfrac{1}{2}(k_1 x^2 + k_2 y^2) + R, \end{aligned}$$

where

$$\lim_{(x,y)\to(0,0)} \frac{R}{x^2 + y^2} = 0.$$

Thus, the curve C is given by

$$k_1 x^2 + k_2 y^2 + 2R = 2\epsilon.$$

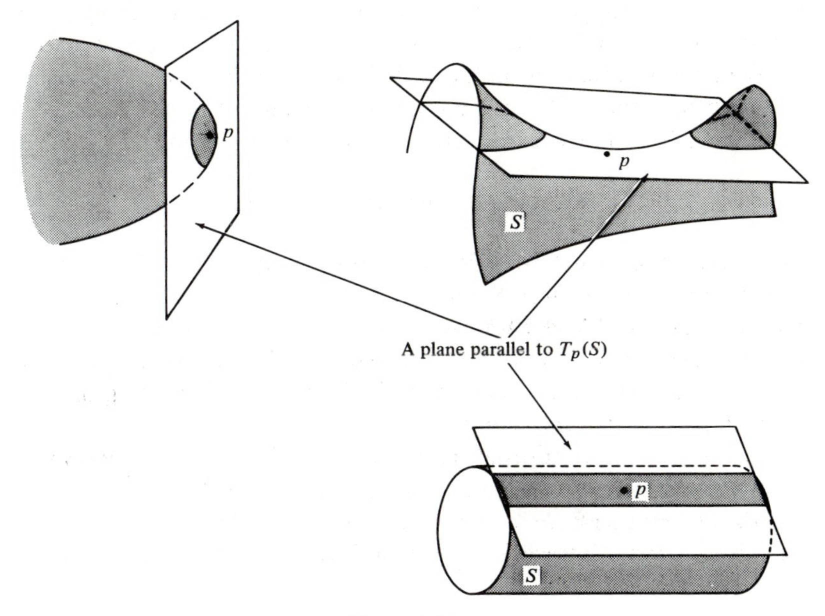

Figure 3-20

Now, if p is not a planar point, we can consider $k_1x^2 + k_2y^2 = 2\epsilon$ as a first-order approximation of C. By using the similarity transformation,

$$x = \bar{x}\sqrt{2\epsilon}, \qquad y = \bar{y}\sqrt{2\epsilon},$$

we have that $k_1x^2 + k_2y^2 = 2\epsilon$ is transformed into the curve

$$k_1\bar{x}^2 + k_2\bar{y}^2 = 1,$$

which is the Dupin indicatrix at p. This means that *if* p *is a nonplanar point, the intersection with* S *of a plane parallel to* $T_p(S)$ *and close to* p *is, in a first-order approximation, a curve similar to the Dupin indicatrix at* p.

If p is a planar point, this interpretation is no longer valid (cf. Exercise 11).

To conclude this section we shall give a geometrical interpretation of the Gaussian curvature in terms of the Gauss map $N: S \to S^2$. Actually this was how Gauss himself introduced this curvature.

To do this, we first need a definition.

Let S and $\bar{S}$ be two oriented regular surfaces. Let $\varphi: S \to \bar{S}$ be a differentiable map and assume that for some $p \in S$, $d\varphi_p$ is nonsingular. We say that φ is *orientation-preserving* at p if given a positive basis $\{w_1, w_2\}$ in $T_p(S)$, then $\{d\varphi_p(w_1), d\varphi_p(w_2)\}$ is a positive basis in $T_{\varphi(p)}(\bar{S})$. If $\{d\varphi_p(w_1), d\varphi_p(w_2)\}$ is not a positive basis, we say that φ is *orientation-reversing* at p.

We now observe that both S and the unit sphere S^2 are embedded in R^3. Thus, an orientation N on S induces an orientation N in S^2. Let $p \in S$ be such that dN_p is nonsingular. Since for a basis $\{w_1, w_2\}$ in $T_p(S)$

$$dN_p(w_1) \wedge dN_p(w_2) = \det(dN_p)(w_1 \wedge w_2) = Kw_1 \wedge w_2,$$

the Gauss map N will be orientation-preserving at $p \in S$ if $K(p) > 0$ and orientation-reversing at $p \in S$ if $K(p) < 0$. Intuitively, this means the following (Fig. 3-21): An orientation of $T_p(S)$ induces an orientation of small closed curves in S around p; the image by N of these curves will have the same or the opposite orientation to the initial one, depending on whether p is an elliptic or hyperbolic point, respectively.

To take this fact into account we shall make the convention that the area of a region contained in a connected neighborhood V, where $K \neq 0$, and the area of its image by N have the same sign if $K > 0$ in V, and opposite signs if $K < 0$ in V (since V is connected, K does not change sign in V).

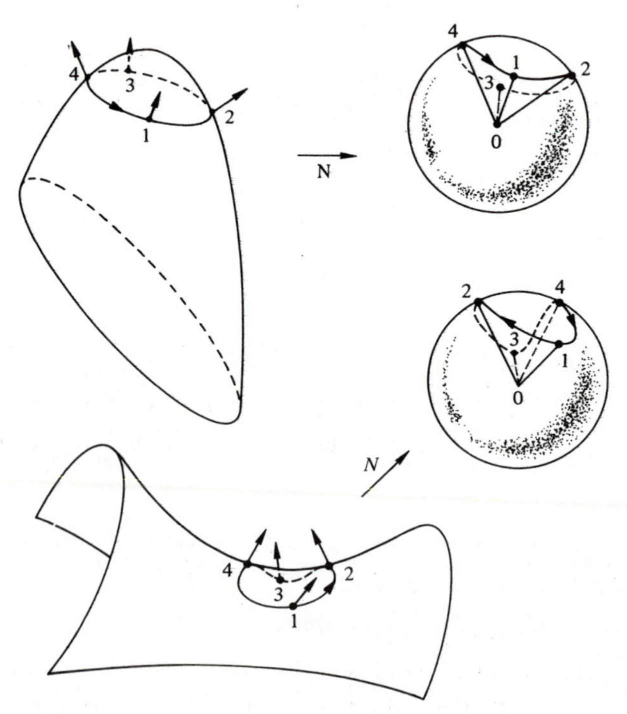

Figure 3-21. The Gauss map preserves orientation at an elliptic point and reverses it at a hyperbolic point.

Now we can state the promised geometric interpretation of the Gaussian curvature K, for $K \neq 0$.

PROPOSITION 2. *Let* p *be a point of a surface* S *such that the Gaussian curvature* $\mathrm{K(p)} \neq 0$, *and let* V *be a connected neighborhood of* p *where* K *does not change sign. Then*

$$\mathrm{K(p)} = \lim_{\mathrm{A} \to 0} \frac{\mathrm{A}'}{\mathrm{A}},$$

where A *is the area of a region* $\mathrm{B} \subset \mathrm{V}$ *containing* p, A′ *is the area of the image of* B *by the Gauss map* $\mathrm{N}\colon \mathrm{S} \longrightarrow \mathrm{S}^2$, *and the limit is taken through a sequence of regions* $\mathrm{B_n}$ *that converges to* p, *in the sense that any sphere around* p *contains all* $\mathrm{B_n}$, *for* n *sufficiently large.*

Proof. The area A of B is given by (cf. Sec. 2-5)

$$A = \iint_R |\mathbf{x}_u \wedge \mathbf{x}_v| \, du \, dv,$$

where $\mathbf{x}(u, v)$ is a parametrization in p, whose coordinate neighborhood contains V (V can be assumed to be sufficiently small) and R is the region in the uv plane corresponding to B. The area A' of $N(B)$ is

$$A' = \iint_R |N_u \wedge N_v| \, du \, dv.$$

Using Eq. (1), the definition of K, and the above convention, we can write

$$A' = \iint_R K|\mathbf{x}_u \wedge \mathbf{x}_v| \, du \, dv. \tag{12}$$

Going to the limit and denoting also by R the area of the region R, we obtain

$$\lim_{A \to 0} \frac{A'}{A} = \lim_{R \to 0} \frac{A'/R}{A/R} = \frac{\lim\limits_{R \to 0} (1/R) \iint_R K|\mathbf{x}_u \wedge \mathbf{x}_v| \, du \, dv}{\lim\limits_{R \to 0} (1/R) \iint_R |\mathbf{x}_u \wedge \mathbf{x}_v| \, du \, dv}$$

$$= \frac{K|\mathbf{x}_u \wedge \mathbf{x}_v|}{|\mathbf{x}_u \wedge \mathbf{x}_v|} = K$$

(notice that we have used the mean value theorem for double integrals), and this proves the proposition. **Q.E.D.**

Remark. Comparing the proposition with the expression of the curvature

$$k = \lim_{s \to 0} \frac{\sigma}{s}$$

of a plane curve C at p (here s is the arc length of a small segment of C containing p, and σ is the arc length of its image in the indicatrix of tangents; cf. Exercise 3 of Sec. 1-5), we see that the Gaussian curvature K is the analogue, for surfaces, of the curvature k of plane curves.

EXERCISES

1. Show that at the origin $(0, 0, 0)$ of the hyperboloid $z = axy$ we have $K = -a^2$ and $H = 0$.

***2.** Determine the asymptotic curves and the lines of curvature of the helicoid $x = v \cos u$, $y = v \sin u$, $z = cu$, and show that its mean curvature is zero.

***3.** Determine the asymptotic curves of the catenoid

$$\mathbf{x}(u, v) = (\cosh v \cos u, \cosh v \sin u, v).$$

4. Determine the asymptotic curves and the lines of curvature of $z = xy$.

5. Consider the parametrized surface (Enneper's surface)

$$\mathbf{x}(u, v) = \left(u - \frac{u^3}{3} + uv^2, v - \frac{v^3}{3} + vu^2, u^2 - v^2\right)$$

and show that

a. The coefficients of the first fundamental form are

$$E = G = (1 + u^2 + v^2)^2, \qquad F = 0.$$

b. The coefficients of the second fundamental form are

$$e = 2, \qquad g = -2, \qquad f = 0.$$

c. The principal curvatures are

$$k_1 = \frac{2}{(1 + u^2 + v^2)^2}, \qquad k_2 = -\frac{2}{(1 + u^2 + v^2)^2}.$$

d. The lines of curvature are the coordinate curves.

e. The asymptotic curves are $u + v = \text{const.}$, $u - v = \text{const.}$

6. (*A Surface with $K \equiv -1$; the Pseudosphere.*)

***a.** Determine an equation for the plane curve C, which is such that the segment of the tangent line between the point of tangency and some line r in the plane, which does not meet the curve, is constantly equal to 1 (this curve is called the *tractrix*; see Fig. 1-9).

b. Rotate the tractrix C about the line r; determine if the "surface" of revolution thus obtained (the *pseudosphere*; see Fig. 3-22) is regular and find out a parametrization in a neighborhood of a regular point.

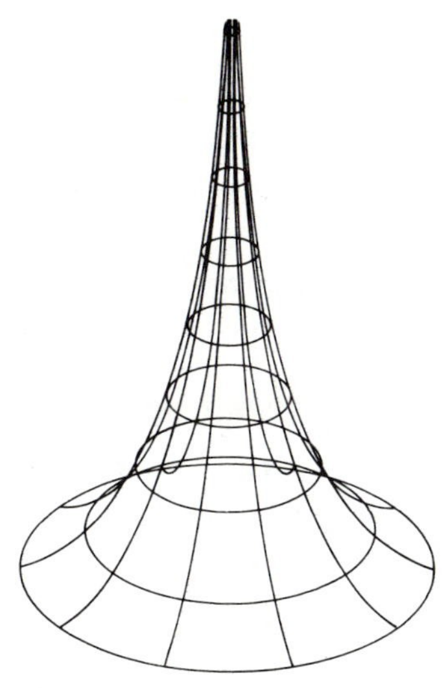

Figure 3-22. The pseudosphere.

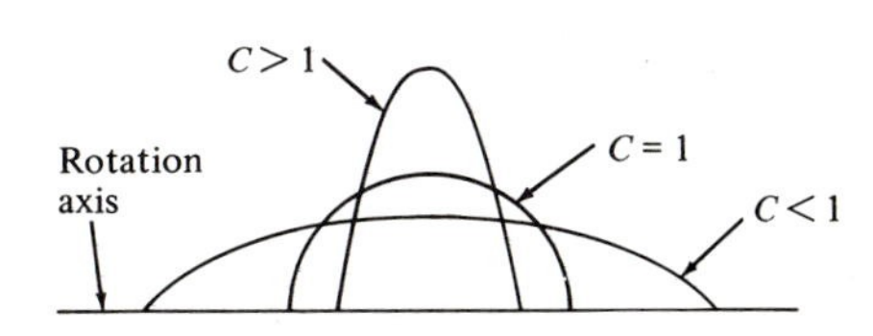

Figure 3-23

c. Show that the Gaussian curvature of any regular point of the pseudosphere is -1.

7. (*Surfaces of Revolution with Constant Curvature.*) $(\varphi(v)\cos u, \varphi(v)\sin u, \psi(v))$ is given as a surface of revolution with constant Gaussian curvature K. To determine the functions φ and ψ, choose the parameter v in such a way that $(\varphi')^2 + (\psi')^2 = 1$ (geometrically, this means that v is the arc length of the generating curve $(\varphi(v), \psi(v))$). Show that

a. φ satisfies $\varphi'' + K\varphi = 0$ and ψ is given by $\psi = \int \sqrt{1-(\varphi')^2}\,dv$; thus, $0 < u < 2\pi$, and the domain of v is such that the last integral makes sense.

b. All surfaces of revolution with constant curvature $K = 1$ which intersect perpendicularly the plane xOy are given by

$$\varphi(v) = C\cos v, \qquad \psi(v) = \int_0^v \sqrt{1 - C^2\sin^2 v}\,dv,$$

where C is a constant ($C = \varphi(0)$). Determine the domain of v and draw a rough sketch of the profile of the surface in the xz plane for the cases $C = 1$, $C > 1$, $C < 1$. Observe that $C = 1$ gives a sphere (Fig. 3-23).

c. All surfaces of revolution with constant curvature $K = -1$ may be given by one of the following types:

1. $\varphi(v) = C\cosh v,$

$$\psi(v) = \int_0^v \sqrt{1 - C^2\sinh^2 v}\,dv.$$

2. $\varphi(v) = C \sinh v,$

$\psi(v) = \int_0^v \sqrt{1 - C^2 \cosh^2 v}\, dv.$

3. $\varphi(v) = e^v,$

$\psi(v) = \int_0^v \sqrt{1 - e^{2v}}\, dv.$

Determine the domain of v and draw a rough sketch of the profile of the surface in the xz plane.

d. The surface of type 3 in part c is the pseudosphere of Exercise 6.

e. The only surfaces of revolution with $K \equiv 0$ are the right circular cylinder, the right circular cone, and the plane.

8. (*Contact of Order ≥ 2 of Surfaces.*) Two surfaces S and $\bar{S}$, with a common point p, have *contact of order* ≥ 2 at p if there exist parametrizations $\mathbf{x}(u, v)$ and $\bar{\mathbf{x}}(u, v)$ in p of S and $\bar{S}$, respectively, such that

$$\mathbf{x}_u = \bar{\mathbf{x}}_u, \qquad \mathbf{x}_v = \bar{\mathbf{x}}_v, \qquad \mathbf{x}_{uu} = \bar{\mathbf{x}}_{uu}, \qquad \mathbf{x}_{uv} = \bar{\mathbf{x}}_{uv}, \qquad \mathbf{x}_{vv} = \bar{\mathbf{x}}_{vv}$$

at p. Prove the following:

***a.** Let S and $\bar{S}$ have contact of order ≥ 2 at p; $\mathbf{x}\colon U \longrightarrow S$ and $\bar{\mathbf{x}}\colon U \longrightarrow \bar{S}$ be arbitrary parametrizations in p of S and $\bar{S}$, respectively; and $f\colon V \subset R^3 \longrightarrow R$ be a differentiable function in a neighborhood V of p in R^3. Then the partial derivatives of order ≤ 2 of $f \circ \bar{\mathbf{x}}\colon U \longrightarrow R$ are zero in $\bar{\mathbf{x}}^{-1}(p)$ if and only if the partial derivatives of order ≤ 2 of $f \circ \mathbf{x}\colon U \longrightarrow R$ are zero in $\mathbf{x}^{-1}(p)$.

***b.** Let S and $\bar{S}$ have contact of order ≥ 2 at p. Let $z = f(x, y)$, $z = \bar{f}(x, y)$ be the equations, in a neighborhood of p, of S and $\bar{S}$, respectively, where the xy plane is the common tangent plane at $p = (0, 0)$. Then the function $f(x, y) - \bar{f}(x, y)$ has all partial derivatives of order ≤ 2, at $(0, 0)$, equal to zero.

c. Let p be a point in a surface $S \subset R^3$. Let $Oxyz$ be a cartesian coordinate system for R^3 such that $O = p$ and the xy plane is the tangent plane of S at p. Show that the paraboloid

$$z = \tfrac{1}{2}(x^2 f_{xx} + 2xy f_{xy} + y^2 f_{yy}), \tag{$*$}$$

obtained by neglecting third- and higher-order terms in the Taylor development around $p = (0, 0)$, has contact of order ≥ 2 at p with S (the surface $(*)$ is called the *osculating paraboloid* of S at p).

***d.** If a paraboloid (the degenerate cases of plane and parabolic cylinder are included) has contact of order ≥ 2 with a surface S at p, then it is the osculating paraboloid of S at p.

e. If two surfaces have contact of order ≥ 2 at p, then the osculating paraboloids of S and $\bar{S}$ at p coincide. Conclude that the Gaussian and mean curvatures of S and $\bar{S}$ at p are equal.

f. The notion of contact of order ≥ 2 is invariant by diffeomorphisms of R^3;

that is, if S and $\bar{S}$ have contact of order ≥ 2 at p and $\varphi\colon R^3 \longrightarrow R^3$ is a diffeomorphism, then $\varphi(S)$ and $\varphi(\bar{S})$ have contact of order ≥ 2 at $\varphi(p)$.

g. If S and $\bar{S}$ have contact of order ≥ 2 at p, then

$$\lim_{r \to 0} \frac{d}{r^2} = 0,$$

where d is the length of the segment cut by the surfaces in a straight line normal to $T_p(S) = T_p(\bar{S})$, which is at a distance r from p.

9. (*Contact of Curves.*) Define contact of order $\geq n$ (n integer ≥ 1) for regular curves in R^3 with a common point p and prove that

a. The notion of contact of order $\geq n$ is invariant by diffeomorphisms.

b. Two curves have contact of order ≥ 1 at p if and only if they are tangent at p.

10. (*Contact of Curves and Surfaces.*) A curve C and a surface S, which have a common point p, have contact of order $\geq n$ (n integer ≥ 1) at p if there exists a curve $\bar{C} \subset S$ passing through p such that C and $\bar{C}$ have contact of order $\geq n$ at p. Prove that

a. If $f(x, y, z) = 0$ is a representation of a neighborhood of p in S and $\alpha(t) = (x(t), y(t), z(t))$ is a parametrization of C in p, with $\alpha(0) = p$, then C and S have contact of order $\geq n$ if and only if

$$f(x(0), y(0), z(0)) = 0, \qquad \frac{df}{dt} = 0, \ldots, \frac{d^n f}{dt^n} = 0,$$

where the derivatives are computed for $t = 0$.

b. If a plane has contact of order ≥ 2 with a curve C at p, then this is the osculating plane of C at p.

c. If a sphere has contact of order ≥ 3 with a curve C at p, and $\alpha(s)$ is a parametrization by arc length of this curve, with $\alpha(0) = p$, then the center of the sphere is given by

$$\alpha(0) + \frac{1}{k}n + \frac{k'}{k^2\tau}b.$$

Such a sphere is called the *osculating sphere* of C at p.

11. Consider the monkey saddle S of Example 2. Construct the Dupin indicatrix at $p = (0, 0, 0)$ using the definition of Sec. 3-2, and compare it with the curve obtained as the intersection of S with a plane parallel to $T_p(S)$ and close to p. Why are they not "approximately similar" (cf. Example 5 of Sec. 3-3)? Go through the argument of Example 5 of Sec. 3-3 and point out where it breaks down.

12. Consider the parametrized surface

$$\mathbf{x}(u, v) = \left(\sin u \cos v, \sin u \sin v, \cos u + \log \tan \frac{u}{2} + \varphi(v)\right),$$

where φ is a differentiable function. Prove that

a. The curves $v =$ const. are contained in planes which pass through the z axis and intersect the surface under a constant angle θ given by

$$\cos\theta = \frac{\varphi'}{\sqrt{1 + (\varphi')^2}}.$$

Conclude that the curves $v =$ const. are lines of curvature of the surface.

b. The length of the segment of a tangent line to a curve $v =$ const., determined by its point of tangency and the z axis, is constantly equal to 1. Conclude that the curves $v =$ const. are tractrices (cf. Exercise 6).

13. Let $F: R^3 \longrightarrow R^3$ be the map (a similarity) defined by $F(p) = cp$, $p \in R^3$, c a positive constant. Let $S \subset R^3$ be a regular surface and set $F(S) = \bar{S}$. Show that $\bar{S}$ is a regular surface, and find formulas relating the Gaussian and mean curvatures, K and H, of S with the Gaussian and mean curvatures, $\bar{K}$ and $\bar{H}$, of $\bar{S}$.

14. Consider the surface obtained by rotating the curve $y = x^3$, $-1 < x < 1$, about the line $x = 1$. Show that the points obtained by rotation of the origin $(0, 0)$ of the curve are planar points of the surface.

***15.** Give an example of a surface which has an isolated parabolic point p (that is, no other parabolic point is contained in some neighborhood of p).

***16.** Show that a surface which is compact (i.e., it is bounded and closed in R^3) has an elliptic point.

17. Define Gaussian curvature for a nonorientable surface. Can you define mean curvature for a nonorientable surface?

18. Show that the Möbius strip of Fig. 3-1 can be parametrized by

$$\mathbf{x}(u, v) = \left(\left(2 - v\sin\frac{u}{2}\right)\sin u, \left(2 - v\sin\frac{u}{2}\right)\cos u, v\cos\frac{u}{2}\right)$$

and that its Gaussian curvature is

$$K = -\frac{1}{\{\frac{1}{4}v^2 + (2 - v\sin(u/2))^2\}^2}.$$

***19.** Obtain the asymptotic curves of the one-sheeted hyperboloid $x^2 + y^2 - z^2 = 1$.

***20.** Determine the umbilical points of the elipsoid

$$\frac{x^2}{a^2} + \frac{y^2}{b^2} + \frac{z^2}{c^2} = 1.$$

***21.** Let S be a surface with orientation N. Let $V \subset S$ be an open set in S and let $f: V \subset S \longrightarrow R$ be any nowhere-zero differentiable function in V. Let v_1 and v_2 be two differentiable (tangent) vector fields in V such that at each point of V, v_1 and v_2 are orthonormal and $v_1 \wedge v_2 = N$.

a. Prove that the Gaussian curvature K of V is given by

$$K = \frac{\langle d(fN)(v_1) \wedge d(fN)(v_2), fN \rangle}{f^3}.$$

The virtue of this formula is that by a clever choice of f we can often simplify the computation of K, as illustrated in part b.

b. Apply the above result to show that if f is the restriction of

$$\sqrt{\frac{x^2}{a^4} + \frac{y^2}{b^4} + \frac{z^2}{c^4}}$$

to the ellipsoid

$$\frac{x^2}{a^2} + \frac{y^2}{b^2} + \frac{z^2}{c^2} = 1,$$

then the Gaussian curvature of the ellipsoid is

$$K = \frac{1}{a^2b^2c^2}\frac{1}{f^4}.$$

22. (*The Hessian*). Let $h\colon S \to R$ be a differentiable function on a surface S, and let $p \in S$ be a critical point of h (i.e., $dh_p = 0$). Let $w \in T_p(S)$ and let

$$\alpha\colon (-\epsilon, \epsilon) \to S$$

be a parametrized curve with $\alpha(0) = p$, $\alpha'(0) = w$. Set

$$H_p h(w) = \left.\frac{d^2(h \circ \alpha)}{dt^2}\right|_{t=0}.$$

a. Let $\mathbf{x}\colon U \to S$ be a parametrization of S at p, and show that (the fact that p is a critical point of h is essential here)

$$H_p h(u'\mathbf{x}_u + v'\mathbf{x}_v) = h_{uu}(p)(u')^2 + 2h_{uv}(p)u'v' + h_{vv}(p)(v')^2.$$

Conclude that $H_p h\colon T_p(S) \to R$ is a well-defined (i.e., it does not depend on the choice of α) quadratic form on $T_p(S)$. $H_p h$ is called the *Hessian* of h at p.

b. Let $h\colon S \to R$ be the height function of S relative to $T_p(S)$; that is, $h(q) = \langle q - p, N(p)\rangle$, $q \in S$. Verify that p is a critical point of h and thus that the Hessian $H_p h$ is well defined. Show that if $w \in T_p(S)$, $|w| = 1$, then

$$H_p h(w) = \text{normal curvature at } p \text{ in the direction of } w.$$

Conclude that *the Hessian at* p *of the height function relative to* $T_p(S)$ *is the second fundamental form of* S *at* p.

23. (*Morse Functions on Surfaces.*) A critical point $p \in S$ of a differentiable function $h\colon S \to R$ is *nondegenerate* if the self-adjoint linear map $A_p h$ associated to the quadratic form $H_p h$ (cf. the appendix to Chap. 3) is nonsingular (here $H_p h$ is the Hessian of h at p; cf. Excercise 22). Otherwise, p is a *degenerate* critical point. A differentiable function on S is a *Morse function* if all its critical points are nondegenerate. Let $h_r: S \subset R^3 \longrightarrow R$ be the distance function from S to r; i.e.,

$$h_r(q) = \sqrt{\langle q - r, q - r \rangle}, \qquad q \in S, \qquad r \in R^3, \qquad r \notin S.$$

a. Show that $p \in S$ is a critical point of h_r if and only if the straight line pr is normal to S at p.

b. Let p be a critical point of $h_r: S \longrightarrow R$. Let $w \in T_p(S)$, $|w| = 1$, and let $\alpha: (-\epsilon, \epsilon) \longrightarrow S$ be a curve parametrized by arc length with $\alpha(0) = p$, $\alpha'(s) = w$. Prove that

$$H_p h_r(w) = \frac{1}{h_r(p)} - k_n,$$

where k_n is the normal curvature at p along the direction of w. Conclude that the orthonormal basis $\{e_1, e_2\}$, where e_1 and e_2 are along the principal directions of $T_p(S)$, diagonalizes the self-adjoint linear map $A_p h_r$. Conclude further that p is a degenerate critical point of h_r if and only if either $h_r(p) = 1/k_1$ or $h_r(p) = 1/k_2$, where k_1 and k_2 are the principal curvatures at p.

c. Show that the set

$$B = \{r \in R^3;\ h_r \text{ is a Morse function}\}$$

is an open and dense set in R^3; here dense in R^3 means that in each neighborhood of a given point of R^3 there exists a point of B (*this shows that on any regular surface there are "many" Morse functions*).

24. (*Local Convexity and Curvature*). A surface $S \subset R^3$ is *locally convex* at a point $p \in S$ if there exists a neighborhood $V \subset S$ of p such that V is contained in one of the closed half-spaces determined by $T_p(S)$ in R^3. If, in addition, V has only one common point with $T_p(S)$, then S is called *strictly locally convex* at p.

a. Prove that S is strictly locally convex at p if the principal curvatures of S at p are nonzero with the same sign (that is, the Gaussian curvature $K(p)$ satisfies $K(p) > 0$).

b. Prove that if S is locally convex at p, then the principal curvatures at p do not have different signs (thus, $K(p) \geq 0$).

c. To show that $K \geq 0$ does not imply local convexity, consider the surface $f(x, y) = x^3(1 + y^2)$, defined in the open set $U = \{(x, y) \in R^2; y^2 < \frac{1}{2}\}$. Show that the Gaussian curvature of this surface is nonnegative on U and yet the surface is not locally convex at $(0, 0) \in U$ (a deep theorem, due to R. Sacksteder, implies that such an example cannot be extended to the entire R^2 if we insist on keeping the curvature nonnegative; cf. Remark 3 of Sec. 5-6).

***d.** The example of part c is also very special in the following local sense. Let p be a point in a surface S, and assume that there exists a neighborhood $V \subset S$

of p such that the principal curvatures on V do not have different signs (this does not happen in the example of part c). Prove that S is locally convex at p.

3-4. Vector Fields†

In this section we shall use the fundamental theorems of ordinary differential equations (existence, uniqueness, and dependence on the initial conditions) to prove the existence of certain coordinate systems on surfaces.

If the reader is willing to assume the results of Corollaries 2, 3, and 4 at the end of this section (which can be understood without reading the section), this material may be omitted on a first reading.

We shall begin with a geometric presentation of the material on differential equations that we intend to use.

A *vector field* in an open set $U \subset R^2$ is a map which assigns to each $q \in U$ a vector $w(q) \in R^2$. The vector field w is said to be *differentiable* if writing $q = (x, y)$ and $w(q) = (a(x, y), b(x, y))$, the functions a and b are differentiable functions in U.

Geometrically, the definition corresponds to assigning to each point $(x, y) \in U$ a vector with coordinates $a(x, y)$ and $b(x, y)$ which vary differentiably with (x, y) (Fig. 3-24).

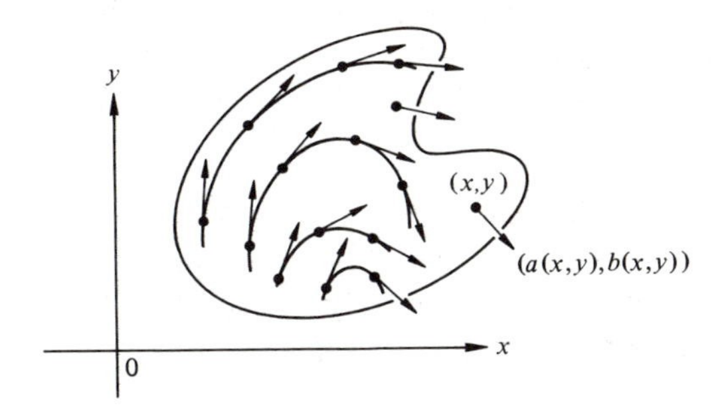

Figure 3-24

In what follows we shall consider only differentiable vector fields.

In Fig. 3-25 some examples of vector fields are shown.

Given a vector field w, it is natural to ask whether there exists a *trajectory* of this field, that is, whether there exists a differentiable parametrized curve $\alpha(t) = (x(t), y(t))$, $t \in I$, such that $\alpha'(t) = w(\alpha(t))$.

For instance, a trajectory, passing through the point (x_0, y_0), of the vector field $w(x, y) = (x, y)$ is the straight line $\alpha(t) = (x_0 e^t, y_0 e^t)$, $t \in R$, and a trajectory of $w(x, y) = (y, -x)$, passing through (x_0, y_0), is the circle $\beta(t) = (r \sin t, r \cos t)$, $t \in R$, $r^2 = x_0^2 + y_0^2$.

---

†This section may be omitted on a first reading.

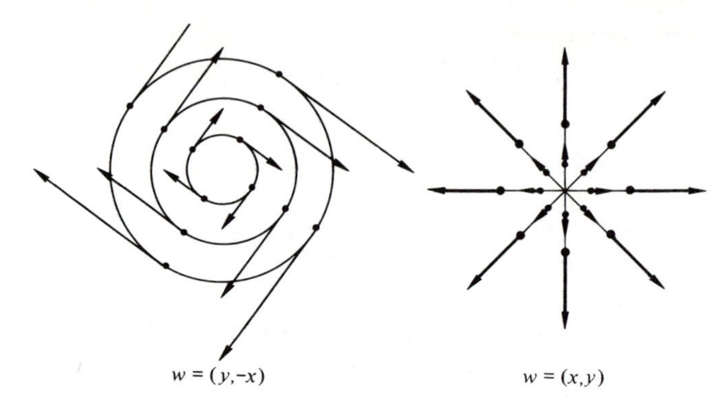

Figure 3-25

In the language of ordinary differential equations, one says that the vector field w determines a system of *differential equations*,

$$\frac{dx}{dt} = a(x, y),$$
$$\frac{dy}{dt} = b(x, y), \tag{1}$$

and that a trajectory of w is a *solution* to Eq. (1).

The fundamental theorem of (local) existence and uniqueness of solutions of Eq. (1) is equivalent to the following statement on trajectories (in what follows, the letters I and J will denote open intervals of the line R, containing the origin $0 \in R$).

THEOREM 1. *Let* w *be a vector field in an open set* $U \subset R^2$. *Given* $p \in U$, *there exists a trajectory* $\alpha: I \to U$ *of* w *(i.e.,* $\alpha'(t) = w(\alpha(t))$, $t \in I$) *with* $\alpha(0) = p$. *This trajectory is unique in the following sense: Any other trajectory* $\beta: J \to U$ *with* $\beta(0) = p$ *agrees with* α *in* $I \cap J$.

An important complement to Theorem 1 is the fact that the trajectory passing through p "varies differentiably with p." This idea can be made precise as follows.

THEOREM 2. *Let* w *be a vector field in an open set* $U \subset R^2$. *For each* $p \in U$ *there exist a neighborhood* $V \subset U$ *of* p, *an interval* I, *and a mapping* $\alpha: V \times I \to U$ *such that*

1. *For a fixed* $q \in V$, *the curve* $\alpha(q, t)$, $t \in I$, *is the trajectory of* w *passing through* q; *that is,*

$$\alpha(q, 0) = q, \qquad \frac{\partial \alpha}{\partial t}(q, t) = w(\alpha(q, t)).$$

2. *α is differentiable.*

Geometrically Theorem 2 means that all trajectories which pass, for $t = 0$, in a certain neighborhood V of p may be "collected" into a single differentiable map. It is in this sense that we say that the trajectories depend differentiably on p (Fig. 3-26).

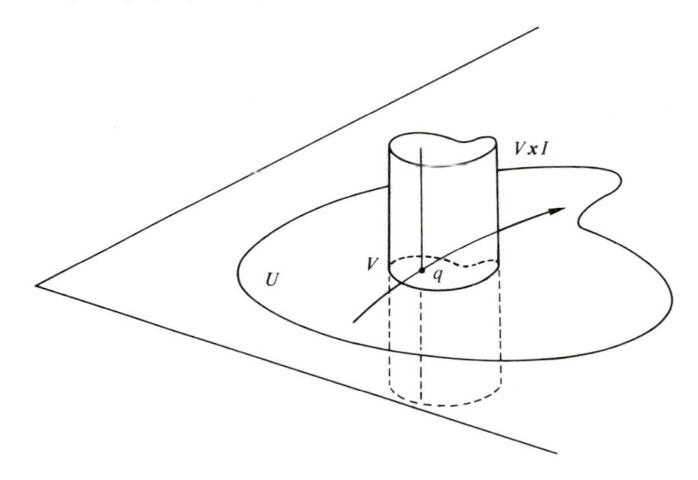

Figure 3-26

The map α is called the *(local) flow* of w at p.

Theorems 1 and 2 will be assumed in this book; for a proof, one can consult, for instance, W. Hurewicz, *Lectures on Ordinary Differential Equations*, M.I.T. Press, Cambridge, Mass., 1958, Chap. 2. For our purposes, we need the following consequence of these theorems.

LEMMA. *Let* w *be a vector field in an open set* $U \subset R^2$ *and let* $p \in U$ *be such that* $w(p) \neq 0$. *Then there exist a neighborhood* $W \subset U$ *of* p *and a differentiable function* $f: W \to R$ *such that* f *is constant along each trajectory of* w *and* $df_q \neq 0$ *for all* $q \in W$.

Proof. Choose a cartesian coordinate system in R^2 such that $p = (0, 0)$ and $w(p)$ is in the direction of the x axis. Let $\alpha: V \times I \to U$ be the local flow at p, $V \subset U$, $t \in I$, and let $\tilde{\alpha}$ be the restriction of α to the rectangle

$$(V \times I) \cap \{(x, y, t) \in R^3; x = 0\}.$$

(See Fig. 3-27.) By the definition of local flow, $d\tilde{\alpha}_p$ maps the unit vector of the t axis into w and maps the unit vector of the y axis into itself. Therefore, $d\tilde{\alpha}_p$ is nonsingular. It follows that there exists a neighborhood $W \subset U$ of p, where $\tilde{\alpha}^{-1}$ is defined and differentiable. The projection of $\tilde{\alpha}^{-1}(x, y)$ onto the y axis is a differentiable function $\xi = f(x, y)$, which has the same

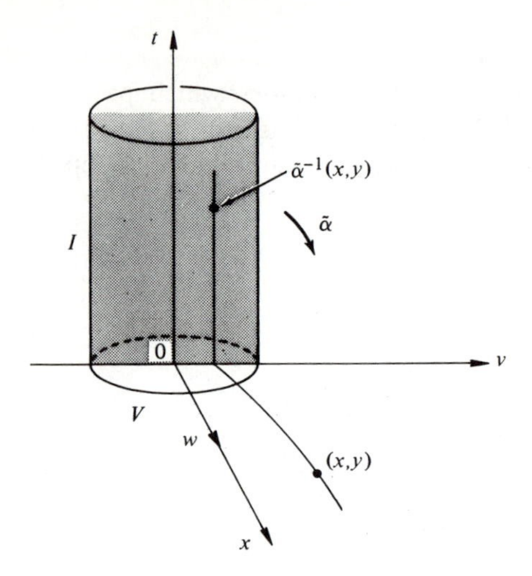

Figure 3-27

value ξ for all points of the trajectory passing through $(0, \xi)$. Since $d\tilde{\alpha}_p$ is nonsingular, W may be taken sufficiently small so that $df_q \neq 0$ for all $q \in W$. f is therefore the required function. **Q.E.D.**

The function f of the above lemma is called a (local) *first integral* of w in a neighborhood of p. For instance, if $w(x, y) = (y, -x)$ is defined in R^2, a first integral $f: R^2 - \{(0, 0)\} \to R$ is $f(x, y) = x^2 + y^2$.

Closely associated with the concept of vector field is the concept of field of directions.

A *field of directions* r in an open set $U \subset R^2$ is a correspondence which assigns to each $p \in U$ a line $r(p)$ in R^2 passing through p. r is said to be *differentiable* at $p \in U$ if there exists a nonzero differentiable vector field w, defined in a neighborhood $V \subset U$ of p, such that for each $q \in V$, $w(q) \neq 0$ is a *basis* of $r(q)$; r is *differentiable in* U if it is differentiable for every $p \in U$.

To each nonzero differentiable vector field w in $U \subset R^2$, there corresponds a differentiable field of directions given by $r(p) =$ line generated by $w(p)$, $p \in U$.

By its very definition, each differentiable field of directions gives rise, locally, to a nonzero differentiable vector field. This, however, is not true globally, as is shown by the field of directions in $R^2 - \{(0, 0)\}$ given by the tangent lines to the curves of Fig. 3-28; any attempt to orient these curves in order to obtain a differentiable nonzero vector field leads to a contradiction.

A regular connected curve $C \subset U$ is an *integral curve* of a field of directions r defined in $U \subset R^2$ if $r(q)$ is the tangent line to C at q for all $q \in C$.

By what has been seen previously, it is clear that given a differentiable field of directions r in an open set $U \subset R^2$, there passes, for each $q \in U$, an integral curve C of r; C agrees locally with the trace of a trajectory through q

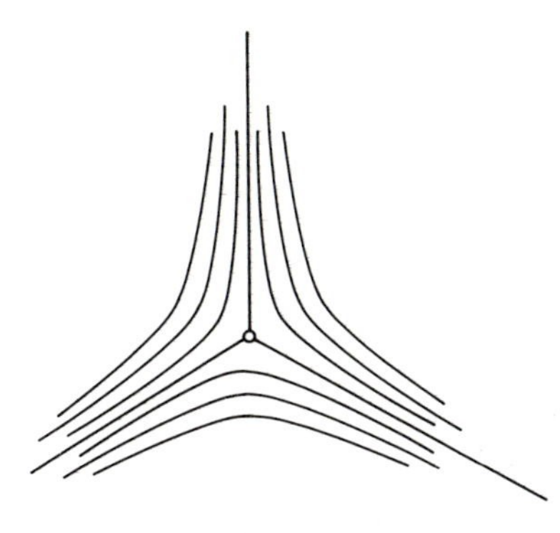

Figure 3-28. A nonorientable field of directions in $R^2 - \{(0, 0)\}$.

of the vector field determined in U by r. In what follows, we shall consider only differentiable fields of directions and shall omit, in general, the word differentiable.

A natural way of describing a field of directions is as follows. We say that two nonzero vectors w_1 and w_2 at $q \in R^2$ are *equivalent* if $w_1 = \lambda w_2$ for some $\lambda \in R$, $\lambda \neq 0$. Two such vectors represent the same straight line passing through q, and, conversely, if two nonzero vectors belong to the same straight line passing through q, they are equivalent. Thus, a field of directions r on an open set $U \subset R^2$ can be given by assigning to each $q \in U$ a pair of real numbers (r_1, r_2) (the coordinates of a nonzero vector belonging to r), where we consider the pairs (r_1, r_2) and $(\lambda r_1, \lambda r_2)$, $\lambda \neq 0$, as equivalent.

In the language of differential equations, a field of directions r is usually given by

$$a(x, y)\frac{dx}{dt} + b(x, y)\frac{dy}{dt} = 0, \tag{2}$$

which simply means that at a point $q = (x, y)$ we associate the line passing through q that contains the vector $(b, -a)$ or any of its nonzero multiples (Fig. 3-29). The trace of the trajectory of the vector field $(b, -a)$ is an integral curve of r. Because the parametrization plays no role in the above considerations, it is often used, instead of Eq. (2), the expression

$$a\,dx + b\,dy = 0$$

with the same meaning as before.

The ideas introduced above belong to the domain of the local facts of R^2, which depend only on the "differentiable structure" of R^2. They can, therefore, be transported to a regular surface, without further difficulties, as follows.

DEFINITION 1. *A* vector field w *in an open set* U $\subset$ S *of a regular surface* S *is a correspondence which assigns to each* p $\in$ U *a vector* w(p) $\in$ $T_p(S)$. *The vector field* w *is* differentiable *at* p $\in$ U *if, for some parametrization*

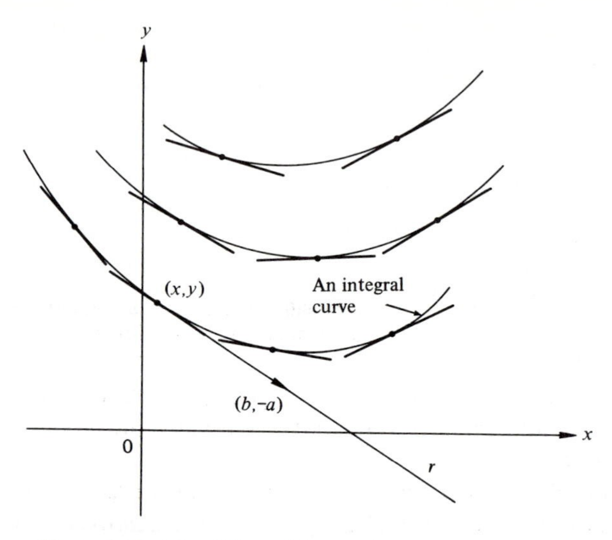

Figure 3-29. The differential equation $adx + bdy = 0$.

$\mathbf{x}(u, v)$ *at* p, *the functions* $a(u, v)$ *and* $b(u, v)$ *given by*

$$w(p) = a(u, v)\mathbf{x}_u + b(u, v)\mathbf{x}_v$$

are differentiable functions at p*; it is clear that this definition does not depend on the choice of* $\mathbf{x}$.

We can define, similarly, trajectories, field of directions, and integral curves. Theorems 1 and 2 and the lemma above extend easily to the present situation; up to a change of R^2 by S, the statements are exactly the same.

Example 1. A vector field in the usual torus T is obtained by parametrizing the meridians of T by arc length and defining $w(p)$ as the velocity vector of the meridian through p (Fig. 3-30). Notice that $|w(p)| = 1$ for all $p \in T$. It is left as an exercise (Exercise 2) to verify that w is differentiable.

Example 2. A similar procedure, this time on the sphere S^2 and using the semimeridians of S^2, yields a vector field w defined in the sphere minus the

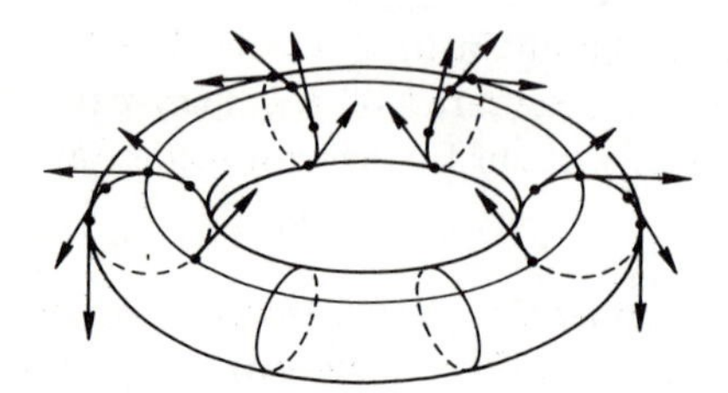

Figure 3-30

two poles N and S. To obtain a vector field defined in the whole sphere, reparametrize all the semimeridians by the same parameter t, $-1 < t < 1$, and define $v(p) = (1 - t^2)w(p)$ for $p \in S^2 - \{N\} \cup \{S\}$ and $v(N) = v(S) = 0$ (Fig. 3-31).

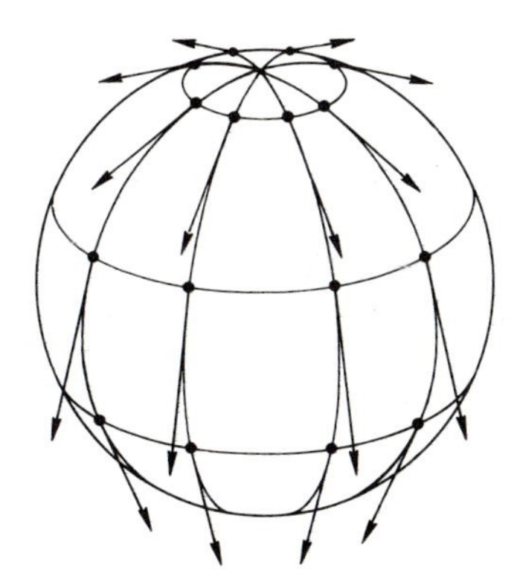

Figure 3-31

Example 3. Let $S = \{(x, y, z) \in R^3; z = x^2 - y^2\}$ be the hyperbolic paraboloid. The intersection with S of the planes $z = \text{const.} \neq 0$ determines a family of curves $\{C_\alpha\}$ such that through each point of $S - \{(0, 0, 0)\}$ there passes one curve C_α. The tangent lines to such curves give a differentiable field of directions r on $S - \{(0, 0, 0)\}$. We want to find a field of directions r' on $S - \{(0, 0, 0)\}$ that is orthogonal to r at each point and to determine the integral curves of r'. r' is called the *orthogonal field* to r, and its integral curves are called the *orthogonal family* of r (cf. Exercise 15, Sec. 2-5).

We begin by parametrizing S by

$$\mathbf{x}(u, v) = (u, v, u^2 - v^2), \qquad u = x, \qquad v = y.$$

The family $\{C_\alpha\}$ is given by $u^2 - v^2 = \text{const.} \neq 0$ (or rather by the image under $\mathbf{x}$ of this set). If $u'\mathbf{x}_u + v'\mathbf{x}_v$ is a tangent vector of a regular parametrization of some curve C_α, we obtain, by differentiating $u^2 - v^2 = \text{const.}$,

$$2uu' - 2vv' = 0.$$

Thus, $(u', v') = (-v, -u)$. It follows that r is given, in the parametrization $\mathbf{x}$, by the pair (v, u) or any of its nonzero multiples.

Now, let $(a(u, v), b(u, v))$ be an expression for the orthogonal field r', in the parametrization $\mathbf{x}$. Since

$$E = 1 + 4u^2, \qquad F = -4uv, \qquad G = 1 + 4v^2,$$

and r' is orthogonal to r at each point, we have

$$Eav + F(bv + au) + Gbu = 0$$

or

$$(1 + 4u^2)av - 4uv(bv + au) + (1 + 4v^2)bu = 0.$$

It follows that

$$va + ub = 0. \tag{3}$$

This determines the pair (a, b) at each point, up to a nonzero multiple, and hence the field r'.

To find the integral curves of r', let $u'\mathbf{x}_u + v'\mathbf{x}_v$ be a tangent vector of some regular parametrization of an integral curve of r'. Then (u', v') satisfies Eq. (3); that is,

$$vu' + uv' = 0$$

or

$$uv = \text{const.}$$

It follows that the orthogonal family of $\{C_\alpha\}$ is given by the intersections with S of the hyperbolic cylinders $xy = \text{const.} \neq 0$.

The main result of this section is the following theorem.

THEOREM. *Let* w_1 *and* w_2 *be two vector fields in an open set* $U \subset S$, *which are linearly independent at some point* $p \in U$. *Then it is possible to parametrize a neighborhood* $V \subset U$ *of* p *in such a way that for each* $q \in V$ *the coordinate lines of this parametrization passing through* q *are tangent to the lines determined by* $w_1(q)$ *and* $w_2(q)$.

Proof. Let W be a neighborhood of p where the first integrals f_1 and f_2 of w_1 and w_2, respectively, are defined. Define a map $\varphi\colon W \to R^2$ by

$$\varphi(q) = (f_1(q), f_2(q)), \qquad q \in W.$$

Since f_1 is constant on the trajectories of w_1 and $(df_1) \neq 0$, we have at p

$$d\varphi_p(w_1) = ((df_1)_p(w_1), (df_2)_p(w_1)) = (0, a),$$

where $a = (df_2)_p(w_1) \neq 0$, since w_1 and w_2 are linearly independent. Similarly,

$$d\varphi_p(w_2) = (b, 0),$$

where $b = (df_1)_p(w_2) \neq 0$.

It follows that $d\varphi_p$ is nonsingular, and hence that φ is a local diffeomorphism. There exists, therefore, a neighborhood $\bar{U} \subset R^2$ of $\varphi(p)$ which is mapped diffeomorphically by $\mathbf{x} = \varphi^{-1}$ onto a neighborhood $V = \mathbf{x}(\bar{U})$ of p; that is, $\mathbf{x}$ is a parametrization of S at p, whose coordinate curves

$$f_1(q) = \text{const.}, \qquad f_2(q) = \text{const.},$$

are tangent at q to the lines determined by $w_1(q)$, $w_2(q)$, respectively. **Q.E.D.**

It should be remarked that the theorem does not imply that the coordinate curves can be so parametrized that their velocity vectors are $w_1(q)$ and $w_2(q)$. The statement of the theorem applies to the coordinate curves as regular (point set) curves; more precisely, we have

COROLLARY 1. *Given two fields of directions* r *and* r' *in an open set* $U \subset S$ *such that at* $p \in U$, $r(p) \neq r'(p)$, *there exists a parametrization* $\mathbf{x}$ *in a neighborhood of* p *such that the coordinate curves of* $\mathbf{x}$ *are the integral curves of* r *and* r'.

A first application of the above theorem is the proof of the existence of an orthogonal parametrization at any point of a regular surface.

COROLLARY 2. *For all* $p \in S$ *there exists a parametrization* $\mathbf{x}(u, v)$ *in a neighborhood* V *of* p *such that the coordinate curves* $u = \text{const.}$, $v = \text{const.}$ *intersect orthogonally for each* $q \in V$ (*such an* $\mathbf{x}$ *is called an orthogonal parametrization*).

Proof. Consider an arbitrary parametrization $\bar{\mathbf{x}}: \bar{U} \to S$ at p, and define two vector fields $w_1 = \bar{\mathbf{x}}_{\bar{u}}$, $w_2 = -(\bar{F}/\bar{E})\bar{\mathbf{x}}_{\bar{u}} + \bar{\mathbf{x}}_{\bar{v}}$ in $\bar{\mathbf{x}}(\bar{U})$, where $\bar{E}, \bar{F}, \bar{G}$ are the coefficients of the first fundamental form in $\bar{\mathbf{x}}$. Since $w_1(q)$, $w_2(q)$ are orthogonal vectors, for each $q \in \bar{\mathbf{x}}(\bar{U})$, an application of the theorem yields the required parametrization. **Q.E.D.**

A second application of the theorem (more precisely, of Corollary 1) is the existence of coordinates given by the asymptotic and principal directions.

As we have seen in Sec. 3-3, the asymptotic curves are solutions of

$$e(u')^2 + 2fu'v' + g(v')^2 = 0.$$

In a neighborhood of a hyperbolic point p, we have $eg - f^2 < 0$, and the left-hand side of the above equation can be decomposed into two distinct linear factors, yielding

$$(Au' + Bv')(Au' + Dv') = 0, \tag{4}$$

where the coefficients are determined by

$$A^2 = e, \qquad A(B + D) = 2f, \qquad BD = g.$$

The above system of equations has real solutions, since $eg - f^2 < 0$. Thus, Eq. (4) gives rise to two equations:

$$Au' + Bv' = 0, \tag{4a}$$

$$Au' + Dv' = 0. \tag{4b}$$

Each of these equations determines a differentiable field of directions (for instance, Eq. (4a) determines the direction r which contains the nonzero vector $(B, -A)$), and at each point of the neighborhood in question the directions given by Eqs. (4a) and (4b) are distinct. By applying Corollary 1, we see that it is possible to parametrize a neighborhood of p in such a way that the coordinate curves are the integral curves of Eqs. (4a) and (4b). In other words,

COROLLARY 3. *Let* $p \in S$ *be a hyperbolic point of* S. *Then it is possible to parametrize a neighborhood of* p *in such a way that the coordinate curves of this parametrization are the asymptotic curves of* S.

Example 4. An almost trivial example, but one which illustrates the mechanism of the above method, is given by the hyperbolic paraboloid $z = x^2 - y^2$. As usual we parametrize the entire surface by

$$\mathbf{x}(u, v) = (u, v, u^2 - v^2).$$

A simple computation shows that

$$e = \frac{2}{(1 + 4u^2 + 4v^2)^{1/2}}, \qquad f = 0, \qquad g = -\frac{2}{(1 + 4u^2 + 4v^2)^{1/2}}.$$

Thus, the equation of the asymptotic curves can be written as

$$\frac{2}{(1 + 4u^2 + 4v^2)^{1/2}}((u')^2 - (v')^2) = 0,$$

which can be factored into two linear equations and give the two fields of directions:

$$r_1: \quad u' + v' = 0,$$

$$r_2: \quad u' - v' = 0.$$

The integral curves of these fields of directions are given by the two families of curves:

$$r_1: \quad u + v = \text{const.},$$

$$r_2: \quad u - v = \text{const.}$$

Now, the functions $f_1(u, v) = u + v$, $f_2(u, v) = u - v$ are clearly first integrals of the vector fields associated to r_1 and r_2, respectively. Thus, by setting

$$\bar{u} = u + v, \qquad \bar{v} = u - v,$$

we obtain a new parametrization for the entire surface $z = x^2 - y^2$ in which the coordinate curves are the asymptotic curves of the surface.

In this particular case, the change of parameters holds for the entire surface. In general, it may fail to be globally one-to-one, even if the whole surface consists only of hyperbolic points.

Similarly, in a neighborhood of a nonumbilical point of S, it is possible to decompose the differential equation of the lines of curvature into distinct linear factors. By an analogous argument we obtain

COROLLARY 4. *Let* $p \in S$ *be a nonumbilical point of* S. *Then it is possible to parametrize a neighborhood of* p *in such a way that the coordinate curves of this parametrization are the lines of curvature of* S.

EXERCISES

1. Prove that the differentiability of a vector field does not depend on the choice of a coordinate system.
2. Prove that the vector field obtained on the torus by parametrizing all its meridians by arc length and taking their tangent vectors (Example 1) is differentiable.
3. Prove that a vector field w defined on a regular surface $S \subset R^3$ is differentiable if and only if it is differentiable as a map $w: S \to R^3$.
4. Let S be a surface and $\mathbf{x}: U \to S$ be a parametrization of S. Then

$$a(u, v)u' + b(u, v)v' = 0,$$

where a and b are differentiable functions, determines a field of directions r on $\mathbf{x}(U)$, namely, the correspondence which assigns to each $\mathbf{x}(u, v)$ the straight line containing the vector $b\mathbf{x}_u - a\mathbf{x}_v$. Show that a necessary and sufficient condition for the existence of an orthogonal field r' on $\mathbf{x}(U)$ (cf. Example 3) is that both functions

$$Eb - Fa, \qquad Fb - Ga$$

are nowhere simultaneously zero (here E, F, and G are the coefficients of the first fundamental form in $\mathbf{x}$) and that r' is then determined by

$$(Eb - Fa)u' + (Fb - Ga)v' = 0.$$

5. Let S be a surface and $\mathbf{x}: U \longrightarrow S$ be a parametrization of S. If $ac - b^2 < 0$, show that

$$a(u, v)(u')^2 + 2b(u, v)u'v' + c(u, v)(v')^2 = 0$$

can be factored into two distinct equations, each of which determines a field of directions on $\mathbf{x}(U) \subset S$. Prove that these two fields of directions are orthogonal if and only if

$$Ec - 2Fb + Ga = 0.$$

6. A straight line r meets the z axis and moves in such a way that it makes a constant angle $\alpha \neq 0$ with the z axis and each of its points describes a helix of pitch $c \neq 0$ about the z axis. The figure described by r is the trace of the parametrized surface (see Fig. 3-32)

$$\mathbf{x}(u, v) = (v \sin \alpha \cos u, v \sin \alpha \sin u, v \cos \alpha + cu).$$

$\mathbf{x}$ is easily seen to be a regular parametrized surface (cf. Exercise 13, Sec. 2-5). Restrict the parameters (u, v) to an open set U so that $\mathbf{x}(U) = S$ is a regular surface (cf. Prop. 2, Sec. 2-3).

a. Find the orthogonal family (cf. Example 3) to the family of coordinate curves $u = \text{const.}$

b. Use the curves $u = \text{const.}$ and their orthogonal family to obtain an orthogonal parametrization for S. Show that in the new parameters $(\bar{u}, \bar{v})$ the coefficients of the first fundamental form are

$$\bar{G} = 1, \qquad \bar{F} = 0, \qquad \bar{E} = \{c^2 + (\bar{v} - c\bar{u} \cos \alpha)^2\} \sin^2 \alpha.$$

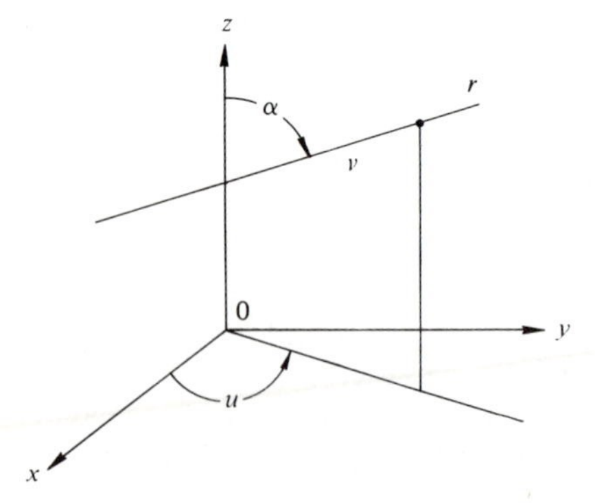

Figure 3-32

7. Define the *derivative* $w(f)$ *of a differentiable function* $f: U \subset S \longrightarrow R$ *relative to a vector field* w *in* U *by*

$$w(f)(q) = \frac{d}{dt}(f \circ \alpha)\bigg|_{t=0}, \qquad q \in U,$$

where $\alpha: I \longrightarrow S$ is a curve such that $\alpha(0) = q$, $\alpha'(0) = w(q)$. Prove that

a. w is differentiable in U if and only if $w(f)$ is differentiable for all differentiable f in U.

b. Let λ and μ be real numbers and $g: U \subset S \longrightarrow R$ be a differentiable function on U; then

$$w(\lambda f + \mu f) = \lambda w(f) + \mu w(f),$$
$$w(fg) = w(f)g + fw(g).$$

8. Show that if w is a differentiable vector field on a surface S and $w(p) \neq 0$ for some $p \in S$, then it is possible to parametrize a neighborhood of p by $\mathbf{x}(u, v)$ in such a way that $\mathbf{x}_u = w$.

9. a. Let $A: V \longrightarrow W$ be a nonsingular linear map of vector spaces V and W of dimension 2 and endowed with inner products $\langle\, ,\, \rangle$ and $(\, ,\,)$, respectively. A is a *similitude* if there exists a real number $\lambda \neq 0$ such that $(Av_1, Av_2) = \lambda\langle v_1, v_2\rangle$ for all vectors $v_1, v_2 \in V$. Assume that A is not a similitude and show that there exists a *unique* pair of orthonormal vectors e_1 and e_2 in V such that Ae_1, Ae_2 are orthogonal in W.

b. Use part a to prove *Tissot's theorem:* Let $\varphi: U_1 \subset S_1 \longrightarrow S_2$ be a diffeomorphism from a neighborhood U_1 of a point p of a surface S_1 into a surface S_2. Assume that the linear map $d\varphi$ is nowhere a similitude. Then it is possible to parametrize a neighborhood of p in S_1 by an orthogonal parametrization $\mathbf{x}_1: U \longrightarrow S_1$ in such a way that $\varphi \circ \mathbf{x}_1 = \mathbf{x}_2: U \longrightarrow S_2$ is also an orthogonal parametrization in a neighborhood of $\varphi(p) \in S_2$.

10. Let T be the torus of Example 6 of Sec. 2-2 and define a map $\varphi: R^2 \longrightarrow T$ by

$$\varphi(u, v) = ((r\cos u + a)\cos v, (r\cos u + a)\sin v, r\sin u),$$

where u and v are the cartesian coordinates of R^2. Let $u = at$, $v = bt$ be a straight line in R^2, passing by $(0, 0) \in R^2$, and consider the curve in T $\alpha(t) = \varphi(at, bt)$. Prove that

a. φ is a local diffeomorphism.

b. The curve $\alpha(t)$ is a regular curve; $\alpha(t)$ is a closed curve if and only if b/a is a rational number.

***c.** If b/a is irrational, the curve $\alpha(t)$ is dense in T; that is, in each neighborhood of a point $p \in T$ there exists a point of $\alpha(t)$.

***11.** Use the local uniqueness of trajectories of a vector field w in $U \subset S$ to prove the following result. Given $p \in U$, there exists a unique trajectory $\alpha: I \longrightarrow U$ of w, with $\alpha(0) = p$, which is *maximal* in the following sense: Any other trajectory $\beta: J \longrightarrow U$, with $\beta(0) = p$, is the restriction of α to J (i.e., $J \subset I$ and $\alpha|J = \beta$).

***12.** Prove that if w is a differentiable vector field on a compact surface S and $\alpha(t)$ is the maximal trajectory of w with $\alpha(0) = p \in S$, then $\alpha(t)$ is defined for all $t \in R$.

13. Construct a differentiable vector field on an open disk of the plane (which is not compact) such that a maximal trajectory $\alpha(t)$ is not defined for all $t \in R$ (this shows that the compactness condition of Exercise 12 is essential).

3-5. Ruled Surfaces and Minimal Surfaces†

In differential geometry one finds quite a number of special cases (surfaces of revolution, parallel surfaces, ruled surfaces, minimal surfaces, etc.) which may either become interesting in their own right (like minimal surfaces), or give a beautiful example of the power and limitations of differentiable methods in geometry. According to the spirit of this book, we have so far treated these special cases in examples and exercises.

It might be useful, however, to present some of these topics in more detail. We intend to do that now. We shall use this section to develop the theory of ruled surfaces and to give an introduction to the theory of minimal surfaces. Throughout the section it will be convenient to use the notion of parametrized surface defined in Sec. 2-3.

If the reader wishes so, the entire section or one of its topics may be omitted. Except for a reference to Sec. A in Example 6 of Sec. B, the two topics are independent and their results will not be used in any essential way in this book.

A. Ruled Surfaces

A (differentiable) *one-parameter family of* (straight) *lines* $\{\alpha(t), w(t)\}$ is a correspondence that assigns to each $t \in I$ a point $\alpha(t) \in R^3$ and a vector $w(t) \in R^3$, $w(t) \neq 0$, so that both $\alpha(t)$ and $w(t)$ depend differentiably on t. For each $t \in I$, the line L_t which passes through $\alpha(t)$ and is parallel to $w(t)$ is called *the line of the family at t*.

Given a one-parameter family of lines $\{\alpha(t), w(t)\}$, the parametrized surface

$$\mathbf{x}(t, v) = \alpha(t) + vw(t), \qquad t \in I, \qquad v \in R,$$

is called the *ruled surface* generated by the family $\{\alpha(t), w(t)\}$. The lines L_t are called the *rulings*, and the curve $\alpha(t)$ is called a *directrix* of the surface $\mathbf{x}$. Sometimes we use the expression ruled surface to mean the trace of $\mathbf{x}$. It should be noticed that we also allow $\mathbf{x}$ to have singular points, that is, points (t, v) where $\mathbf{x}_t \wedge \mathbf{x}_v = 0$.

Example 1. The simplest examples of ruled surfaces are the tangent surfaces to a regular curve (cf. Example 4, Sec. 2-3), the cylinders and the cones. A *cylinder* is a ruled surface generated by a one-parameter family of lines $\{\alpha(t), w(t)\}$, $t \in I$, where $\alpha(I)$ is contained in a plane P and $w(t)$ is parallel

†This section may be omitted on a first reading.

to a fixed direction in R^3 (Fig. 3-33(a)). A *cone* is a ruled surface generated by a family $\{\alpha(t), w(t)\}$, $t \in I$, where $\alpha(I) \subset P$ and the rulings L_t all pass through a point $p \notin P$ (Fig. 3-33(b)).

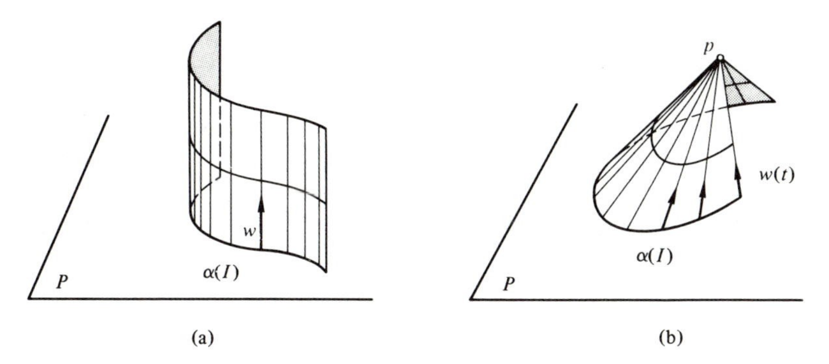

Figure 3-33

Example 2. Let S^1 be the unit circle $x^2 + y^2 = 1$ in the xy plane, and let $\alpha(s)$ be a parametrization of S^1 by arc length. For each s, let $w(s) = \alpha'(s) + e_3$, where e_3 is the unit vector of the z axis (Fig. 3-34). Then

$$\mathbf{x}(s, v) = \alpha(s) + v(\alpha'(s) + e_3)$$

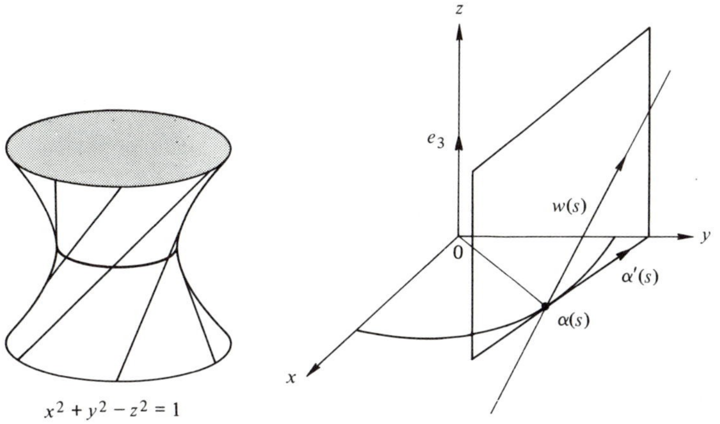

Figure 3-34. $x^2 + y^2 - z^2 = 1$ as a ruled surface.

is a ruled surface. It can be put into a more familiar form if we write

$$\mathbf{x}(s, v) = (\cos s - v \sin s, \sin s + v \cos s, v)$$

and notice that $x^2 + y^2 - z^2 = 1 + v^2 - v^2 = 1$. This shows that the trace of $\mathbf{x}$ is a hyperboloid of revolution.

It is interesting to observe that if we take $w(s) = -\alpha'(s) + e_3$, we again obtain the same surface. This shows that the hyperboloid of revolution has two sets of rulings.

We have defined ruled surfaces in such a way that allows the appearance of singularities. This is necessary if we want to include tangent surfaces and cones. We shall soon show, at least for ruled surfaces that satisfy some reasonable condition, that the singularities of such a surface (if any) will concentrate along a curve of this surface.

We shall now start the study of general ruled surfaces. We can assume, without loss of generality, that $|w(t)| = 1$, $t \in I$. To be able to develop the theory, we need the nontrivial assumption that $w'(t) \neq 0$ for all $t \in I$. If the zeros of $w'(t)$ are isolated, we can divide our surface into pieces such that the theory can be applied to each of them. However, if the zeros of $w'(t)$ have cluster points, the situation may become complicated and will not be treated here.

The assumption $w'(t) \neq 0$, $t \in I$, is usually expressed by saying that the ruled surface $\mathbf{x}$ is *noncylindrical*.

Unless otherwise stated, we shall assume that

$$\mathbf{x}(t, v) = \alpha(t) + vw(t) \tag{1}$$

is a noncylindrical ruled surface with $|w(t)| = 1$, $t \in I$. Notice that the assumption $|w(t)| \equiv 1$ implies that $\langle w(t), w'(t) \rangle = 0$ for all $t \in I$.

We first want to find a parametrized curve $\beta(t)$ such that $\langle \beta'(t), w'(t) \rangle = 0$, $t \in I$, and $\beta(t)$ lies on the trace of $\mathbf{x}$; that is,

$$\beta(t) = \alpha(t) + u(t)w(t), \tag{2}$$

for some real-valued function $u = u(t)$. Assuming the existence of such a curve β, one obtains

$$\beta' = \alpha' + u'w + uw';$$

hence, since $\langle w, w' \rangle = 0$,

$$0 = \langle \beta', w' \rangle = \langle \alpha', w' \rangle + u\langle w', w' \rangle.$$

It follows that $u = u(t)$ is given by

$$u = -\frac{\langle \alpha', w' \rangle}{\langle w', w' \rangle} \tag{3}$$

Thus, if we define $\beta(t)$ by Eqs. (2) and (3), we obtain the required curve.

We shall now show that the curve β does not depend on the choice of the directrix α for the ruled surface. β is then called the *line of striction*, and its points are called the *central points* of the ruled surface.

To prove our claim, let $\bar{\alpha}$ be another directrix of the ruled surface; that is, let, for all (t, v),

$$\mathbf{x}(t, v) = \alpha(t) + vw(t) = \bar{\alpha}(t) + sw(t) \tag{4}$$

for some function $s = s(t)$. Then, from Eqs. (2) and (3) we obtain

$$\beta - \bar{\beta} = (\alpha - \bar{\alpha}) + \frac{\langle \bar{\alpha}' - \alpha', w' \rangle}{\langle w', w' \rangle} w,$$

where $\bar{\beta}$ is the line of striction corresponding to $\bar{\alpha}$. On the other hand, Eq. (4) implies that

$$\alpha - \bar{\alpha} = (s - v)w(t).$$

Thus,

$$\beta - \bar{\beta} = \left\{ (s - v) + \frac{\langle (v - s)w', w' \rangle}{\langle w', w' \rangle} \right\} w = 0,$$

since $\langle w, w' \rangle = 0$. This proves our claim.

We now take the line of striction as the directrix of the ruled surface and write it as follows:

$$\mathbf{x}(t, u) = \beta(t) + uw(t). \tag{5}$$

With this choice, we have

$$\mathbf{x}_t = \beta' + uw', \qquad \mathbf{x}_u = w$$

and

$$\mathbf{x}_t \wedge \mathbf{x}_u = \beta' \wedge w + uw' \wedge w.$$

Since $\langle w', w \rangle = 0$ and $\langle w', \beta' \rangle = 0$, we conclude that $\beta' \wedge w = \lambda w'$ for some function $\lambda = \lambda(t)$. Thus,

$$\begin{aligned} |\mathbf{x}_t \wedge \mathbf{x}_u|^2 &= |\lambda w' + uw' \wedge w|^2 \\ &= \lambda^2 |w'|^2 + u^2 |w'|^2 = (\lambda^2 + u^2)|w'|^2. \end{aligned}$$

It follows that the only singular points of the ruled surface (5) are along the line of striction $u = 0$, and they will occur if and only if $\lambda(t) = 0$. Observe also that

$$\lambda = \frac{(\beta', w, w')}{|w'|^2},$$

where, as usual, (β', w, w') is a short for $\langle \beta' \wedge w, w' \rangle$.

Let us compute the Gaussian curvature of the surface (5) at its regular points. Since

$$\mathbf{x}_{tt} = \beta'' + uw'', \qquad \mathbf{x}_{tu} = w', \qquad \mathbf{x}_{uu} = 0,$$

we have, for the coefficients of the second fundamental form,

$$g = 0, \qquad f = \frac{(\mathbf{x}_t, \mathbf{x}_u, \mathbf{x}_{ut})}{|\mathbf{x}_t \wedge \mathbf{x}_u|} = \frac{(\beta', w, w')}{|\mathbf{x}_t \wedge \mathbf{x}_u|^2};$$

hence (since $g = 0$ we do not need the value of e to compute K),

$$K = \frac{eg - f^2}{EG - F^2} = -\frac{\lambda^2 |w'|^4}{(\lambda^2 + u^2)^2 |w'|^4} = -\frac{\lambda^2}{(\lambda^2 + u^2)^2}. \tag{6}$$

This shows that, at regular points, *the Gaussian curvature K of a ruled surface satisfies $K \leq 0$, and K is zero only along those rulings which meet the line of striction at a singular point.*

Equation (6) allows us to give a geometric interpretation of the (regular) central points of a ruled surface. Indeed, the points of a ruling, except perhaps the central point, are regular points of the surface. If $\lambda \neq 0$, the function $|K(u)|$ is a continuous function on the ruling and, by Eq. (6), the central point is characterized by the fact that $|K(u)|$ has a maximum there.

For another geometrical interpretation of the line of striction see Exercise 4.

We also remark that the curvature K takes up the same values at points on a ruling that are symmetric relative to the central point (this justifies the name central).

The function $\lambda(t)$ is called the *distribution parameter* of $\mathbf{x}$. Since the line of striction is independent of the choice of the directrix, it follows that the same holds for λ. If $\mathbf{x}$ is regular, we have the following interpretation of λ. The normal vector to the surface at (t, u) is

$$N(t, u) = \frac{\mathbf{x}_t \wedge \mathbf{x}_u}{|\mathbf{x}_t \wedge \mathbf{x}_u|} = \frac{\lambda w' + uw' \wedge w}{\sqrt{\lambda^2 + u^2}\,|w'|}.$$

On the other hand ($\lambda \neq 0$),

$$N(t, 0) = \frac{w'}{|w'|}.$$

Therefore, if θ is the angle formed by $N(t, u)$ and $N(t, 0)$,

$$\tan \theta = \frac{u}{\lambda}. \tag{7}$$

Thus, if θ is the angle which the normal vector at a point of a ruling makes with the normal vector at the central point of this ruling, then $\tan\theta$ *is proportional to the distance between these two points, and the coefficient of proportionality is the inverse of the distribution parameter.*

Example 3. Let S be the hyperbolic paraboloid $z = kxy$, $k \neq 0$. To show that S is a ruled surface, we observe that the lines $y = z/tk$, $x = t$, for each $t \neq 0$ belong to S. If we take the intersection of this family of lines with the plane $z = 0$, we obtain the curve $x = t$, $y = 0$, $z = 0$. Taking this curve as directrix and vectors $w(t)$ parallel to the lines $y = z/tk$, $x = t$, we obtain

$$\alpha(t) = (t, 0, 0), \qquad w(t) = \left(0, \frac{1}{k}, t\right).$$

This gives a ruled surface (Fig. 3-35)

$$\mathbf{x}(t, v) = \alpha(t) + vw(t) = \left(t, \frac{v}{k}, vt\right), \qquad t \in R, v \in R,$$

the trace of which clearly agrees with S.

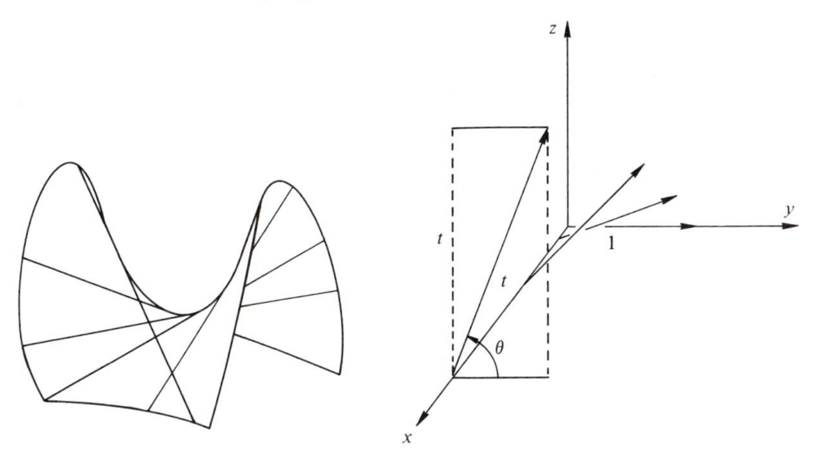

Figure 3-35. $z = xy$ as a ruled surface.

Since $\alpha'(t) = (1, 0, 0)$, we obtain that the line of striction is α itself. The distribution parameter is

$$\lambda = \frac{1 + k^2t^2}{k^2}.$$

We also remark that the tangent of the angle θ which $w(t)$ makes with $w(0)$ is $\tan\theta = tk$.

The last remark leads to an interesting general property of a ruled surface. If we consider the family of normal vectors along a ruling of a regular ruled surface, this family generates another ruled surface. By Eq. (7) and the last remark, the latter surface is exactly the hyperbolic paraboloid $x = kxy$, where $1/k$ is the value of the distribution parameter at the chosen ruling.

Among the ruled surfaces, the developables play a distinguished role. Let us start again with an arbitrary ruled surface (not necessarily noncylindrical)

$$\mathbf{x}(t, v) = \alpha(t) + vw(t), \tag{8}$$

generated by the family $\{\alpha(t), w(t)\}$ with $|w(t)| \equiv 1$. The surface (8) is said to be *developable* if

$$(w, w', \alpha') \equiv 0. \tag{9}$$

To find a geometric interpretation for condition (9), we shall compute the Gaussian curvature of a developable surface at a regular point. A computation entirely similar to the one made to obtain Eq. (6) gives

$$g = 0, \qquad f = \frac{(w, w', \alpha')}{|\mathbf{x}_t \wedge \mathbf{x}_v|^2}.$$

By condition (9), $f \equiv 0$; hence,

$$K = \frac{eg - f^2}{EG - F^2} \equiv 0.$$

This implies that, *at regular points, the Gaussian curvature of a developable surface is identically zero.*

For another geometric interpretation of a developable surface, see Exercise 6.

We can now distinguish two nonexhaustive cases of developable surfaces:

1. $w(t) \wedge w'(t) \equiv 0$. This implies that $w'(t) \equiv 0$. Thus, $w(t)$ is constant and the ruled surface is a cylinder over a curve obtained intersecting the cylinder with a plane normal to $w(t)$.
2. $w(t) \wedge w'(t) \neq 0$ for all $t \in I$. In this case $w'(t) \neq 0$ for all $t \in I$. Thus, the surface is noncylindrical, and we can apply our previous work. Thus, we can determine the line of striction (2) and check that the distribution parameter

$$\lambda = \frac{(\beta', w, w')}{|w'|^2} \equiv 0. \tag{10}$$

Therefore, the line of striction will be the locus of singular points of the developable surface. If $\beta'(t) \neq 0$ for all $t \in I$, it follows from Eq. (10) and the fact that $\langle \beta', w' \rangle \equiv 0$ that w is parallel to β'. Thus, the ruled surface is the tangent surface of β. If $\beta'(t) = 0$ for all $t \in I$, then the line of striction is a point, and the ruled surface is a cone with vertex at this point.

Of course, the above cases do not exhaust all possibilities. As usual, if there is a clustering of zeros of the functions involved, the analysis may become rather complicated. At any rate, away from these cluster points, a developable surface is a union of pieces of cylinders, cones, and tangent surfaces.

As we have seen, at regular points, the Gaussian curvature of a developable surface is identically zero. In Sec. 5-8 we shall prove a sort of global converse to this which implies that a regular surface $S \subset R^3$ which is closed as a subset of R^3 and has zero Gaussian curvature is a cylinder.

Example 4. (*The Envelope of the Family of Tangent Planes Along a Curve of a Surface*). Let S be a regular surface and $\alpha = \alpha(s)$ a curve on S parametrized by arc length. Assume that α is nowhere tangent to an asymptotic direction. Consider the ruled surface

$$\mathbf{x}(s, v) = \alpha(s) + v\frac{N(s) \wedge N'(s)}{|N'(s)|}, \tag{11}$$

where by $N(s)$ we denote the unit normal vector of S restricted to the curve $\alpha(s)$ (since $\alpha'(s)$ is not an asymptotic direction, $N'(s) \neq 0$ for all s). We shall show that $\mathbf{x}$ is a developable surface which is regular in a neighborhood of $v = 0$ and is tangent to S along $v = 0$. Before that, however, let us give a geometric interpretation of the surface $\mathbf{x}$.

Consider the family $\{T_{\alpha(s)}(S)\}$ of tangent planes to the surface S along the curve $\alpha(s)$. If Δs is small, the two planes $T_{\alpha(s)}(S)$ and $T_{\alpha(s+\Delta s)}(S)$ of the family will intersect along a straight line parallel to the vector

$$\frac{N(s) \wedge N(s + \Delta s)}{\Delta s}.$$

If we let Δs go to zero, this straight line will approach a limiting position parallel to the vector

$$\begin{aligned}\lim_{\Delta s \to 0} \frac{N(s) \wedge N(s + \Delta s)}{\Delta s} &= \lim_{\Delta s \to 0} N(s) \wedge \frac{(N(s + \Delta s) - N(s))}{\Delta s} \\ &= N(s) \wedge N'(s).\end{aligned}$$

This means intuitively that the rulings of **x** are the limiting positions of the intersection of neighboring planes of the family $\{T_{\alpha(s)}(S)\}$. **x** is called the *envelope of the family of tangent planes of* S *along* $\alpha(s)$ (Fig. 3-36).

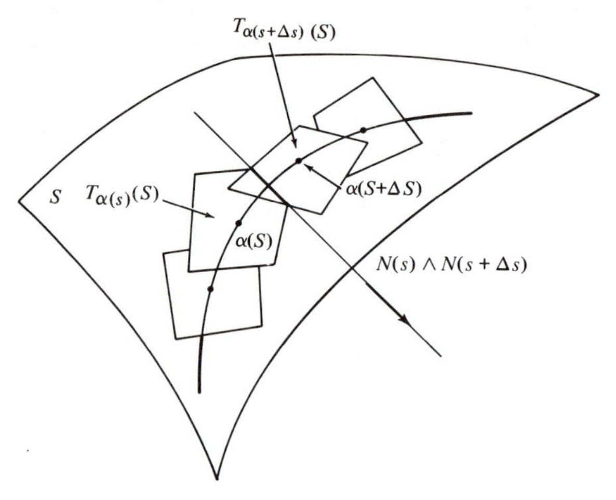

Figure 3-36

For instance, if α is a parametrization of a parallel of a sphere S^2, then the envelope of tangent planes of S^2 along α is either a cylinder, if the parallel is an equator, or a cone, if the parallel is not an equator (Fig. 3-37).

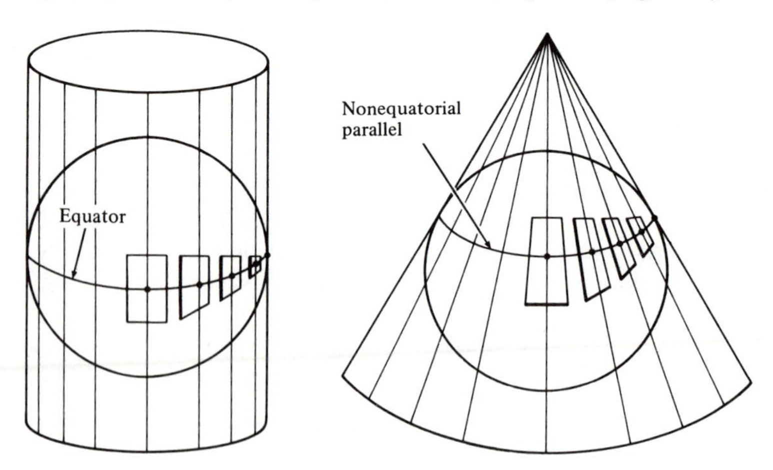

Figure 3-37. Envelopes of families of tangent planes along parallels of a sphere.

To show that **x** is a developable surface, we shall check that condition (9) holds for **x**. In fact, by a straightforward computation, we obtain

$$\left\langle \frac{N \wedge N'}{|N'|} \wedge \left(\frac{N \wedge N'}{|N'|}\right)', \alpha' \right\rangle = \left\langle \frac{N \wedge N'}{|N'|} \wedge \frac{(N \wedge N')'}{|N'|}, \alpha' \right\rangle$$
$$= \frac{1}{|N'|^2}\langle\langle N \wedge N', N''\rangle N, \alpha'\rangle = 0.$$

This proves our claim.

We shall now prove that $\mathbf{x}$ is regular in a neighborhood of $v = 0$ and that it is tangent to S along α. In fact, at $v = 0$, we have

$$\mathbf{x}_s \wedge \mathbf{x}_v = \alpha' \wedge \frac{(N \wedge N')}{|N'|} = \langle N', \alpha'\rangle \frac{N}{|N'|} = -\langle N, \alpha''\rangle \frac{N}{|N'|}$$
$$= -\frac{(k_n N)}{|N'|},$$

where $k_n = k_n(s)$ is the normal curvature of α. Since $k_n(s)$ is nowhere zero, this shows that $\mathbf{x}$ is regular in a neighborhood of $v = 0$ and that the unit normal vector of $\mathbf{x}$ at $\mathbf{x}(s, 0)$ agrees with $N(s)$. Thus, $\mathbf{x}$ is tangent to S along $v = 0$, and this completes the proof of our assertions.

We shall summarize our conclusions as follows. *Let* $\alpha(s)$ *be a curve parametrized by arc length on a surface* S *and assume that* α *is nowhere tangent to an asymptotic direction. Then the envelope* (9) *of the family of tangent planes to* S *along* α *is a developable surface, regular in a neighborhood of* $\alpha(s)$ *and tangent to* S *along* $\alpha(s)$.

B. Minimal Surfaces

A regular parametrized surface is called *minimal* if its mean curvature vanishes everywhere. A regular surface $S \subset R^3$ is *minimal* if each of its parametrizations is minimal.

To explain why we use the word minimal for such surfaces, we need to introduce the notion of a variation. Let $\mathbf{x}: U \subset R^2 \to R^3$ be a regular parametrized surface. Choose a bounded domain $D \subset U$ (cf. Sec. 2-5) and a differentiable function $h: \bar{D} \to R$, where $\bar{D}$ is the union of the domain D with its boundary ∂D. The *normal variation* of $\mathbf{x}(\bar{D})$, determined by h, is the map (Fig. 3-38) given by,

$$\varphi: \bar{D} \times (\epsilon, \epsilon) \longrightarrow R^3$$
$$\varphi(u, v, t) = \mathbf{x}(u, v) + th(u, v)N(u, v), \qquad (u, v) \in \bar{D}, t \in (-\epsilon, \epsilon).$$

For each fixed $t \in (-\epsilon, \epsilon)$, the map $\mathbf{x}^t: D \to R^3$

$$\mathbf{x}^t(u, v) = \varphi(u, v, t)$$

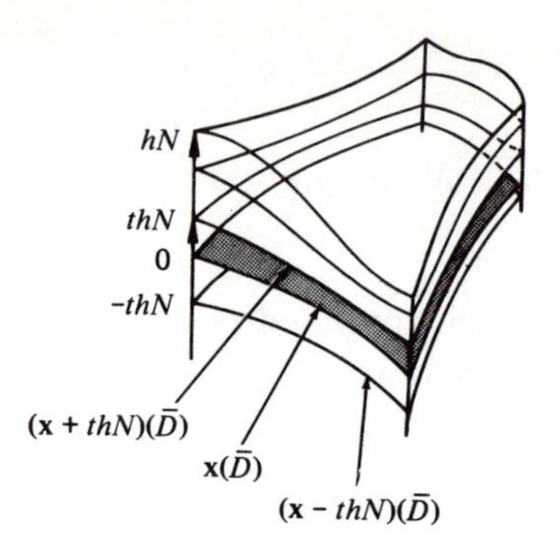

Figure 3-38. A normal variation of $\mathbf{x}(D)$.

is a parametrized surface with

$$\frac{\partial \mathbf{x}^t}{\partial u} = \mathbf{x}_u + thN_u + th_uN,$$

$$\frac{\partial \mathbf{x}^t}{\partial v} = \mathbf{x}_v + thN_v + th_vN.$$

Thus, if we denote by E^t, F^t, G^t the coefficients of the first fundamental form of $\mathbf{x}^t$, we obtain

$$\begin{aligned}
E^t &= E + th(\langle x_u, N_u\rangle + \langle \mathbf{x}_u, N_u\rangle) + t^2h^2\langle N_u, N_u\rangle + t^2h_uh_u,\\
F^t &= F + th(\langle \mathbf{x}_u, N_v\rangle + \langle \mathbf{x}_v, N_u\rangle) + t^2h^2\langle N_u, N_v\rangle + t^2h_uh_v,\\
G^t &= G + th(\langle \mathbf{x}_v, N_v\rangle + \langle \mathbf{x}_v, N_v\rangle) + t^2h^2\langle N_v, N_v\rangle + t^2h_vh_v.
\end{aligned}$$

By using the fact that

$$\langle \mathbf{x}_u, N_u\rangle = -e, \qquad \langle \mathbf{x}_u, N_v\rangle + \langle \mathbf{x}_v, N_u\rangle = -2f, \qquad \langle \mathbf{x}_v, N_v\rangle = -g$$

and that the mean curvature H is (Sec. 3-3, Eq. (5))

$$H = \frac{1}{2}\frac{Eg - 2fF + Ge}{EG - F^2},$$

we obtain

$$\begin{aligned}
E^tG^t - (F^t)^2 &= EG - F^2 - 2th(Eg - 2Ff + Ge) + R\\
&= (EG - F^2)(1 - 4thH) + R,
\end{aligned}$$

where $\lim_{t\to 0}(R/t) = 0$.

It follows that if ϵ is sufficiently small, $\mathbf{x}^t$ is a regular parametrized surface. Furthermore, the area $A(t)$ of $\mathbf{x}^t(\bar{D})$ is

$$A(t) = \int_{\bar{D}} \sqrt{E^t G^t - (F^t)^2}\, du\, dv$$
$$= \int_{\bar{D}} \sqrt{1 - 4thH + \bar{R}}\, \sqrt{EG - F^2}\, du\, dv,$$

where $\bar{R} = R/(EG - F^2)$. It follows that if ϵ is small, A is a differentiable function and its derivative at $t = 0$ is

$$A'(0) = -\int_{\bar{D}} 2hH\sqrt{EG - F^2}\, du\, dv \tag{12}$$

We are now prepared to justify the use of the word minimal in connection with surfaces with vanishing mean curvature.

PROPOSITION 1. *Let* $\mathbf{x}: U \to R^3$ *be a regular parametrized surface and let* $D \subset U$ *be a bounded domain in* U. *Then* **x** *is minimal if and only if* $A'(0) = 0$ *for all such* D *and all normal variations of* $\mathbf{x}(\bar{D})$.

Proof. If **x** is minimal, $H \equiv 0$ and the condition is clearly satisfied. Conversely, assume that the condition is satisfied and that $H(q) \neq 0$ for some $q \in D$. Choose $h: \bar{D} \to R$ such that $h(q) = H(q)$ and h is identically zero outside a small neighborhood of q. Then $A'(0) < 0$ for the variation determined by this h, and that is a contradiction. **Q.E.D.**

Thus, any bounded region $\mathbf{x}(\bar{D})$ of a minimal surface **x** is a critical point for the area function of any normal variation of $\mathbf{x}(\bar{D})$. It should be noticed that this critical point may not be a minimum and that this makes the word minimal seem somewhat awkward. It is, however, a time-honored terminology which was introduced by Lagrange (who first defined a minimal surface) in 1760.

Minimal surfaces are usually associated with soap films that can be obtained by dipping a wire frame into a soap solution and withdrawing it carefully. If the experiment is well performed, a soap film is obtained that has the same frame as a boundary. It can be shown by physical considerations that the film will assume a position where at its regular points the mean curvature is zero. In this way we can "manufacture" beautiful minimal surfaces, such as the one in Fig. 3-39.

Remark 1. It should be pointed out that not all soap films are minimal surfaces according to our definition. We have assumed minimal surfaces to be regular (we could have assumed some isolated singular points, but to go beyond that would make the treatment much less elementary). However, soap films can be formed, for instance, using a cube as a frame (Fig. 3-40), that have singularities along lines.

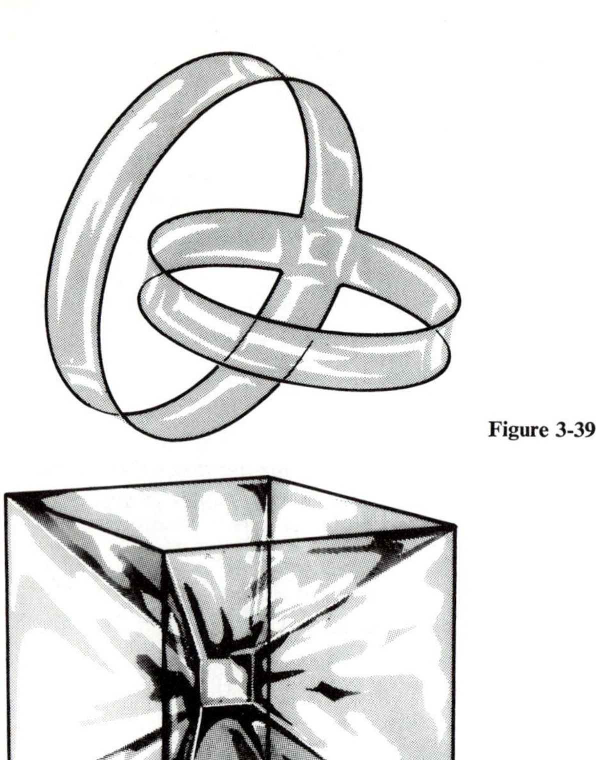

Figure 3-39

Figure 3-40

Remark 2. The connection between minimal surfaces and soap films motivated the celebrated Plateau's problem (Plateau was a Belgian physicist who made careful experiments with soap films around 1850). The problem can be roughly described as follows: *to prove that for each closed curve* $C \subset R^3$ *there exists a surface* S *of minimum area with* C *as boundary*. To make the problem precise (which curves and surfaces are allowed and what is meant by C being a boundary of S) is itself a nontrivial part of the problem. A version of Plateau's problem was solved simultaneously by Douglas and Radó in 1930. Further versions (and generalizations of the problem for higher dimensions) have inspired the creation of mathematical entities which include at least as many things as soap-like films. We refer the interested reader to the Chap. 2 of Lawson [20] (references are at the end of the book) for further details and a recent bibliography of Plateau's problem.

It will be convenient to introduce, for an arbitrary parametrized regular surface, the *mean curvature vector* defined by $\mathbf{H} = HN$. The geometrical meaning of the direction of $\mathbf{H}$ can be obtained from Eq. (12). Indeed, if we choose $h = H$, we have, for this particular variation,

$$A'(0) = -2\int_{\bar{D}} \langle \mathbf{H}, \mathbf{H} \rangle \sqrt{EG - F^2}\, du\, dv < 0.$$

This means that *if we deform* $\mathbf{x}(\bar{D})$ *in the direction of the vector* $\mathbf{H}$, *the area is initially decreasing.*

The mean curvature vector has another interpretation which we shall now pursue, since it has important implications for the theory of minimal surfaces.

A regular parametrized surface $\mathbf{x} = \mathbf{x}(u, v)$ is said to be *isothermal* if $\langle \mathbf{x}_u, \mathbf{x}_u \rangle = \langle \mathbf{x}_v, \mathbf{x}_v \rangle$ and $\langle \mathbf{x}_u, \mathbf{x}_v \rangle = 0$.

PROPOSITION 2. *Let* $\mathbf{x} = \mathbf{x}(u, v)$ *be a regular parametrized surface and assume that* $\mathbf{x}$ *is isothermal. Then*

$$\mathbf{x}_{uu} + \mathbf{x}_{vv} = 2\lambda^2 \mathbf{H},$$

where $\lambda^2 = \langle \mathbf{x}_u, \mathbf{x}_u \rangle = \langle \mathbf{x}_v, \mathbf{x}_v \rangle$.

Proof. Since $\mathbf{x}$ is isothermal, $\langle \mathbf{x}_u, \mathbf{x}_u \rangle = \langle \mathbf{x}_v, \mathbf{x}_v \rangle$ and $\langle \mathbf{x}_u, \mathbf{x}_v \rangle = 0$. By differentiation, we obtain

$$\langle \mathbf{x}_{uu}, \mathbf{x}_u \rangle = \langle \mathbf{x}_{vu}, \mathbf{x}_v \rangle = -\langle \mathbf{x}_u, \mathbf{x}_{vv} \rangle.$$

Thus,

$$\langle \mathbf{x}_{uu} + \mathbf{x}_{vv}, \mathbf{x}_u \rangle = 0.$$

Similarly,

$$\langle \mathbf{x}_{uu} + \mathbf{x}_{vv}, \mathbf{x}_v \rangle = 0.$$

It follows that $\mathbf{x}_{uu} + \mathbf{x}_{vv}$ is parallel to N. Since $\mathbf{x}$ is isothermal,

$$H = \frac{1}{2}\frac{g + e}{\lambda^2}.$$

Thus,

$$2\lambda^2 H = g + e = \langle N, \mathbf{x}_{uu} + \mathbf{x}_{vv} \rangle;$$

hence,

$$\mathbf{x}_{uu} + \mathbf{x}_{vv} = 2\lambda^2 \mathbf{H}.$$ **Q.E.D.**

The Laplacian Δf of a differentiable function $f\colon U \subset R^2 \to R$ is defined by $\Delta f = (\partial^2 f/\partial u^2) + (\partial^2 f/\partial v^2)$, $(u, v) \in U$. We say that f is *harmonic* in U if $\Delta f = 0$. From Prop. 2, we obtain

COROLLARY: *Let* $\mathbf{x}(u, v) = (x(u, v), y(u, v), z(u, v))$ *be a parametrized surface and assume that* $\mathbf{x}$ *is isothermal. Then* $\mathbf{x}$ *is minimal if and only if its coordinate functions* x, y, z *are harmonic.*

Example 5. *The catenoid*, given by

$$x(u, v) = (a \cosh v \cos u, a \cosh v \sin u, av),$$
$$0 < u < 2\pi, \qquad -\infty < v < \infty.$$

This is the surface generated by rotating the catenary $y = a \cosh(z/a)$ about the z axis (Fig. 3-41). It is easily checked that $E = G = a^2 \cosh^2 v$, $F = 0$, and that $\mathbf{x}_{uu} + \mathbf{x}_{vv} = 0$. Thus, the catenoid is a minimal surface. It can be characterized as the only surface of revolution which is minimal.

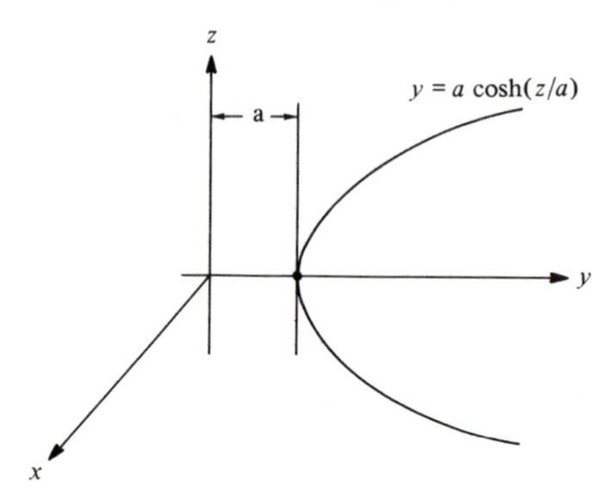

Figure 3-41

The last assertion can be proved as follows. We want to find a curve $y = f(x)$ such that, when rotated about the x axis, it describes a minimal surface. Since the parallels and the meridians of a surface of revolution are lines of curvature of the surface (Sec. 3-3, Example 4), we must have that the curvature of the curve $y = f(x)$ is the negative of the normal curvature of the circle generated by the point $f(x)$ (both are principal curvatures). Since the curvature of $y = f(x)$ is

$$\frac{y''}{(1 + (y')^2)^{3/2}}$$

and the normal curvature of the circle is the projection of its usual curvature $(= 1/y)$ over the normal N to the surface (see Fig. 3-42), we obtain

$$\frac{y''}{(1 + (y')^2)^{3/2}} = -\frac{1}{y} \cos \varphi.$$

But $-\cos \varphi = \cos \theta$ (see Fig. 3-42), and since $\tan \theta = y'$, we obtain

$$\frac{y''}{(1 + (y')^2)^{3/2}} = \frac{1}{y} \frac{1}{(1 + (y')^2)^{1/2}}$$

as the equation to be satisfied by the curve $y = f(x)$.

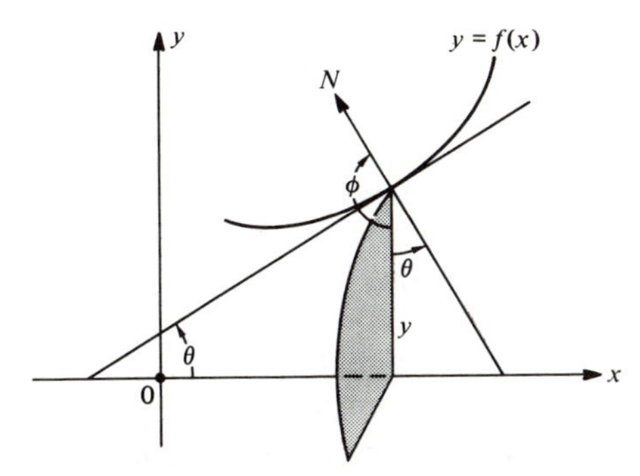

Figure 3-42

Clearly, there exists a point x where $f'(x) \neq 0$. Let us work in a neighborhood of this point where $f' \neq 0$. Multiplying both members of the above equation by $2y'$, we obtain,

$$\frac{2y''y'}{1 + (y')^2} = \frac{2y'}{y}.$$

Setting $1 + (y')^2 = z$ (hence, $2y''y' = z'$), we have

$$\frac{z'}{z} = \frac{2y'}{y},$$

which, by integration, gives (k is a constant)

$$\log z = \log y^2 + \log k^2 = \log(yk)^2$$

or

$$1 + (y')^2 = z = (yk)^2.$$

The last expression can be written

$$\frac{k\,dy}{\sqrt{(yk)^2 - 1}} = k\,dx,$$

which, again by integration, gives (c is a constant)

$$\cosh^{-1}(yk) = kx + c$$

or

$$y = \frac{1}{k}\cosh(kx + c).$$

Thus, in the neighborhood of a point where $f' \neq 0$, the curve $y = f(x)$ is a catenary. But then y' can only be zero at $x = 0$, and if the surface is to be connected, it is by continuity a catenoid, as we claimed.

Example 6 (*The Helicoid*). (cf. Example 3, Sec. 2-5.)

$$\mathbf{x}(u, v) = (a \sinh v \cos u, a \sinh v \sin u, au).$$

It is easily checked that $E = G = a^2 \cosh^2 v$, $F = 0$, and $\mathbf{x}_{uu} + \mathbf{x}_{vv} = 0$. Thus, the helicoid is a minimal surface. It has the additional property that it is the only minimal surface, other than the plane, which is also a ruled surface.

We can give a proof of the last assertion if we assume that the zeros of the Gaussian curvature of a minimal surface are isolated (for a proof, see, for instance, the survey of Osserman quoted at the end of this section, p. 76). Granted this, we shall proceed as follows.

Assume that the surface is not a plane. Then in some neighborhood W of the surface the Gaussian curvature K is strictly negative. Since the mean curvature is zero, W is covered by two families of asymptotic curves which intersect orthogonally. Since the rulings are asymptotic curves and the surface is not a plane, we can choose a point $q \in W$ such that the asymptotic curve, other than the ruling, passing through q has nonzero torsion at q. Since the osculating plane of an asymptotic curve is the tangent plane to the surface, there is a neighborhood $V \subset W$ such that the rulings of V are principal normals to the family of twisted asymptotic curves (Fig. 3-43). It is an interesting exercise in curves to prove that this can occur if and only if the twisted curves are circular helices (cf. Exercise 18, Sec. 1-5). Thus, V is a part of a helicoid. Since the torsion of a circular helix is constant, we easily see that the whole surface is part of a helicoid, as we claimed.

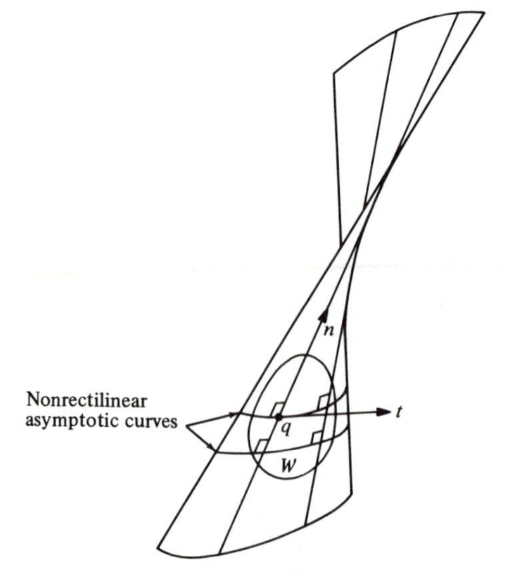

Figure 3-43

The helicoid and the catenoid were discovered in 1776 by Meusnier, who also proved that Lagrange's definition of minimal surfaces as critical points of a variational problem is equivalent to the vanishing of the mean curvature. For a long time, they were the only known examples of minimal surfaces. Only in 1835 did Scherk find further examples, one of which is described in Example 8. In Exercise 14, we shall describe an interesting connection between the helicoid and the catenoid.

Example 7 (*Enneper's Minimal Surface*). Enneper's surface is the parametrized surface

$$\mathbf{x}(u, v) = \left(u - \frac{u^3}{3} + uv^2, v - \frac{v^3}{3} + vu^2, u^2 - v^2\right), \qquad (u, v) \in R^2,$$

which is easily seen to be minimal (Fig. 3-44). Notice that by changing (u, v) into $(-v, u)$ we change, in the surface, (x, y, z) into $(-y, x, -z)$. Thus, if we perform a positive rotation of $\pi/2$ about the z axis and follow it by a symmetry in the xy plane, the surface remains invariant.

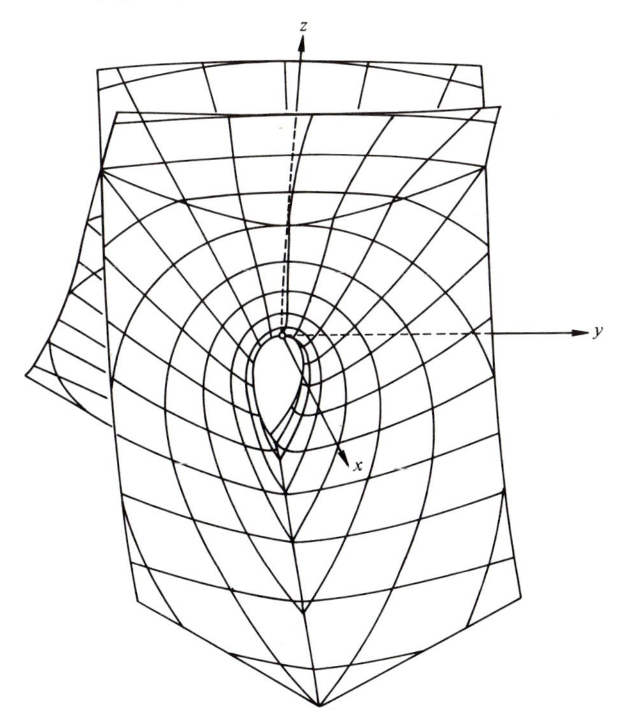

Figure 3-44. Enneper's surface. Reproduced, with modifications, from K. Leichtweiss, "Minimalflächen im Grossen," *Überblicke Math.* 2 (1969), 7–49, Fig. 4, with permission.

An interesting feature of Enneper's surface is that it has self-intersections. This can be shown by setting $u = \rho \cos \theta$, $v = \rho \sin \theta$ and writing

$$\mathbf{x}(\rho, \theta) = \left(\rho \cos \theta - \frac{\rho^3}{3} \cos 3\theta, \rho \sin \theta + \frac{\rho^3}{3} \sin 3\theta, \rho^2 \cos 2\theta\right) \cdot$$

Thus, if $\mathbf{x}(\rho_1, \theta_1) = \mathbf{x}(\rho_2, \theta_2)$, a straightforward computation shows that

$$\begin{aligned} x^2 + y^2 &= \rho_1^2 + \frac{\rho_1^6}{9} - \cos 4\theta \frac{2\rho_1^4}{3} \\ &= \left(\rho_1 + \frac{\rho_1^3}{3}\right)^2 - \frac{4}{3}(\rho_1^2 \cos 2\theta_1)^2 \\ &= \left(\rho_2 + \frac{\rho_2^3}{3}\right)^2 - \frac{4}{3}(\rho_2^2 \cos 2\theta_2)^2. \end{aligned}$$

Hence, since $\rho_1^2 \cos^2 2\theta_1 = \rho_2^2 \cos^2 2\theta_2$, we obtain

$$\rho_1 + \frac{\rho_1^3}{3} = \rho_2 + \frac{\rho_2^3}{3},$$

which implies that $\rho_1 = \rho_2$. It follows that $\cos 2\theta_1 = \cos 2\theta_2$.

If, for instance, $\rho_1 = \rho_2$ and $\theta_1 = 2\pi - \theta_2$, we obtain from

$$y(\rho_1, \theta_1) = y(\rho_2, \theta_2)$$

that $y = -y$. Hence, $y = 0$; that is, the points (ρ_1, θ_2) and (ρ_2, θ_2) belong to the curve $\sin \theta + (\rho^2/3) \sin 3\theta = 0$. Clearly, for each point (ρ, θ) belonging to this curve, the point $(\rho, 2\pi - \theta)$ also belongs to it, and

$$x(\rho, \theta) = x(\rho, 2\pi - \theta), \; z(\rho, \theta) = z(\rho, 2\pi - \theta).$$

Thus, the intersection of the surface with the plane $y = 0$ is a curve along which the surface intersects itself.

Similarly, it can be shown that the intersection of the surface with the plane $x = 0$ is also a curve of self-intersection (this corresponds to the case $\rho_1 = \rho_2$, $\theta_1 = \pi - \theta_2$). It is easily seen that they are the only self-intersections of Enneper's surface.

I want to thank Alcides Lins Neto for having worked out this example in order to draw a first sketch of Fig. 3-44.

Before going into the next example, we shall establish a useful relation between minimal surfaces and analytic functions of a complex variable. Let $\mathbb{C}$ denote the complex plane, which is, as usual, identified with R^2 by setting $\zeta = u + iv$, $\zeta \in \mathbb{C}$, $(u, v) \in R^2$. We recall that a function $f\colon U \subset \mathbb{C} \to \mathbb{C}$

is *analytic* when, by writing

$$f(\zeta) = f_1(u, v) + if_2(u, v),$$

the real functions f_1 and f_2 have continuous partial derivatives of first order which satisfy the so-called Cauchy-Riemann equations:

$$\frac{\partial f_1}{\partial u} = \frac{\partial f_2}{\partial v}, \qquad \frac{\partial f_1}{\partial v} = -\frac{\partial f_2}{\partial u}.$$

Now let $\mathbf{x}: U \subset R^2 \longrightarrow R^3$ be a regular parametrized surface and define complex functions $\varphi_1, \varphi_2, \varphi_3$ by

$$\varphi_1(\zeta) = \frac{\partial x}{\partial u} - i\frac{\partial x}{\partial v}, \qquad \varphi_2(\zeta) = \frac{\partial y}{\partial u} - i\frac{\partial y}{\partial v}, \qquad \varphi_3(\zeta) = \frac{\partial z}{\partial u} - i\frac{\partial z}{\partial v},$$

where x, y, and z are the component functions of $\mathbf{x}$.

LEMMA. $\mathbf{x}$ *is isothermal if and only if* $\varphi_1^2 + \varphi_2^2 + \varphi_3^2 \equiv 0$. *If this last condition is satisfied,* $\mathbf{x}$ *is minimal if and only if* φ_1, φ_2, *and* φ_3 *are analytic functions.*

Proof. By a simple computation, we obtain that

$$\varphi_1^2 + \varphi_2^2 + \varphi_3^2 = E - G + 2iF,$$

whence the first part of the lemma. Furthermore, $\mathbf{x}_{uu} + \mathbf{x}_{vv} = 0$ if and only if

$$\frac{\partial}{\partial u}\left(\frac{\partial x}{\partial v}\right) = -\frac{\partial}{\partial v}\left(\frac{\partial x}{\partial v}\right),$$
$$\frac{\partial}{\partial u}\left(\frac{\partial y}{\partial u}\right) = -\frac{\partial}{\partial v}\left(\frac{\partial y}{\partial v}\right),$$
$$\frac{\partial}{\partial u}\left(\frac{\partial z}{\partial u}\right) = -\frac{\partial}{\partial v}\left(\frac{\partial z}{\partial v}\right),$$

which give one-half of the Cauchy-Riemann equations for $\varphi_1, \varphi_2, \varphi_3$. Since the other half is automatically satisfied, we conclude that $\mathbf{x}_{uu} + \mathbf{x}_{vv} = 0$ if and only if φ_1, φ_2, and φ_3 are analytic. **Q.E.D.**

Example 8 (*Scherk's Minimal Surface*). This is given by

$$\mathbf{x}(u, v) = \left(\arg\frac{\zeta + i}{\zeta - i}, \arg\frac{\zeta + 1}{\zeta - 1}, \log\left|\frac{\zeta^2 + 1}{\zeta^2 - 1}\right|\right),$$
$$\zeta \neq \pm 1, \zeta \neq \pm i,$$

where $\zeta = u + iv$, and $\arg \zeta$ is the angle that the real axis makes with ζ.

We easily compute that

$$\arg\frac{\zeta+i}{\zeta-i}=\tan^{-1}\frac{2u}{u^2+v^2-1},$$

$$\arg\frac{\zeta+1}{\zeta-1}=\tan^{-1}\frac{-2v}{u^2+v^2-1}$$

$$\log\left|\frac{\zeta^2+1}{\zeta^2-1}\right|=\frac{1}{2}\log\frac{(u^2-v^2+1)^2+4u^2v^2}{(u^2-v^2-1)^2+4u^2v^2};$$

hence,

$$\varphi_1=\frac{\partial x}{\partial u}-i\frac{\partial x}{\partial v}=-\frac{2}{1+\zeta^2},\qquad \varphi_2=-\frac{2i}{1-\zeta^2},\qquad \varphi_3=\frac{4\zeta}{1-\zeta^4}.$$

Since $\varphi_1^2+\varphi_2^2+\varphi_3^2\equiv 0$ and φ_1, φ_2, and φ_3 are analytic, $\mathbf{x}$ is an isothermal parametrization of a minimal surface.

It is easily seen from the expressions of x, y, and z that

$$z=\log\frac{\cos y}{\cos x}.$$

This representation shows that Scherk's surface is defined on the chessboard pattern of Fig. 3-45 (except at the vertices of the squares, where the surface is actually a vertical line).

Minimal surfaces are perhaps the best-studied surfaces in differential geometry, and we have barely touched the subject. A very readable introduction can be found in R. Osserman, *A Survey of Minimal Surfaces*, Van Nostrand Mathematical Studies, Van Nostrand Reinhold, New York, 1969. The theory has developed into a rich branch of differential geometry in which interesting and nontrivial questions are still being investigated. It has deep connections with analytic functions of complex variables and partial differential equations. As a rule, the results of the theory have the charming quality that they are easy to visualize and very hard to prove. To convey to the reader some flavor of the subject we shall close this brief account by stating without proof one striking result.

THEOREM (Osserman). *Let* $S\subset R^3$ *be a regular, closed (as a subset of* R^3*) minimal surface in* R^3 *which is not a plane. Then the image of the Gauss map* $N\colon S\to S^2$ *is* dense in the sphere S^2 (*that is, arbitrarily close to any point of* S^2 *there is a point of* $N(S)\subset S^2$).

A proof of this theorem can be found in Osserman's survey, quoted above. Actually, the theorem is somewhat stronger in that it applies to complete surfaces, a concept to be defined in Sec. 5-3.

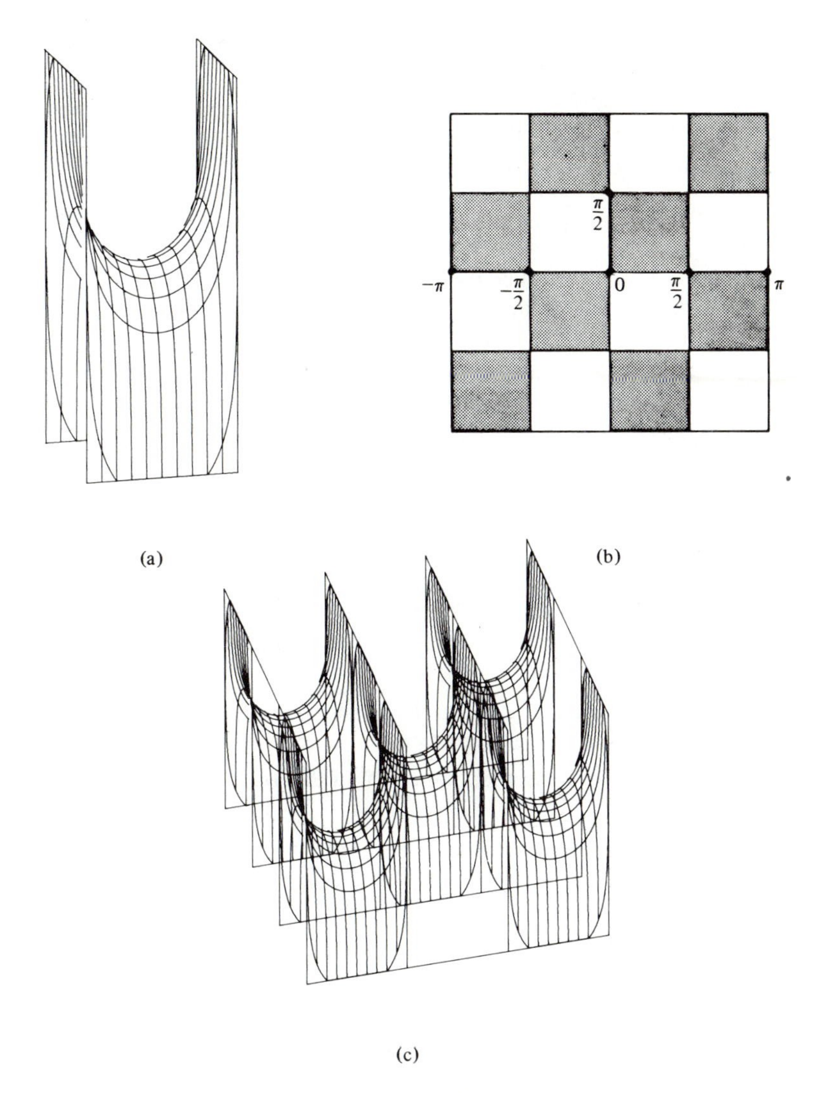

Figure 3-45. Scherk's surface.

EXERCISES

1. Show that the helicoid (cf. Example 3, Sec. 2-5) is a ruled surface, its line of striction is the z axis, and its distribution parameter is constant.

2. Show that on the hyperboloid of revolution $x^2 + y^2 - z^2 = 1$, the parallel of least radius is the line of striction, the rulings meet is under a constant angle, and the distribution parameter is constant.

3. Let $\alpha: I \longrightarrow S \subset R^3$ be a curve on a regular surface S and consider the ruled surface generated by the family $\{\alpha(t), N(t)\}$, where $N(t)$ is the normal to the surface at $\alpha(t)$. Prove that $\alpha(I) \subset S$ is a line of curvature in S if and only if this ruled surface is developable.

4. Assume that a noncylindrical ruled surface

$$x(t, v) = \alpha(t) + vw(t), \qquad |w| = 1,$$

is regular. Let $w(t_1)$, $w(t_2)$ be the directions of two rulings of $\mathbf{x}$ and let $\mathbf{x}(t_1, v_1)$, $\mathbf{x}(t_2, v_2)$ be the feet of the common perpendicular to these two rulings. As $t_2 \longrightarrow t_1$, these points tend to a point $\mathbf{x}(t_1, \bar{v})$. To determine $(t_1, \bar{v})$ prove the following:

a. The unit vector of the common perpendicular converges to a unit vector tangent to the surface at $(t_1, \bar{v})$. Conclude that, at $(t_1, \bar{v})$,

$$\langle w' \wedge w, N \rangle = 0.$$

b. $\bar{v} = -(\langle \alpha', w' \rangle / \langle w', w' \rangle)$.

Thus, $(t_1, \bar{v})$ is the central point of the ruling through t_1, and this gives another interpretation of the line of striction (assumed nonsingular).

5. A *right conoid* is a ruled surface whose rulings L_t intersect perpendicularly at fixed axis r which does not meet the directrix $\alpha: I \longrightarrow R^3$.

a. Find a parametrization for the right conoid and determine a condition that implies it to be noncylindrical.

b. Given a noncylindrical right conoid, find the line of striction and the distribution parameter.

6. Let

$$\mathbf{x}(t, v) = \alpha(t) + vw(t)$$

be a developable surface. Prove that at a regular point we have

$$\langle N_v, \mathbf{x}_v \rangle = \langle N_v, \mathbf{x}_t \rangle = 0.$$

Conclude that *the tangent plane of a developable surface is constant along* (the regular points of) *a fixed ruling.*

7. Let S be a regular surface and let $C \subset S$ be a regular curve on S, nowhere tangent to an asymptotic direction. Consider the envelope of the family of tangent planes of S along C. Prove that the direction of the ruling that passes through a point $p \in C$ is conjugate to the tangent direction of C at p.

8. Show that if $C \subset S^2$ is a parallel of a unit sphere S^2, then the envelope of tangent planes of S^2 along C is either a cylinder, if C is an equator, or a cone, if C is not an equator.

9. (*Focal Surfaces.*) Let S be a regular surface without parabolic or umbilical points. Let $\mathbf{x}: U \longrightarrow S$ be a parametrization of S such that the coordinate curves are lines of curvature (if U is small, this is no restriction. cf. Corollary 4, Sec. 3-4). The parametrized surfaces

$$\mathbf{y}(u, v) = \mathbf{x}(u, v) + \rho_1 N(u, v),$$
$$\mathbf{z}(u, v) = \mathbf{x}(u, v) + \rho_2 N(u, v),$$

where $\rho_1 = 1/k_1$, $\rho_2 = 1/k_2$, are called *focal surfaces* of $\mathbf{x}(U)$ (or *surfaces of centers* of $\mathbf{x}(U)$; this terminology comes from the fact that $\mathbf{y}(u, v)$, for instance, is the center of the osculating circle (cf. Sec. 1-6, Exercise 2) of the normal section at $\mathbf{x}(u, v)$ corresponding to the principal curvature k_1). Prove that

a. If $(k_1)_u$ and $(k_2)_v$ are nowhere zero, then $\mathbf{y}$ and $\mathbf{z}$ are regular parametrized surfaces.

b. At the regular points, the directions on a focal surface corresponding to the principal directions on $\mathbf{x}(U)$ are conjugate. That means, for instance, that $\mathbf{y}_u$ and $\mathbf{y}_v$ are conjugate vectors in $\mathbf{y}(U)$ for all $(u, v) \in U$.

c. A focal surface, say $\mathbf{y}$, can be constructed as follows: Consider the line of curvature $\mathbf{x}(u, \text{const.})$ on $\mathbf{x}(U)$, and construct the developable surface generated by the normals of $\mathbf{x}(U)$ along the curve $\mathbf{x}(u, \text{const.})$ (cf. Exercise 3). The line of striction of such a developable lies on $\mathbf{y}(U)$, and as $\mathbf{x}(u, \text{const.})$ describes $\mathbf{x}(U)$, this line describes $\mathbf{y}(U)$ (Fig. 3-46).

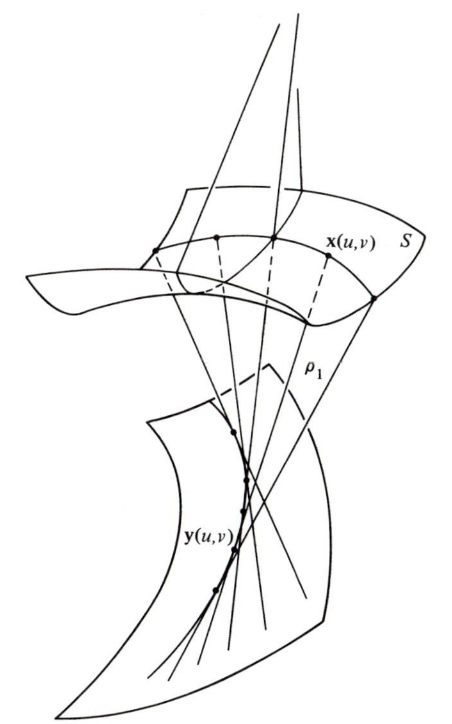

Figure 3-46. Construction of a focal surface.

10. Example 4 can be generalized as follows. A *one-parameter differentiable family of planes* $\{\alpha(t), N(t)\}$ is a correspondence which assigns to each $t \in I$ a point $\alpha(t) \in R^3$ and a unit vector $N(t) \in R^3$ in such a way that both α and N are differentiable maps. A family $\{\alpha(t), N(t)\}$, $t \in I$, is said to be a *family of tangent planes* if $\alpha'(t) \neq 0$, $N'(t) \neq 0$, and $\langle \alpha'(t), N(t) \rangle = 0$ for all $t \in I$.

a. Give proof that a differentiable one-parameter family of tangent planes $\{\alpha(t), N(t)\}$ determines a differentiable one-parameter family of lines $\{\alpha(t), (N \wedge N')/|N'|\}$ which generates a developable surface

$$\mathbf{x}(t, v) = \alpha(t) + v\frac{N \wedge N'}{|N'|}. \qquad (*)$$

The surface $(*)$ is called the *envelope of the family* $\{\alpha(t), N(t)\}$.

b. Prove that if $\alpha'(t) \wedge (N(t) \wedge N'(t)) \neq 0$ for all $t \in I$, then the envelope $(*)$ is regular in a neighborhood of $v = 0$, and the unit normal vector of $\mathbf{x}$ at $(t, 0)$ is $N(t)$.

c. Let $\alpha = \alpha(s)$ be a curve in R^3 parametrized by arc length. Assume that the curvature $k(s)$ and the torsion $\tau(s)$ of α are nowhere zero. Prove that the family of osculating planes $\{\alpha(s), b(s)\}$ is a one-parameter differentiable family of tangent planes and that the envelope of this family is the tangent surface to $\alpha(s)$ (cf. Example 5, Sec. 2-3).

11. Let $\mathbf{x} = \mathbf{x}(u, v)$ be a regular parametrized surface. A *parallel surface* to $\mathbf{x}$ is a parametrized surface

$$\mathbf{y}(u, v) = \mathbf{x}(u, v) + aN(u, v),$$

where a is a constant.

a. Prove that $\mathbf{y}_u \wedge \mathbf{y}_v = (1 - 2Ha + Ka^2)(\mathbf{x}_u \wedge \mathbf{x}_v)$, where K and H are the Gaussian and mean curvatures of $\mathbf{x}$, respectively.

b. Prove that at the regular points, the Gaussian curvature of $\mathbf{y}$ is

$$\frac{K}{1 - 2Ha + Ka^2}$$

and the mean curvature of $\mathbf{y}$ is

$$\frac{H - Ka}{1 - 2Ha + Ka^2}.$$

c. Let a surface $\mathbf{x}$ have constant mean curvature equal to $c \neq 0$ and consider the parallel surface to $\mathbf{x}$ at a distance $1/2c$. Prove that this parallel surface has constant Gaussian curvature equal to $4c^2$.

12. Prove that there are no compact (i.e., bounded and closed in R^3) minimal surfaces.

13. a. Let S be a regular surface without umbilical points. Prove that S is a minimal surface if and only if the Gauss map $N: S \longrightarrow S^2$ satisfies, for all $p \in S$

and all $w_1, w_2 \in T_p(S)$,

$$\langle dN_p(w_1), dN_p(w_2)\rangle_{N(p)} = \lambda(p)\langle w_1, w_2\rangle_p,$$

where $\lambda(p) \neq 0$ is a number which depends only on p.

b. Let $\mathbf{x}\colon U \longrightarrow S^2$ be a parametrization of the unit sphere S^2 by $(\theta, \bar{\varphi}) \in U$, where θ is the colatitude (cf. Example 1, Sec. 2-2) and $\bar{\varphi}$ is the arc length of the parallel determined by θ. Consider a neighborhood V of a point p of the minimal surface S in part a such that $N\colon S \longrightarrow S^2$ restricted to V is a diffeomorphism (since $K(p) = \det(dN_p) \neq 0$, such a V exists by the inverse function theorem). Prove that the parametrization $y = N^{-1} \circ \mathbf{x}\colon U \longrightarrow S$ is isothermal (*this gives a way of introducing isothermal parametrizations on minimal surfaces without planar points*).

14. When two differentiable functions $f, g\colon U \subset R^2 \longrightarrow R$ satisfy the Cauchy-Riemann equations

$$\frac{\partial f}{\partial u} = \frac{\partial g}{\partial v}, \qquad \frac{\partial f}{\partial v} = -\frac{\partial g}{\partial u},$$

they are easily seen to be harmonic; in this situation, f and g are said to be *harmonic conjugate*. Let $\mathbf{x}$ and $\mathbf{y}$ be isothermal parametrizations of minimal surfaces such that their component functions are pairwise harmonic conjugate; then $\mathbf{x}$ and $\mathbf{y}$ are called *conjugate minimal surfaces*. Prove that

a. The helicoid and the catenoid are conjugate minimal surfaces.

b. Given two conjugate minimal surfaces, $\mathbf{x}$ and $\mathbf{y}$, the surface

$$\mathbf{z} = (\cos t)\mathbf{x} + (\sin t)\mathbf{y} \tag{$*$}$$

is again minimal for all $t \in R$.

c. All surfaces of the one-parameter family $(*)$ have the same fundamental form: $E = \langle \mathbf{x}_u, \mathbf{x}_u\rangle = \langle \mathbf{y}_v, \mathbf{y}_v\rangle$, $F = 0$, $G = \langle \mathbf{x}_v, \mathbf{x}_v\rangle = \langle \mathbf{y}_u, \mathbf{y}_u\rangle$.

Thus, any two conjugate minimal surfaces can be joined through a one-parameter family of minimal surfaces, and the first fundamental form of this family is independent of t.

Appendix Self-Adjoint Linear Maps and Quadratic Forms

In this appendix, V will denote a vector space of dimension 2, endowed with an inner product $\langle\ ,\ \rangle$. All that follows can be easily extended to a finite n-dimensional vector space, but for the sake of simplicity, we shall treat only the case $n = 2$.

We say that a linear map $A: V \to V$ is *self-adjoint* if $\langle Av, w\rangle = \langle v, Aw\rangle$ for all $v, w \in V$.

Notice that if $\{e_1, e_2\}$ is an orthonormal basis for V and (α_{ij}), $i, j = 1, 2$, is the matrix of A relative to that basis, then

$$\langle Ae_i, e_j\rangle = \alpha_{ij} = \langle e_i, Ae_j\rangle = \langle Ae_j, e_i\rangle = \alpha_{ji};$$

that is, the matrix (α_{ij}) is symmetric.

To each self-adjoint linear map we associate a map $B: V \times V \to R$ defined by

$$B(v, w) = \langle Av, w\rangle.$$

B is clearly bilinear; that is, it is linear in both v and w. Moreover, the fact that A is self-adjoint implies that $B(v, w) = B(w, v)$; that is, B is a bilinear symmetric form in V.

Conversely, if B is a bilinear symmetric form in V, we can define a linear map $A: V \to V$ by $\langle Av, w\rangle = B(v, w)$ and the symmetry of B implies that A is self-adjoint.

On the other hand, to each symmetric, bilinear form B in V, there corresponds a quadratic form Q in V given by

$$Q(v) = B(v, v), \qquad v \in V,$$

and the knowledge of Q determines B completely, since

$$B(u, v) = \tfrac{1}{2}[Q(u + v) - Q(u) - Q(v)].$$

Thus, a one-to-one correspondence is established between quadratic forms in V and self-adjoint linear maps of V.

The goal of this appendix is to prove that (see the theorem below) given a self-adjoint linear map $A: V \longrightarrow V$, there exists an orthonormal basis for V such that relative to that basis the matrix of A is a diagonal matrix. Furthermore, the elements on the diagonal are the maximum and the minimum of the corresponding quadratic form restricted to the unit circle of V.

LEMMA. *If the function* $Q(x, y) = ax^2 + 2bxy + cy^2$, *restricted to the unit circle* $x^2 + y^2 = 1$, *has a maximum at the point* $(1, 0)$, *then* $b = 0$.

Proof. Parametrize the circle $x^2 + y^2 = 1$ by $x = \cos t$, $y = \sin t$, $t \in (0 - \epsilon, 2\pi + \epsilon)$. Thus, Q, restricted to that circle, becomes a function of t:

$$Q(t) = a \cos^2 t + 2b \cos t \sin t + c \sin^2 t.$$

Since Q has a maximum at the point $(1, 0)$ we have

$$\left(\frac{dQ}{dt}\right)_{t=0} = 2b = 0.$$

Hence, $b = 0$ as we wished. **Q.E.D.**

PROPOSITION. *Given a quadratic form* Q *in* V, *there exists an orthonormal basis* $\{e_1, e_2\}$ *of* V *such that if* $v \in V$ *is given by* $v = xe_1 + ye_2$, *then*

$$Q(v) = \lambda_1 x^2 + \lambda_2 y^2,$$

where λ_1 *and* λ_2 *are the maximum and minimum, respectively, of* Q *on the unit circle* $|v| = 1$.

Proof. Let λ_1 be the maximum of Q on the unit circle $|v| = 1$, and let e_1 be a unit vector with $Q(e_1) = \lambda_1$. Such an e_1 exists by continuity of Q on the compact set $|v| = 1$. Let e_2 be a unit vector that is orthogonal to e_1, and set $\lambda_2 = Q(e_2)$. We shall show that the basis $\{e_1, e_2\}$ satisfies the conditions of the proposition.

Let B be the symmetric bilinear form that is associated to Q and set $v = xe_1 + ye_2$. Then

$$\begin{aligned} Q(v) = B(v, v) &= B(xe_1 + ye_2, xe_1 + ye_2) \\ &= \lambda_1 x^2 + 2bxy + \lambda_2 y^2, \end{aligned}$$

where $b = B(e_1, e_2)$. By the lemma, $b = 0$, and it only remains to prove that λ_2 is the minimum of Q in the circle $|v| = 1$. This is immediate because, for any $v = xe_1 + ye_2$ with $x^2 + y^2 = 1$, we have that

$$Q(v) = \lambda_1 x^2 + \lambda_2 y^2 \geq \lambda_2(x^2 + y^2) = \lambda_2,$$

since $\lambda_2 \leq \lambda_1$. **Q.E.D.**

We say that a vector $v \neq 0$ is an *eigenvector* of a linear map $A: V \rightarrow V$ if $Av = \lambda v$ for some real number λ; λ is then called an *eigenvalue* of A.

THEOREM. *Let* $\mathrm{A: V \rightarrow V}$ *be a self-adjoint linear map. Then there exists an orthonormal basis* $\{\mathrm{e}_1, \mathrm{e}_2\}$ *of* V *such that* $\mathrm{A(e_1)} = \lambda_1 \mathrm{e}_1$, $\mathrm{A(e_2)} = \lambda_2 \mathrm{e}_2$ *(that is,* e_1 *and* e_2 *are eigenvectors, and* λ_1, λ_2 *are eigenvalues of* A*). In the basis* $\{\mathrm{e}_1, \mathrm{e}_2\}$*, the matrix of* A *is clearly diagonal and the elements* λ_1, λ_2, $\lambda_1 \geq \lambda_2$*, on the diagonal are the maximum and the minimum, respectively, of the quadratic form* $\mathrm{Q(v)} = \langle \mathrm{Av, v} \rangle$ *on the unit circle of* V.

Proof. Consider the quadratic form $Q(v) = \langle Av, v \rangle$. By the proposition above, there exists an orthonormal basis $\{e_1, e_2\}$ of V, with $Q(e_1) = \lambda_1$, $Q(e_2) = \lambda_2 \leq \lambda_1$, where λ_1 and λ_2 are the maximum and minimum, respectively, of Q in the unit circle. It remains, therefore, to prove that

$$A(e_1) = \lambda_1 e_1, \qquad A(e_2) = \lambda_2(e_2).$$

Since $B(e_1, e_2) = \langle Ae_1, e_2 \rangle = 0$ (by the lemma) and $e_2 \neq 0$, we have that either Ae_1 is parallel to e_1 or $Ae_1 = 0$. If Ae_1 is parallel to e_1, then $Ae_1 = \alpha e_1$, and since $\langle Ae_1, e_1 \rangle = \lambda_1 = \langle \alpha e_1, e_1 \rangle = \alpha$, we conclude that $Ae_1 = \lambda_1 e_1$; if $Ae_1 = 0$, then $\lambda_1 = \langle Ae_1, e_1 \rangle = 0$, and $Ae_1 = 0 = \lambda_1 e_1$. Thus, we have in any case that $Ae_1 = \lambda_1 e_1$.

Now using the fact that

$$B(e_1, e_2) = \langle Ae_2, e_1 \rangle = 0$$

and that

$$\langle Ae_2, e_2 \rangle = \lambda_2,$$

we can prove in the same way that $Ae_2 = \lambda_2 e_2$. **Q.E.D.**

Remark. The extension of the above results to an n-dimensional vector space, $n > 2$, requires only the following precaution. In the previous proposition, we choose the maximum $\lambda_1 = Q(e_1)$ of Q in the unit sphere, and then show that Q restricts to a quadratic form Q_1 in the subspace V_1 orthogonal to e_1. We choose for $\lambda_2 = Q_1(e_2)$ the maximum of Q_1 in the unit sphere of V_1, and so forth.

4 The Intrinsic Geometry of Surfaces

4-1. Introduction

In Chap. 2 we introduced the first fundamental form of a surface S and showed how it can be used to compute simple metric concepts on S (length, angle, area, etc.). The important point is that such computations can be made without "leaving" the surface, once the first fundamental form is known. Because of this, these concepts are said to be intrinsic to the surface S.

The geometry of the first fundamental form, however, does not exhaust itself with the simple concepts mentioned above. As we shall see in this chapter, many important local properties of a surface can be expressed only in terms of the first fundamental form. The study of such properties is called the *intrinsic geometry* of the surface. This chapter is dedicated to intrinsic geometry.

In Sec. 4-2 we shall define the notion of isometry, which essentially makes precise the intuitive idea of two surfaces having "the same" first fundamental forms.

In Sec. 4-3 we shall prove the celebrated Gauss formula that expresses the Gaussian curvature K as a function of the coefficients of the first fundamental form and its derivatives. This means that K is an intrinsic concept, a very striking fact if we consider that K was defined using the second fundamental form.

In Sec. 4-4 we shall start a systematic study of intrinsic geometry. It turns out that the subject can be unified through the concept of covariant derivative of a vector field on a surface. This is a generalization of the usual derivative of a vector field on the plane and plays a fundamental role throughout the chapter.

Section 4-5 is devoted to the Gauss-Bonnet theorem both in its local and global versions. This is probably the most important theorem of this book. Even in a short course, one should make an effort to reach Sec. 4-5.

In Sec. 4-6 we shall define the exponential map and use it to introduce two special coordinate systems, namely, the normal coordinates and the geodesic polar coordinates.

In Sec. 4-7 we shall take up some delicate points on the theory of geodesics which were left aside in the previous sections. For instance, we shall prove the existence, for each point p of a surface S, of a neighborhood of p in S which is a normal neighborhood of all its points (the definition of normal neighborhood is given in Sec. 4-6). This result and a related one are used in Chap. 5; however, it is probably convenient to assume them and omit Sec. 4-7 on a first reading. We shall also prove the existence of convex neighborhoods, but this is used nowhere else in the book.

4-2. Isometries; Conformal Maps

Examples 1 and 2 of Sec. 2-5 display an interesting peculiarity. Although the cylinder and the plane are distinct surfaces, their first fundamental forms are "equal" (at least in the coordinate neighborhoods that we have considered). This means that insofar as intrinsic metric questions are concerned (length, angle, area), the plane and the cylinder behave locally in the same way. (This is intuitively clear, since by cutting a cylinder along a generator we may unroll it onto a part of a plane.) In this chapter we shall see that many other important concepts associated to a regular surface depend only on the first fundamental form and should be included in the category of intrinsic concepts. It is therefore convenient that we formulate in a precise way what is meant by two regular surfaces having equal first fundamental forms.

S and $\bar{S}$ will always denote regular surfaces.

DEFINITION 1. *A diffeomorphism* $\varphi\colon S \to \bar{S}$ *is an* isometry *if for all* $p \in S$ *and all pairs* $w_1, w_2 \in T_p(S)$ *we have*

$$\langle w_1, w_2 \rangle_p = \langle d\varphi_p(w_1), d\varphi_p(w_2) \rangle_{\varphi(p)}.$$

The surfaces S *and* $\bar{S}$ *are then said to be* isometric.

In other words, a diffeomorphism φ is an isometry if the differential $d\varphi$ preserves the inner product. It follows that, $d\varphi$ being an isometry,

$$I_p(w) = \langle w, w \rangle_p = \langle d\varphi_p(w), d\varphi_p(w) \rangle_{\varphi(p)} = I_{\varphi(p)}(d\varphi_p(w))$$

for all $w \in T_p(S)$. Conversely, if a diffeomorphism φ preserves the first fundamental form, that is,

$$I_p(w) = I_{\varphi(p)}(d\varphi_p(w)) \qquad \text{for all } w \in T_p(S),$$

then

$$\begin{aligned} 2\langle w_1, w_2 \rangle &= I_p(w_1 + w_2) - I_p(w_1) - I_p(w_2) \\ &= I_{\varphi(p)}(d\varphi_p(w_1 + w_2)) - I_{\varphi(p)}(d\varphi_p(w_1)) - I_{\varphi(p)}(d\varphi_p(w_2)) \\ &= 2\langle d\varphi_p(w_1), d\varphi_p(w_2) \rangle, \end{aligned}$$

and φ is, therefore, an isometry.

DEFINITION 2. *A map* $\varphi\colon \mathrm{V} \longrightarrow \bar{\mathrm{S}}$ *of a neighborhood* V *of* $\mathrm{p} \in \mathrm{S}$ *is a* local isometry *at* p *if there exists a neighborhood* $\bar{\mathrm{V}}$ *of* $\varphi(\mathrm{p}) \in \bar{\mathrm{S}}$ *such that* $\varphi\colon \mathrm{V} \longrightarrow \bar{\mathrm{V}}$ *is an isometry. If there exists a local isometry into* $\bar{\mathrm{S}}$ *at every* $\mathrm{p} \in \mathrm{S}$, *the surface* S *is said to be* locally isometric *to* $\bar{\mathrm{S}}$. S *and* $\bar{\mathrm{S}}$ *are* locally isometric *if* S *is locally isometric to* $\bar{\mathrm{S}}$ *and* $\bar{\mathrm{S}}$ *is locally isometric to* S.

It is clear that if $\varphi\colon S \longrightarrow \bar{S}$ is a diffeomorphism and a local isometry for every $p \in S$, then φ is an isometry (globally). It may, however, happen that two surfaces are locally isometric without being (globally) isometric, as shown in the following example.

Example 1. Let φ be a map of the coordinate neighborhood $\bar{\mathbf{x}}(U)$ of the cylinder given in Example 2 of Sec. 2-5 into the plane $\mathbf{x}(R^2)$ of Example 1 of Sec. 2-5, defined by $\varphi = \mathbf{x} \circ \bar{\mathbf{x}}^{-1}$ (we have changed $\mathbf{x}$ to $\bar{\mathbf{x}}$ in the parametrization of the cylinder). Then φ is a local isometry. In fact, each vector w, tangent to the cylinder at a point $p \in \bar{\mathbf{x}}(U)$, is tangent to a curve $\bar{\mathbf{x}}(u(t), v(t))$, where $(u(t), v(t))$ is a curve in $U \subset R^2$. Thus, w can be written as

$$w = \bar{\mathbf{x}}_u u' + \bar{\mathbf{x}}_v v'.$$

On the other hand, $d\varphi(w)$ is tangent to the curve

$$\varphi(\bar{\mathbf{x}}(u(t), v(t))) = \mathbf{x}(u(t), v(t)).$$

Thus, $d\varphi(w) = \mathbf{x}_u u' + \mathbf{x}_v v'$. Since $E = \bar{E}, F = \bar{F}, G = \bar{G}$, we obtain

$$\begin{aligned} I_p(w) &= \bar{E}(u')^2 + 2\bar{F}u'v' + \bar{G}(v')^2 \\ &= E(u')^2 + 2Fu'v' + G(v')^2 = I_{\varphi(p)}(d\varphi_p(w)), \end{aligned}$$

as we claimed. It follows that the cylinder $x^2 + y^2 = 1$ is locally isometric to a plane.

The isometry cannot be extended to the entire cylinder because the cylinder is not even homeomorphic to a plane. A rigorous proof of the last assertion would take us too far afield, but the following intuitive argument may give an idea of the proof. Any simple closed curve in the plane can be shrunk continuously into a point without leaving the plane (Fig. 4-1). Such a property would certainly be preserved under a homeomorphism. But a parallel of the cylinder (Fig. 4-1) does not have that property, and this contradicts the existence of a homeomorphism between the plane and the cylinder.

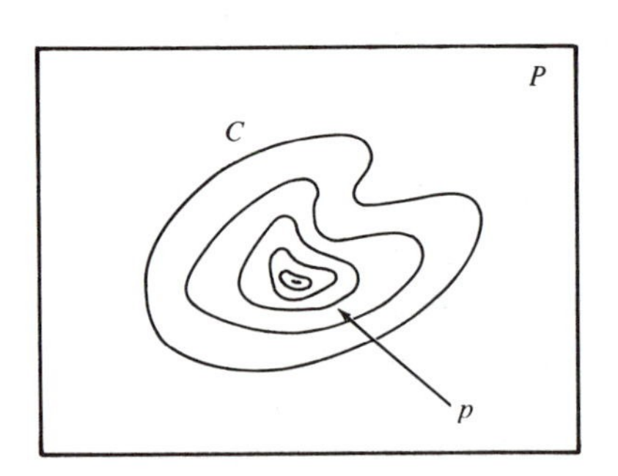

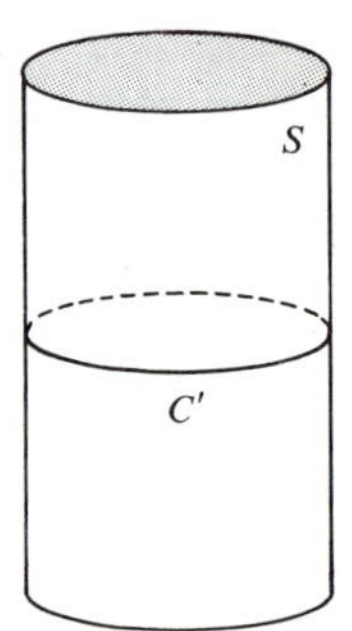

Figure 4-1. $C \subset P$ can be shrunk continuously into p without leaving P. The same does not hold for $C' \subset S$.

Before presenting further examples, we shall generalize the argument given above to obtain a criterion for local isometry in terms of local coordinates.

PROPOSITION 1. *Assume the existence of parametrizations* $\mathbf{x}: U \to S$ *and* $\bar{\mathbf{x}}: U \to \bar{S}$ *such that* $E = \bar{E}$, $F = \bar{F}$, $G = \bar{G}$ *in* U. *Then the map* $\varphi = \bar{\mathbf{x}} \circ \mathbf{x}^{-1}: \mathbf{x}(U) \to \bar{S}$ *is a local isometry.*

Proof. Let $p \in \mathbf{x}(U)$ and $w \in T_p(S)$. Then w is tangent to a curve $\mathbf{x}(\alpha(t))$ at $t = 0$, where $\alpha(t) = (u(t), v(t))$ is a curve in U; thus, w may be written $(t = 0)$

$$w = \mathbf{x}_u u' + \mathbf{x}_v v'.$$

By definition, the vector $d\varphi_p(w)$ is the tangent vector to the curve $\bar{\mathbf{x}} \circ \mathbf{x}^{-1} \circ \bar{\mathbf{x}}(\alpha(t))$, i.e., to the curve $\bar{\mathbf{x}}(\alpha(t))$ at $t = 0$ (Fig. 4-2). Thus,

$$d\varphi_p(w) = \bar{\mathbf{x}}_u u' + \bar{\mathbf{x}}_v v'.$$

Since

$$I_p(w) = E(u')^2 + 2Fu'v' + G(v')^2,$$
$$I_{\varphi(p)}(d\varphi_p(w)) = \bar{E}(u')^2 + 2\bar{F}u'v' + \bar{G}(v')^2,$$

we conclude that $I_p(w) = I_{\varphi(p)}(d\varphi_p(w))$ for all $p \in \mathbf{x}(U)$ and all $w \in T_p(S)$; hence, φ is a local isometry. **Q.E.D.**

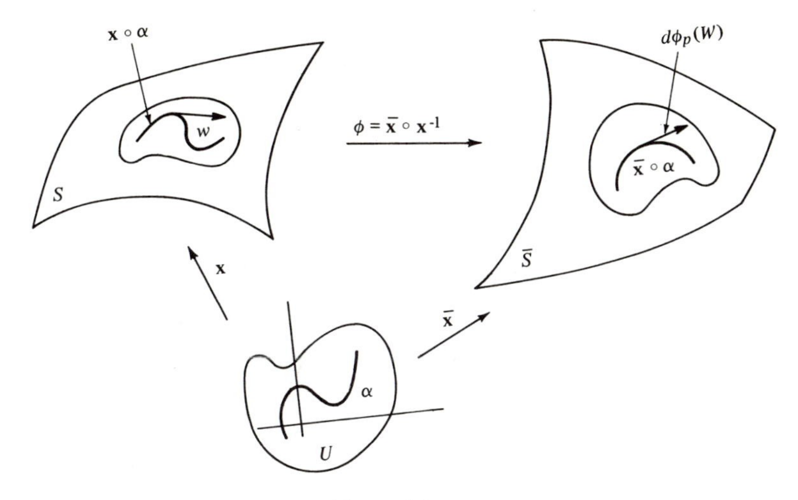

Figure 4-2

Example 2. Let S be a surface of revolution and let

$$\mathbf{x}(v, v) = (f(v) \cos u, f(v) \sin u, g(v)),$$
$$a \le v \le b, \qquad 0 < u < 2\pi, \qquad f(v) > 0,$$

be a parametrization of S (cf. Example 4, Sec. 2-3). The coefficients of the first fundamental form of S in the parametrization $\mathbf{x}$ are given by

$$E = (f(v))^2, \qquad F = 0, \qquad G = (f'(v))^2 + g'(v))^2.$$

In particular, the surface of revolution of the *catenary*,

$$x = a \cosh v, \qquad z = av, \qquad -\infty < v < \infty,$$

has the following parametrization:

$$\mathbf{x}(u, v) = (a \cosh v \cos u, a \cosh v \sin u, av),$$
$$0 < u < 2\pi, \qquad -\infty < v < \infty,$$

relative to which the coefficients of the first fundamental form are

$$E = a^2 \cosh^2 v, \qquad F = 0, \qquad G = a^2(1 + \sinh^2 v) = a^2 \cosh^2 v.$$

This surface of revolution is called the *catenoid* (see Fig. 4-3). We shall show that the catenoid is locally isometric to the helicoid of Example 3, Sec. 2-5.

A parametrization for the helicoid is given by

$$\bar{\mathbf{x}}(\bar{u}, \bar{v}) = (\bar{v} \cos \bar{u}, \bar{v} \sin \bar{u}, a\bar{u}), \qquad 0 < \bar{u} < 2\pi, -\infty < \bar{v} < \infty.$$

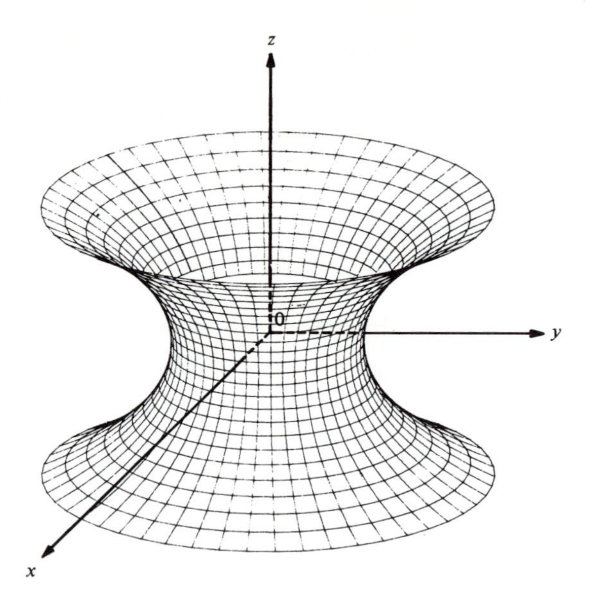

Figure 4-3. The catenoid.

Let us make the following change of parameters:

$$\bar{u} = u, \qquad \bar{v} = a \sinh v, \qquad 0 < u < 2\pi, \ -\infty < v < \infty,$$

which is possible since the map is clearly one-to-one, and the Jacobian

$$\frac{\partial(\bar{u}, \bar{v})}{\partial(u, v)} = a \cosh v$$

is nonzero everywhere. Thus, a new parametrization of the helicoid is

$$\bar{\mathbf{x}}(u, v) = (a \sinh v \cos u, a \sinh v \sin u, au),$$

relative to which the coefficients of the first fundamental form are given by

$$E = a^2 \cosh^2 v, \qquad F = 0, \qquad G = a^2 \cosh^2 v.$$

Using Prop. 1, we conclude that the catenoid and the helicoid are locally isometric.

Figure 4-4 gives a geometric idea of how the isometry operates; it maps "one turn" of the helicoid (coordinate neighborhood corresponding to $0 < u < 2\pi$) into the catenoid minus one meridian.

Remark 1. The isometry between the helicoid and the catenoid has already appeared in Chap. 3 in the context of minimal surfaces; cf. Exercise 14, Sec. 3-5.

Example 3. We shall prove that the one-sheeted cone (minus the vertex)

$$z = +k\sqrt{x^2 + y^2}, \qquad (x, y) \neq (0, 0),$$

is locally isometric to a plane. The idea is to show that a cone minus a generator can be "rolled" onto a piece of a plane.

Let $U \subset R^2$ be the open set given in polar coordinates (ρ, θ) by

$$0 < \rho < \infty, \qquad 0 < \theta < 2\pi \sin \alpha,$$

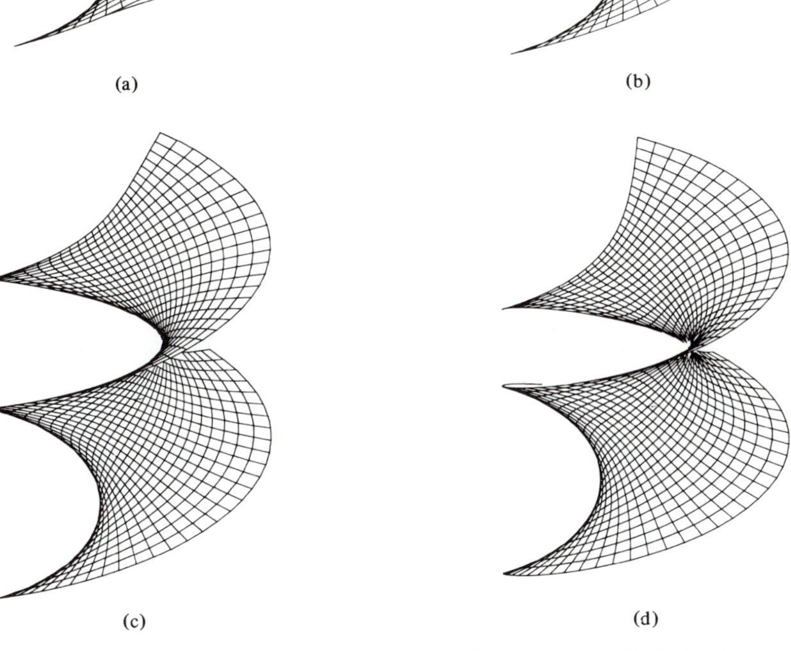

Figure 4-4. Isometric deformation of helicoid to catenoid. (a) Phase 1. (b) Phase 2. (c) Phase 3. (d) Phase 4.

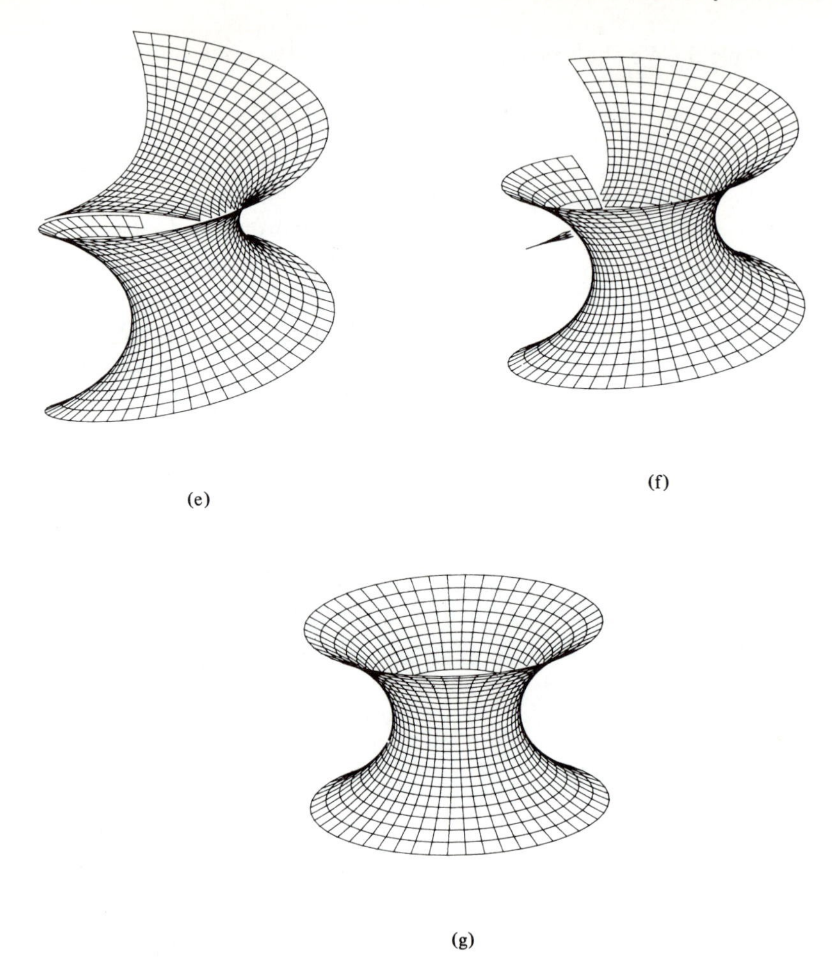

Figure 4-4. (e) Phase 5. (f) Phase 6. (g) Phase 7.

where 2α $(0 < 2\alpha < \pi)$ is the angle at the vertex of the cone (i.e., where cotan $\alpha = k$), and let $F\colon U \longrightarrow R^3$ be the map (Fig. 4-5)

$$F(\rho, \theta) = \left(\rho \sin\alpha \cos\left(\frac{\theta}{\sin\alpha}\right), \rho \sin\alpha \sin\left(\frac{\theta}{\sin\alpha}\right), \rho\cos\alpha\right).$$

It is clear that $F(U)$ is contained in the cone because

$$k\sqrt{x^2 + y^2} = \operatorname{cotan}\alpha\sqrt{\rho^2 \sin^2\alpha} = \rho\cos\alpha = z.$$

Furthermore, when θ describes the interval $(0, 2\pi\sin\alpha)$, $\theta/\sin\alpha$ describes the

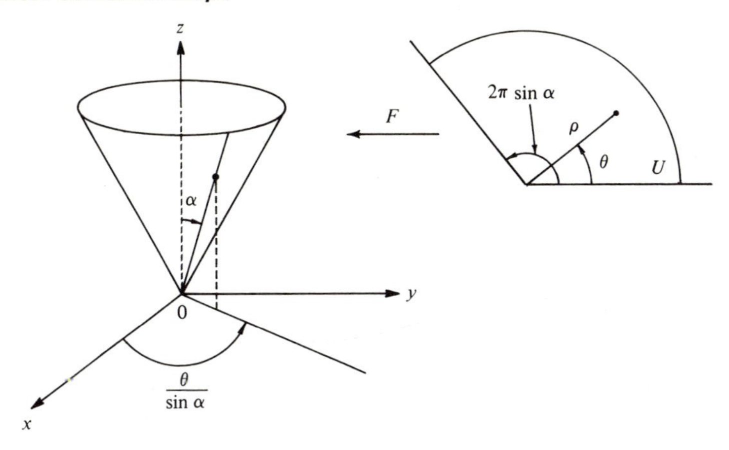

Figure 4-5

interval $(0, 2\pi)$. Thus, all points of the cone except the generator $\theta = 0$ are covered by $F(U)$.

It is easily checked that F and dF are one-to-one in U; therefore, F is a diffeomorphism of U onto the cone minus a generator.

We shall now show that F is an isometry. In fact, U may be thought of as a regular surface, parametrized by

$$\bar{\mathbf{x}}(\rho, \theta) = (\rho \cos \theta, \rho \sin \theta, 0), \qquad 0 < \rho < \infty,\ 0 < \theta < 2\pi \sin \alpha.$$

The coefficients of the first fundamental form of U in this parametrization are

$$\bar{E} = 1, \qquad \bar{F} = 0, \qquad \bar{G} = \rho^2.$$

On the other hand, the coefficientes of the first fundamental form of the cone in the parametrization $F \circ \mathbf{x}$ are

$$E = 1, \qquad F = 0, \qquad G = \rho^2.$$

From Prop. 1 we conclude that F is a local isometry, as we wished.

Remark 2. The fact that we can compute lengths of curves on a surface S by using only its first fundamental form allows us to introduce a notion of "intrinsic" distance for points in S. Roughly speaking, we define the (intrinsic) *distance* $d(p, q)$ between two points of S as the infimum of the length of curves on S joining p and q. (We shall go into that in more detail in Sec. 5-3.) This distance is clearly greater than or equal to the distance $\|p - q\|$ of p to q *as points in* R^3 (Fig. 4-6). We shall show in Exercise 3 that the distance d is invariant under isometries; that is, if $\varphi: S \longrightarrow \bar{S}$ is an isometry, then $d(p, q) = d(\varphi(p), \varphi(q))$, $p, q \in S$.

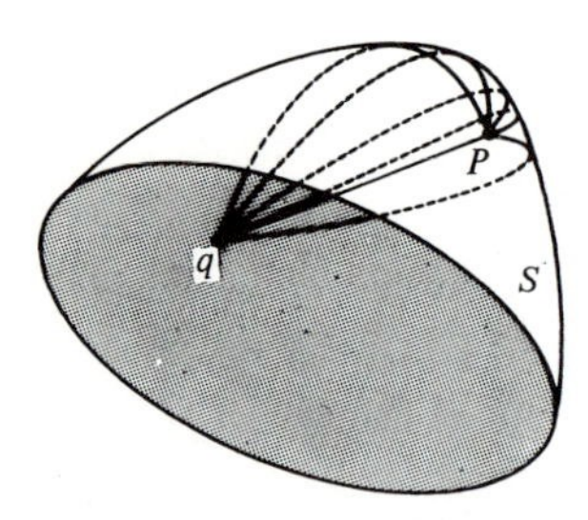

Figure 4-6

The notion of isometry is the natural concept of equivalence for the metric properties of regular surfaces. In the same way as diffeomorphic surfaces are equivalent from the point of view of differentiability, so isometric surfaces are equivalent from the metric viewpoint.

It is possible to define further types of equivalence in the study of surfaces. From our point of view, diffeomorphisms and isometries are the most important. However, when dealing with problems associated with analytic functions of complex variables, it is important to introduce the conformal equivalence, which we shall now discuss briefly.

DEFINITION 3. *A diffeomorphism* $\varphi\colon S \to \bar{S}$ *is called a* conformal map *if for all* $p \in S$ *and all* $v_1, v_2 \in T_p(S)$ *we have*

$$\langle d\varphi_p(v_1), d\varphi_p(v_2)\rangle = \lambda^2(p)\langle v_1, v_2\rangle_p,$$

where λ^2 *is a nowhere-zero differentiable function on* S; *the surfaces* S *and* $\bar{S}$ *are then said to be* conformal. *A map* $\varphi\colon V \to \bar{S}$ *of a neighborhood* V *of* $p \in S$ *into* $\bar{S}$ *is a* local conformal map *at* p *if there exists a neighborhood* $\bar{V}$ *of* $\varphi(p)$ *such that* $\varphi\colon V \to \bar{V}$ *is a conformal map. If for each* $p \in S$, *there exists a local conformal map at* p, *the surface* S *is said to be* locally conformal *to* $\bar{S}$.

The geometric meaning of the above definition is that the angles (but not necessarily the lengths) are preserved by conformal maps. In fact, let $\alpha\colon I \to S$ and $\beta\colon I \to S$ be two curves in S which intersect at, say, $t = 0$. Their angle θ at $t = 0$ is given by

$$\cos\theta = \frac{\langle \alpha', \beta'\rangle}{|\alpha'||\beta'|}, \qquad 0 < \theta < \pi.$$

A conformal map $\varphi\colon S \to \bar{S}$ maps these curves into curves $\varphi\circ\alpha\colon I \to \bar{S}$, $\varphi\circ\beta\colon I \to \bar{S}$, which intersect for $t = 0$, making an angle $\bar{\theta}$ given by

$$\cos\bar{\theta} = \frac{\langle d\varphi(\alpha'), d\varphi(\beta')\rangle}{|d\varphi(\alpha')||d\varphi(\beta')|} = \frac{\lambda^2\langle \alpha', \beta'\rangle}{\lambda^2|\alpha'||\beta'|} = \cos\theta,$$

as we claimed. It is not hard to prove that this property characterizes the locally conformal maps (Exercise 14).

The following proposition is the analogue of Prop. 1 for conformal maps, and its proof is also left as an exercise.

PROPOSITION 2. *Let* $\mathbf{x}: U \longrightarrow S$ *and* $\bar{\mathbf{x}}: U \longrightarrow \bar{S}$ *be parametrizations such that* $E = \lambda^2\bar{E}$, $F = \lambda^2\bar{F}$, $G = \lambda^2\bar{G}$ *in* U, *where* λ^2 *is a nowhere-zero differentiable function in* U. *Then the map* $\varphi = \bar{\mathbf{x}}\circ\mathbf{x}^{-1}: \mathbf{x}(U) \longrightarrow \bar{S}$ *is a local conformal map.*

Local conformality is easily seen to be an equivalence relation; that is, if S_1 is locally conformal to S_2 and S_2 is locally conformal to S_3, then S_1 is locally conformal to S_3.

The most important property of conformal maps is given by the following theorem, which we shall not prove.

THEOREM. *Any two regular surfaces are locally conformal.*

The proof is based on the possibility of parametrizing a neighborhood of any point of a regular surface in such a way that the coefficients of the first fundamental form are

$$E = \lambda^2(u, v) > 0, \qquad F = 0, \qquad G = \lambda^2(u, v).$$

Such a coordinate system is called *isothermal.* Once the existence of an isothermal coordinate system of a regular surface S is assumed, S is clearly locally conformal to a plane, and by composition locally conformal to any other surface.

The proof that there exist isothermal coordinate systems on any regular surface is delicate and will not be taken up here. The interested reader may consult L. Bers, *Riemann Surfaces*, New York University, Institute of Mathematical Sciences, New York, 1957–1958, pp. 15–35.

Remark 3. Isothermal parametrizations already appeared in Chap. 3 in the context of minimal surfaces; cf. Prop. 2 and Exercise 13 of Sec. 3-5.

EXERCISES

1. Let $F: U \subset R^2 \longrightarrow R^3$ be given by

$$F(u, v) = (u \sin \alpha \cos v,\ u \sin \alpha \sin v,\ u \cos \alpha),$$
$$(u, v) \in U = \{(u, v) \in R^2;\ u > 0\}, \qquad \alpha = \text{const.}$$

a. Prove that F is a local diffeomorphism of U onto a cone C with the vertex at the origin and 2α as the angle of the vertex.

b. Is F a local isometry?

2. Prove the following "converse" of Prop. 1: Let $\varphi: S \longrightarrow \bar{S}$ be an isometry and $\mathbf{x}: U \longrightarrow S$ a parametrization at $p \in S$; then $\bar{\mathbf{x}} = \varphi \circ \mathbf{x}$ is a parametrization at $\varphi(p)$ and $E = \bar{E}$, $F = \bar{F}$, $G = \bar{G}$.

***3.** Show that a diffeomorphism $\varphi: S \longrightarrow \bar{S}$ is an isometry if and only if the arc length of any parametrized curve in S is equal to the arc length of the image curve by φ.

4. Use the stereographic projection (cf. Exercise 16, Sec. 2-2) to show that the sphere is locally conformal to a plane.

5. Let $\alpha_1: I \longrightarrow R^3$, $\alpha_2: I \longrightarrow R^3$ be regular parametrized curves, where the parameter is the arc length. Assume that the curvatures k_1 of α_1 and k_2 of α_2 satisfy $k_1(s) = k_2(s) \neq 0$, $s \in I$. Let

$$\mathbf{x}_1(s, v) = \alpha_1(s) + v\alpha_1'(s),$$
$$\mathbf{x}_2(s, v) = \alpha_2(s) + v\alpha_2'(s)$$

be their (regular) tangent surfaces (cf. Example 5, Sec. 2-3) and let V be a neighborhood of (t_0, s_0) such that $\mathbf{x}_1(V) \subset R^3$, $\mathbf{x}_2(V) \subset R^3$ are regular surfaces (cf. Prop. 2, Sec. 2-3). Prove that $\mathbf{x}_1 \circ \mathbf{x}_2^{-1}: \mathbf{x}_2(V) \longrightarrow \mathbf{x}_1(V)$ is an isometry.

***6.** Let $\alpha: I \longrightarrow R^3$ be a regular parametrized curve with $k(t) \neq 0$, $t \in I$. Let $\mathbf{x}(t, v)$ be its tangent surface. Prove that, for each $(t_0, v_0) \in I \times (R - \{0\})$, there exists a neighborhood V of (t_0, v_0) such that $\mathbf{x}(V)$ is isometric to an open set of the plane (*thus, tangent surfaces are locally isometric to planes*).

7. Let V and W be (finite-dimensional) vector spaces with inner products denoted by $\langle\ ,\ \rangle$ and let $F: V \longrightarrow W$ be a linear map. Prove that the following conditions are equivalent:

a. $\langle F(v_1), F(v_2)\rangle = \langle v_1, v_2\rangle$ for all $v_1, v_2 \in V_1$.

b. $|F(v)| = |v|$ for all $v \in V$.

c. If $\{v_1, \ldots, v_n\}$ is an orthonormal basis in V, then $\{F(v_1), \ldots, F(v_n)\}$ is an orthonormal basis in W.

d. There exists an orthonormal basis $\{v_1, \ldots, v_n\}$ in V such that $\{F(v_1), \ldots, F(v_n)\}$ is an orthonormal basis in W.

If any of these conditions is satisfied, F is called a *linear isometry* of V into W. (When $W = V$, a linear isometry is often called an *orthogonal transformation*.)

***8.** Let $G: R^3 \longrightarrow R^3$ be a map such that

$$|G(p) - G(q)| = |p - q| \qquad \text{for all } p, q \in R^3$$

(that is, G is a *distance-preserving* map). Prove that there exists $p_0 \in R^3$ and a linear isometry (cf. Exercise 7) F of the vector space R^3 such that

$$G(p) = F(p) + p_0 \qquad \text{for all } p \in R^3.$$

9. Let S_1, S_2, and S_3 be regular surfaces. Prove that

a. If $\varphi: S_1 \longrightarrow S_2$ is an isometry, then $\varphi^{-1}: S_2 \longrightarrow S_1$ is also an isometry.

b. If $\varphi: S_1 \longrightarrow S_2$, $\psi: S_2 \longrightarrow S_3$ are isometries, then $\psi \circ \varphi: S_1 \longrightarrow S_3$ is an isometry.

This implies that the isometries of a regular surface S constitute in a natural way a group, called the *group of isometries* of S.

10. Let S be a surface of revolution. Prove that the rotations about its axis are isometries of S.

***11. a.** Let $S \subset R^3$ be a regular surface and let $F: R^3 \longrightarrow R^3$ be a distance-preserving map of R^3 (see Exercise 8) such that $F(S) \subset S$. Prove that the restriction of F to S is an isometry of S.

b. Use part a to show that the group of isometries (see Exercise 10) of the unit sphere $x^2 + y^2 + z^2 = 1$ is contained in the group of orthogonal linear transformations of R^3 (it is actually equal; see Exercise 23, Sec. 4-4).

c. Give an example to show that there are isometries $\varphi: S_1 \longrightarrow S_2$ which cannot be extended into distance-preserving maps $F: R^3 \longrightarrow R^3$.

***12.** Let $C = \{(x, y, z) \in R^3; x^2 + y^2 = 1\}$ be a cylinder. Construct an isometry $\varphi: C \longrightarrow C$ such that the set of fixed points of φ, i.e., the set $\{p \in C; \varphi(p) = p\}$, contains exactly two points.

13. Let V and W be (finite-dimensional) vector spaces with inner products $\langle\ ,\ \rangle$. Let $G: V \longrightarrow W$ be a linear map. Prove that the following conditions are equivalent:

a. There exists a real constant $\lambda \neq 0$ such that

$$\langle G(v_1), G(v_2)\rangle = \lambda^2\langle v_1, v_2\rangle \qquad \text{for all } v_1, v_2 \in V.$$

b. There exists a real constant $\lambda > 0$ such that

$$|G(v)| = \lambda|v| \qquad \text{for all } v \in V.$$

c. There exists an orthonormal basis $\{v_1, \ldots, v_n\}$ of V such that $\{G(v_1), \ldots, G(v_n)\}$ is an orthogonal basis of W and, also, the vectors $G(v_i)$, $i = 1, \ldots, n$, have the same (nonzero) length.

If any of these conditions is satisfied, G is called a *linear conformal map* (or a *similitude*).

14. We say that a differentiable map $\varphi: S_1 \longrightarrow S_2$ *preserves angles* when for every $p \in S_1$ and every pair $v_1, v_2 \in T_p(S_1)$ we have

$$\cos(v_1, v_2) = \cos(d\varphi_p(v_1), d\varphi_p(v_2)).$$

Prove that φ is locally conformal if and only if it preserves angles.

15. Let $\varphi: R^2 \longrightarrow R^2$ be given by $\varphi(x, y) = (u(x, y), v(x, y))$, where u and v are differentiable functions that satisfy the Cauchy-Riemann equations

$$u_x = v_y, \qquad u_y = -v_x.$$

Show that φ is a local conformal map from $R^2 - Q$ into R^2, where $Q = \{(x, y) \in R^2; u_x^2 + u_y^2 = 0\}$.

16. Let $\mathbf{x}: U \subset R^2 \to R^3$, where

$$U = \{(\theta, \varphi) \in R^2; 0 < \theta < \pi, 0 < \varphi < 2\pi\},$$
$$\mathbf{x}(\theta, \varphi) = (\sin\theta\cos\varphi, \sin\theta\sin\varphi, \cos\theta),$$

be a parametrization of the unit sphere S^2. Let

$$\log\tan\tfrac{1}{2}\theta = u, \qquad \varphi = v,$$

and show that a new parametrization of the coordinate neighborhood $\mathbf{x}(U) = V$ can be given by

$$\mathbf{y}(u, v) = (\operatorname{sech} u\cos v, \operatorname{sech} u\sin v, \tanh u).$$

Prove that in the parametrization $\mathbf{y}$ the coefficients of the first fundamental form are

$$E = G = \operatorname{sech}^2 u, \qquad F = 0.$$

Thus, $\mathbf{y}^{-1}: V \subset S^2 \to R^2$ is a conformal map which takes the meridians and parallels of S^2 into straight lines of the plane. This is called *Mercator's projection.*

***17.** Consider a triangle on the unit sphere so that its sides are made up of segments of loxodromes (i.e., curves which make a constant angle with the meridians; cf. Example 4, Sec. 2-5), and do not contain poles. Prove that the sum of the interior angles of such a triangle is π.

18. A diffeomorphism $\varphi: S \to \bar{S}$ is said to be *area-preserving* if the area of any region $R \subset S$ is equal to the area of $\varphi(R)$. Prove that if φ is area-preserving and conformal, then φ is an isometry.

19. Let $S^2 = \{(x, y, z) \in R^3; x^2 + y^2 + z^2 = 1\}$ be the unit sphere and $C = \{(x, y, z) \in R^3; x^2 + y^2 = 1\}$ be the circumscribed cylinder. Let

$$\varphi: S^2 - \{(0, 0, 1) \cup (0, 0, -1)\} = M \to C$$

be a map defined as follows. For each $p \in M$, the line passing through p and perpendicular to $0z$ meets $0z$ at the point q. Let l be the half-line starting from q and containing p (Fig. 4-7). By definition, $\varphi(p) = C \cap l$.

Prove that φ is an area-preserving diffeomorphism.

20. Let $\mathbf{x}: U \subset R^2 \to S$ be the parametrization of a surface of revolution S:

$$\mathbf{x}(u, v) = (f(v)\cos u, f(v)\sin u, g(v)), \qquad f(v) > 0,$$
$$U = \{(u, v) \in R^2; 0 < u < 2\pi, a < v < b\}.$$

a. Show that the map $\varphi: U \to R^2$ given by

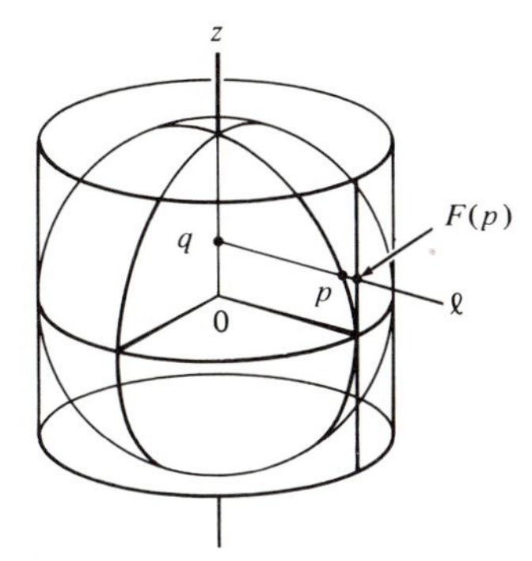

Figure 4-7

$$\varphi(u, v) = \left(u, \int \frac{\sqrt{(f'(v))^2 + (g'(v))^2}}{f(v)}\, dv\right)$$

is a local diffeomorphism.

b. Use part a to prove that a surface of revolution S is locally conformal to a plane in such a way that each local conformal map $\theta\colon V \subset S \longrightarrow R^2$ takes the parallels and the meridians of the neighborhood V into an orthogonal system of straight lines in $\theta(V) \subset R^2$. (Notice that this generalizes Mercator's projection of Exercise 16.)

c. Show that the map $\psi\colon U \longrightarrow R^2$ given by

$$\psi(u, v) = \left(u, \int f(v)\sqrt{(f'(v))^2 + (g'(v))^2}\, dv\right)$$

is a local diffeomorphism.

d. Use part c to prove that for each point p of a surface of revolution S there exists a neighborhood $V \subset S$ and a map $\bar{\theta}\colon V \longrightarrow R^2$ of V into a plane that is area-preserving.

4-3. The Gauss Theorem and the Equations of Compatibility

The properties of Chap. 3 were obtained from the study of the variation of the tangent plane in a neighborhood of a point. Proceeding with the analogy with curves, we are going to assign to each point of a surface a trihedron (the analogue of Frenet's trihedron) and study the derivatives of its vectors.

S will denote, as usual, a regular, orientable, and oriented surface. Let $\mathbf{x}\colon U \subset R^2 \longrightarrow S$ be a parametrization in the orientation of S. It is possible to assign to each point of $\mathbf{x}(U)$ a natural trihedron given by the vectors $\mathbf{x}_u$, $\mathbf{x}_v$, and N. The study of this trihedron will be the subject of this section.

By expressing the derivatives of the vectors $\mathbf{x}_u$, $\mathbf{x}_v$, and N in the basis $\{\mathbf{x}_u, \mathbf{x}_v, N\}$, we obtain

$$\begin{aligned}
\mathbf{x}_{uu} &= \Gamma_{11}^1 \mathbf{x}_u + \Gamma_{11}^2 \mathbf{x}_v + L_1 N, \\
\mathbf{x}_{uv} &= \Gamma_{12}^1 \mathbf{x}_u + \Gamma_{12}^2 \mathbf{x}_v + L_2 N, \\
\mathbf{x}_{vu} &= \Gamma_{21}^1 \mathbf{x}_u + \Gamma_{21}^2 \mathbf{x}_v + \bar{L}_2 N, \\
\mathbf{x}_{vv} &= \Gamma_{22}^1 \mathbf{x}_u + \Gamma_{22}^2 \mathbf{x}_v + L_3 N, \\
N_u &= a_{11} \mathbf{x}_u + a_{21} \mathbf{x}_v, \\
N_v &= a_{12} \mathbf{x}_u + a_{22} \mathbf{x}_v,
\end{aligned} \tag{1}$$

where the a_{ij}, $i, j = 1, 2$, were obtained in Chap. 3 and the other coefficients are to be determined. The coefficients Γ_{ij}^k, $i, j, k = 1, 2$, are called the Christoffel symbols of S in the parametrization $\mathbf{x}$. Since $\mathbf{x}_{uv} = \mathbf{x}_{vu}$, we conclude that $\Gamma_{12}^1 = \Gamma_{21}^1$ and $\Gamma_{12}^2 = \Gamma_{21}^2$; that is, the *Christoffel symbols* are symmetric relative to the lower indices.

By taking the inner product of the first four relations in (1) with N, we immediately obtain $L_1 = e$, $L_2 = \bar{L}_2 = f$, $L_3 = g$, where e, f, g are the coefficients of the second fundamental form of S.

To determine the Christoffel symbols, we take the inner product of the first four relations with $\mathbf{x}_u$ and $\mathbf{x}_v$, obtaining the system

$$\begin{aligned}
&\begin{cases}
\Gamma_{11}^1 E + \Gamma_{11}^2 F = \langle \mathbf{x}_{uu}, \mathbf{x}_u \rangle = \frac{1}{2} E_u, \\
\Gamma_{11}^1 F + \Gamma_{11}^2 G = \langle \mathbf{x}_{uu}, \mathbf{x}_v \rangle = F_u - \frac{1}{2} E_v,
\end{cases} \\
&\begin{cases}
\Gamma_{12}^1 E + \Gamma_{12}^2 F = \langle \mathbf{x}_{uv}, \mathbf{x}_u \rangle = \frac{1}{2} E_v, \\
\Gamma_{12}^1 F + \Gamma_{12}^2 G = \langle \mathbf{x}_{uv}, \mathbf{x}_v \rangle = \frac{1}{2} G_u,
\end{cases} \\
&\begin{cases}
\Gamma_{22}^1 E + \Gamma_{22}^2 F = \langle \mathbf{x}_{vv}, \mathbf{x}_u \rangle = F_v - \frac{1}{2} G_u, \\
\Gamma_{22}^1 F + \Gamma_{22}^2 G = \langle \mathbf{x}_{vv}, \mathbf{x}_v \rangle = \frac{1}{2} G_v.
\end{cases}
\end{aligned} \tag{2}$$

Note that the above equations have been grouped into three pairs of equations and that for each pair the determinant of the system is $EG - F^2 \neq 0$. Thus, it is possible to solve the above system and *to compute the Christoffel symbols in terms of the coefficients of the first fundamental form, E, F, G, and their derivatives*. We shall not obtain the explicit expressions of the Γ_{ij}^k, since it is easier to work in each particular case with the system (2). (See Example 1 below.) However, the following consequence of the fact that we can solve the system (2) is very important: *All geometric concepts and properties expressed in terms of the Christoffel symbols are invariant under isometries.*

Example 1. We shall compute the Christoffel symbols for a surface of revolution parametrized by (cf. Example 4, Sec. 2-3)

$$\mathbf{x}(u, v) = (f(v)\cos u, f(v)\sin u, g(v)), \qquad f(v) \neq 0.$$

Since

$$E = (f(v))^2, \qquad F = 0, \qquad G = (f'(v))^2 + (g'(v))^2,$$

we obtain

$$\begin{aligned} E_u &= 0, \qquad & E_v &= 2ff', \\ F_u &= F_v = 0, \qquad & G_u &= 0, \\ & & G_v &= 2(f'f'' + g'g''), \end{aligned}$$

where prime denotes derivative with respect to v. The first two equations of the system (2) then give

$$\Gamma_{11}^1 = 0, \qquad \Gamma_{11}^2 = -\frac{ff'}{(f')^2 + (g')^2}.$$

Next, the second pair of equations in system (2) yield

$$\Gamma_{12}^1 = \frac{ff'}{f^2}, \qquad \Gamma_{12}^2 = 0.$$

Finally, from the last two equations in system (2) we obtain

$$\Gamma_{22}^1 = 0, \qquad \Gamma_{22}^2 = \frac{f'f'' + g'g''}{(f')^2 + (g')^2}.$$

As we have just seen, the expressions of the derivatives of $\mathbf{x}_u$, $\mathbf{x}_v$, and N in the basis $\{\mathbf{x}_u, \mathbf{x}_v, N\}$ involve only the knowledge of the coefficients of the first and second fundamental forms of S. A way of obtaining relations between these coefficients is to consider the expressions

$$\begin{aligned} (\mathbf{x}_{uu})_v - (\mathbf{x}_{uv})_u &= 0, \\ (\mathbf{x}_{vv})_u - (\mathbf{x}_{vu})_v &= 0, \\ N_{uv} - N_{vu} &= 0. \end{aligned} \tag{3}$$

By introducing the values of (1), we may write the above relations in the form

$$\begin{aligned} A_1\mathbf{x}_u + B_1\mathbf{x}_v + C_1N &= 0, \\ A_2\mathbf{x}_u + B_2\mathbf{x}_v + C_2N &= 0, \\ A_3\mathbf{x}_u + B_3\mathbf{x}_v + C_3N &= 0, \end{aligned} \tag{3a}$$

where A_i, B_i, C_i, $i = 1, 2, 3$, are functions of E, F, G, e, f, g and of their derivatives. Since the vectors $\mathbf{x}_u$, $\mathbf{x}_v$, N are linearly independent, (3a) implies that there exist nine relations:

$$A_i = 0, \qquad B_i = 0, \qquad C_i = 0, \qquad i = 1, 2, 3.$$

As an example, we shall determine the relations $A_1 = 0$, $B_1 = 0$, $C_1 = 0$. By using the values of (1), the first of the relations (3) may be written

$$\begin{aligned}\Gamma_{11}^1\mathbf{x}_{uv} + \Gamma_{11}^2\mathbf{x}_{vv} + eN_v + (\Gamma_{11}^1)_v\mathbf{x}_u + (\Gamma_{11}^2)_v\mathbf{x}_v + e_vN \\ = \Gamma_{12}^1\mathbf{x}_{uu} + \Gamma_{12}^2\mathbf{x}_{vu} + fN_u + (\Gamma_{12}^1)_u\mathbf{x}_u + (\Gamma_{12}^2)_u\mathbf{x}_v + f_uN.\end{aligned} \tag{4}$$

By using (1) again and equating the coefficients of $\mathbf{x}_v$, we obtain

$$\begin{aligned}\Gamma_{11}^1\Gamma_{12}^2 + \Gamma_{11}^2\Gamma_{22}^2 + ea_{22} + (\Gamma_{11}^2)_v \\ = \Gamma_{12}^1\Gamma_{11}^2 + \Gamma_{12}^2\Gamma_{12}^2 + fa_{21} + (\Gamma_{12}^2)_u.\end{aligned}$$

Introducing the values of a_{ij} already computed (cf. Sec. 3-3) it follows that

$$\begin{aligned}(\Gamma_{12}^2)_u - (\Gamma_{11}^2)_v + \Gamma_{12}^1\Gamma_{11}^2 + \Gamma_{12}^2\Gamma_{12}^2 - \Gamma_{11}^2\Gamma_{22}^2 - \Gamma_{11}^1\Gamma_{12}^2 \\ = -E\frac{eg - f^2}{EG - F^2} \\ = -EK.\end{aligned} \tag{5}$$

At this point it is convenient to interrupt our computations in order to draw attention to the fact that the above equation proves the following theorem, due to K. F. Gauss.

THEOREMA EGREGIUM (Gauss). *The Gaussian curvature* K *of a surface is invariant by local isometries.*

In fact if $\mathbf{x}: U \subset R^2 \longrightarrow S$ is a parametrization at $p \in S$ and if $\varphi: V \subset S \longrightarrow S$, where $V \subset \mathbf{x}(U)$ is a neighborhood of p, is a local isometry at p, then $\mathbf{y} = \mathbf{x}\circ\varphi$ is a parametrization of S at $\varphi(p)$. Since φ is an isometry, the coefficients of the first fundamental form in the parametrizations $\mathbf{x}$ and $\mathbf{y}$ agree at corresponding points q and $\varphi(q)$, $q \in V$; thus, the corresponding Christoffel symbols also agree. By Eq. (5), K can be computed at a point as a function of the Christoffel symbols in a given parametrization at the point. It follows that $K(q) = K(\varphi(q))$ for all $q \in V$.

The above expression, which yields the value of K in terms of the coefficients of the first fundamental form and its derivatives, is known as the *Gauss formula*. It was first proved by Gauss in a famous paper [1].

The Gauss theorem is considered, by the extension of its consequences, one of the most important facts of differential geometry. For the moment we shall mention only the following corollary.

As was proved in Sec. 4-2, a catenoid is locally isometric to a helicoid. It follows from the Gauss theorem that the Gaussian curvatures are equal at corresponding points, a fact which is geometrically nontrivial.

Actually, it is a remarkable fact that a concept such as the Gaussian curvature, the definition of which made essential use of the position of a surface in the space, does not depend on this position but only on the metric structure (first fundamental form) of the surface.

We shall see in the next section that many other concepts of differential geometry are in the same setting as the Gaussian curvature; that is, they depend only on the first fundamental form of the surface. It thus makes sense to talk about a geometry of the first fundamental form, which we call intrinsic geometry, since it may be developed without any reference to the space that contains the surface (once the first fundamental form is given).

†With an eye to a further geometrical result we come back to our computations. By equating the coefficients of $\mathbf{x}_u$ in (4), we see that the relation $A_1 = 0$ may be written in the form

$$(\Gamma_{12}^1)_u - (\Gamma_{11}^1)_v + \Gamma_{12}^2\Gamma_{12}^1 - \Gamma_{11}^2\Gamma_{22}^1 = FK. \tag{5a}$$

By equating also in Eq. (4) the coefficients of N, we obtain $C_1 = 0$ in the form

$$e_v - f_u = e\Gamma_{12}^1 + f(\Gamma_{12}^2 - \Gamma_{11}^1) - g\Gamma_{11}^2. \tag{6}$$

Observe that relation (5a) is (when $F \neq 0$) merely another form of the Gauss formula (5).

By applying the same process to the second expression of (3), we obtain that both the equations $A_2 = 0$ and $B_2 = 0$ give again the Gauss formula (5). Furthermore, $C_2 = 0$ is given by

$$f_v - g_u = e\Gamma_{22}^1 + f(\Gamma_{22}^2 - \Gamma_{12}^1) - g\Gamma_{12}^2. \tag{6a}$$

Finally, the same process can be applied to the last expression of (3), yielding that $C_3 = 0$ is an identity and that $A_3 = 0$ and $B_3 = 0$ are again Eqs. (6) and (6a). Equations (6) and (6a) are called *Mainardi-Codazzi equations.*

The Gauss formula and the Mainardi-Codazzi equations are known under the name of *compatibility equations* of the theory of surfaces.

---

†The rest of this section will not be used until Chap. 5. If omitted, Exercises 7 and 8 should also be omitted.

A natural question is whether there exist further relations of compatibility between the first and the second fundamental forms besides those already obtained. The theorem stated below shows that the answer is negative. In other words, by successive derivations or any other process we would obtain no further relations among the coefficients E, F, G, e, f, g and their derivatives. Actually, the theorem is more explicit and asserts that the knowledge of the first and second fundamental forms determines a surface locally. More precisely,

THEOREM (Bonnet). *Let* E, F, G, e, f, g *be differentiable functions, defined in an open set* $V \subset R^2$, *with* $E > 0$ *and* $G > 0$. *Assume that the given functions satisfy formally the Gauss and Mainardi-Codazzi equations and that* $EG - F^2 > 0$. *Then, for every* $q \in V$ *there exists a neighborhood* $U \subset V$ *of* q *and a diffeomorphism* $\mathbf{x}: U \to \mathbf{x}(U) \subset R^3$ *such that the regular surface* $\mathbf{x}(U) \subset R^3$ *has* E, F, G *and* e, f, g *as coefficients of the first and second fundamental forms, respectively. Furthermore, if* U *is connected and if*

$$\bar{\mathbf{x}}: U \to \bar{\mathbf{x}}(U) \subset R^3$$

is another diffeomorphism satisfying the same conditions, then there exist a translation T *and a proper linear orthogonal transformation* ρ *in* R^3 *such that* $\bar{\mathbf{x}} = T \circ \rho \circ \mathbf{x}$.

A proof of this theorem may be found in the appendix to Chap. 4.

For later use, it is convenient to observe how the Mainardi-Codazzi equations simplify when the coordinate neighborhood contains no umbilical points and the coordinate curves are lines of curvature ($F = 0 = f$). Then, Eqs. (6) and (6a) may be written

$$e_v = e\Gamma^1_{12} - g\Gamma^2_{11}, \qquad g_u = g\Gamma^2_{12} - e\Gamma^1_{22}.$$

By taking into consideration that $F = 0$ implies that

$$\Gamma^2_{11} = -\frac{1}{2}\frac{E_v}{G}, \qquad \Gamma^1_{12} = \frac{1}{2}\frac{E_v}{E},$$

$$\Gamma^1_{22} = -\frac{1}{2}\frac{G_u}{E}, \qquad \Gamma^2_{12} = \frac{1}{2}\frac{G_u}{G},$$

we conclude that the Mainardi-Codazzi equations take the following form:

$$e_v = \frac{E_v}{2}\left(\frac{e}{E} + \frac{g}{G}\right), \tag{7}$$

$$g_u = \frac{G_u}{2}\left(\frac{e}{E} + \frac{g}{G}\right). \tag{7a}$$

EXERCISES

1. Show that if $\mathbf{x}$ is an orthogonal parametrization, that is, $F = 0$, then

$$K = -\frac{1}{2\sqrt{EG}}\left\{\left(\frac{E_v}{\sqrt{EG}}\right)_v + \left(\frac{G_u}{\sqrt{EG}}\right)_u\right\}.$$

2. Show that if $\mathbf{x}$ is an isothermal parametrization, that is, $E = G = \lambda(u, v)$ and $F = 0$, then

$$K = -\frac{1}{2\lambda}\Delta(\log \lambda),$$

where $\Delta\varphi$ denotes the Laplacian $(\partial^2\varphi/\partial u^2) + (\partial^2\varphi/\partial v^2)$ of the function φ. Conclude that when $E = G = (u^2 + v^2 + c)^{-2}$ and $F = 0$, then $K = \text{const.} = 4c$.

3. Verify that the surfaces

$$\mathbf{x}(u, v) = (u \cos v, u \sin v, \log u),$$
$$\bar{\mathbf{x}}(u, v) = (u \cos v, u \sin v, v),$$

have equal Gaussian curvature at the points $\mathbf{x}(u, v)$ and $\bar{\mathbf{x}}(u, v)$ but that the mapping $\bar{\mathbf{x}} \circ \mathbf{x}^{-1}$ is not an isometry. This shows that the "converse" of the Gauss theorem is not true.

4. Show that no neighborhood of a point in a sphere may be isometrically mapped into a plane.

5. If the coordinate curves form a Tchebyshef net (cf. Exercises 7 and 8, Sec. 2-5), then $E = G = 1$ and $F = \cos\theta$. Show that in this case

$$K = -\frac{\theta_{uv}}{\sin\theta}.$$

6. Use Bonnet's theorem to show that there exists no surface $\mathbf{x}(u, v)$ such that $E = G = 1$, $F = 0$ and $e = 1$, $g = -1$, $f = 0$.

7. Does there exist a surface $\mathbf{x} = \mathbf{x}(u, v)$ with $E = 1$, $F = 0$, $G = \cos^2 u$ and $e = \cos^2 u$, $f = 0$, $g = 1$?

8. Compute the Christoffel symbols for an open set of the plane

a. In cartesian coordinates.

b. In polar coordinates.

Use the Gauss formula to compute K in both cases.

9. Justify why the surfaces below are not pairwise locally isometric:

a. Sphere.

b. Cylinder.

c. Saddle $z = x^2 - y^2$.

4-4. Parallel Transport. Geodesics.

We shall now proceed to a systematic exposition of the intrinsic geometry. To display the intuitive meaning of the concepts, we shall often give definitions and interpretations involving the space exterior to the surface. However, we shall prove in each case that the concepts to be introduced depend only on the first fundamental form.

We shall start with the definition of covariant derivative of a vector field, which is the analogue for surfaces of the usual differentiation of vectors in the plane. We recall that a (*tangent*) *vector field* in an open set $U \subset S$ of a regular surface S is a correspondence w that assigns to each $p \in U$ a vector $w(p) \in T_p(S)$. The vector field w is *differentiable* at p if, for some parametrization $\mathbf{x}(u, v)$ in p, the components a and b of $w = a\mathbf{x}_u + b\mathbf{x}_v$ in the basis $\{\mathbf{x}_u, \mathbf{x}_v\}$ are differentiable functions at p. w is differentiable in U if it is differentiable for every $p \in U$.

DEFINITION 1. *Let* w *be a differentiable vector field in an open set* $\mathrm{U} \subset \mathrm{S}$ *and* $\mathrm{p} \in \mathrm{U}$. *Let* $\mathrm{y} \in \mathrm{T_p(S)}$. *Consider a parametrized curve*

$$\alpha\colon (-\epsilon, \epsilon) \longrightarrow \mathrm{U},$$

with $\alpha(0) = \mathrm{p}$ *and* $\alpha'(0) = \mathrm{y}$, *and let* $\mathrm{w(t)}$, $\mathrm{t} \in (-\epsilon, \epsilon)$, *be the restriction of the vector field* w *to the curve* α. *The vector obtained by the normal projection of* $(\mathrm{dw/dt})(0)$ *onto the plane* $\mathrm{T_p(S)}$ *is called the* covariant derivative *at* p *of the vector field* w *relative to the vector* y. *This covariant derivative is denoted by* $(\mathrm{Dw/dt})(0)$ *or* $(\mathrm{D_y w})(\mathrm{p})$ (*Fig. 4-8*).

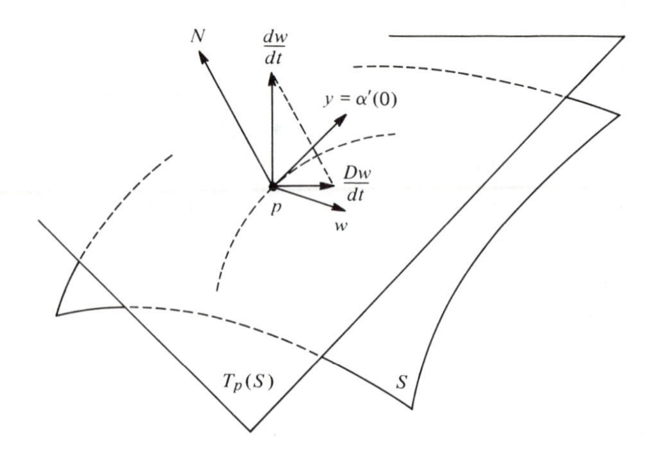

Figure 4-8. The covariant derivative.

The above definition makes use of the normal vector of S and of a particular curve α, tangent to y at p. To show that covariant differentiation is a concept of the intrinsic geometry and that it does not depend on the choice of the curve α, we shall obtain its expression in terms of a parametrization $\mathbf{x}(u, v)$ of S in p.

Let $\mathbf{x}(u(t), v(t)) = \alpha(t)$ be the expression of the curve α and let

$$
\begin{aligned}
w(t) &= a(u(t), v(t))\mathbf{x}_u + b(u(t), v(t))\mathbf{x}_v \\
&= a(t)\mathbf{x}_u + b(t)\mathbf{x}_v,
\end{aligned}
$$

be the expression of $w(t)$ in the parametrization $\mathbf{x}(u, v)$. Then

$$
\frac{dw}{dt} = a(\mathbf{x}_{uu}u' + \mathbf{x}_{uv}v') + b(\mathbf{x}_{vu}u' + \mathbf{x}_{vv}v') + a'\mathbf{x}_u + b'\mathbf{x}_v,
$$

where prime denotes the derivative with respect to t.

Since Dw/dt is the component of dw/dt in the tangent plane, we use the expressions in (1) of Sec. 4-1 for $\mathbf{x}_{uu}$, $\mathbf{x}_{uv}$, and $\mathbf{x}_{vv}$ and, by dropping the normal component, we obtain

$$
\begin{aligned}
\frac{Dw}{dt} &= (a' + \Gamma^1_{11}au' + \Gamma^1_{12}av' + \Gamma^1_{12}bu' + \Gamma^1_{22}bv')\mathbf{x}_u \\
&\quad + (b' + \Gamma^2_{11}au' + \Gamma^2_{12}av' + \Gamma^2_{12}bu' + \Gamma^2_{22}bv')\mathbf{x}_v.
\end{aligned}
\tag{1}
$$

Expression (1) shows that Dw/dt depends only on the vector $(u', v') = y$ and not on the curve α. Furthermore, the surface makes its appearance in Eq. (1) through the Christoffel symbols, that is, through the first fundamental form. Our assertions are, therefore, proved.

If, in particular, S is a plane, we know that it is possible to find a parametrization in such a way that $E = G = 1$ and $F = 0$. A quick inspection of the equations that give the Christoffel symbols shows that in this case the Γ^k_{ij} become zero. In this case, it follows from Eq. (1) that the covariant derivative agrees with the usual derivative of vectors in the plane (this can also be seen geometrically from Def. 1). The covariant derivative is, therefore, a generalization of the usual derivative of vectors in the plane.

Another consequence of Eq. (1) is that the definition of covariant derivative may be extended to a vector field which is defined only at the points of a parametrized curve. To make this point clear, we need some definitions.

DEFINITION 2. *A* parametrized curve $\alpha\colon [0, l] \to S$ *is the restriction to* $[0, l]$ *of a differentiable mapping of* $(0 - \epsilon, l + \epsilon)$, $\epsilon > 0$, *into* S. *If* $\alpha(0) = p$ *and* $\alpha(l) = q$, *we say that* α joins p *to* q. α *is* regular *if* $\alpha'(t) \neq 0$ *for* $t \in [0, l]$.

In what follows it will be convenient to use the notation $[0, l] = I$ whenever the specification of the end point l is not necessary.

DEFINITION 3. *Let* $\alpha: I \longrightarrow S$ *be a parametrized curve in* S. *A* vector field w along α *is a correspondence that assigns to each* $t \in I$ *a vector*

$$w(t) \in T_{\alpha(t)}(S).$$

The vector field w *is* differentiable *at* $t_0 \in I$ *if for some parametrization* $\mathbf{x}(u, v)$ *in* $\alpha(t_0)$ *the components* $a(t)$, $b(t)$ *of* $w(t) = a\mathbf{x}_u + b\mathbf{x}_v$ *are differentiable functions of* t *at* t_0. w *is* differentiable *in* I *if it is differentiable for every* $t \in I$.

An example of a (differentiable) vector field along α is given by the field $\alpha'(t)$ of the tangent vectors of α (Fig. 4-9).

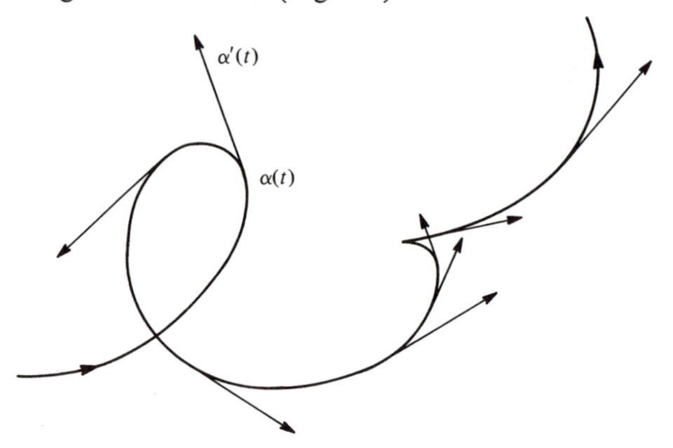

Figure 4-9. The field of tangent vectors along a curve α.

DEFINITION 4. *Let* w *be a differentiable vector field along* $\alpha: I \longrightarrow S$. *The expression* (1) *of* $(Dw/dt)(t)$, $t \in I$, *is well defined and is called the* covariant derivative *of* w *at* t.

From a point of view external to the surface, in order to obtain the covariant derivative of a field w along $\alpha: I \longrightarrow S$ at $t \in I$ we take the usual derivative $(dw/dt)(t)$ of w in t and project this vector orthogonally onto the tangent plane $T_{\alpha(t)}(S)$. It follows that when two surfaces are tangent along a parametrized curve α the covariant derivative of a field w along α is the same for both surfaces.

If $\alpha(t)$ is a curve on S, we can think of it as the trajectory of a point which is moving on the surface. $\alpha'(t)$ is then the speed and $\alpha''(t)$ the acceleration of α. The covariant derivative $D\alpha'/dt$ of the field $\alpha'(t)$ is the tangential component of the acceleration $\alpha''(t)$. Intuitively $D\alpha'/dt$ is the acceleration of the point $\alpha(t)$ "as seen from the surface S."

DEFINITION 5. *A vector field* w *along a parametrized curve* $\alpha: I \to S$ *is said to be* parallel *if* $Dw/dt = 0$ *for every* $t \in I$.

In the particular case of the plane, the notion of parallel field along a parametrized curve reduces to that of a constant field along the curve; that is, the length of the vector and its angle with a fixed direction are constant (Fig. 4-10). Those properties are partially reobtained on any surface as the following proposition shows.

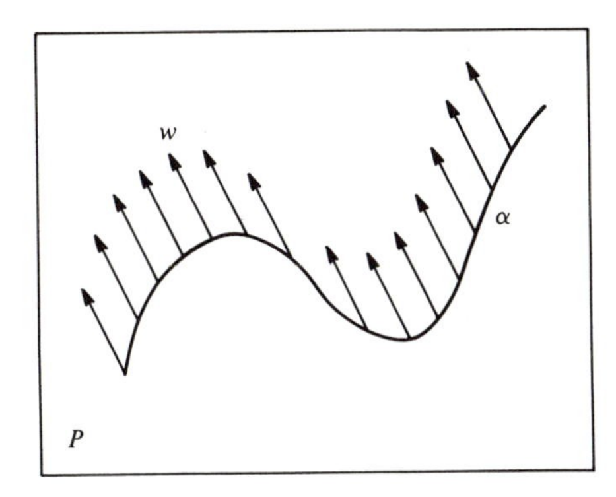

Figure 4-10

PROPOSITION 1. *Let* w *and* v *be parallel vector fields along* $\alpha: I \to S$. *Then* $\langle w(t), v(t) \rangle$ *is constant. In particular,* $|w(t)|$ *and* $|v(t)|$ *are constant, and the angle between* v(t) *and* w(t) *is constant.*

Proof. To say that the vector field w is parallel along α means that dw/dt is normal to the plane which is tangent to the surface at $\alpha(t)$; that is,

$$\langle v(t), w'(t) \rangle = 0, \qquad t \in I.$$

On the other hand, $v'(t)$ is also normal to the tangent plane at $\alpha(t)$. Thus,

$$\langle v(t), w(t) \rangle' - \langle v'(t), w(t) \rangle + \langle v(t), w'(t) \rangle = 0;$$

that is, $\langle v(t), w(t) \rangle =$ constant. **Q.E.D.**

Of course, on an arbitrary surface parallel fields may look strange to our R^3 intuition. For instance, the tangent vector field of a meridian (parametrized by arc length) of a unit sphere S^2 is a parallel field on S^2 (Fig. 4-11). In fact, since the meridian is a great circle on S^2, the usual derivative of such a field is normal to S^2. Thus, its covariant derivative is zero.

The following proposition shows that there exist parallel vector fields along a parametrized curve $\alpha(t)$ and that they are completely determined by their values at a point t_0.

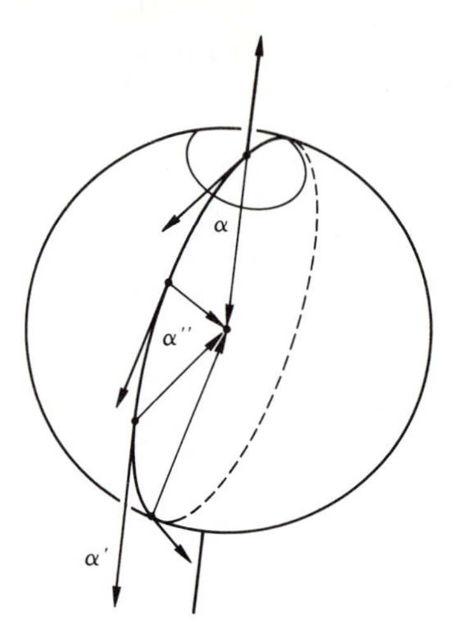

Figure 4-11. Parallel field on a sphere.

PROPOSITION 2. *Let* $\alpha: I \longrightarrow S$ *be a parametrized curve in* S *and let* $w_0 \in T_{\alpha(t_0)}(S)$, $t_0 \in I$. *Then there exists a unique parallel vector field* $w(t)$ *along* $\alpha(t)$, *with* $w(t_0) = w_0$.

An elementary proof of Prop. 2 will be given later in this section. Those who are familiar with the material of Sec. 3-6 will notice, however, that the proof is an immediate consequence of the theorem of existence and uniqueness of differential equations.

Proposition 2 allows us to talk about parallel transport of a vector along a parametrized curve.

DEFINITION 6. *Let* $\alpha: I \longrightarrow S$ *be a parametrized curve and* $w_0 \in T_{\alpha(t_0)}(S)$, $t_0 \in I$. *Let* w *be the parallel vector field along* α, *with* $w(t_0) = w_0$. *The vector* $w(t_1)$, $t_1 \in I$, *is called the* parallel transport *of* w_0 *along* α *at the point* t_1.

It should be remarked that if $\alpha: I \longrightarrow S$, $t \in I$, is regular, then the parallel transport does not depend on the parametrization of $\alpha(I)$. As a matter of fact, if $\beta: J \longrightarrow S$, $\sigma \in J$ is another regular parametrization for $\alpha(I)$, it follows from Eq. (1) that

$$\frac{Dw}{d\sigma} = \frac{Dw}{dt}\frac{dt}{d\sigma}, \qquad t \in I, \sigma \in I.$$

Since $dt/d\sigma \neq 0$, $w(t)$ is parallel if and only if $w(\sigma)$ is parallel.

Proposition 1 contains an interesting property of the parallel transport. Fix two points $p, q \in S$ and a parametrized curve $\alpha: I \longrightarrow S$ with $\alpha(0) = p$, $\alpha(1) = q$. Denote by $P_\alpha: T_p(S) \longrightarrow T_q(S)$ the map that assigns to each $v \in T_p(S)$ its parallel transport along α at q. Proposition 1 says that this map is an isometry.

Another interesting property of the parallel transport is that if two surfaces S and $\bar{S}$ are tangent along a parametrized curve α and w_0 is a vector of $T_{\alpha(t_0)}(S) = T_{\alpha(t_0)}(\bar{S})$, then $w(t)$ is the parallel transport of w_0 relative to the surface S if and only if $w(t)$ is the parallel transport of w_0 relative to $\bar{S}$. Indeed, the covariant derivative Dw/dt of w is the same for both surfaces. Since the parallel transport is unique, the assertion follows.

The above property will allow us to give a simple and instructive example of parallel transport.

Example 1. Let C be a parallel of colatitude φ (see Fig. 4-12) of an oriented unit sphere and let w_0 be a unit vector, tangent to C at some point p of C. Let us determine the parallel transport of w_0 along C, parametrized by arc length s, with $s = 0$ at p.

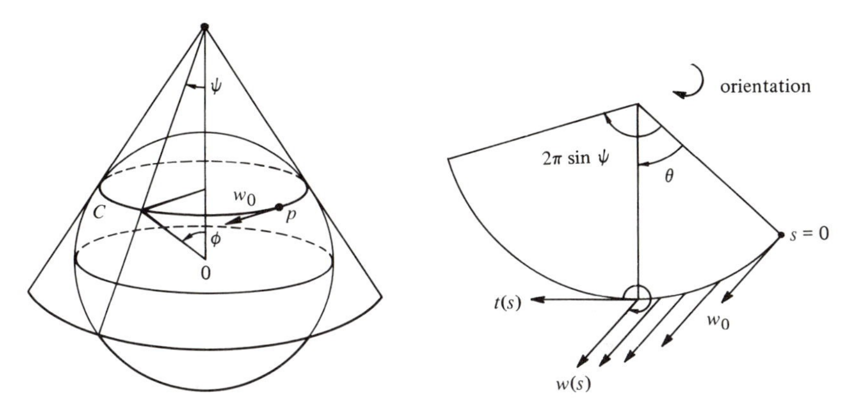

Figure 4-12 Figure 4-13

Consider the cone which is tangent to the sphere along C. The angle ψ at the vertex of this cone is given by $\psi = (\pi/2) - \varphi$. By the above property, the problem reduces to the determination of the parallel transport of w_0, along C, relative to the tangent cone.

The cone minus one generator is, however, isometric to an open set $U \subset R^2$ (cf. Example 3, Sec. 4-2), given in polar coordinates by

$$0 < \rho < +\infty, \qquad 0 < \theta < 2\pi \sin \psi.$$

Since in the plane the parallel transport coincides with the usual notion, we obtain, for a displacement s of p, corresponding to the central angle θ (see Fig. 4-13) that the oriented angle formed by the tangent vector $t(s)$ with the parallel transport $w(s)$ is given by $2\pi - \theta$.

It is sometimes convenient to introduce the notion of a "broken curve," which can be expressed as follows.

DEFINITION 7. *A map* $\alpha: [0, l] \to S$ *is a* parametrized piecewise regular curve *if* α *is continuous and there exists a subdivision*

$$0 = t_0 < t_1 < \cdots < t_k < t_{k+1} = l$$

of the interval $[0, l]$ *in such a way that the restriction* $\alpha | [t_i, t_{i+1}]$, $i = 0, \ldots, k$, *is a parametrized regular curve. Each* $\alpha | [t_i, t_{i+1}]$ *is called* a regular arc *of* α.

The notion of parallel transport can be easily extended to parametrized piecewise regular curves. If, say, the initial value w_0 lies in the interval $[t_i, t_{i+1}]$, we perform the parallel transport in the regular arc $\alpha | [t_i, t_{i+1}]$ as usual; if $t_{i+1} \neq l$, we take $w(t_{i+1})$ as the initial value for the parallel transport in the next arc $\alpha | [t_{i+1}, t_{i+2}]$, and so forth.

Example 2.† The previous example is a particular case of an interesting geometric construction of the parallel transport. Let C be a regular curve on a surface S and assume that C is nowhere tangent to an asymptotic direction. Consider the envelope of the family of tangent planes of S along C (cf. Example 4, Sec. 3-5). In a neighborhood of C, this envelope is a regular surface Σ which is tangent to S along C. (In Example 1, Σ can be taken as a ribbon around C on the cone which is tangent to the sphere along C.) Thus, the parallel transport along C of any vector $w \in T_p(S)$, $p \in S$, is the same whether we consider it relative to S or to Σ. Furthermore, Σ is a developable surface; hence, its Gaussian curvature is identically zero.

Now, we shall prove later in this book (Sec. 4-6, theorem of Minding) that a surface of zero Gaussian curvature is locally isometric to a plane. Thus, we can map a neighborhood $V \subset \Sigma$ of p into a plane P by an isometry $\varphi: V \to P$. To obtain the parallel transport of w along $V \cap C$, we take the usual parallel transport in the plane of $d\varphi_p(w)$ along $\varphi(C)$ and pull it back to Σ by $d\varphi$ (Fig. 4-14).

This gives a geometric construction for the parallel transport along small arcs of C. We leave it as an exercise to show that this construction can be extended stepwise to a given arc of C. (Use the Heine-Borel theorem and proceed as in the case of broken curves.)

The parametrized curves $\gamma: I \to R^2$ of a plane along which the field of their tangent vectors $\gamma'(t)$ is parallel are precisely the straight lines of that plane. The parametrized curves that satisfy an analogous condition for a surface are called geodesics.

---

†This example uses the material on ruled surfaces of Sec. 3–5.

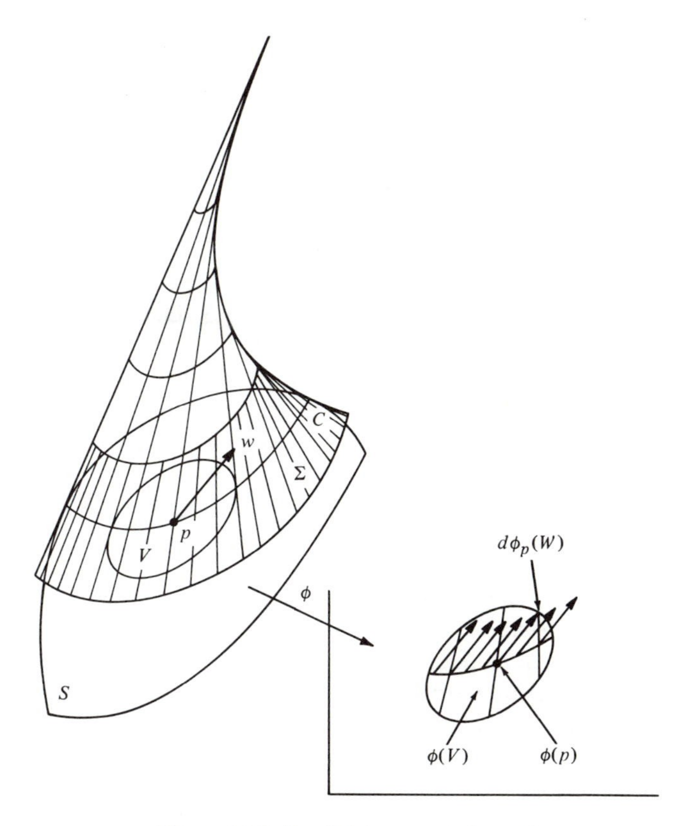

Figure 4-14. Parallel transport along C.

DEFINITION 8. *A nonconstant, parametrized curve* $\gamma\colon I \longrightarrow S$ *is said to be* geodesic *at* $t \in I$ *if the field of its tangent vectors* $\gamma'(t)$ *is parallel along* γ *at* t; *that is,*

$$\frac{D\gamma'(t)}{dt} = 0;$$

γ *is a* parametrized geodesic *if it is geodesic for all* $t \in I$.

By Prop. 1, we obtain immediately that $|\gamma'(t)| = \text{const.} = c \neq 0$. Therefore, we may introduce the arc length $s = ct$ as a parameter, and we conclude that the parameter t of a parametrized geodesic γ is proportional to the arc length of γ.

Observe that a parametrized geodesic may admit self-intersections. (Example 6 will illustrate this; see Fig. 4-20.) However, its tangent vector is never zero, and thus the parametrization is regular.

The notion of geodesic is clearly local. The previous considerations allow us to extend the definition of geodesic to subsets of S that are regular curves.

DEFINITION 8a. *A regular connected curve* C *in* S *is said to be a* geodesic *if, for every* p $\in$ S, *the parametrization* $\alpha(s)$ *of a coordinate neighborhood of* p *by the arc length* s *is a parametrized geodesic; that is,* $\alpha'(s)$ *is a parallel vector field along* $\alpha(s)$.

Observe that every straight line contained in a surface satisfies Def. 8a.

From a point of view exterior to the surface S, Def. 8a is equivalent to saying that $\alpha''(s) = kn$ is normal to the tangent plane, that is, parallel to the normal to the surface. In other words, a regular curve $C \subset S$ $(k \neq 0)$ is a geodesic if and only if its principal normal at each point $p \in C$ is parallel to the normal to S at p.

The above property can be used to identify some geodesics geometrically, as shown in the examples below.

Example 3. The great circles of a sphere S^2 are geodesics. Indeed, the great circles C are obtained by intersecting the sphere with a plane that passes through the center O of the sphere. The principal normal at a point $p \in C$ lies in the direction of the line that connects p to O because C is a circle of center O. Since S^2 is a sphere, the normal lies in the same direction, which verifies our assertion.

Later in this section we shall prove the general fact that for each point $p \in S$ and each direction in $T_p(S)$ there exists exactly one geodesic $C \subset S$ passing through p and tangent to this direction. For the case of the sphere, through each point and tangent to each direction there passes exactly one great circle, which, as we proved before, is a geodesic. Therefore, by uniqueness, the great circles are the only geodesics of a sphere.

Example 4. For the right circular cylinder over the circle $x^2 + y^2 = 1$, it is clear that the circles obtained by the intersection of the cylinder with planes that are normal to the axis of the cylinder are geodesics. That is so because the principal normal to any of its points is parallel to the normal to the surface at this point.

On the other hand, by the observation after Def. 8a the straight lines of the cylinder (generators) are also geodesics.

To verify the existence of other geodesics on the cylinder C we shall consider a parametrization (cf. Example 2, Sec. 2-5)

$$\mathbf{x}(u, v) = (\cos u, \sin u, v)$$

of the cylinder in a point $p \in C$, with $\mathbf{x}(0, 0) = p$. In this parametrization, a neighborhood of p in C is expressed by $\mathbf{x}(u(s), v(s))$, where s is the arc length of C. As we saw previously (cf. Example 1, Sec. 4-2), $\mathbf{x}$ is a local isometry which maps a neighborhood U of $(0, 0)$ of the uv plane into the cylinder. Since the condition of being a geodesic is local and invariant by isometries, the curve $(u(s), v(s))$ must be a geodesic in U passing through $(0, 0)$. But the geodesics of the plane are the straight lines. Therefore, excluding the cases already obtained,

$$u(s) = as, \qquad v(s) = bs, \qquad a^2 + b^2 = 1.$$

It follows that when a regular curve C (which is neither a circle or a line) is a geodesic of the cylinder it is locally of the form (Fig. 4-15)

$$(\cos as, \sin as, bs),$$

and thus it is a helix. In this way, all the geodesics of a right circular cylinder are determined.

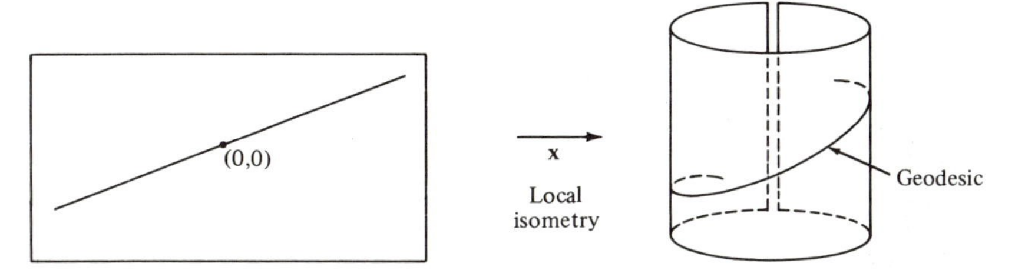

Figure 4-15. Geodesics on a cylinder.

Observe that given two points on a cylinder which are not in a circle parallel to the xy plane, it is possible to connect them through an infinite number of helices. This fact means that two points of a cylinder may in general be connected through an infinite number of geodesics, in contrast to the situation in the plane. Observe that such a situation may occur only with geodesics that make a "complete turn," since the cylinder minus a generator is isometric to a plane (Fig. 4-16).

Proceeding with the analogy with the plane, we observe that the lines, that is, the geodesics of a plane, are also characterized as regular curves of curvature zero. Now, the curvature of an oriented plane curve is given by the absolute value of the derivative of the unit vector field tangent to the curve, associated to a sign which denotes the concavity of the curve in relation to the orientation of the plane (cf. Sec. 1-5, Remark 1). To take the sign into consideration, it is convenient to introduce the following definition.

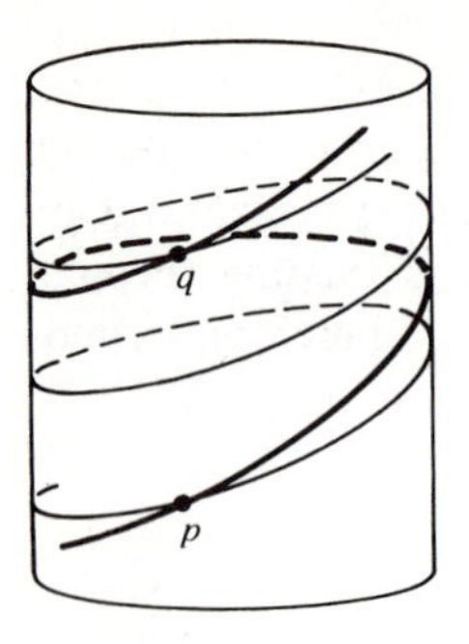

Figure 4-16. Two geodesics on a cylinder joining p and q.

DEFINITION 9. *Let* w *be a differentiable field of unit vectors along a parametrized curve* $\alpha: I \longrightarrow S$ *on an oriented surface* S. *Since* $w(t)$, $t \in I$, *is a unit vector field,* $(dw/dt)(t)$ *is normal to* $w(t)$, *and therefore*

$$\frac{Dw}{dt} = \lambda(N \wedge w(t)).$$

The real number $\lambda = \lambda(t)$, *denoted by* $[Dw/dt]$, *is called the* algebraic value of the covariant derivative *of* w *at* t.

Observe that the sign of $[Dw/dt]$ depends on the orientation of S and that $[Dw/dt] = \langle dw/dt, N \wedge w \rangle$.

We should also make the general remark that, from now on, the orientation of S will play an essential role in the concepts to be introduced. The careful reader will have noticed that the definitions of parallel transport and geodesic do not depend on the orientation of S. In constrast, the geodesic curvature, to be defined below, changes its sign with a change of orientation of S.

We shall now define, for a curve in a surface, a concept which is an analogue of the curvature of plane curves.

DEFINITION 10. *Let* C *be an oriented regular curve contained on an oriented surface* S, *and let* $\alpha(s)$ *be a parametrization of* C, *in a neighborhood of* $p \in S$, *by the arc length* s. *The algebraic value of the covariant derivative* $[D\alpha'(s)/ds] = k_g$ *of* $\alpha'(s)$ *at* p *is called the* geodesic curvature *of* C *at* p.

The geodesics which are regular curves are thus characterized as curves whose geodesic curvature is zero.

From a point of view external to the surface, the absolute value of the geodesic curvature k_g of C at p is the absolute value of the tangential component of the vector $\alpha''(s) = kn$, where k is the curvature of C at p and n is the normal vector of C at p. Recalling that the absolute value of the normal component of the vector kn is the absolute value of the normal curvature

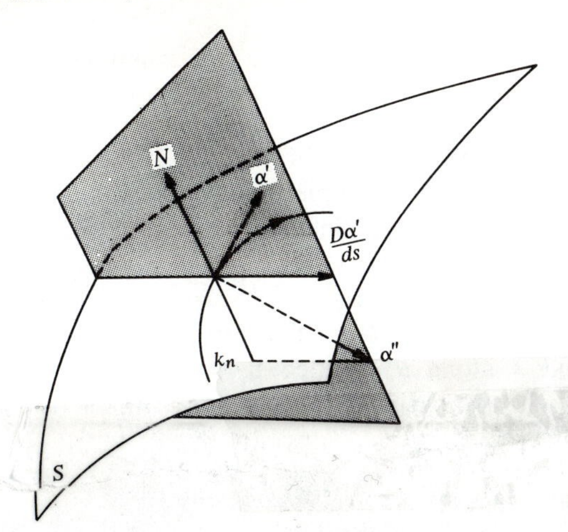

Figure 4-17

k_n of $C \subset S$ in p, we have immediately (Fig. 4-17)

$$k^2 = k_g^2 + k_n^2.$$

For instance, the absolute value of the geodesic curvature k_g of a parallel C of colatitude φ in a unit sphere S^2 can be computed from the relation (see Fig. 4-18)

$$\sin^2 \varphi = k_n^2 + k_g^2 = \sin^4 \varphi + k_g^2;$$

that is,

$$k_g^2 = \sin^2 \varphi(1 - \sin^2 \varphi) = \tfrac{1}{4} \sin^2 2\varphi.$$

The sign of k_g depends on the orientations of S^2 and C.

A further consequence of that external interpretation is that when two surfaces are tangent along a regular curve C, the absolute value of the geodesic curvature of C is the same relatively to any of the two surfaces.

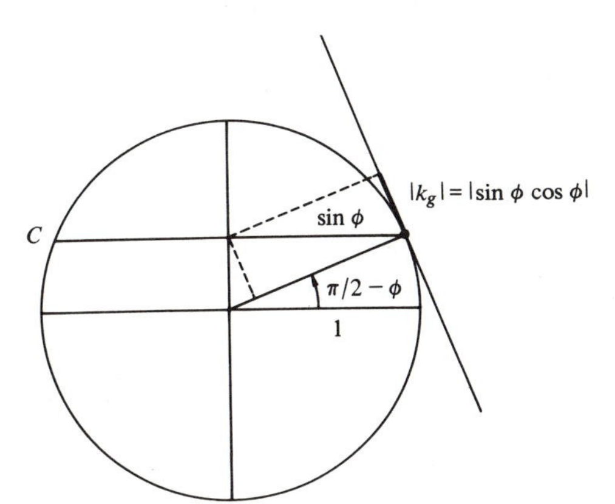

Figure 4-18. Geodesic curvature of a parallel on a unit sphere.

Remark. The geodesic curvature of $C \subset S$ changes sign when we change the orientation of either C or S.

We shall now obtain an expression for the algebraic value of the covariant derivative (Prop. 3 below). For that we need some preliminaries.

Let v and w be two differentiable vector fields along the parametrized curve $\alpha: I \to S$, with $|v(t)| = |w(t)| = 1$, $t \in I$. We want to define a differentiable function $\varphi: I \to R$ in such a way that $\varphi(t)$, $t \in I$, is a determination of the angle from $v(t)$ to $w(t)$ in the orientation of S. For that, we consider the differentiable vector field $\bar{v}$ along α, defined by the condition that $\{v(t), \bar{v}(t)\}$ is an orthonormal positive basis for every $t \in I$. Thus, $w(t)$ may be expressed as

$$w(t) = a(t)v(t) + b(t)\bar{v}(t),$$

where a and b are differentiable functions in I and $a^2 + b^2 = 1$.

Lemma 1 below shows that by fixing a determination φ_0 of the angle from $v(t_0)$ to $w(t_0)$ it is possible to "extend it" differentiably in I, and this yields the desired function.

LEMMA 1. *Let* a *and* b *be differentiable functions in* I *with* $a^2 + b^2 = 1$ *and* φ_0 *be such that* $a(t_0) = \cos \varphi_0$, $b(t_0) = \sin \varphi_0$. *Then the differentiable function*

$$\varphi = \varphi_0 + \int_{t_0}^{t} (ab' - ba')\, dt$$

is such that $\cos \varphi(t) = a(t)$, $\sin \varphi(t) = b(t)$, $t \in I$, *and* $\varphi(t_0) = \varphi_0$.

Proof. It suffices to show that the function

$$(a - \cos \varphi)^2 + (b - \sin \varphi)^2 = 2 - 2(a \cos \varphi + b \sin \varphi)$$

is identically zero, or that

$$A = a \cos \varphi + b \sin \varphi = 1.$$

By using the fact that $aa' = -bb'$ and the definition of φ, we easily obtain

$$\begin{aligned} A' &= -a(\sin \varphi)\varphi' + b(\cos \varphi)\varphi' + a' \cos \varphi + b' \sin \varphi \\ &= -b'(\sin \varphi)(a^2 + b^2) - a'(\cos \varphi)(a^2 + b^2) \\ &\quad + a' \cos \varphi + b' \sin \varphi = 0. \end{aligned}$$

Therefore, $A(t) = \text{const.}$, and since $A(t_0) = 1$, the lemma is proved. **Q.E.D.**

We may now relate the covariant derivative of two unit vector fields along a curve to the variation of the angle that they form.

LEMMA 2. *Let* v *and* w *be two differentiable vector fields along the curve* $\alpha: I \rightarrow S$, *with* $|w(t)| = |v(t)| = 1$, $t \in I$. *Then*

$$\left[\frac{Dw}{dt}\right] - \left[\frac{Dv}{dt}\right] = \frac{d\varphi}{dt},$$

where φ *is one of the differentiable determinations of the angle from* v *to* w, *as given by Lemma 1.*

Proof. We introduce the vectors $\bar{v} = N \wedge v$ and $\bar{w} = N \wedge w$. Then

$$w = (\cos \varphi)v + (\sin \varphi)\bar{v}, \tag{2}$$

$$\begin{aligned} \bar{w} = N \wedge w &= (\cos \varphi)N \wedge v + (\sin \varphi)N \wedge \bar{v} \\ &= (\cos \varphi)\bar{v} - (\sin \varphi)v. \end{aligned} \tag{3}$$

By differentiating (2) with respect to t, we obtain

$$w' = -(\sin \varphi)\varphi' v + (\cos \varphi)v' + (\cos \varphi)\varphi' \bar{v} + (\sin \varphi)\bar{v}'.$$

By taking the inner product of the last relation with $\bar{w}$, using (3), and observing that $\langle v, \bar{v} \rangle = 0$, $\langle v, v' \rangle = 0$, we conclude that

$$\begin{aligned} \langle w', \bar{w} \rangle &= (\sin^2 \varphi)\varphi' + (\cos^2 \varphi)\langle v', \bar{v} \rangle + (\cos^2 \varphi)\varphi' - (\sin^2 \varphi)\langle \bar{v}', v \rangle \\ &= \varphi' + (\cos^2 \varphi)\langle v', \bar{v} \rangle - (\sin^2 \varphi)\langle \bar{v}', v \rangle. \end{aligned}$$

On the other hand, since $\langle v, \bar{v} \rangle = 0$, that is,

$$\langle v', \bar{v} \rangle = -\langle v, \bar{v}' \rangle,$$

we conclude that

$$\langle w', \bar{w} \rangle = \varphi' + (\cos^2 \varphi + \sin^2 \varphi)\langle v', \bar{v} \rangle = \varphi' + \langle v', \bar{v} \rangle.$$

It follows that

$$\left[\frac{Dw}{dt}\right] = \langle w', \bar{w} \rangle = \varphi' + \langle v', \bar{v} \rangle = \frac{d\varphi}{dt} + \left[\frac{Dv}{dt}\right],$$

since

$$\langle w', \bar{w} \rangle = \left\langle \frac{dw}{dt}, \bar{w} \right\rangle = \left[\frac{Dw}{dt}\right]\langle N \wedge w, \bar{w} \rangle = \left[\frac{Dw}{dt}\right],$$

which concludes the proof of the lemma. **Q.E.D.**

An immediate consequence of the above lemma is the following observation. Let C be a regular oriented curve on S, $\alpha(s)$ a parametrization by the arc length s of C at $p \in C$, and $v(s)$ a parallel field along $\alpha(s)$. Then, by taking $w(s) = \alpha'(s)$, we obtain

$$k_g(s) = \left[\frac{D\alpha'(s)}{ds}\right] = \frac{d\varphi}{ds}.$$

In other words, *the geodesic curvature is the rate of change of the angle that the tangent to the curve makes with a parallel direction along the curve.* In the case of the plane, the parallel direction is fixed and the geodesic curvature reduces to the usual curvature.

We are now able to obtain the promised expression for the algebraic value of the covariant derivative. Whenever we speak of a parametrization of an oriented surface, this parametrization is assumed to be compatible with the given orientation.

PROPOSITION 3. *Let* $\mathbf{x}(u, v)$ *be an orthogonal parametrization (that is,* $F = 0$*) of a neighborhood of an oriented surface* S*, and* $w(t)$ *be a differentiable field of unit vectors along the curve* $\mathbf{x}(u(t), v(t))$*. Then*

$$\left[\frac{Dw}{dt}\right] = \frac{1}{2\sqrt{EG}}\left\{G_u\frac{dv}{dt} - E_v\frac{du}{dt}\right\} + \frac{d\varphi}{dt},$$

where $\varphi(t)$ *is the angle from* $\mathbf{x}_u$ *to* $w(t)$ *in the given orientation.*

Proof. Let $e_1 = \mathbf{x}_u/\sqrt{E}$, $e_2 = \mathbf{x}_v/\sqrt{G}$ be the unit vectors tangent to the coordinate curves. Observe that $e_1 \wedge e_2 = N$, where N is the given orientation of S. By using Lemma 2, we may write

$$\left[\frac{Dw}{dt}\right] = \left[\frac{De_1}{dt}\right] + \frac{d\varphi}{dt},$$

where $e_1(t) = e_1(u(t), v(t))$ is the field e_1 restricted to the curve $\mathbf{x}(u(t), v(t))$. Now

$$\left[\frac{De_1}{dt}\right] = \left\langle\frac{de_1}{dt}, N \wedge e_1\right\rangle = \left\langle\frac{de_1}{dt}, e_2\right\rangle = \langle(e_1)_u, e_2\rangle\frac{du}{dt} + \langle(e_1)_v, e_2\rangle\frac{dv}{dt}.$$

On the other hand, since $F = 0$, we have

$$\langle \mathbf{x}_{uu}, \mathbf{x}_v\rangle = -\tfrac{1}{2}E_v,$$

and therefore

$$\langle(e_1)_u, e_2\rangle = \left\langle\left(\frac{\mathbf{x}_u}{\sqrt{E}}\right)_u, \frac{\mathbf{x}_v}{\sqrt{G}}\right\rangle = -\frac{1}{2}\frac{E_v}{\sqrt{EG}}.$$

Similarly,

$$\langle(e_1)_v, e_2\rangle = \frac{1}{2}\frac{G_u}{\sqrt{EG}}.$$

By introducing these relations in the expression of $[Dw/dt]$, we finally obtain

$$\left[\frac{Dw}{dt}\right] = \frac{1}{2\sqrt{EG}}\left\{G_u\frac{dv}{dt} - E_v\frac{du}{dt}\right\} + \frac{d\varphi}{dt},$$

which completes the proof. **Q.E.D.**

As an application of Prop. 3, we shall prove the existence and uniqueness of the parallel transport (Prop. 2).

Proof of Prop. 2. Let us assume initially that the parametrized curve $\alpha: I \longrightarrow S$ is contained in a coordinate neighborhood of an orthogonal parametrization $\mathbf{x}(u, v)$. Then, with the notations of Prop. 3, the condition of parallelism for the field w becomes

$$\frac{d\varphi}{dt} = -\frac{1}{2\sqrt{EG}}\left\{G_u\frac{dv}{dt} - E_v\frac{du}{dt}\right\} = B(t).$$

Denoting by φ_0 a determination of the oriented angle from $\mathbf{x}_u$ to w_0, the field w is entirely determined by

$$\varphi = \varphi_0 + \int_{t_0}^{t} B(t)\, dt,$$

which proves the existence and uniqueness of w in this case.

If $\alpha(I)$ is not contained in a coordinate neighborhood, we shall use the compactness of I to divide $\alpha(I)$ into a finite number of parts, each contained in a coordinate neighborhood. By using the uniqueness of the first part of the proof in the nonempty intersections of these pieces, it is easy to extend the result to the present case. **Q.E.D.**

A further application of Prop. 3 is the following expression for the geodesic curvature, known as *Liouville's formula.*

PROPOSITION 4 (Liouville). *Let* $\alpha(s)$ *be a parametrization by arc length of a neighborhood of a point* $p \in S$ *of a regular oriented curve* C *on an oriented surface* S. *Let* $\mathbf{x}(u, v)$ *be an orthogonal parametrization of* S *in* p *and* $\varphi(s)$ *be the angle that* $\mathbf{x}_u$ *makes with* $\alpha'(s)$ *in the given orientation. Then*

$$k_g = (k_g)_1 \cos\varphi + (k_g)_2 \sin\varphi + \frac{d\varphi}{ds},$$

where $(k_g)_1$ *and* $(k_g)_2$ *are the geodesic curvatures of the coordinate curves* $v = const.$ *and* $u = const.$ *respectively.*

Proof. By setting $w = \alpha'(s)$ in Prop. 3, we obtain

$$k_g = \frac{1}{2\sqrt{EG}}\left\{G_u\frac{dv}{ds} - E_v\frac{du}{ds}\right\} + \frac{d\varphi}{ds}.$$

Along the coordinate curve $v = \text{const.}$ $u = u(s)$, we have $dv/ds = 0$ and $du/ds = 1/\sqrt{E}$; therefore,

$$(k_g)_1 = -\frac{E_v}{2E\sqrt{G}}.$$

Similarly,

$$(k_g)_2 = \frac{G_u}{2G\sqrt{E}}.$$

By introducing these relations in the above formula for k_g, we obtain

$$k_g = (k_g)_1\sqrt{E}\,\frac{du}{ds} + (k_g)_2\sqrt{G}\,\frac{dv}{ds} + \frac{d\varphi}{ds}.$$

Since

$$\sqrt{E}\,\frac{du}{ds} = \left\langle \alpha'(s), \frac{\mathbf{x}_u}{\sqrt{E}}\right\rangle = \cos\varphi \quad \text{and} \quad \sqrt{G}\,\frac{dv}{ds} = \sin\varphi,$$

we finally arrive at

$$k_g = (k_g)_1\cos\varphi + (k_g)_2\sin\varphi + \frac{d\varphi}{ds},$$

as we wished. **Q.E.D.**

We shall now introduce the equations of a geodesic in a coordinate neighborhood. For that, let $\gamma\colon I \to S$ be a parametrized curve of S and let $\mathbf{x}(u, v)$ be a parametrization of S in a neighborhood V of $\gamma(t_0)$, $t_0 \in I$. Let $J \subset I$ be an open interval containing t_0 such that $\gamma(J) \subset V$. Let $\mathbf{x}(u(t), v(t))$, $t \in J$, be the expression of $\gamma\colon J \to S$ in the parametrization $\mathbf{x}$. Then, the tangent vector field $\gamma'(t)$, $t \in J$, is given by

$$w = u'(t)\mathbf{x}_u + v'(t)\mathbf{x}_v.$$

Therefore, the fact that w is parallel is equivalent to the system of differential equations

$$\begin{aligned} u'' + \Gamma^1_{11}(u')^2 + 2\Gamma^1_{12}u'v' + \Gamma^1_{22}(v')^2 &= 0,\\ v'' + \Gamma^2_{11}(u')^2 + 2\Gamma^2_{12}u'v' + \Gamma^2_{22}(v')^2 &= 0, \end{aligned} \qquad \textbf{(4)}$$

obtained from Eq. (1) by making $a = u'$ and $b = v'$, and equating to zero the coefficients of $\mathbf{x}_u$ and $\mathbf{x}_v$.

In other words, $\gamma: I \longrightarrow S$ is a geodesic if and only if system (4) is satisfied for every interval $J \subset I$ such that $\gamma(J)$ is contained in a coordinate neighborhood. The system (4) is known as the *differential equations of the geodesics of* S.

An important consequence of the fact that the geodesics are characterized by the system (4) is the following proposition.

PROPOSITION 5. *Given a point* $\mathrm{p} \in \mathrm{S}$ *and a vector* $\mathrm{w} \in \mathrm{T_p(S)}$, $\mathrm{w} \neq 0$, *there exist an* $\epsilon > 0$ *and a unique parametrized geodesic* $\gamma: (-\epsilon, \epsilon) \longrightarrow \mathrm{S}$ *such that* $\gamma(0) = \mathrm{p}$, $\gamma'(0) = \mathrm{w}$.

In Sec. 4-5 we shall show how Prop. 5 may be derived from theorems on vector fields.

Remark. The reason for taking $w \neq 0$ in Prop. 5 comes from the fact that we have excluded the constant curves in the definition of parametrized geodesics (cf. Def. 8).

We shall use the rest of this section to give some geometrical applications of the differential equations (4). This material can be omitted if the reader wants to do so. In this case, Exercises 18, 20, and 21 should also be omitted.

Example 5. We shall use system (4) to study locally the geodesics of a surface of revolution (cf. Example 4, Sec. 2-3) with the parametrization

$$x = f(v) \cos u, \qquad y = f(v) \sin u, \qquad z = g(v).$$

By Example 1 of Sec. 4-1, the Christoffel symbols are given by

$$\Gamma_{11}^1 = 0, \qquad \Gamma_{11}^2 = -\frac{ff'}{(f')^2 + (g')^2}, \qquad \Gamma_{12}^1 = \frac{ff'}{f^2},$$

$$\Gamma_{12}^2 = 0, \qquad \Gamma_{22}^1 = 0, \qquad \Gamma_{22}^2 = \frac{f'f'' + g'g''}{(f')^2 + (g')^2}.$$

With the values above, system (4) becomes

$$\begin{aligned} u'' + \frac{2ff'}{f^2}u'v' &= 0, \\ v'' - \frac{ff'}{(f')^2 + (g')^2}(u')^2 + \frac{f'f'' + f'g''}{(f')^2 + (g')^2}(v')^2 &= 0. \end{aligned} \tag{4a}$$

We are going to obtain some conclusions from these equations.

First, as expected, the meridians $u = \text{const.}$ and $v = v(s)$, parametrized by arc length s, are geodesics. Indeed, the first equation of (4a) is trivially satisfied by $u = \text{const.}$ The second equation becomes

$$v'' + \frac{f'f'' + g'g''}{(f')^2 + (g')^2}(v')^2 = 0.$$

Since the first fundamental form along the meridian $u = \text{const.}$ $v = v(s)$ yields

$$((f')^2 + (g')^2)(v')^2 = 1,$$

we conclude that

$$(v')^2 = \frac{1}{(f')^2 + (g')^2}.$$

Therefore, by derivation,

$$2v'v'' = -\frac{2(f'f'' + g'g'')}{((f')^2 + (g')^2)^2}v' = -\frac{2(f'f'' + g'g'')}{(f')^2 + (g')^2}(v')^3,$$

or, since $v' \neq 0$,

$$v'' = -\frac{f'f'' + g'g''}{(f')^2 + (g')^2}(v')^2;$$

that is, along the meridian the second of the equations (4a) is also satisfied, which shows that in fact the meridians are geodesics.

Now we are going to determine which parallels $v = \text{const.}$ $u = u(s)$, parametrized by arc length, are geodesics. The first of the equations (4a) gives $u' = \text{const.}$ and the second becomes

$$\frac{ff'}{(f')^2 + (g')^2}(u')^2 = 0.$$

In order that the parallel $v = \text{const.}$, $u = u(s)$ be a geodesic it is necessary that $u' \neq 0$. Since $(f')^2 + (g')^2 \neq 0$ and $f \neq 0$, we conclude from the above equation that $f' = 0$.

In other words, a necessary condition for a parallel of a surface of revolution to be a geodesic is that such a parallel be generated by the rotation of a point of the generating curve where the tangent is parallel to the axis of revolution (Fig. 4-19). This condition is clearly sufficient, since it implies that the normal line of the parallel agrees with the normal line to the surface (Fig. 4-19).

We shall obtain for further use an interesting geometric consequence from the first of the equations (4a), known as Clairaut's relation. Observe that the first of the equations (4a) may be written as

$$(f^2u')' = f^2u'' + 2ff'u'v' = 0;$$

hence,

$$f^2u' = \text{const.} = c.$$

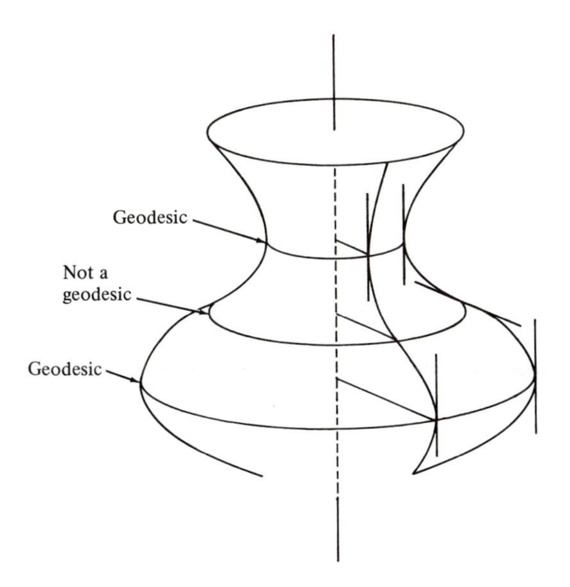

Figure 4-19

On the other hand, the angle θ, $0 \leq \theta \leq \pi/2$, of a geodesic with a parallel that intersects it is given by

$$\cos\theta = \frac{|\langle \mathbf{x}_u, \mathbf{x}_u u' + \mathbf{x}_v v' \rangle|}{|\mathbf{x}_u|} = |fu'|,$$

where $\{\mathbf{x}_u, \mathbf{x}_v\}$ is the associated basis to the given parametrization. Since $f = r$ is the radius of the parallel at the intersection point, we obtain *Clairaut's relation*:

$$r\cos\theta = \text{const.} = |c|.$$

In the next example we shall show how useful this relation is. See also Exercises 18, 20, and 21.

Finally, we shall show that system (4a) may be integrated by means of primitives. Let $u = u(s)$, $v = v(s)$ be a geodesic parametrized by arc length, which we shall assume not to be a meridian or a parallel of the surface. The first of the equations (4a) is then written as $f^2 u' = \text{const.} = c \neq 0$.

Observe initially that the first fundamental form along $(u(s), v(s))$,

$$1 = f^2\left(\frac{du}{ds}\right)^2 + ((f')^2 + (g')^2)\left(\frac{dv}{ds}\right)^2, \tag{5}$$

together with the first of the equations (4a), is equivalent to the second of the equations (4a). In fact, by substituting $f^2 u' = c$ in Eq. (5), we obtain

$$\left(\frac{dv}{ds}\right)^2((f')^2 + (g')^2) = -\frac{c^2}{f^2} + 1;$$

hence, by differentiating with respect to s,

$$2\frac{dv}{ds}\frac{d^2v}{ds^2}((f')^2 + (g')^2) + \left(\frac{dv}{ds}\right)^2(2f'f'' + 2g'g'')\frac{dv}{ds} = \frac{2ff'c^2}{f^4}\frac{dv}{ds},$$

which is equivalent to the second equation of (4a), since $(u(s), v(s))$ is not a parallel. (Of course the geodesic may be tangent to a parallel which is not a geodesic and then $v'(s) = 0$. However, Clairaut's relation shows that this happens only at isolated points.)

On the other hand, since $c \neq 0$ (because the geodesic is not a meridian), we have $u'(s) \neq 0$. It follows that we may invert $u = u(s)$, obtaining $s = s(u)$, and therefore $v = v(s(u))$. By multiplying Eq. (5) by $(ds/du)^2$, we obtain

$$\left(\frac{ds}{du}\right)^2 = f^2 + ((f')^2 + (g')^2)\left(\frac{dv}{ds}\frac{ds}{du}\right)^2,$$

or, by using the fact that $(ds/du)^2 = f^4/c^2$,

$$f^2 = c^2 + c^2\frac{(f')^2 + (g')^2}{f^2}\left(\frac{dv}{du}\right)^2,$$

that is,

$$\frac{dv}{du} = \frac{1}{c}f\sqrt{\frac{f^2 - c^2}{(f')^2 + (g')^2}};$$

hence,

$$u = c\int\frac{1}{f}\sqrt{\frac{(f')^2 + (g')^2}{f^2 - c^2}}\,dv + \text{const.} \tag{6}$$

which is the equation of a segment of a geodesic of a surface of revolution which is neither a parallel nor a meridian.

Example 6. We are going to show that any geodesic of a paraboloid of revolution $z = x^2 + y^2$ which is not a meridian intersects itself an infinite number of times.

Let p_0 be a point of the paraboloid and let P_0 be the parallel of radius r_0 passing through p_0. Let γ be a parametrized geodesic passing through p_0 and making an angle θ_0 with P_0. Since, by Clairaut's relation,

$$r\cos\theta = \text{const.} = |c|, \qquad 0 \leq \theta \leq \frac{\pi}{2},$$

we conclude that θ increases with r.

Therefore, if we follow in the direction of the increasing parallels, θ increases. It may happen that in some revolution surfaces γ approaches asymptotically a meridian. We shall show in a while that such is not the case with a paraboloid of revolution. That is, the geodesic γ intersects all the meridians, and therefore it makes an infinite number of turns around the paraboloid.

On the other hand, if we follow the direction of decreasing parallels, the angle θ decreases and approaches the value 0, which corresponds to a parallel of radius $|c|$ (observe that if $\theta_0 \neq 0$, $|c| < r$). We shall prove later in this book that no geodesic of a surface of revolution can be asymptotic to a parallel which is not itself a geodesic (Sec. 4-7). Since no parallel of the paraboloid is a geodesic, the geodesic γ is actually tangent to the parallel of radius $|c|$ at the point p_1. Because 1 is a maximum for $\cos\theta$, the value of r will increase starting from p_1. We are, therefore, in the same situation as before. The geodesic will go around the paraboloid an infinite number of turns, in the direction of the increasing r's, and it will clearly intersect the other branch infinitely often (Fig. 4-20).

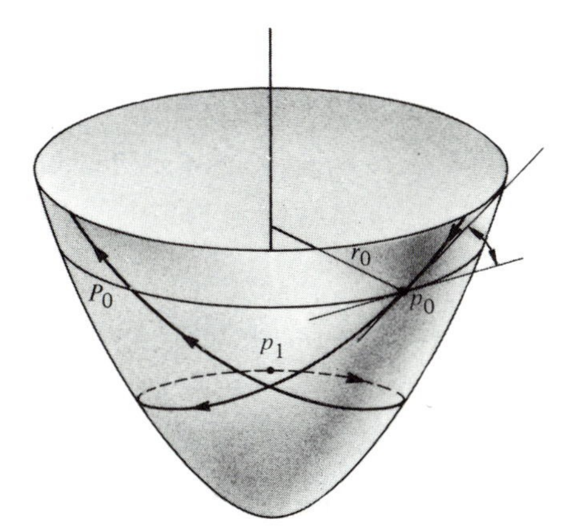

Figure 4-20

Observe that if $\theta_0 = 0$, the initial situation is that of the point p_1.

It remains to show that when r increases, the geodesic γ meets all the meridians of the paraboloid. Observe initially that the geodesic cannot be tangent to a meridian. Otherwise, it would coincide with the meridian by the uniqueness part of Prop. 5. Since the angle θ increases with r, if γ did not cut all the meridians, it would approach asymptotically a meridian, say M.

Let us assume that this is the case and let us choose a system of local coordinates for the paraboloid $z = x^2 + y^2$, given by

$$x = v\cos u, \qquad y = v\sin u, \qquad z = v^2,$$
$$0 < v < +\infty, \qquad 0 < u < 2\pi,$$

in such a way that the corresponding coordinate neighborhood contains M as $u = u_0$. By hypothesis $u \longrightarrow u_0$ when $v \longrightarrow \infty$. On the other hand, the equation of the geodesic γ in this coordinate system is given by (cf. Eq. (6), Example 5 and choose an orientation on γ such that $c > 0$)

$$u = c \int \frac{1}{v} \sqrt{\frac{1 + 4v^2}{v^2 - c^2}}\, dv + \text{const.} > c \int \frac{dv}{v} + \text{const.},$$

since

$$\frac{1 + 4v^2}{v^2 - c^2} > 1.$$

It follows from the above inequality that as $v \longrightarrow \infty$, u increases beyond any value, which contradicts the fact that γ approaches M asymptotically. Therefore, γ intersects all the meridians, and this completes the proof of the assertion made at the beginning of this example.

EXERCISES

1. a. Show that if a curve $C \subset S$ is both a line of curvature and a geodesic, then C is a plane curve.

b. Show that if a (nonrectilinear) geodesic is a plane curve, then it is a line of curvature.

c. Give an example of a line of curvature which is a plane curve and not a geodesic.

2. Prove that a curve $C \subset S$ is both an asymptotic curve and a geodesic if and only if C is a (segment of a) straight line.

3. Show, without using Prop. 5, that the straight lines are the only geodesics of a plane.

4. Let v and w be vector fields along a curve $\alpha\colon I \longrightarrow S$. Prove that

$$\frac{d}{dt}\langle v(t), w(t)\rangle = \left\langle \frac{Dv}{dt}, w(t)\right\rangle + \left\langle v(t), \frac{Dw}{dt}\right\rangle.$$

5. Consider the torus of revolution generated by rotating the circle

$$(x - a)^2 + z^2 = r^2, y = 0,$$

about the z axis ($a > r > 0$). The parallels generated by the points $(a + r, 0)$, $(a - r, 0)$, (a, r) are called the *maximum parallel*, the *minimum parallel*, and the *upper parallel*, respectively. Check which of these parallels is

a. A geodesic.

b. An asymptotic curve.

c. A line of curvature.

***6.** Compute the geodesic curvature of the upper parallel of the torus of Exercise 5.

7. Intersect the cylinder $x^2 + y^2 = 1$ with a plane passing through the x axis and making an angle θ with the xy plane.

a. Show that the intersecting curve is an ellipse C.

b. Compute the absolute value of the geodesic curvature of C in the cylinder at the points where C meets their axes.

***8.** Show that if all the geodesics of a connected surface are plane curves, then the surface is contained in a plane or a sphere.

***9.** Consider two meridians of a sphere C_1 and C_2 which make an angle φ at the point p_1. Take the parallel transport of the tangent vector w_0 of C_1, along C_1 and C_2, from the initial point p_1 to the point p_2 where the two meridians meet again, obtaining, respectively, w_1 and w_2. Compute the angle from w_1 to w_2.

***10.** Show that the geodesic curvature of an oriented curve $C \subset S$ at a point $p \in C$ is equal to the curvature of the plane curve obtained by projecting C onto the tangent plane $T_p(S)$ along the normal to the surface at p.

11. State precisely and prove: The algebraic value of the covariant derivative is invariant under orientation-preserving isometries.

***12.** We say that a set of regular curves on a surface S is a *differentiable family of curves* on S if the tangent lines to the curves of the set make up a differentiable field of directions (see Sec. 3-4). Assume that a surface S admits two differentiable orthogonal families of geodesics. Prove that the Gaussian curvature of S is zero.

***13.** Let V be a connected neighborhood of a point p of a surface S, and assume that the parallel transport between any two points of V does not depend on the curve joining these two points. Prove that the Gaussian curvature of V is zero.

14. Let S be an oriented regular surface and let $\alpha: I \longrightarrow S$ be a curve parametrized by arc length. At the point $p = \alpha(s)$ consider the three unit vectors (the *Darboux trihedron*) $T(s) = \alpha'(s)$, $N(s) =$ the normal vector to S at p, $V(s) = N(s) \wedge T(s)$. Show that

$$\frac{dT}{ds} = 0 + aV + bN,$$

$$\frac{dV}{ds} = -aT + 0 + cN,$$

$$\frac{dN}{ds} = -bT - cV + 0,$$

where $a = a(s)$, $b = b(s)$, $c = c(s)$, $s \in I$. The above formulas are the analogues of Frenet's formulas for the trihedron T, V, N. To establish the geometrical meaning of the coefficients, prove that

a. $c = -\langle dN/ds, V\rangle$; conclude from this that $\alpha(I) \subset S$ is a line of curvature if and only if $c \equiv 0$ (c is called the *geodesic torsion* of α; cf. Exercise 19, Sec. 3-2).

b. b is the normal curvature of $\alpha(I) \subset S$ at p.

c. a is the geodesic curvature of $\alpha(I) \subset S$ at p.

15. Let p_0 be a pole of a unit sphere S^2 and q, r be two points on the corresponding equator in such a way that the meridians p_0q and p_0r make an angle θ at p_0. Consider a unit vector v tangent to the meridian p_0q at p_0, and take the parallel transport of v along the closed curve made up by the meridian p_0q, the parallel qr, and the meridian rp_0 (Fig. 4-21).

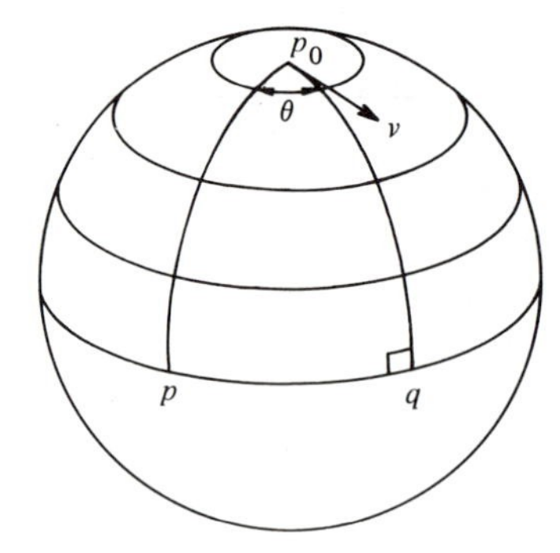

Figure 4-21

a. Determine the angle of the final position of v with v.

b. Do the same thing when the points p, q instead of being on the equator are taken on a parallel of colatitude φ (cf. Example 1).

***16.** Let p be a point of an oriented surface S and assume that there is a neighborhood of p in S all points of which are parabolic. Prove that the (unique) asymptotic curve through p is an open segment of a straight line. Give an example to show that the condition of having a neighborhood of parabolic points is essential.

17. Let $\alpha: I \longrightarrow S$ be a curve parametrized by arc length s, with nonzero curvature. Consider the parametrized surface (Sec. 2-3)

$$\mathbf{x}(s, v) = \alpha(s) + vb(s), \qquad s \in I,\ -\epsilon < v < \epsilon,\ \epsilon > 0,$$

where b is the binormal vector of α. Prove that if ϵ is small, $\mathbf{x}(I \times (-\epsilon, \epsilon)) = S$ is a regular surface over which $\alpha(I)$ is a geodesic (*thus, every curve is a geodesic on the surface generated by its binormals*).

***18.** Consider a geodesic which starts at a point p in the upper part ($z > 0$) of a hyperboloid of revolution $x^2 + y^2 - z^2 = 1$ and makes an angle θ with the parallel passing through p in such a way that $\cos\theta = 1/r$, where r is the distance from p to the z axis. Show that by following the geodesic in the direction of decreasing parallels, it approaches asymptotically the parallel $x^2 + y^2 = 1$, $z = 0$ (Fig. 4-22).

***19.** Show that when the differential equations (4) of the geodesics are referred to the arc length then the second equation of (4) is, except for the coordinate curves, a consequence of the first equation of (4).

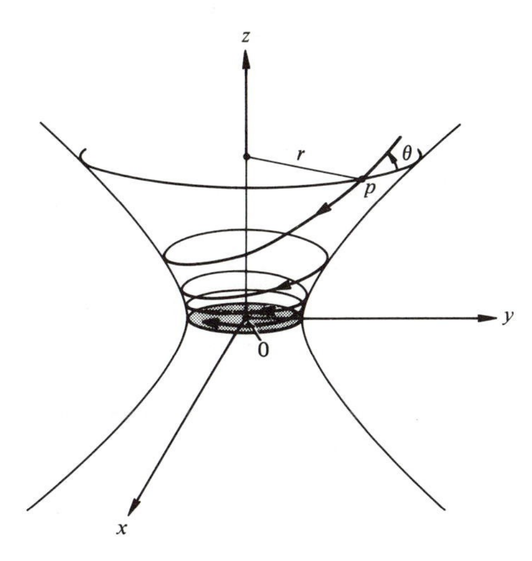

Figure 4-22

***20.** Let T be a torus of revolution which we shall assume to be parametrized by (cf. Example 6, Sec. 2-2)

$$\mathbf{x}(u, v) = ((r \cos u + a) \cos v, (r \cos u + a) \sin v, r \sin u).$$

Prove that

a. If a geodesic is tangent to the parallel $u = \pi/2$, then it is entirely contained in the region of T given by

$$-\frac{\pi}{2} \leq u \leq \frac{\pi}{2}.$$

b. A geodesic that intersects the parallel $u = 0$ under an angle θ $(0 < \theta < \pi/2)$ also intersects the parallel $u = \pi$ if

$$\cos \theta < \frac{a - r}{a + r}.$$

21. *Surfaces of Liouville* are those surfaces for which it is possible to obtain a system of local coordinates $\mathbf{x}(u, v)$ such that the coefficients of the first fundamental form are written in the form

$$E = G = U + V, \qquad F = 0,$$

where $U = U(u)$ is a function of u alone and $V = V(v)$ is a function of v alone. Observe that the surfaces of Liouville generalize the surfaces of revolution and prove that (cf. Example 5)

a. The geodesics of a surface of Liouville may be obtained by primitivation in the form

$$\int \frac{du}{\sqrt{U - c}} = \pm \int \frac{dv}{\sqrt{V + c}} + c_1,$$

where c and c_1 are constants that depend on the initial conditions.

b. If θ, $0 \leq \theta \leq \pi/2$, is the angle which a geodesic makes with the curve $v = \text{const.}$, then

$$U \sin^2 \theta - V \cos^2 \theta = \text{const.}$$

(Notice that this is the analogue of Clairaut's relation for the surfaces of Liouville.)

22. Let $S^2 = \{(x, y, z) \in R^3; x^2 + y^2 + z^2 = 1\}$ and let $p \in S^2$. For each piecewise regular parametrized curve $\alpha\colon [0, l] \to S^2$ with $\alpha(0) = \alpha(l) = p$, let $P_\alpha\colon T_p(S^2) \to T_p(S^2)$ be the map which assigns to each $v \in T_p(S^2)$ its parallel transport along α back to p. By Prop. 1, P_α is an isometry. Prove that for every rotations R of $T_p(S)$ there exists an α such that $R = P_\alpha$.

23. Show that the isometries of the unit sphere

$$S^2 = \{(x, y, z) \in R^3; x^2 + y^2 + z^2 = 1\}$$

are the restrictions to S^2 of the linear orthogonal transformations of R^3.

4-5. The Gauss-Bonnet Theorem and Its Applications

In this section, we shall present the Gauss-Bonnet theorem and some of its consequences. The geometry involved in this theorem is fairly simple, and the difficulty of its proof lies in certain topological facts. These facts will be presented without proofs.

The Gauss-Bonnet theorem is probably the deepest theorem in the differential geometry of surfaces. A first version of this theorem was presented by Gauss in a famous paper [1] and deals with geodesic triangles on surfaces (that is, triangles whose sides are arcs of geodesics). Roughly speaking, it asserts that the excess over π of the sum of the interior angles $\varphi_1, \varphi_2, \varphi_3$ of a geodesic triangle T is equal to the integral of the Gaussian curvature K over T; that is (Fig. 4-23),

$$\sum_{i=1}^{3} \varphi_i - \pi = \iint_T K\, d\sigma.$$

For instance, if $K \equiv 0$, we obtain that $\sum \varphi_i = \pi$, an extension of Thales' theorem of high school geometry to surfaces of zero curvature. Also, if $K \equiv 1$, we obtain that $\sum \varphi_i - \pi = \text{area}\,(T) > 0$. Thus, on a unit sphere, the sum of the interior angles of any geodesic triangle is greater than π, and the

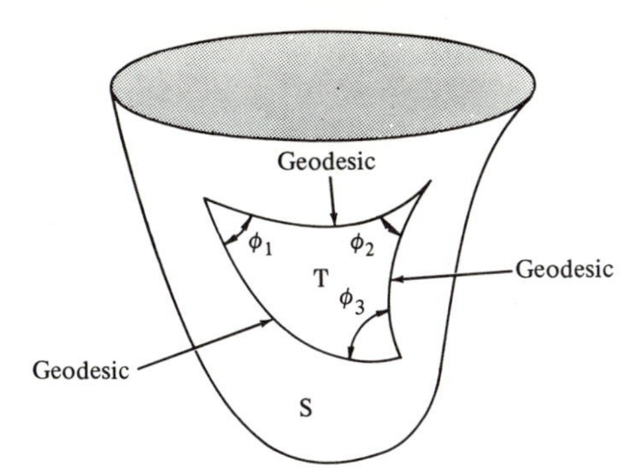

Figure 4-23. A geodesic triangle.

excess over π is exactly the area of T. Similarly, on the pseudosphere (Exercise 6, Sec. 3-3) the sum of the interior angles of any geodesic triangle is smaller than π (Fig. 4-24).

The extension of the theorem to a region bounded by a nongeodesic simple curve (see Eq. (1) below) is due to O. Bonnet. To extend it even further, say, to compact surfaces, some topological considerations will come into play. Actually, one of the most important features of the Gauss-Bonnet theorem is that it provides a remarkable relation between the topology of a compact surface and the integral of its curvature (see Corollary 2 below).

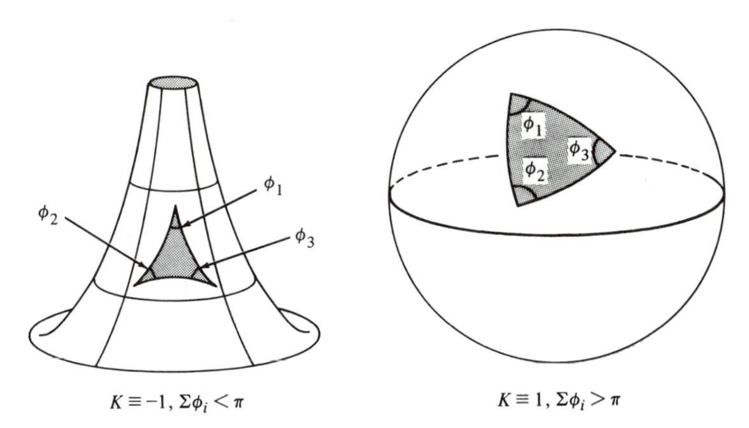

Figure 4-24

We shall now begin the details of a local version of the Gauss-Bonnet theorem. We need a few definitions.

Let $\alpha: [0, l] \longrightarrow S$ be a continuous map from the closed interval $[0, l]$ into the regular surface S. We say that α is a *simple, closed, piecewise regular, parametrized curve if*

1. $\alpha(0) = \alpha(l)$.

2. $t_1 \neq t_2$, $t_1, t_2 \in [0, l)$, implies that $\alpha(t_1) \neq \alpha(t_2)$.
3. There exists a subdivision

$$0 = t_0 < t_1 < \cdots < t_k < t_{k+1} = l,$$

of $[0, l]$ such that α is differentiable and regular in each $[t_i, t_{i+1}]$, $i = 0, \ldots, k$.

Intuitively, this means that α is a closed curve (condition 1) without self-intersections (condition 2), which fails to have a well-defined tangent line only at a finite number of points (condition 3).

The points $\alpha(t_i)$, $i = 0, \ldots, k$, are called the *vertices* of α and the traces $\alpha([t_i, t_{i+1}])$ are called the *regular arcs* of α. It is usual to call the trace $\alpha([0, l])$ of α a *closed piecewise regular curve.*

By the condition of regularity, for each vertex $\alpha(t_i)$ there exist the limit from the left, i.e., for $t < t_i$

$$\lim_{t \to t_i} \alpha'(t) = \alpha'(t_i - 0) \neq 0,$$

and the limit from the right, i.e., for $t > t_i$,

$$\lim_{t \to t_i} \alpha'(t) = \alpha'(t_i + 0) \neq 0.$$

Assume now that S is oriented and let $|\theta_i|$, $0 < |\theta_i| \leq \pi$, be the smallest determination of the angle from $\alpha'(t_i - 0)$ to $\alpha'(t_i + 0)$. If $|\theta_i| \neq \pi$, we give θ_i the sign of the determinant $(\alpha'(t_i - 0), \alpha'(t_i + 0), N)$. This means that if the vertex $\alpha(t_i)$ is not a "cusp" (Fig. 4-25), the sign of θ_i is given by the orientation of S. The signed angle θ_i, $-\pi < \theta_i < \pi$, is called the *external angle* at the vertex $\alpha(t_i)$.

In the case that the vertex $\alpha(t_i)$ is a cusp, that is, $|\theta_i| = \pi$, we choose the sign of θ_i as follows. By the condition of regularity, we can see that there exists a number $\epsilon' > 0$ such that the determinant $(\alpha'(t_i - \epsilon), \alpha'(t_i + \epsilon), N)$

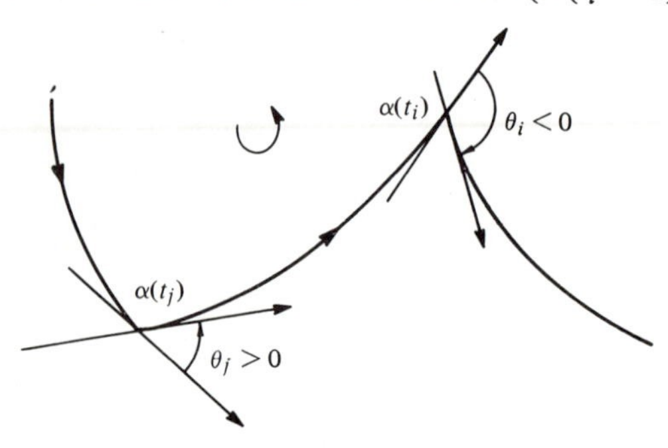

Figure 4-25

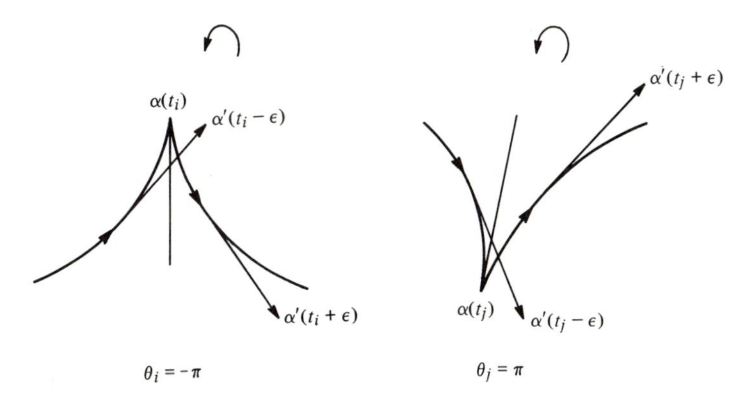

Figure 4-26. The sign of the external angle in the case of a cusp.

does not change sign for all $0 < \epsilon < \epsilon'$. We give θ_i the sign of this determinant (Fig. 4-26).

Let $\mathbf{x}: U \subset R^2 \to S$ be a parametrization compatible with the orientation of S. Assume further that U is homeomorphic to an open disk in the plane.

Let α: $[0, l] \to \mathbf{x}(U) \subset S$ be a simple closed, piecewise regular, parametrized curve, with vertices $\alpha(t_i)$ and external angles θ_i, $i = 0, \ldots, k$.

Let φ_i: $[t_i, t_{i+1}] \to R$ be differentiable functions which measure at each $t \in [t_i, t_{i+1}]$ the positive angle from $\mathbf{x}_u$ to $\alpha'(t)$ (cf. Lemma 1, Sec. 4-4).

The first topological fact that we shall present without proof is the following.

THEOREM (of Turning Tangents). *With the above notation*

$$\sum_{i=0}^{k} (\varphi_i(t_{i+1}) - \varphi_i(t_i)) + \sum_{i=0}^{k} \theta_i = \pm 2\pi,$$

where the sign plus or minus depends on the orientation of α.

The theorem states that the total variation of the angle of the tangent vector to α with a given direction plus the "jumps" at the vertices is equal to 2π.

An elegant proof of this theorem has been given by H. Hopf, *Compositio Math.* 2 (1935), 50–62. For the case where α has no vertices, Hopf's proof can be found in Sec. 5-7 (Theorem 2) of this book.

Before stating the local version of the Gauss-Bonnet theorem we still need some terminology.

Let S be an oriented surface. A region $R \subset S$ (union of a connected open set with its boundary) is called a *simple region* if R is homeomorphic to a disk and the boundary ∂R of R is the trace of a simple, closed, piecewise regular, parametrized curve $\alpha: I \to S$. We say then that α is *positively oriented* if for each $\alpha(t)$, belonging to a regular arc, the positive orthogonal basis $\{\alpha'(t), h(t)\}$ satisfies the condition that $h(t)$ "points toward" R; more precisely, for any curve $\beta: I \to R$ with $\beta(0) = \alpha(t)$ and $\beta'(0) \neq \alpha'(t)$, we have that $\langle \beta'(0), h(t) \rangle > 0$. Intuitively, this means that if one is walking on the curve α in the positive direction and with one's head pointing to N, then the region R remains to the left (Fig. 4-27). It can be shown that one of the two possible orientations of α makes it positively oriented.

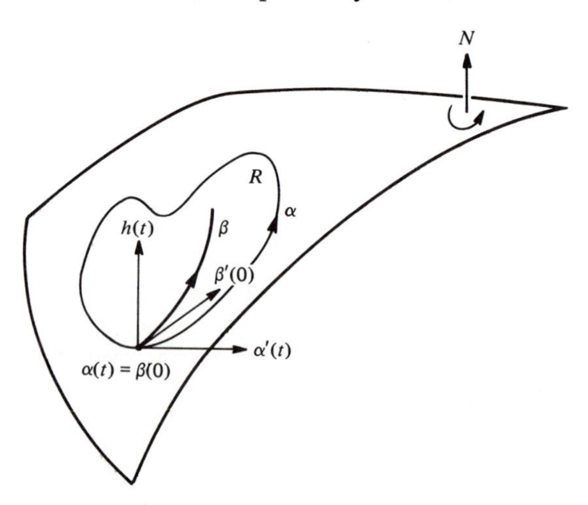

Figure 4-27. A positively oriented boundary curve.

Now let $\mathbf{x}: U \subset R^2 \to S$ be a parametrization of S compatible with its orientation and let $R \subset \mathbf{x}(U)$ be a bounded region of S. If f is a differentiable function on S, then it is easily seen that the integral

$$\iint_{\mathbf{x}^{-1}(R)} f(u, v)\sqrt{EG - F^2}\, du\, dv$$

does not depend on the parametrization $\mathbf{x}$, chosen in the class of orientation of $\mathbf{x}$. (The proof is the same as in the definition of area; cf. Sec. 2-5.) This integral has, therefore, a geometrical meaning and is called *the integral of f over the region R*. It is usual to denote it by

$$\iint_R f\, d\sigma.$$

With these definitions, we now state the

GAUSS-BONNET THEOREM (Local). *Let $\mathbf{x}: U \to S$ be an orthogonal parametrization (that is, $F = 0$), of an oriented surface S, where $U \subset R^2$ is homeomorphic to an open disk and $\mathbf{x}$ is compatible with the orientation of S.*

Let $R \subset \mathbf{x}(U)$ *be a simple region of* S *and let* $\alpha\colon I \longrightarrow S$ *be such that* $\partial R = \alpha(I)$. *Assume that* α *is positively oriented, parametrized by arc length* s, *and let* $\alpha(s_0), \ldots, \alpha(s_k)$ *and* $\theta_0, \ldots, \theta_k$ *be, respectively, the vertices and the external angles of* α. *Then*

$$\sum_{i=0}^{k} \int_{s_i}^{s_{i+1}} k_g(s)\, ds + \iint_R K\, d\sigma + \sum_{i=0}^{k} \theta_i = 2\pi, \qquad (1)$$

where $k_g(s)$ *is the geodesic curvature of the regular arcs of* α *and* K *is the Gaussian curvature of* S.

Remark. The restriction that the region R be contained in the image set of an orthogonal parametrization is needed only to simplify the proof. As we shall see later (Corollary 1 of the global Gauss-Bonnet theorem) the above result still holds for any simple region of a regular surface. This is quite plausible, since Eq. (1) does not involve in any way a particular parametrization.†

Proof. Let $u = u(s)$, $v = v(s)$ be the expression of α in the parametrization $\mathbf{x}$. By using Prop. 3 of Sec. 4-4, we have

$$k_g(s) = \frac{1}{2\sqrt{EG}}\left\{G_u \frac{dv}{ds} - E_v \frac{du}{ds}\right\} + \frac{d\varphi_i}{ds},$$

where $\varphi_i = \varphi_i(s)$ is a differentiable function which measures the positive angle from $\mathbf{x}_u$ to $\alpha'(s)$ in $[s_i, s_{i+1}]$. By integrating the above expression in every interval $[s_i, s_{i+1}]$ and adding up the results,

$$\sum_{i=0}^{k} \int_{s_i}^{s_{i+1}} k_g(s)\, ds = \sum_{i=0}^{k} \int_{s_i}^{s_{i+1}} \left(\frac{G_u}{2\sqrt{EG}}\frac{dv}{ds} - \frac{E_v}{2\sqrt{EG}}\frac{du}{ds}\right) ds$$
$$+ \sum_{i=0}^{k} \int_{s_i}^{s_{i+1}} \frac{d\varphi_i}{ds}\, ds.$$

Now we use the Gauss-Green theorem in the uv plane which states the following: *If* $P(u, v)$ *and* $Q(u, v)$ *are differentiable functions in a simple region* $A \subset R^2$, *the boundary of which is given by* $u = u(s)$, $v = v(s)$, *then*

$$\sum_{i=0}^{k} \int_{s_i}^{s_{i+1}} \left(P\frac{du}{ds} + Q\frac{dv}{ds}\right) ds = \iint_A \left(\frac{\partial Q}{\partial u} - \frac{\partial P}{\partial v}\right) du\, dv.$$

---

†If the truth of this assertion is assumed, applications 2 and 6 given below can be presented now.

It follows that

$$\sum_{i=0}^{k}\int_{s_i}^{s_{i+1}} k_g(s)\,ds = \iint_{\mathbf{x}^{-1}(R)} \left\{\left(\frac{E_v}{2\sqrt{EG}}\right)_v + \left(\frac{G_u}{2\sqrt{EG}}\right)_u\right\} du\,dv + \sum_{i=0}^{k}\int_{s_i}^{s_{i+1}} \frac{d\varphi_i}{ds}\,ds.$$

From the Gauss formula for $F = 0$ (cf. Exercise 1, Sec. 4-3), we know that

$$\iint_{\mathbf{x}^{-1}(R)} \left\{\left(\frac{E_v}{2\sqrt{EG}}\right)_v + \left(\frac{G_u}{2\sqrt{EG}}\right)\right\} du\,dv = -\iint_{\mathbf{x}^{-1}(R)} K\sqrt{EG}\,du\,dv = -\iint_R K\,d\sigma.$$

On the other hand, by the theorem of turning tangents,

$$\sum_{i=0}^{k}\int_{s_i}^{s_{i+1}} \frac{d\varphi_i}{ds}\,ds = \sum_{i=0}^{k}(\varphi_i(s_{i+1}) - \varphi_i(s_i)) = \pm 2\pi - \sum_{i=0}^{k}\theta_i.$$

Since the curve α is positively oriented, the sign should be plus, as can easily be seen in the particular case of the circle in a plane.

By putting these facts together, we obtain

$$\sum_{i=0}^{k}\int_{s_i}^{s_{i+1}} k_g(s)\,ds + \iint_R K\,d\sigma + \sum_{i=0}^{k}\theta_i = 2\pi. \qquad \text{Q.E.D.}$$

Before going into a global version of the Gauss-Bonnet theorem, we would like to show how the techniques used in the proof of this theorem may also be used to give an interpretation of the Gaussian curvature in terms of parallelism.

To do that, let $\mathbf{x}: U \to S$ be an orthogonal parametrization at a point $p \in S$, and let $R \subset \mathbf{x}(U)$ be a simple region without vertices, containing p in its interior. Let $\alpha: [0, l] \to \mathbf{x}(U)$ be a curve parametrized by arc length s such that the trace of α is the boundary of R. Let w_0 be a unit vector tangent to S at $\alpha(0)$ and let $w(s)$, $s \in [0, l]$, be the parallel transport of w_0 along α (Fig. 4-28). By using Prop. 3 of Sec. 4-4 and the Gauss-Green theorem in the uv plane, we obtain

$$0 = \int_0^l \left[\frac{Dw}{ds}\right] ds = \int_0^l \frac{1}{2\sqrt{EG}}\left\{G_u\frac{dv}{ds} - E_v\frac{du}{ds}\right\} ds + \int_0^l \frac{d\varphi}{ds}$$

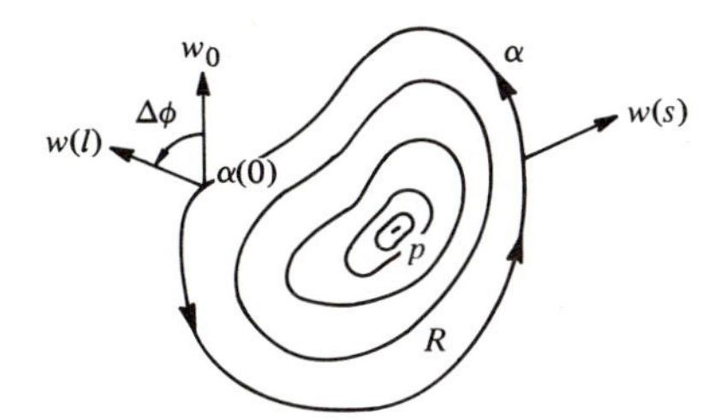

Figure 4-28

$$= -\iint_R K\,d\sigma + \varphi(l) - \varphi(0),$$

where $\varphi = \varphi(s)$ is a differentiable determination of the angle from $\mathbf{x}_u$ to $w(s)$. It follows that $\varphi(l) - \varphi(0) = \Delta\varphi$ is given by

$$\Delta\varphi = \iint_R K\,d\sigma. \tag{2}$$

Now, $\Delta\varphi$ does not depend on the choice of w_0, and it follows from the expression above that $\Delta\varphi$ does not depend on the choice of $\alpha(0)$ either. By taking the limit (in the sense of Prop. 2, Sec. 3-3)

$$\lim_{R\to p} \frac{\Delta\varphi}{A(R)} = K(p),$$

where $A(R)$ denotes the area of the region R, we obtain the desired interpretation of K.

To globalize the Gauss-Bonnet theorem, we need further topological preliminaries.

Let S be a regular surface. A region $R \subset S$ is said to be *regular* if R is compact and its boundary ∂R is the finite union of (simple) closed piecewise regular curves which do not intersect (the region in Fig. 4-29(a) is regular, but that in Fig. 4-29(b) is not). For convenience, we shall consider a compact surface as a regular region, the boundary of which is empty.

A simple region which has only three vertices with external angles $\alpha_i \neq 0$, $i = 1, 2, 3$, is called a *triangle*.

A *triangulation* of a regular region $R \subset S$ is a finite family $\mathfrak{J}$ of triangles T_i, $i = 1, \dots, n$, such that

1. $\bigcup_{i=1}^n T_i = R$.
2. If $T_i \cap T_j \neq \phi$, then $T_i \cap T_j$ is either a common edge of T_i and T_j or a common vertex of T_i and T_j.

Given a triangulation $\mathfrak{J}$ of a regular region $R \subset S$ of a surface S, we shall denote by F the number of triangles (faces), by E the number of sides (edges),

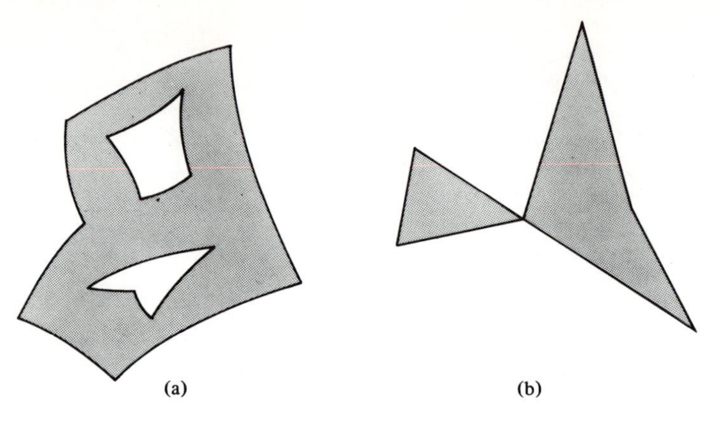

Figure 4-29

and by V the number of vertices of the triangulation. The number

$$F - E + V = \chi$$

is called the *Euler-Poincaré characteristic* of the triangulation.

The following propositions are presented without proofs. An exposition of these facts may be found, for instance, in L. Ahlfors and L. Sario, *Riemann Surfaces*, Princeton University Press, Princeton, N.J., 1960, Chap. 1.

PROPOSITION 1. *Every regular region of a regular surface admits a triangulation.*

PROPOSITION 2. *Let* S *be an oriented surface and* $\{\mathbf{x}_\alpha\}$, $\alpha \in$ A, *a family of parametrizations compatible with the orientation of* S. *Let* R $\subset$ S *be a regular region of* S. *Then there is a triangulation* $\mathfrak{J}$ *of* R *such that every triangle* T $\in \mathfrak{J}$ *is contained in some coordinate neighborhood of the family* $\{\mathbf{x}_\alpha\}$. *Furthermore, if the boundary of every triangle of* $\mathfrak{J}$ *is positively oriented, adjacent triangles determine opposite orientations in the common edge (Fig. 4-30).*

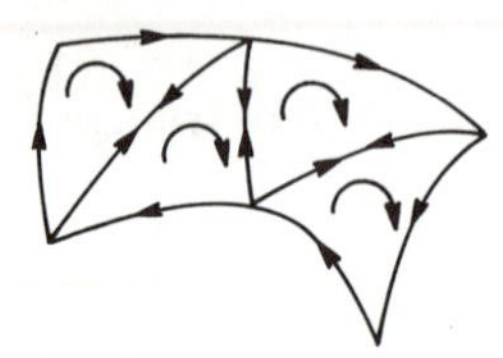

Figure 4-30

PROPOSITION 3. *If* R $\subset$ S *is a regular region of a surface* S, *the Euler-Poincaré characteristic does not depend on the triangulation of* R. *It is convenient, therefore, to denote it by* χ(R).

The latter proposition shows that the Euler-Poincaré characteristic is a topological invariant of the regular region R. For the sake of the applications of the Gauss-Bonnet theorem, we shall mention the important fact that this invariant allows a topological classification of the compact surfaces in R^3.

It should be observed that a direct computation shows that the Euler-Poincaré characteristic of the sphere is 2, that of the torus (sphere with one "handle"; see Fig. 4-31) is zero, that of the double torus (sphere with two

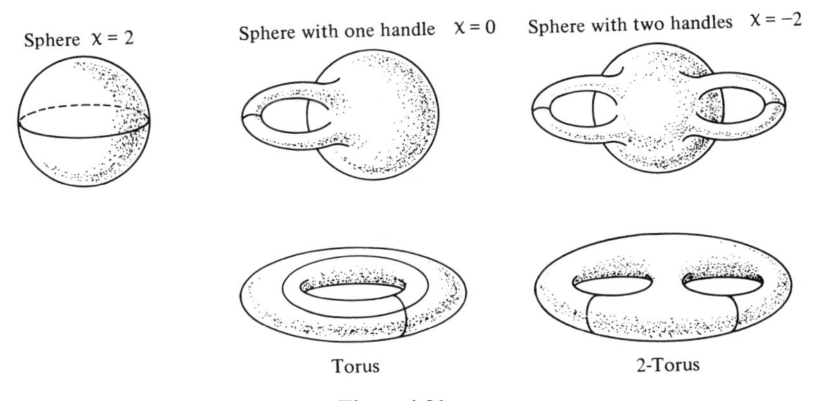

Figure 4-31

handles) is -2, and, in general, that of the n-torus (sphere with n handles) is $-2(n-1)$.

The following proposition shows that this list exhausts all compact surfaces in R^3.

PROPOSITION 4. *Let* $S \subset R^3$ *be a compact connected surface; then one of the values* $2, 0, -2, \ldots, -2n, \ldots$ *is assumed by the Euler-Poincaré characteristic* $\chi(S)$. *Furthermore, if* $S' \subset R^3$ *is another compact surface and* $\chi(S) = \chi(S')$, *then* S *is homeomorphic to* S'.

In other words, every compact connected surface $S \subset R^3$ is homeomorphic to a sphere with a certain number g of handles. The number

$$g = \frac{2 - \chi(S)}{2}$$

is called the *genus* of S.

Finally, let $R \subset S$ be a regular region of an oriented surface S and let $\mathfrak{J}$ be a triangulation of R such that every triangle $T_j \in \mathfrak{J}$, $j = 1, \ldots, k$, is contained in a coordinate neighborhood $\mathbf{x}_j(U_j)$ of a family of parametrizations $\{\mathbf{x}_\alpha\}$, $\alpha \in A$, compatible with the orientation of S. Let f be a differentiable function on S. The following proposition shows that it makes sense to talk about the integral of f over the region R.

PROPOSITION 5. *With the above notation, the sum*

$$\sum_{j=1}^{k} \iint_{\mathbf{x}_j^{-1}(T_j)} f(u_j, v_j)\sqrt{E_jG_j - F_j^2}\, du_j\, dv_j$$

does not depend on the triangulation $\mathfrak{J}$ *or on the family* $\{\mathbf{x}_j\}$ *of parametrizations of* S.

This sum has, therefore, a geometrical meaning and is called *the integral of* f *over the regular region* R. It is usually denoted by

$$\iint_R f\, d\sigma.$$

We are now in a position to state and prove the

GLOBAL GAUSS-BONNET THEOREM. *Let* $R \subset S$ *be a regular region of an oriented surface and let* $C_1, \ldots, C_n$ *be the closed, simple, piecewise regular curves which form the boundary* ∂R *of* R. *Suppose that each* C_i *is positively oriented and let* $\theta_1, \ldots, \theta_p$ *be the set of all external angles of the curves* $C_1, \ldots, C_n$. *Then*

$$\sum_{i=1}^{n} \int_{C_i} k_g(s)\, ds + \iint_R K\, d\sigma + \sum_{l=1}^{p} \theta_l = 2\pi\chi(R),$$

where s *denotes the arc length of* C_i, *and the integral over* C_i *means the sum of integrals in every regular arc of* C_i.

Proof. Consider a triangulation $\mathfrak{J}$ of the region R such that every triangle T_j is contained in a coordinate neighborhood of a family of orthogonal parametrizations compatible with the orientation of S. Such a triangulation exists by Prop. 2. Furthermore, if the boundary of every triangle of $\mathfrak{J}$ is positively oriented, we obtain opposite orientations in the edges which are common to adjacent triangles (Fig. 4-32).

By applying to every triangle the local Gauss-Bonnet theorem and adding up the results we obtain, using Prop. 5 and the fact that each "interior" side is described twice in opposite orientations,

$$\sum_{i} \int_{C_i} k_g(s)\, ds + \iint_R K\, d\sigma + \sum_{j,k=1}^{F,3} \theta_{jk} = 2\pi F,$$

where F denotes the number of triangles of $\mathfrak{J}$, and $\theta_{j1}, \theta_{j2}, \theta_{j3}$ are the external angles of the triangle T_j.

We shall now introduce the *interior* angles of the triangle T_j, given by $\varphi_{jk} = \pi - \theta_{jk}$. Thus,

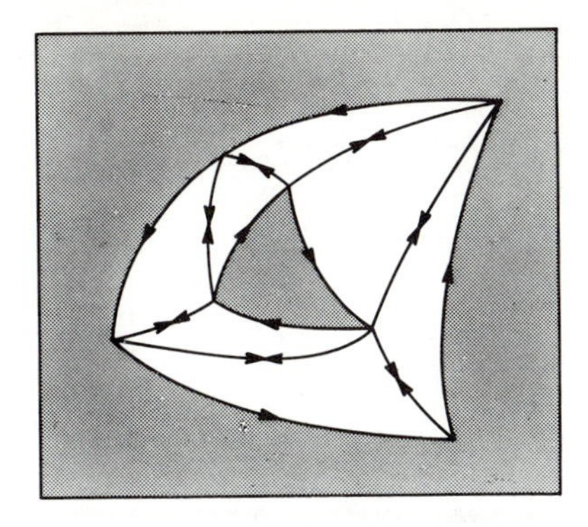

Figure 4-32

$$\sum_{j,k} \theta_{jk} = \sum_{j,k} \pi - \sum_{j,k} \varphi_{jk} = 3\pi F - \sum_{j,k} \varphi_{jk}.$$

We shall use the following notation:

$$E_e = \text{number of external edges of } \mathfrak{J},$$
$$E_i = \text{number of internal edges of } \mathfrak{J},$$
$$V_e = \text{number of external vertices of } \mathfrak{J},$$
$$V_i = \text{number of internal vertices of } \mathfrak{J}.$$

Since the curves C_i are closed, $E_e = V_e$. Furthermore, it is easy to show by induction that

$$3F = 2E_i + E_e$$

and therefore that

$$\sum_{j,k} \theta_{jk} = 2\pi E_i + \pi E_e - \sum_{j,k} \varphi_{jk}.$$

We observe now that the external vertices may be either vertices of some curve C_i or vertices introduced by the triangulation. We set $V_e = V_{ec} + V_{et}$, where V_{ec} is the number of vertices of the curves C_i and V_{et} is the number of external vertices of the triangulation which are not vertices of some curve C_i. Since the sum of angles around each internal vertex is 2π, we obtain

$$\sum_{j,k} \theta_{jk} = 2\pi E_i + \pi E_e - 2\pi V_i - \pi V_{et} - \sum_{l} (\pi - \theta_l).$$

By adding πE_e to and subtracting it from the expression above and taking into consideration that $E_e = V_e$, we conclude that

$$\begin{aligned}\sum_{j,k} \theta_{jk} &= 2\pi E_i + 2\pi E_e - 2\pi V_i - \pi V_e - \pi V_{et} - \pi V_{ec} + \sum_{l} \theta_l \\ &= 2\pi E - 2\pi V + \sum_{l} \theta_l.\end{aligned}$$

By putting things together, we finally obtain

$$\sum_{i=1}^{n} \int_{C_i} k_g(s)\, ds + \iint_R K\, d\sigma + \sum_{l=1}^{p} \theta_l = 2\pi(F - E + V)$$
$$= 2\pi\chi(R). \qquad \textbf{Q.E.D.}$$

Since the Euler-Poincaré characteristic of a simple region is clearly 1, we obtain (cf. Remark 1)

COROLLARY 1. *If* R *is a simple region of* S, *then*

$$\sum_{i=0}^{k} \int_{s_i}^{s_{i+1}} k_g(s)\, ds + \iint_R K\, d\sigma + \sum_{i=0}^{k} \theta_i = 2\pi.$$

By taking into account the fact that a compact surface may be considered as a region with empty boundary, we obtain

COROLLARY 2. *Let* S *be an orientable compact surface; then*

$$\iint_S K\, d\sigma = 2\pi\chi(S).$$

Corollary 2 is most striking. We have only to think of all possible shapes of a surface homeomorphic to a sphere to find it very surprising that in each case the curvature function distributes itself in such a way that the "total curvature," i.e., $\iint K\, d\sigma$, is the same for all cases.

We shall present some applications of the Gauss-Bonnet theorem below. For these applications (and for the exercises at the end of the section), it is convenient to assume a basic fact of the topology of the plane (the Jordan curve theorem) which we shall use in the following form: *Every piecewise regular curve in the plane* (*thus without self-intersections*) *is the boundary of a simple region.*

1. *A compact surface of positive curvature is homeomorphic to a sphere.* The Euler-Poincaré characteristic of such a surface is positive and the sphere is the only compact surface of R^3 which satisfies this condition.

2. *Let* S *be an orientable surface of negative or zero curvature. Then two geodesics* γ_1 *and* γ_2 *which start from a point* $p \in S$ *cannot meet again at a point* $q \in S$ *in such a way that the traces of* γ_1 *and* γ_2 *constitute the boundary of a simple region* R *of* S.

Assume that the contrary is true. By the Gauss-Bonnet theorem (R is simple)

$$\iint_R K\, d\sigma + \theta_1 + \theta_2 = 2\pi,$$

where θ_1 and θ_2 are the external angles of the region R. Since the geodesics γ_1 and γ_2 cannot be mutually tangent, we have $\theta_i < \pi$, $i = 1, 2$. On the other hand, $K \leq 0$, whence the contradiction.

When $\theta_1 = \theta_2 = 0$, the traces of the geodesics γ_1 and γ_2 constitute a simple closed geodesic of S (that is, a closed regular curve which is a geodesic). It follows that on a surface of zero or negative curvature, there exists no simple closed geodesic which is a boundary of a simple region of S.

3. *Let* S *be a surface homeomorphic to a cylinder with Gaussian curvature* K < 0. *Then* S *has at most one simple closed geodesic.*

Suppose that S contains one simple closed geodesic Γ. By application 2, and since there is a homeomorphism φ of S with a plane P minus one point $q \in P$, $\varphi(\Gamma)$ is the boundary of a simple region of P containing q.

Assume now that S contains another simple closed geodesic Γ'. We claim that Γ' does not intersect Γ. Otherwise, the arcs of $\varphi(\Gamma)$ and $\varphi(\Gamma')$ between two "consecutive" intersection points, r_1 and r_2, would be the boundary of a simple region, contradicting application 2 (see Fig. 4-33). By the above argu-

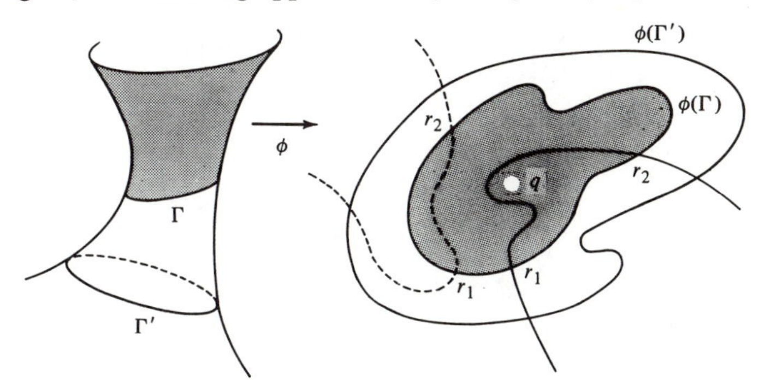

Figure 4-33

ment, $\varphi(\Gamma')$ is again the boundary of a simple region R of P containing q, the interior of which is homeomorphic to a cylinder. Thus, $\chi(R) = 0$. On the other hand, by the Gauss-Bonnet theorem,

$$\iint_{\varphi^{-1}(R)} K\, d\sigma = 2\pi\chi(R) = 0,$$

which is a contradiction, since $K < 0$.

4. *If there exist two simple closed geodesics* Γ_1 *and* Γ_2 *on a compact surface* S *of positive curvature, then* Γ_1 *and* Γ_2 *intersect.*

By application 1, S is homeomorphic to a sphere. If Γ_1 and Γ_2 do not intersect, then the set formed by Γ_1 and Γ_2 is the boundary of a region R, the Euler-Poincaré characteristic of which is $\chi(R) = 0$. By the Gauss-Bonnet theorem,

$$\iint_R K\, d\sigma = 0,$$

which is a contradiction, since $K > 0$.

5. We shall prove the following result, due to Jacobi: *Let* $\alpha: I \to R^3$ *be a closed, regular, parametrized curve with nonzero curvature. Assume that the curve described by the normal vector* $n(s)$ *in the unit sphere* S^2 *(the normal indicatrix) is simple. Then* $n(I)$ *divides* S^2 *in two regions with equal areas.*

We may assume that α is parametrized by arc length. Let $\bar{s}$ denote the arc length of the curve $n = n(s)$ on S^2. The geodesic curvature $\bar{k}_g$ of $n(s)$ is

$$\bar{k}_g = \langle \ddot{n}, n \wedge \dot{n} \rangle,$$

where the dots denote differentiation with respect to $\bar{s}$. Since

$$\dot{n} = \frac{dn}{ds}\frac{ds}{d\bar{s}} = (-kt - \tau b)\frac{ds}{d\bar{s}},$$

$$\ddot{n} = (-kt - \tau b)\frac{d^2 s}{d\bar{s}^2} + (-k't - \tau' b)\left(\frac{ds}{d\bar{s}}\right)^2 - (k^2 + \tau^2)n\left(\frac{ds}{d\bar{s}}\right)^2,$$

and

$$\left(\frac{ds}{d\bar{s}}\right)^2 = \frac{1}{k^2 + \tau^2},$$

we obtain

$$\bar{k}_g = \langle n \wedge \dot{n}, \ddot{n} \rangle = \frac{ds}{d\bar{s}}\langle (kb - \tau t), \ddot{n} \rangle = \left(\frac{ds}{d\bar{s}}\right)^3 (-k\tau' - k'\tau)$$

$$= -\frac{\tau' k - k\tau'}{k^2 + \tau^2}\frac{ds}{d\bar{s}} = -\frac{d}{ds}\tan^{-1}\left(\frac{\tau}{k}\right)\frac{ds}{d\bar{s}}.$$

Thus, by applying the Gauss-Bonnet theorem to one of the regions R bounded by $n(I)$ and using the fact that $K \equiv 1$, we obtain

$$2\pi = \int_R K\, d\sigma + \int_{\partial R} \bar{k}_g\, d\bar{s} = \int_R d\sigma = \text{area of } R.$$

Since the area of S^2 is 4π, the result follows.

6. Let T be a geodesic triangle (that is, the sides of T are geodesics) in an oriented surface S. Let $\theta_1, \theta_2, \theta_3$ be the external angles of T and let $\varphi_1 = \pi - \theta_1$, $\varphi_2 = \pi - \theta_2$, $\varphi_3 = \pi - \theta_3$ be its interior angles. By the Gauss-Bonnet theorem,

$$\iint_T K\, d\sigma + \sum_{i=1}^{3} \theta_i = 2\pi.$$

Thus,

$$\iint_T K\,d\sigma = 2\pi - \sum_{i=1}^{3} (\pi - \varphi_i) = -\pi + \sum_{i=1}^{3} \varphi_i.$$

It follows that the sum of the interior angles, $\sum_{i=1}^{3} \varphi_i$, of a geodesic triangle is

1. *Equal to* π if $K = 0$.
2. *Greater than* π if $K > 0$.
3. *Smaller than* π if $K < 0$.

Furthermore, the difference $\sum_{i=1}^{3} \varphi_i - \pi$ (the *excess* of T) is given precisely by $\iint_T K\,d\sigma$. If $K \neq 0$ on T, this is the area of the image $N(T)$ of T by the Gauss map $N: S \longrightarrow S^2$ (cf. Eq. (12), Sec. 3-3). This was the form in which Gauss himself stated his theorem: *The excess of a geodesic triangle* T *is equal to the area of its spherical image* N(T).

The above fact is related to a historical controversy about the possibility of proving Euclid's fifth axiom (the axiom of the parallels), from which it follows that the sum of the interior angles of any triangle is equal to π. By considering the geodesics as straight lines, it is possible to show that the surfaces of constant negative curvature constitute a (local) model of a geometry where Euclid's axioms hold, except for the fifth and the axiom which guarantees the possibility of extending straight lines indefinitely. Actually, Hilbert showed that there does not exist in R^3 a surface of constant negative curvature, the geodesics of which can be extended indefinitely (the pseudosphere of Exercise 6, Sec. 3-3, has an edge of singular points). Therefore, the surfaces of R^3 with constant negative Gaussian curvature do not yield a model to test the independence of the fifth axiom alone. However, by using the notion of abstract surface, it is possible to bypass this inconvenience and to build a model of geometry where *all* of Euclid's axioms but the fifth are valid. This axiom is, therefore, independent of the others.

In Secs. 5-10 and 5-11, we shall prove the result of Hilbert just quoted and shall describe the abstract model of a noneuclidean geometry.

7. *Vector fields on surfaces.*† Let v be a differentiable vector field on an oriented surface S. We say that $p \in S$ is a *singular point* of v if $v(p) = 0$. The singular point p is *isolated* if there exists a neighborhood V of p in S such that v has no singular points in V other than p.

To each isolated singular point p of a vector field v, we shall associate an

†This application requires the material of Sec. 3-4. If omitted, then Exercises 6–9 of this section should also be omitted.

integer, the index of v, defined as follows. Let $\mathbf{x}: U \longrightarrow S$ be an orthogonal parametrization at $p = \mathbf{x}(0, 0)$ compatible with the orientation of S, and let $\alpha: [0, l] \longrightarrow S$ be a simple, closed, piecewise regular parametrized curve such that $\alpha([0, l]) \subset \mathbf{x}(U)$ is the boundary of a simple region R containing p as its only singular point. Let $v = v(t)$, $t \in [0, l]$, be the restriction of v along α, and let $\varphi = \varphi(t)$ be some differentiable determination of the angle from $\mathbf{x}_u$ to $v(t)$, given by Lemma 1 of Sec. 4-4 (which can easily be extended to piecewise regular curves). Since α is closed, there is an integer I defined by

$$2\pi I = \varphi(l) - \varphi(0) = \int_0^l \frac{d\varphi}{dt}\, dt.$$

I is called the *index* of v at p.

We must show that this definition is independent of the choices made, the first one being the parametrization $\mathbf{x}$. Let $w_0 \in T_{\alpha(0)}(S)$ and let $w(t)$ be the parallel transport of w_0 along α. Let $\psi(t)$ be a differentiable determination of the angle from $\mathbf{x}_u$ to $w(t)$. Then, as we have seen in the interpretation of K in terms of parallel transport (cf. Eq. (2)),

$$\psi(l) - \psi(0) = \iint_R K\, d\sigma.$$

By subtracting the above relations, we obtain

$$\iint_R K\, d\sigma - 2\pi I = (\psi - \varphi)(l) - (\psi - \varphi)(0) = \Delta(\psi - \varphi) \tag{3}$$

Sincc $\psi - \varphi$ does not depend on $\mathbf{x}_u$, the index I is independent of the parametrization $\mathbf{x}$.

The proof that the index does not depend on the choice of α is more technical (although rather intuitive) and we shall only sketch it.

Let α_0 and α_1 be two curves as in the definition of index and let us show that the index of v is the same for both curves. We first suppose that the traces of α_0 and α_1 do not intersect. Then there is a homeomorphism of the region bounded by the traces of α_0 and α_1 onto a region of the plane bounded by two concentric circles C_0 and C_1 (a ring). Since we can obtain a family of concentric circles C_t which depend continuously on t and deform C_0 into C_1, we obtain a family of curves α_t, which depend continuously on t and deform α_0 into α_1 (Fig. 4-34). Denote by I_t the index of v computed with the curve α_t. Now, since the index is an integral, I_t depends continuously on t, $t \in [0, 1]$. Being an integer, I_t is constant under this deformation, and $I_0 = I_1$, as we wished. If the traces of α_0 and α_1 intersect, we choose a curve sufficiently small so that its trace has no intersection with both α_0 and α_1 and then apply the previous result.

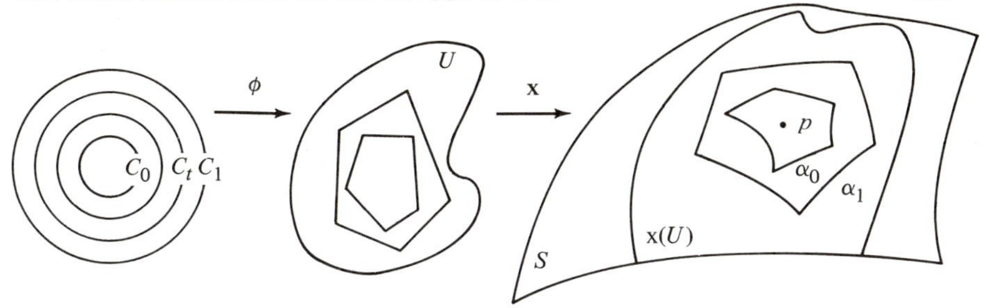

Figure 4-34

It should be noticed that the definition of index can still be applied when p is not a singular point of v. It turns out, however, that the index is then zero. This follows from the fact that, since I does not depend on $\mathbf{x}_u$, we can choose $\mathbf{x}_u$ to be v itself; thus, $\varphi(t) \equiv 0$.

In Fig. 4-35 we show some examples of indices of vector fields in the xy plane which have (0, 0) as a singular point. The curves that appear in the drawings are the trajectories of the vector fields.

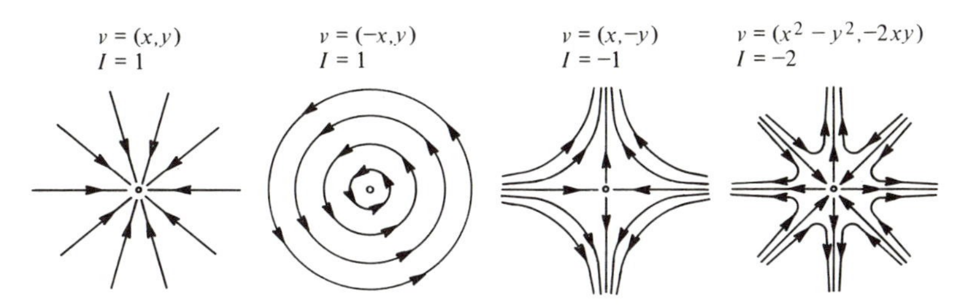

Figure 4-35

Now, let $S \subset R^3$ be an oriented, compact surface and v a differentiable vector field with only isolated singular points. We remark that they are finite in number. Otherwise, by compactness (cf. Sec. 2-7, Property 1), they have a limit point which is a nonisolated singular point. Let $\{\mathbf{x}_\alpha\}$ be a family of orthogonal parametrizations compatible with the orientation of S. Let $\mathfrak{I}$ be a triangulation of S such that

1. Every triangle $T \in \mathfrak{I}$ is contained in some coordinate neighborhood of the family $\{\mathbf{x}_\alpha\}$.
2. Every $T \in \mathfrak{I}$ contains at most one singular point.
3. The boundary of every $T \in \mathfrak{I}$ contains no singular points and is positively oriented.

If we apply Eq. (1) to every triangle $T \in \mathfrak{I}$, sum up the results, and take into account that the edge of each $T \in \mathfrak{I}$ appears twice with opposite orientations, we obtain

$$\iint_S K\,d\sigma - 2\pi \sum_{i=1}^{k} I_i = 0,$$

where I_i is the index of the singular point p_i, $i = 1, \ldots, k$. Joining this with the Gauss-Bonnet theorem (cf. Corollary 2), we finally arrive at

$$\sum I_i = \frac{1}{2\pi} \iint_S K\,d\sigma = \chi(S).$$

Thus, we have proved the following:

POINCARÉ'S THEOREM. *The sum of the indices of a differentiable vector field* v *with isolated singular points on a compact surface* S *is equal to the Euler-Poincaré characteristic of* S.

This is a remarkable result. It implies that $\sum I_i$ does not depend on v but only on the topology of S. For instance, in any surface homeomorphic to a sphere, all vector fields with isolated singularities must have the sum of their indices equal to 2. In particular, no such surface can have a differentiable vector field without singular points.

EXERCISES

1. Let $S \subset R^3$ be a regular, compact, orientable surface which is not homeomorphic to a sphere. Prove that there are points on S where the Gaussian curvature is positive, negative, and zero.

2. Let T be a torus of revolution. Describe the image of the Gauss map of T and show, without using the Gauss-Bonnet theorem, that

$$\iint_T K\,d\sigma = 0.$$

Compute the Euler-Poincaré characteristic of T and check the above result with the Gauss-Bonnet theorem.

3. Let $S \subset R^3$ be a regular surface homeomorphic to a sphere. Let $\Gamma \subset S$ be a simple closed geodesic in S, and let A and B be the regions of S which have Γ as a common boundary. Let $N\colon S \longrightarrow S^2$ be the Gauss map of S. Prove that $N(A)$ and $N(B)$ have the same area.

4. Compute the Euler-Poincaré characteristic of

a. An ellipsoid.

***b.** The surface $S = \{(x, y, z) \in R^3; x^2 + y^4 + z^6 = 1\}$.

5. Let C be a parallel of colatitude φ on an oriented unit sphere S^2, and let w_0 be a unit vector tangent to C at a point $p \in C$ (cf. Example 1, Sec. 4-4). Take the parallel transport of w_0 along C and show that its position, after a complete turn, makes an angle $\Delta\varphi = 2\pi(1 - \cos\varphi)$ with the initial position w_0. Check that

$$\lim_{R\to p} \frac{\Delta\varphi}{A} = 1 = \text{curvature of } S^2,$$

where A is the area of the region R of S^2 bounded by C.

6. Show that (0, 0) is an isolated singular point and compute the index at (0, 0) of the following vector fields in the plane:

***a.** $v = (x, y)$.

b. $v = (-x, y)$.

c. $v = (x, -y)$.

***d.** $v = (x^2 - y^2, -2xy)$.

e. $v = (x^3 - 3xy^2, y^3 - 3x^2y)$.

7. Can it happen that the index of a singular point is zero? If so, give an example.

8. Prove that an orientable compact surface $S \subset R^3$ has a differentiable vector field without singular points if and only if S is homeomorphic to a torus.

9. Let C be a regular closed curve on a sphere S^2. Let v be a differentiable vector field on S^2 such that the trajectories of v are never tangent to C. Prove that each of the two regions determined by C contains at least one singular point of v.

4-6. The Exponential Map. Geodesic Polar Coordinates

In this section we shall introduce some special coordinate systems with an eye toward their geometric applications. The natural way of introducing such coordinates is by means of the exponential map, which we shall now describe.

As we learned in Sec. 4-4, Prop. 5, given a point p of a regular surface S and a nonzero vector $v \in T_p(S)$ there exists a unique parametrized geodesic $\gamma: (-\epsilon, \epsilon) \to S$, with $\gamma(0) = p$ and $\gamma'(0) = v$. To indicate the dependence of this geodesic on the vector v, it is convenient to denote it by $\gamma(t, v) = \gamma$.

LEMMA 1. *If the geodesic* $\gamma(t, v)$ *is defined for* $t \in (-\epsilon, \epsilon)$, *then the geodesic* $\gamma(t, \lambda v)$, $\lambda \in R$, $\lambda \neq 0$, *is defined for* $t \in (-\epsilon/\lambda, \epsilon/\lambda)$, *and* $\gamma(t, \lambda v) = \gamma(\lambda t, v)$.

Proof. Let $\alpha: (-\epsilon/\lambda, \epsilon/\lambda) \to S$ be a parametrized curve defined by $\alpha(t) = \gamma(\lambda t)$. Then $\alpha(0) = \gamma(0)$, $\alpha'(0) = \lambda\gamma'(0)$, and, by the linearity of D (cf. Eq. (1), Sec. 4-2),

$$D_{\alpha'(t)}\alpha'(t) = \lambda^2 D_{\gamma'(t)}\gamma'(t) = 0.$$

If follows that α is a geodesic with initial conditions $\gamma(0)$, $\lambda\gamma'(0)$, and by uniqueness

$$\alpha(t) = \gamma(t, \lambda v) = \gamma(\lambda t, v) \qquad \textbf{Q.E.D.}$$

Intuitively, Lemma 1 means that since the speed of a geodesic is constant, we can go over its trace within a prescribed time by adjusting our speed appropriately.

We shall now introduce the following notation. If $v = T_p(S)$, $v \neq 0$, is such that $\gamma(|v|, v/|v|) = \gamma(1, v)$ is defined, we set

$$\exp_p(v) = \gamma(1, v) \quad \text{and} \quad \exp_p(0) = p.$$

Geometrically, the construction corresponds to laying off (if possible) a length equal to $|v|$ along the geodesic that passes through p in the direction of v; the point of S thus obtained is denoted by $\exp_p(v)$ (Fig. 4-36).

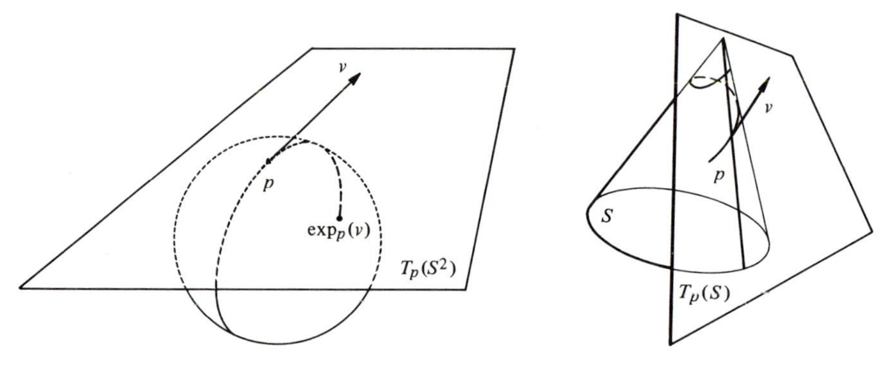

Figure 4-36 **Figure 4-37**

For example, $\exp_p(v)$ is defined on the unit sphere S^2 for every $v \in T_p(S^2)$. The points of the circles of radii $\pi, 3\pi, \ldots, (2n+1)\pi$ are mapped into the point q, the antipodal point of p. The points of the circles of radii 2π, $4\pi, \ldots, 2n\pi$ are mapped back into p.

On the other hand, on the regular surface C formed by the one-sheeted cone minus the vertex, $\exp_p(v)$ is not defined for a vector $v \in T_p(C)$ in the direction of the meridian that connects p to the vertex, when $|v| \geq d$ and d is the distance from p to the vertex (Fig. 4-37).

If, in the example of the sphere, we remove from S^2 the antipodal point of p, then $\exp_p(v)$ is defined only in the interior of a disk of $T_p(S^2)$ of radius π and center in the origin.

The important point is that $\exp_p$ is always defined and differentiable in some neighborhood of p.

PROPOSITION 1. *Given* $p \in S$ *there exists an* $\epsilon > 0$ *such that* $\exp_p$ *is defined and differentiable in the interior* B_ϵ *of a disk of radius* ϵ *of* $T_p(S)$, *with center in the origin.*

Proof. It is clear that for every direction of $T_p(S)$ it is possible, by Lemma 1, to take v sufficiently small so that the interval of definition of $\gamma(t, v)$ contains 1, and thus $\gamma(1, v) = \exp_p(v)$ is defined. To show that this reduction can be made uniformly in all directions, we need the theorem of the dependence of a geodesic on its initial conditions (see Sec. 4-7) in the following form: *Given* $p \in S$ *there exist numbers* $\epsilon_1 > 0, \epsilon_2 > 0$ *and a differentiable map*

$$\gamma\colon (-\epsilon_2, \epsilon_2) \times B_{\epsilon_1} \longrightarrow S$$

such that, for $v \in B_{\epsilon_1}$, $v \neq 0$, $t \in (-\epsilon_2, \epsilon_2)$, *the curve* $\gamma(t, v)$ *is the geodesic of* S *with* $\gamma(0, p) = p$, $\gamma'(0, v) = v$, *and for* $v = 0$, $\gamma(t, 0) = p$.

From this statement and Lemma 1, our assertion follows. In fact, since $\gamma(t, v)$ is defined for $|t| < \epsilon_2$, $|v| < \epsilon_1$, we obtain, by setting $\lambda = \epsilon_2/2$ in Lemma 1, that $\gamma(t, (\epsilon_2/2)v)$ is defined for $|t| < 2$, $|v| < \epsilon_1$. Therefore, by taking a disk $B_\epsilon \subset T_p(S)$, with center at the origin and radius $\epsilon < \epsilon_1\epsilon_2/2$, we have that $\gamma(1, w) = \exp_p w$, $w \in B_\epsilon$, is defined. The differentiability of $\exp_p$ in B_ϵ follows from the differentiability of γ. **Q.E.D.**

An important complement to this result is the following:

PROPOSITION 2. $\exp_p\colon B_\epsilon \subset T_p(S) \longrightarrow S$ *is a diffeomorphism in a neighborhood* $U \subset B_\epsilon$ *of the origin* 0 *of* $T_p(S)$.

Proof. We shall show that the differential $d(\exp_p)$ is nonsingular at $0 \in T_p(S)$. To do this, we identify the space of tangent vectors to $T_p(S)$ at 0 with $T_p(S)$ itself. Consider the curve $\alpha(t) = tv$, $v \in T_p(S)$. It is obvious that $\alpha(0) = 0$ and $\alpha'0) = v$. The curve $(\exp_p \circ \alpha)(t) = \exp_p(tv)$ has at $t = 0$ the tangent vector

$$\frac{d}{dt}(\exp_p(tv))\Big|_{t=0} = \frac{d}{dt}(\gamma(t, v))\Big|_{t=0} = v.$$

It follows that

$$(d\exp_p)_0(v) = v,$$

which shows that $d\exp_p$ is nonsingular at 0. By applying the inverse function theorem (cf. Prop. 3, Sec. 2-4), we complete the proof of the proposition. **Q.E.D.**

It is convenient to call $V \subset S$ a *normal neighborhood* of $p \in S$ if V is the image $V = \exp_p(U)$ of a neighborhood U of the origin of $T_p(S)$ restricted to which $\exp_p$ is a diffeomorphism.

Since the exponential map at $p \in S$ is a diffeomorphism on U, it may be used to introduce coordinates in V. Among the coordinate systems thus introduced, the most usual are

1. The *normal coordinates* which correspond to a system of rectangular coordinates in the tangent plane $T_p(S)$.
2. The *geodesic polar coordinates* which correspond to polar coordinates in the tangent plane $T_p(S)$ (Fig. 4-38).

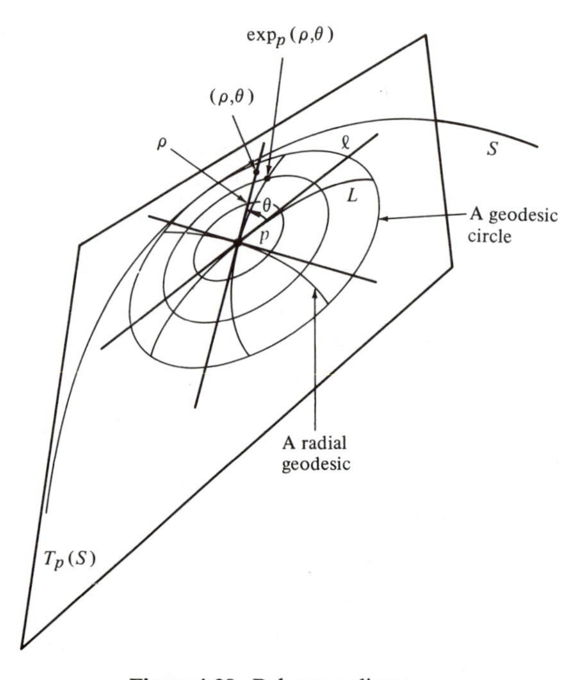

Figure 4-38 Polar coordinates.

We shall first study the normal coordinates, which are obtained by choosing in the plane $T_p(S)$, $p \in S$, two orthogonal unit vectors e_1 and e_2. Since $\exp_p: U \longrightarrow V \subset S$ is a diffeomorphism, it satisfies the conditions for a parametrization in p. If $q \in V$, then $q = \exp_p(w)$, where $w = ue_1 + ve_2 \in U$, and we say that q has coordinates (u, v). It is clear that the normal coordinates thus obtained depend on the choice of e_1, e_2.

In a system of normal coordinates centered in p, the geodesics that pass through p are the images by $\exp_p$ of the lines $u = at, v = bt$ which pass through the origin of $T_p(S)$. Observe also that at p the coefficients of the first fundamental form in such a system are given by $E(p) = G(p) = 1$, $F(p) = 0$.

Now we shall proceed to the geodesic polar coordinates. Choose in the plane $T_p(S)$, $p \in S$, a system of polar coordinates (ρ, θ), where ρ is the polar radius and θ, $0 < \theta < 2\pi$, is the polar angle, the pole of which is the origin 0 of $T_p(S)$. Observe that the polar coordinates in the plane are not defined in the closed half-line l which corresponds to $\theta = 0$. Set $\exp_p(l) = L$. Since $\exp_p\colon U - l \to V - L$ is still a diffeomorphism, we may parametrize the points of $V - L$ by the coordinates (ρ, θ), which are called geodesic polar coordinates.

We shall use the following terminology. The images by $\exp_p\colon U \to V$ of circles in U centered in 0 will be called *geodesic circles* of V, and the images of $\exp_p$ of the lines through 0 will be called *radial geodesics* of V. In $V - L$ these are the curves $\rho =$ const. and $\theta =$ const., respectively.

We shall now determine the coefficients of the first fundamental form in a system of geodesic polar coordinates.

PROPOSITION 3. *Let* $\mathbf{x}\colon \mathrm{U} - l \to \mathrm{V} - \mathrm{L}$ *be a system of geodesic polar coordinates* (ρ, θ). *Then the coefficients* $\mathrm{E} = \mathrm{E}(\rho, \theta)$, $\mathrm{F} = \mathrm{F}(\rho, \theta)$, *and* $\mathrm{G} = \mathrm{G}(\rho, \theta)$ *of the first fundamental form satisfy the conditions*

$$\mathrm{E} = 1, \qquad \mathrm{F} = 0, \qquad \lim_{\rho \to 0} \mathrm{G} = 0, \qquad \lim_{\rho \to 0} (\sqrt{\mathrm{G}})_\rho = 1.$$

Proof. By definition of the exponential map, ρ measures the arc length along the curve $\theta =$ const. It follows immediately that $E = 1$.

By introducing in the differential equation of a geodesic (Eq. (4), Sec. 4-4) the fact that $\theta =$ const. is a geodesic, we conclude that $\Gamma_{11}^2 = 0$. By using the first of the relations (2) of Sec. 4-3 that define the Christoffel symbols, we obtain

$$0 = \tfrac{1}{2}E_\rho = \Gamma_{11}^1 E = \Gamma_{11}^1.$$

By introducing this relation in the second of the equations (2) of Sec. 4-3, we conclude that $F_\rho = 0$, and, therefore, $F(\rho, \theta)$ does not depend on ρ.

For each $q \in V$, we shall denote by $\alpha(\sigma)$ the geodesic circle that passes through q, where $\sigma \in [0, 2\pi]$ (if $q = p$, $\alpha(\sigma)$ is the constant curve $\alpha(\sigma) = p$). We shall denote by $\gamma(s)$, where s is the arc length of γ, the radial geodesic that passes through q. With this notation we may write

$$F(\rho, \theta) = \left\langle \frac{d\alpha}{d\sigma}, \frac{d\gamma}{ds} \right\rangle.$$

The coefficient $F(\rho, \theta)$ is not defined at p. However, if we fix the radial geodesic $\theta =$ const., the second member of the above equation is defined for every point of this geodesic. Since at p, $\alpha(\sigma) = p$, that is, $d\alpha/d\sigma = 0$, we obtain

$$\lim_{\rho \to 0} F(\rho, \theta) = \lim_{\rho \to 0} \left\langle \frac{d\alpha}{d\sigma}, \frac{d\gamma}{ds} \right\rangle = 0.$$

Together with the fact that F does not depend on ρ, this implies that $F = 0$.

To prove the last assertion of the proposition, we choose a system of normal coordinates $(\bar{u}, \bar{v})$ in p in such a way that the change of coordinates is given by

$$\bar{u} = \rho \cos \theta, \qquad \bar{v} = \rho \sin \theta, \qquad \rho \neq 0, \quad 0 < \theta < 2\pi.$$

By recalling that

$$\sqrt{EG - F^2} = \sqrt{\bar{E}\bar{G} - \bar{F}^2}\,\frac{\partial(\bar{u}, \bar{v})}{\partial(\rho, \theta)},$$

where $\partial(\bar{u}, \bar{v})/\partial(\rho, \theta)$ is the Jacobian of the change of coordinates and $\bar{E}, \bar{F}, \bar{G}$, are the coefficients of the first fundamental form in the normal coordinates $(\bar{u}, \bar{v})$, we have

$$\sqrt{G} = \rho\sqrt{\bar{E}\bar{G} - \bar{F}^2}, \qquad \rho \neq 0. \tag{1}$$

Since at p, $\bar{E} = \bar{G} = 1$, $\bar{F} = 0$ (the normal coordinates are defined at p), we conclude that

$$\lim_{\rho \to 0} \sqrt{G} = 0, \qquad \lim_{\rho \to 0}(\sqrt{G})_\rho = 1,$$

which concludes the proof of the proposition. **Q.E.D.**

Remark 1. The geometric meaning of the fact that $F = 0$ is that in a normal neighborhood the family of geodesic circles is orthogonal to the family of radial geodesics. This fact is known as the *Gauss lemma.*

We shall now present some geometrical applications of the geodesic polar coordinates.

First, we shall study the surfaces of constant Gaussian curvature. Since in a polar system $E = 1$ and $F = 0$, the Gaussian curvature K can be written

$$K = -\frac{(\sqrt{G})_{\rho\rho}}{\sqrt{G}}.$$

This expression may be considered as the differential equation which $\sqrt{G}(\rho, \theta)$ should satisfy if we want the surface to have (in the coordinate neighborhood in question) curvature $K(\rho, \theta)$. If K is constant, the above expression, or, equivalently,

$$(\sqrt{G})_{\rho\rho} + K\sqrt{G} = 0, \tag{2}$$

is a linear differential equation of second order with constant coefficients. We shall prove

THEOREM (Minding). *Any two regular surfaces with the same constant Gaussian curvature are locally isometric. More precisely, let* S_1, S_2 *be two*

regular surfaces with the same constant curvature K. *Choose points* $p_1 \in S_1$, $p_2 \in S_2$, *and orthonormal basis* $\{e_1, e_2\} \in T_{p_1}(S_1)$, $\{f_1, f_2\} \in T_{p_2}(S_2)$. *Then there exist neighborhoods* V_1 *of* p_1, V_2 *of* p_2 *and an isometry* $\psi: V_1 \rightarrow V_2$ *such that* $d\psi(e_1) = f_1$, $d\psi(e_2) = f_2$.

Proof. Let us first consider Eq. (2) and study separately the cases (1) $K = 0$, (2) $K > 0$, and (3) $K < 0$.

1. If $K = 0$, $(\sqrt{G})_{\rho\rho} = 0$. Thus, $(\sqrt{G})_\rho = g(\theta)$, where $g(\theta)$ is a function of θ. Since

$$\lim_{\rho \to 0} (\sqrt{G})_\rho = 1,$$

we conclude that $(\sqrt{G})_\rho \equiv 1$. Therefore, $\sqrt{G} = \rho + f(\theta)$, where $f(\theta)$ is a function of θ. Since

$$f(\theta) = \lim_{\rho \to 0} \sqrt{G} = 0,$$

we finally have, in this case,

$$E = 1, \qquad F = 0, \qquad G(\rho, \theta) = \rho^2.$$

2. If $K > 0$, the general solution of Eq. (2) is given by

$$\sqrt{G} = A(\theta) \cos(\sqrt{K}\rho) + B(\theta) \sin(\sqrt{K}\rho),$$

where $A(\theta)$ and $B(\theta)$ are functions of θ. That this expression is a solution of Eq. (2) is easily verified by differentiation.

Since $\lim_{\rho \to 0} \sqrt{G} = 0$, we obtain $A(\theta) = 0$. Thus,

$$(\sqrt{G})_\rho = B(\theta)\sqrt{K} \cos(\sqrt{K}\rho),$$

and since $\lim_{\rho \to 0} (\sqrt{G})_\rho = 1$, we conclude that

$$B(\theta) = \frac{1}{\sqrt{K}}.$$

Therefore, in this case,

$$E = 1, \qquad F = 0, \qquad G = \frac{1}{K} \sin^2(\sqrt{K}\rho).$$

3. Finally, if $K < 0$, the general solution of Eq. (2) is

$$\sqrt{G} = A(\theta) \cosh(\sqrt{-K}\rho) + B(\theta) \sinh(\sqrt{-K}\rho).$$

By using the initial conditions, we verify that in this case

$$E = 1, \qquad F = 0, \qquad G = \frac{1}{-K} \sinh^2(\sqrt{-K}\rho).$$

We are now prepared to prove Minding's theorem. Let V_1 and V_2 be normal neighborhoods of p_1 and p_2, respectively. Let φ be the linear isometry of $T_{p_1}(S_1)$ onto $T_{p_2}(S_2)$ given by $\varphi(e_1) = f_1$, $\varphi(e_2) = f_2$. Take a polar coordinate system (ρ, θ) in $T_{p_1}(S_1)$ with axis l and set $L_1 = \exp_{p_1}(l)$, $L_2 = \exp_{p_2}(\varphi(l))$. Let $\psi\colon V_1 \longrightarrow V_2$ be defined by

$$\psi = \exp_{p_2} \circ \varphi \circ \exp_{p_1}^{-1}.$$

We claim that ψ is the required isometry.

In fact, the restriction $\bar{\psi}$ of ψ to $V_1 - L_1$ maps a polar coordinate neighborhood with coordinates (ρ, θ) centered in p_1 into a polar coordinate neighborhood with coordinates (ρ, θ) centered in p_2. By the above study of Eq. (2), the coefficients of the first fundamental forms at corresponding points are equal. By Prop. 1 of Sec. 4-2, $\bar{\psi}$ is an isometry. By continuity, ψ still preserves inner products at points of L_1 and thus is an isometry. It is immediate to check that $d\psi(e_1) = f_1$, $d\psi(e_2) = f_2$, and this concludes the proof.

Q.E.D.

Remark 2. In the case that K is not constant but maintains its sign, the expression $\sqrt{G}K = -(\sqrt{G})_{\rho\rho}$ has a nice intuitive meaning. Consider the arc length $L(\rho)$ of the curve $\rho =$ const. between two close geodesics $\theta = \theta_0$ and $\theta = \theta_1$:

$$L(\rho) = \int_{\theta_0}^{\theta_1} \sqrt{G(\rho, \theta)}\, d\theta.$$

Assume that $K < 0$. Since

$$\lim_{\rho \to 0} (\sqrt{G})_\rho = 1 \quad \text{and} \quad (\sqrt{G})_{\rho\rho} = -K\sqrt{G} > 0,$$

the function $L(\rho)$ behaves as in Fig. 4-39(a). This means that $L(\rho)$ increases with ρ; that is, as ρ increases, the geodesics $\theta = \theta_0$ and $\theta = \theta_1$ get farther and farther apart (of course, we must remain in the coordinate neighborhood in question).

On the other hand, if $K > 0$, $L(\rho)$ behaves as in Fig. 4-39(b). The geodesics

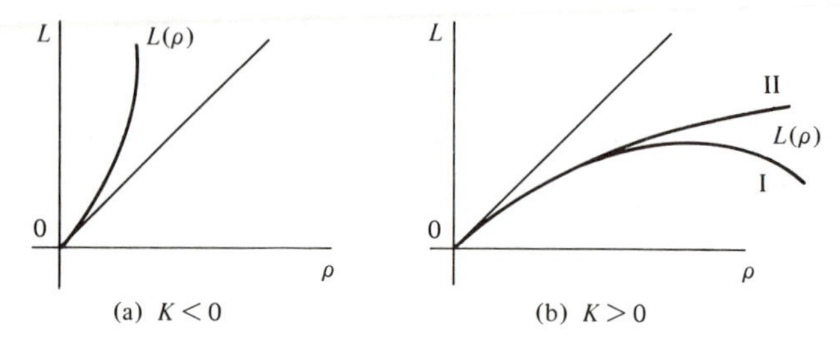

Figure 4-39. Spreading of close geodesics in a normal neighborhood.

$\theta = \theta_0$ and $\theta = \theta_1$ may (case I) or may not (case II) come closer together after a certain value of ρ, and this depends on the Gaussian curvature. For instance, in the case of a sphere two geodesics which leave from a pole start coming closer together after the equator (Fig. 4-40).

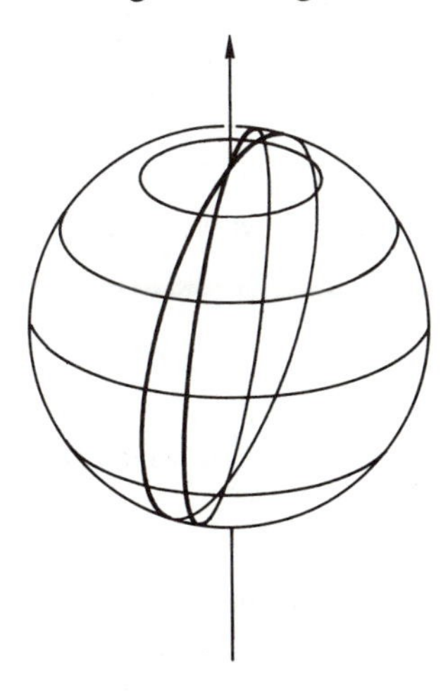

Figure 4-40

In Chap. 5 (Secs. 5-4 and 5-5) we shall come back to this subject and shall make this observation more precise.

Another application of the geodesic polar coordinates consists of a geometrical interpretation of the Gaussian curvature K.

To do this, we first observe that the expression of K in geodesic polar coordinates (ρ, θ), with center $p \in S$, is given by

$$K = -\frac{(\sqrt{G})_{\rho\rho}}{\sqrt{G}},$$

and therefore

$$\frac{\partial^3(\sqrt{G})}{\partial \rho^3} = -K(\sqrt{G})_\rho - K_\rho(\sqrt{G}).$$

Thus, recalling that

$$\lim_{\rho \to 0} \sqrt{G} = 0,$$

we obtain

$$-K(p) = \lim_{\rho \to 0} \frac{\partial^3 \sqrt{G}}{\partial \rho^3}.$$

On the other hand, by defining $\sqrt{G}$ and its successive derivatives with respect to ρ at p by its limit values (cf. Eq. (1)), we may write

$$\sqrt{G}(\rho, \theta) = \sqrt{G}(0, \theta) + \rho(\sqrt{G})_\rho(0, \theta) + \frac{\rho^2}{2!}(\sqrt{G})_{\rho\rho}(0, \theta) + \frac{\rho^3}{3!}(\sqrt{G})_{\rho\rho\rho}(0, \theta) + R(\rho, \theta)$$

where

$$\lim_{\rho \to 0} \frac{R(\rho, \theta)}{\rho^3} = 0,$$

uniformly in θ. By substituting in the above expression the values already known, we obtain

$$\sqrt{G}(\rho, \theta) = \rho - \frac{\rho^3}{3!}K(p) + R.$$

With this value for $\sqrt{G}$, we compute the arc length L of a geodesic circle of radius $\rho = r$:

$$L = \lim_{\epsilon \to 0} \int_{0+\epsilon}^{2\pi-\epsilon} \sqrt{G}(r, \theta)\, d\theta = 2\pi r - \frac{\pi}{3} r^3 K(p) + R_1,$$

where

$$\lim_{r \to 0} \frac{R_1}{r^3} = 0.$$

It follows that

$$K(p) = \lim_{r \to 0} \frac{3}{\pi} \frac{2\pi r - L}{r^3},$$

which gives an intrinsic interpretation of $K(p)$ in terms of the radius r of a geodesic circle $S_r(p)$ around p and the arc lengths L and $2\pi r$ of $S_r(P)$ and $\exp_p^{-1}(S_r(p))$, respectively.

An interpretation of $K(p)$ involving the area of the region bounded by $S_r(p)$ is easily obtained by the above process (see Exercise 3).

As a last application of the geodesic polar coordinates, we shall study some minimal properties of geodesics. A fundamental property of a geodesic is the fact that, locally, it minimizes arc length. More precisely, we have

PROPOSITION 4. *Let* p *be a point on a surface* S. *Then, there exists a neighborhood* $W \subset S$ *of* p *such that if* $\gamma\colon I \to W$ *is a parametrized geodesic with* $\gamma(0) = p$, $\gamma(t_1) = q$, $t_1 \in I$, *and* $\alpha\colon[0, t_1] \to S$ *is a parametrized regular curve joining* p *to* q, *we have*

$$l_\gamma \leq l_\alpha,$$

where l_α *denotes the length of the curve* α. *Moreover, if* $l_\gamma = l_\alpha$, *then the trace of* α *coincides with the trace of* α *between* p *and* q.

Proof. Let V be a normal neighborhood of p, and let $\bar{W}$ be the closed region bounded by a geodesic circle of radius r contained in V. Let (ρ, θ) be geodesic polar coordinates in $\bar{W} - L$ centered in p such that $q \in L$.

Suppose first that $\alpha((0, t_1)) \subset \bar{W} - L$, and set $\alpha(t) = (\rho(t), \theta(t))$. Observe initially that

$$\sqrt{(\rho')^2 + G(\theta')^2} \geq \sqrt{(\rho')^2},$$

and equality holds if and only if $\theta' \equiv 0$; that is, $\theta = \text{const.}$ Therefore, the length $l_\alpha(\epsilon)$ of α between ϵ and $t_1 - \epsilon$ satisfies

$$\begin{aligned} l_\alpha(\epsilon) &= \int_\epsilon^{t_1-\epsilon} \sqrt{(\rho')^2 + G(\theta')^2}\, dt \geq \int_\epsilon^{t_1-\epsilon} \sqrt{(\rho')^2}\, dt \\ &\geq \int_\epsilon^{t_1-\epsilon} \rho'\, dt = l_\gamma - 2\epsilon, \end{aligned}$$

and equality holds if and only if $\theta = \text{const.}$ and $\rho' > 0$. By making $\epsilon \to 0$ in the expression above, we obtain that $l_\alpha \geq l_\gamma$ and that equality holds if and only if α is the radial geodesic $\theta = \text{const.}$ with a parametrization $\rho = \rho(t)$, where $\rho'(t) > 0$. It follows that if $l_\alpha = l_\gamma$, then the traces of α and γ between p and q coincide.

Suppose now that $\alpha((0, t_1))$ intersects L, and assume that this occurs for the first time at, say, $\alpha(t_2)$. Then, by the previous argument, $l_\alpha \geq l_\gamma$ between t_0 and t_2, and $l_\alpha = l_\gamma$ implies that the traces of α and γ coincide. Since $\alpha([0, t_1])$ and L are compact, there exists a $\bar{t} \geq t_2$ such that either $\alpha(\bar{t})$ is the last point where $\alpha((0, t_1))$ intersects L or $\alpha([\bar{t}, t_1]) \subset L$ (Fig. 4-41). In any case, applying the above case, the conclusions of the proposition follow.

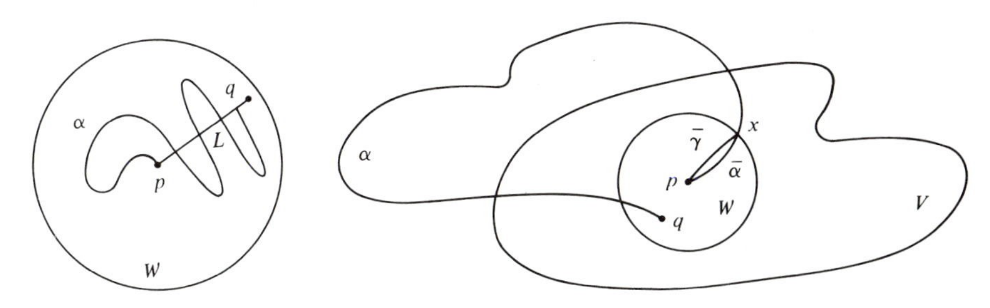

Figure 4-41

Figure 4-42

Suppose finally that $\alpha([0, t_1])$ is not entirely contained in $\bar{W}$. Let $t_0 \in [0, t_1]$ be the first value for which $\alpha(t_0) = x$ belongs to the boundary of $\bar{W}$. Let $\bar{\gamma}$ be the radial geodesic px and let $\bar{\alpha}$ be the restriction of the curve α to the interval $[0, t_0]$. It is clear then that $l_\alpha \geq l_{\bar{\alpha}}$ (see Fig. 4-42).

By the previous argument, $l_{\bar{\alpha}} \geq l_{\bar{\gamma}}$. Since q is a point in the interior of $\bar{W}$, $l_{\bar{\gamma}} > l_\gamma$. We conclude that $l_\alpha > l_\gamma$, which ends the proof. **Q.E.D.**

Remark 3. For simplicity, we have proved the proposition for regular curves. However, it still holds for piecewise regular curves (cf. Def. 7, Sec. 4-4); the proof is entirely analogous and will be left as an exercise.

Remark 4. The proof also shows that the converse of the last assertion of Prop. 4 holds true. However, this converse does not generalize to piecewise regular curves.

The previous proposition is not true globally, as is shown by the example of the sphere. Two nonantipodal points of a sphere may be connected by two meridians of unequal lengths and only the smaller one satisfies the conclusions of the proposition. In other words, a geodesic, if sufficiently extended, may not be the shortest path between its end points. The following proposition shows, however, that when a regular curve is the shortest path between any two of its points, this curve is necessarily a geodesic.

PROPOSITION 5. *Let* $\alpha: I \to S$ *be a regular parametrized curve with a parameter proportional to arc length. Suppose that the arc length of* α *between any two points* $t, \tau \in I$, *is smaller than or equal to the arc length of any regular parametrized curve joining* $\alpha(t)$ *to* $\alpha(\tau)$. *Then* α *is a geodesic.*

Proof. Let $t_0 \in I$ be an arbitrary point of I and let W be the neighborhood of $\alpha(t_0) = p$ given by Prop. 4. Let $q = \alpha(t_1) \in W$. From the case of equality in Prop. 4, it follows that α is a geodesic in (t_0, t_1). Otherwise α would have, between t_0 and t_1, a length greater than the radial geodesic joining $\alpha(t_0)$ to $\alpha(t_1)$, a contradiction to the hypothesis. Since α is regular, we have, by continuity, that α still is a geodesic in t_0. **Q.E.D.**

EXERCISES

1. Prove that on a surface of constant curvature the geodesic circles have constant geodesic curvature.

2. Show that the equations of the geodesics in geodesic polar coordinates ($E = 1$, $F = 0$) are given by

$$\rho'' - \frac{1}{2}G_\rho(\theta')^2 = 0$$

$$\theta'' + \frac{G_\rho}{G}\rho'\theta' + \frac{1}{2}\frac{G_\theta}{G}(\theta')^2 = 0.$$

3. If p is a point of a regular surface S, prove that

$$K(p) = \lim_{r \to 0} \frac{12}{\pi} \frac{\pi r^2 - A}{r^4},$$

where $K(p)$ is the Gaussian curvature of S at p, r is the radius of a geodesic circle $S_r(p)$ centered in p, and A is the area of the region bounded by $S_r(p)$.

4. Show that in a system of normal coordinates centered in p, all the Christoffel symbols are zero at p.

5. For which of the pair of surfaces given below does there exist a local isometry?

a. Torus of revolution and cone.

b. Cone and sphere.

c. Cone and cylinder.

6. Let S be a surface, let p be a point of S, and let $S^1(p)$ be a geodesic circle around p, sufficiently small to be contained in a normal neighborhood. Let r and s be two points of $S^1(p)$, and C be an arc of $S^1(p)$ between r and s. Consider the curve $\exp_p^{-1}(C) \subset T_p(S)$. Prove that $S^1(p)$ can be chosen sufficiently small so that

a. If $K > 0$, then $l(\exp_p^{-1}(C)) > l(C)$, where $l(\ \ )$ denotes the arc length of the corresponding curve.

b. If $K < 0$, then $l(\exp_p^{-1}(C)) < l(C)$.

7. Let (ρ, θ) be a system of geodesic polar coordinates ($E = 1, F = 0$) on a surface, and let $\gamma(\rho(s), \theta(s))$ be a geodesic that makes an angle $\varphi(s)$ with the curves $\theta = \text{const.}$ For definiteness, the curves $\theta = \text{const.}$ are oriented in the sense of increasing ρ's and φ is measured from $\theta = \text{const.}$ to γ in the orientation given by the parametrization (ρ, θ). Show that

$$\frac{d\varphi}{ds} + (\sqrt{G})_\rho \frac{d\theta}{ds} = 0.$$

***8.** (*Gauss Theorem on the Sum of the Internal Angles of a "Small" Geodesic Triangle.*) Let Δ be a geodesic triangle (that is, its sides are segments of geodesics) on a surface S. Assume that Δ is sufficiently small to be contained in a normal neighborhood of some of its vertices. Prove directly (i.e., without using the Gauss-Bonnet theorem) that

$$\iint_\Delta K\, dA = \left(\sum_{i=1}^{3} \alpha_i\right) - \pi,$$

where K is the Gaussian curvature of S, and $0 < \alpha_i < \pi$, $i = 1, 2, 3$, are the internal angles of the triangle Δ.

9. (*A Local Isoperimetric Inequality for Geodesic Circles.*) Let $p \in S$ and let $S_r(p)$ be a geodesic circle of center p and radius r. Let L be the arc length of $S_r(p)$ and A be the area of the region bounded by $S_r(p)$. Prove that

$$4\pi A - L^2 = \pi^2 r^4 K(p) + R,$$

where $K(p)$ is the Gaussian curvature of S at p and

$$\lim_{r \to 0} \frac{R}{r^4} = 0.$$

Thus, if $K(p) > 0$ (or < 0) and r is small, $4\pi A - L^2 > 0$ (or < 0). (Compare the isoperimetric inequality of Sec. 1-7.)

10. Let S be a connected surface and let $\varphi, \psi: S \to S$ be two isometries of S. Assume that there exists a point $p \in S$ such that $\varphi(p) = \psi(p)$ and $d\varphi_p(v) = d\psi_p(v)$ for all $v \in T_p(S)$. Prove that $\varphi(q) = \psi(q)$ for all $q \in S$.

11. (*Free Mobility of Small Geodesic Triangles.*) Let S be a surface of constant Gaussian curvature. Choose points $p_1, p_1' \in S$ and let V, V' be normal neighborhoods of p_1, p_1', respectively. Choose geodesic triangles p_1, p_2, p_3 in V (geodesic means that the sides $\widehat{p_1p_2}$, $\widehat{p_2p_3}$, $\widehat{p_3p_1}$ are geodesic arcs) and p_1', p_2', p_3' in V' in such a way that

$$l(p_1, p_2) = l(p_1', p_2'),$$
$$l(p_2, p_3) = l(p_2', p_3'),$$
$$l(p_3, p_1) = l(p_3', p_1')$$

(here l denotes the length of a geodesic arc). Show that there exists an isometry $\theta: V \to V'$ which maps the first triangle onto the second. (This is the local version, for surfaces of constant curvature, of the theorem of high school geometry that any two triangles in the plane with equal corresponding sides are congruent.)

12. A diffeomorphism $\varphi: S_1 \to S_2$ is said to be a *geodesic mapping* if for every geodesic $C \subset S_1$ of S_1, the regular curve $\varphi(C) \subset S_2$ is a geodesic of S_2. If U is a neighborhood of $p \in S_1$, then $\varphi: U \to S_2$ is said to be a *local geodesic mapping* in p if there exists a neighborhood V of $\varphi(p)$ in S_2 such that $\varphi: U \to V$ is a geodesic mapping.

a. Show that if $\varphi: S_1 \to S_2$ is both a geodesic and a conformal mapping, then φ is a *similarity*; that is,

$$\langle v, w\rangle_p = \lambda\langle d\varphi_p(v), d\varphi_p(w)\rangle, \qquad p \in S_1, v, w \in T_p(S_1),$$

where λ is constant.

b. Let $S^2 = \{(x, y, z) \in R^3; x^2 + y^2 + z^2 = 1\}$ be the unit sphere, $S^- = \{(x, y, z) \in S^2; z < 0\}$ be its lower hemisphere, and P be the plane $z = -1$. Prove that the map (central projection) $\varphi: S^- \to P$ which takes a point $p \in S^-$ to the intersection of P with the line that connects p to the center of S^2 is a geodesic mapping.

***c.** Show that a surface of constant curvature admits a local geodesic mapping into the plane for every $p \in S$.

13. (*Beltrami's Theorem.*) In Exercise 12, part c, it was shown that a surface S of constant curvature K admits a local geodesic mapping in the plane for every $p \in S$. To prove the converse (Beltrami's theorem)—*If a regular connected surface* S *admits for every* p $\in$ S *a local geodesic mapping into the plane, then* S *has constant curvature*, the following assertions should be proved:

a. If $v = v(u)$ is a geodesic, in a coordinate neighborhood of a surface parametrized by (u, v), which does not coincide with $u = \text{const.}$, then

$$\frac{d^2v}{du^2} = \Gamma_{22}^1\left(\frac{dv}{du}\right)^3 + (2\Gamma_{12}^1 - \Gamma_{22}^2)\left(\frac{dv}{du}\right)^2 + (\Gamma_{11}^1 - 2\Gamma_{12}^2)\frac{dv}{du} - \Gamma_{11}^2.$$

***b.** If S admits a local geodesic mapping $\varphi\colon V \longrightarrow R^2$ of a neighborhood V of a point $p \in S$ into the plane R^2, then it is possible to parametrize the neighborhood V by (u, v) in such a way that

$$\Gamma_{22}^1 = \Gamma_{11}^2 = 0, \qquad \Gamma_{22}^2 = 2\Gamma_{12}^1, \qquad \Gamma_{11}^1 = 2\Gamma_{12}^2.$$

***c.** If there exists a geodesic mapping of a neighborhood V of $p \in S$ into a plane, then the curvature K in V satisfies the relations

$$KE = \Gamma_{12}^2\Gamma_{12}^2 - (\Gamma_{12}^2)_u \qquad (a)$$

$$KF = \Gamma_{12}^1\Gamma_{12}^2 - (\Gamma_{12}^2)_v \qquad (b)$$

$$KG = \Gamma_{12}^1\Gamma_{12}^1 - (\Gamma_{12}^1)_v \qquad (c)$$

$$KF = \Gamma_{12}^2\Gamma_{12}^1 - (\Gamma_{12}^1)_u \qquad (d)$$

***d.** If there exists a geodesic mapping of a neighborhood V of $p \in S$ into a plane, then the curvature K in V is constant.

e. Use the above, and a standard argument of connectedness, to prove Beltrami's theorem.

14. (*The Holonomy Group.*) Let S be a regular surface and $p \in S$. For each piecewise regular parametrized curve $\alpha\colon [0, l] \longrightarrow S$ with $\alpha(0) = \alpha(l) = p$, let $P_\alpha\colon T_p(S) \longrightarrow T_p(S)$ be the map which assigns to each $v \in T_p(S)$ its parallel transport along α back to p. By Prop. 1 of Sec. 4-4, P_α is a linear isometry of $T_p(S)$. If $\beta\colon [l, \bar{l}]$ is another piecewise regular parametrized curve with $\beta(l) = \beta(\bar{l}) = p$, define the curve $\beta \circ \alpha\colon [0, l + \bar{l}] \longrightarrow S$ by running successively first α and then β; that is, $\beta \circ \alpha(s) = \alpha(s)$ if $s \in [0, l]$, and $\beta \circ \alpha(s) = \beta(s)$ if $s \in [l, \bar{l}]$.

a. Consider the set

$$H_p(S) = \{P_\alpha\colon T_p(S) \longrightarrow T_p(S);\ \text{all } \alpha \text{ joining } p \text{ to } p\},$$

where α is piecewise regular. Define in this set the operation $P_\beta \circ P_\alpha = P_{\beta\circ\alpha}$; that is, $P_\beta \circ P_\alpha$ is the usual composition of performing first P_α and then P_β. Prove that, with this operation, $H_p(S)$ is a group (actually, a subgroup of the group of linear isometries of $T_p(S)$). $H_p(S)$ is called the *holonomy group* of S at p.

b. Show that the holonomy group at any point of a surface with $K \equiv 0$ reduces to the identity.

c. Prove that if S is connected, the holonomy groups $H_p(S)$ and $H_q(S)$ at two arbitrary points $p, q \in S$ are isomorphic. Thus, we can talk about *the* (abstract) *holonomy group of a surface.*

d. Prove that the holonomy group of a sphere is isomorphic to the group of 2×2 rotation matrices (cf. Exercise 22, Sec. 4-4).

4-7. Further Properties of Geodesics; Convex Neighborhoods†

In this section we shall show how certain facts on geodesics (in particular, Prop. 5 of Sec. 4-4) follow from the general theorem of existence, uniqueness, and dependence on the initial condition of vector fields.

The geodesics in a parametrization $\mathbf{x}(u, v)$ are given by the system

$$\begin{aligned} u'' + \Gamma^1_{11}(u')^2 + 2\Gamma^1_{12}u'v' + \Gamma^1_{22}(v')^2 &= 0, \\ v'' + \Gamma^2_{11}(u')^2 + 2\Gamma^2_{12}u'v' + \Gamma^2_{22}(v')^2 &= 0, \end{aligned} \tag{1}$$

where the Γ^k_{ij} are functions of the local coordinates u and v. By setting $u' = \xi$ and $v' = \eta$, we may write the above system in the general form

$$\begin{aligned} \xi' &= F_1(u, v, \xi, \eta), \\ \eta' &= F_2(u, v, \xi, \eta), \\ u' &= F_3(u, v, \xi, \eta), \\ v' &= F_4(u, v, \xi, \eta), \end{aligned} \tag{2}$$

where $F_3(u, v, \xi, \eta) = \xi$, $F_4(u, v, \xi, \eta) = \eta$.

It is convenient to use the following notation: (u, v, ξ, η) will denote a point of R^4 which will be thought of as the cartesian product $R^4 = R^2 \times R^2$; (u, v) will denote a point of the first factor and (ξ, η) a point of the second factor.

The system (2) is equivalent to a vector field in an open set of R^4 which is defined in a way entirely analogous to vector fields in R^2 (cf. Sec. 3-4). The theorem of existence and uniqueness of trajectories (Theorem 1, Sec. 3-4) still holds in this case (actually, the theorem holds for R^n; cf. S. Lang, *Analysis* I, Addison-Wesley, Reading, Mass., 1968, pp. 383–386) and is stated as follows:

Given the system (2) *in an open set* $\mathrm{U} \subset \mathrm{R}^4$ *and given a point*

$$(\mathrm{u}_0, \mathrm{v}_0, \xi_0, \eta_0) \in \mathrm{U}$$

there exists a unique trajectory $\alpha\colon (-\epsilon, \epsilon) \to \mathrm{U}$ *of Eq.* (2), *with*

$$\alpha(0) = (\mathrm{u}_0, \mathrm{v}_0, \xi_0, \eta_0).$$

†This section may be omitted on a first reading. Propositions 1 and 2 (the statements of which can be understood without reading the section) are, however, used in Chap. 5.

To apply this result to a regular surface S, we should observe that, given a parametrization $\mathbf{x}(u, v)$ in $p \in S$, of coordinate neighborhood V, the set of pairs (q, v), $q \in V$, $v \in T_q(S)$, may be identified to an open set $V \times R^2 = U \subset R^4$. For that, we identify each $T_q(S)$, $q \in V$, with R^2 by means of the basis $\{\mathbf{x}_u, \mathbf{x}_v\}$. Whenever we speak about differentiability and continuity in the set of pairs (q, v) we mean the differentiability and continuity induced by this identification.

Assuming the above theorem, the proof of Prop. 5 of Sec. 4-4 is trivial. Indeed, the equations of the geodesics in the parametrization $\mathbf{x}(u, v)$ in $p \in S$ yield a system of the form (2) in $U \subset R^4$. The fundamental theorem implies then that given a point $q = (u_0, v_0) \in V$ and a nonzero tangent vector $v = (\xi_0, \eta_0) \in T_q(S)$ there exists a unique parametrized geodesic

$$\gamma = \pi \circ \alpha : (-\epsilon, \epsilon) \longrightarrow V$$

in V (where $\pi(q, v) = q$ is the projection $V \times R^2 \longrightarrow V$).

The theorem of the dependence on the initial conditions for the vector field defined by Eq. (2) is also important. It is essentially the same as that for the vector fields of R^2: *Given a point* $\mathrm{p} = (\mathrm{u}_0, \mathrm{v}_0, \xi_0, \eta_0) \in \mathrm{U}$, *there exist a neighborhood* $\mathrm{V} = \mathrm{V}_1 \times \mathrm{V}_2$ *of* p (*where* V_1 *is a neighborhood of* $(\mathrm{u}_0, \mathrm{v}_0)$ *and* V_2 *is a neighborhood of* (ξ_0, η_0)), *an open interval* I, *and a differentiable mapping* $\alpha : \mathrm{I} \times \mathrm{V}_1 \times \mathrm{V}_2 \longrightarrow \mathrm{U}$ *such that, fixed* $(\mathrm{u}, \mathrm{v}, \xi, \eta) = (\mathrm{q}, \mathrm{v}) \in \mathrm{V}$, *then* $\alpha(\mathrm{t}, \mathrm{q}, \mathrm{v})$, $\mathrm{t} \in \mathrm{I}$, *is the trajectory of* (2) *passing through* (q, v).

To apply this statement to a regular surface S, we introduce a parametrization in $p \in S$, with coordinate neighborhood V, and identify, as above, the set of pairs (q, v), $q \in V$, $v \in T_q(S)$, with $V \times R^2$. Taking as the initial condition the pair $(p, 0)$, we obtain an interval $(-\epsilon_2, \epsilon_2)$, a neighborhood $V_1 \subset V$ of p in S, a neighborhood V_2 of the origin in R^2, and a differentiable map

$$\gamma : (-\epsilon_2, \epsilon_2) \times V_1 \times V_2 \longrightarrow V$$

such that if $(q, v) \in V_1 \times V_2$, $v \neq 0$, the curve

$$t \longrightarrow \gamma(t, q, v), \qquad t \in (-\epsilon_2, \epsilon_2),$$

is the geodesic of S satisfying $\gamma(0, q, v) = q$, $\gamma'(0, q, v) = v$, and if $v = 0$, this curve reduces to the point q. Here $\gamma = \pi \circ \alpha$, where $\pi(q, v) = q$ is the projection $U = V \times R^2 \longrightarrow V$ and α is the map given above.

Back in the surface, the set $V_1 \times V_2$ is of the form

$$\{(q, v), p \in V_1, v \in V_q(0) \subset T_q(S)\},$$

where $V_q(0)$ denotes a neighborhood of the origin in $T_q(S)$. Thus, if we restrict

γ to $(-\epsilon_2, \epsilon_2) \times \{p\} \times V_2$, we can choose $\{p\} \times V_2 = B_{\epsilon_1} \subset T_p(S)$, and obtain

THEOREM 1. *Given* $p \in S$ *there exist numbers* $\epsilon_1 > 0, \epsilon_2 > 0$ *and a differentiable map*

$$\gamma\colon (-\epsilon_2, \epsilon_2) \times B_{\epsilon_1} \longrightarrow S, \qquad B_{\epsilon_1} \subset T_p(S)$$

such that for $v \in B_{\epsilon_1}, v \neq 0, t \in (-\epsilon_2, \epsilon_2)$ *the curve* $t \longrightarrow \gamma(t, v)$ *is the geodesic of* S *with* $\gamma(0, v) = p, \gamma'(0, v) = v$, *and for* $v = 0, \gamma(t, 0) = p$.

This result was used in the proof of Prop. 1 of Sec. 4-6.

The above theorem corresponds to the case where p is fixed. To handle the general case, let us denote by $B_r(q)$ the domain bounded by a (small) geodesic circle of radius r and center q, and by $\bar{B}_r(q)$ the union of $B_r(q)$ with its boundary.

Let $\epsilon > 0$ be such that $\bar{B}_\epsilon(p) \subset V_1$. Let $B_{\delta(q)}(0) \subset \bar{V}_q(0)$ be the largest open disk in the set $\bar{V}_q(0)$ formed by the union of $V_q(0)$ with its limit points, and set $\epsilon_1 = \inf \delta(q), q \in \bar{B}_\epsilon(p)$. Clearly, $\epsilon_1 > 0$. Thus, the set

$$\mathfrak{U} = \{(q, v); q \in B_\epsilon(p), v \in B_{\epsilon_1}(0) \subset T_q(S)\}$$

is contained in $V_1 \times V_2$, and we obtain

THEOREM 1a. *Given* $p \in S$, *there exist positive numbers* $\epsilon, \epsilon_1, \epsilon_2$ *and a differentiable map*

$$\gamma\colon (-\epsilon_2, \epsilon_2) \times \mathfrak{U} \longrightarrow S,$$

where

$$\mathfrak{U} = \{(q, v); q \in B_\epsilon(p), v \in B_{\epsilon_1}(0) \subset T_q(S)\},$$

such that $\gamma(t, q, 0) = q$, *and for* $v \neq 0$ *the curve*

$$t \longrightarrow \gamma(t, q, v), \qquad t \in (-\epsilon_2, \epsilon_2)$$

is the geodesic of S *with* $\gamma(0, q, v) = q, \gamma'(0, q, v) = v$.

Let us apply Theorem 1a to obtain the following refinement of the existence of normal geodesics.

PROPOSITION 1. *Given* $p \in S$ *there exist a neighborhood* W *of* p *in* S *and a number* $\delta > 0$ *such that for every* $q \in W$, exp_q *is a diffeomorphism on* $B_\delta(0) \subset T_q(S)$ *and* $exp_q\,(B_\delta(0)) \supset W$; *that is,* W *is a normal neighborhood of all its points.*

Proof. Let V be a coordinate neighborhood of p. Let $\epsilon, \epsilon_1, \epsilon_2$ and $\gamma: (-\epsilon_2, \epsilon_2) \times \mathcal{U} \longrightarrow V$ be as in Theorem 1a. By choosing $\epsilon_1 < \epsilon_2$, we can make sure that, for $(q, v) \in \mathcal{U}$, $\exp_q(v) = \gamma(|v|, q, v)$ is well defined. Thus, we can define a differentiable map $\varphi: \mathcal{U} \longrightarrow V \times V$ by

$$\varphi(q, v) = (q, \exp_q(v)).$$

We first show that $d\varphi$ is nonsingular at $(p, 0)$. For that, we investigate how φ transforms the curves in $\mathcal{U}$ given by

$$t \longrightarrow (p, tw), \qquad t \longrightarrow (\alpha(t), 0),$$

where $w \in T_p(S)$ and $\alpha(t)$ is a curve in S with $\alpha(0) = p$. Observe that the tangent vectors of these curves at $t = 0$ are $(0, w)$ and $(\alpha'(0), 0)$, respectively. Thus,

$$d\varphi_{(p,0)}(0, w) = \frac{d}{dt}(p, \exp_p(wt))\bigg|_{t=0} = (0, w),$$

$$d\varphi_{(p,0)}(\alpha'(0), 0) = \frac{d}{dt}(\alpha(t), \exp_{\alpha(t)}(0))\bigg|_{t=0} = (\alpha'(0), \alpha'(0)),$$

and $d\varphi_{(p,0)}$ takes linearly independent vectors into linearly independent vectors. Hence, $d\varphi_{(p,0)}$ is nonsingular.

It follows that we can apply the inverse function theorem, and conclude the existence of a neighborhood $\mathcal{V}$ of $(p, 0)$ in $\mathcal{U}$ such that φ maps $\mathcal{V}$ diffeomorphically onto a neighborhood of (p, p) in $V \times V$. Let $U \subset B_\epsilon(p)$ and $\delta > 0$ be such that

$$\mathcal{V} = \{(q, v) \in \mathcal{U};\ q \in U, v \in B_\delta(0) \subset T_q(S)\}.$$

Finally, let $W \subset U$ be a neighborhood of p such that $W \times W \subset \varphi(\mathcal{V})$.

We claim that δ and W thus obtained satisfy the statement of the theorem. In fact, since φ is a diffeomorphism in $\mathcal{V}$, $\exp_q$ is a diffeomorphism in $B_\delta(0)$, $q \in W$. Furthermore, if $q \in W$, then

$$\varphi(\{q\} \times B_\delta(0)) \supset \{q\} \times W,$$

and, by definition of φ, $\exp_q(B_\delta(0)) \supset W$. **Q.E.D.**

Remark 1. From the previous proposition, it follows that given two points $q_1, q_2 \in W$ there exists a unique geodesic γ of length less than δ joining q_1 and q_2. Furthermore, the proof also shows that γ "depends differentiably" on q_1 and q_2 in the following sense: Given $(q_1, q_2) \in W \times W$, a unique $v \in T_{q_1}(S)$ is determined (precisely, the v given by $\varphi^{-1}(q_1, q_2) = (q_1, v)$) which depends differentiably on (q_1, q_2) and is such that $\gamma'(0) = v$.

One of the applications of the previous result consists of proving that a curve which locally minimizes arc length cannot be "broken." More precisely, we have

PROPOSITION 2. *Let* $\alpha: I \to S$ *be a parametrized, piecewise regular curve such that in each regular arc the parameter is proportional to the arc length. Suppose that the arc length between any two of its points is smaller than or equal to the arc length of any parametrized regular curve joining these points. Then* α *is a geodesic; in particular,* α *is regular everywhere.*

Proof. Let $0 = t_0 \leq t_1 \leq \cdots \leq t_k \leq t_{k+1} = l$ be a subdivision of $[0, l] = I$ in such a way that $\alpha \mid [t_i, t_{i+1}]$, $i = 0, \ldots, k$, is regular. By Prop. 5 of Sec. 4-6, α is geodesic at the points of (t_i, t_{i+1}). To prove that α is geodesic in t_i, consider the neighborhood W, given by Prop. 1, of $\alpha(t_i)$. Let $q_1 = \alpha(t_i - \epsilon), q_2 = \alpha(t_i + \epsilon)$, $\epsilon > 0$, be two points of W, and let γ be the radial geodesic of $B_\delta(q_1)$ joining q_1 to q_2 (Fig. 4-43). By Prop. 4 of Sec. 4-6, extended to the piecewise regular curves, $l(\gamma) \leq l(\alpha)$ between q_1 and q_2. Together with the hypothesis of the proposition, this implies that $l(\gamma) = l(\alpha)$. Thus, again by Prop. 4 of Sec. 4-6, the traces of γ and α coincide. Therefore, α is geodesic in t_1, which ends the proof. **Q.E.D.**

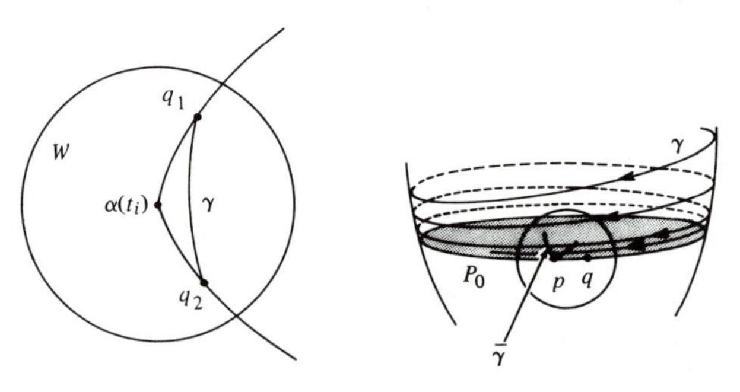

Figure 4-43 **Figure 4-44**

In Example 6 of Sec. 4-4 we have used the following fact: *A geodesic* $\gamma(t)$ *of a surface of revolution cannot be asymptotic to a parallel* P_0 *which is not itself a geodesic.* As a further application of Prop. 1, we shall sketch a proof of this fact (the details can be filled in as an exercise).

Assume the contrary to the above statement, and let p be a point in the parallel P_0. Let W and δ be the neighborhood and the number given by Prop. 1, and let $q \in P_0 \cap W, q \neq p$. Because $\gamma(t)$ is asymptotic to P_0, the point p is a limit of points $\gamma(t_i)$, where $\{t_i\} \to \infty$, and the tangents of γ at t_i converge to the tangent of P_0 at p. By Remark 1, the geodesic $\bar{\gamma}(t)$ with length smaller than δ joining p to q must be tangent to P_0 at p. By Clairaut's relation (cf. Example 5, Sec. 4-4), a small arc of $\bar{\gamma}(t)$ around p will be in the region of W where $\gamma(t)$ lies. It follows that, sufficiently close to p, there is a pair of points in W joined by two geodesics of length smaller than δ (see Fig. 4-44). This is a contradiction and proves our claim.

One natural question about Prop. 1 is whether the geodesic of length less than δ which joins two points q_1, q_2 of W is contained in W. If this is the case for every pair of points in W, we say that W is *convex*.

We say that a parametrized geodesic joining two points is *minimal* if its length is smaller than or equal to that of any other parametrized piecewise regular curve joining these two points.

When W is convex, we have by Prop. 4 (see also Remark 3) of Sec. 4-6 that the geodesic γ joining $q_1 \in W$ to $q_2 \in W$ is minimal. Thus, in this case, we may say that any two points of W are joined by a unique minimal geodesic in W. In general, however, W is not convex.

We shall now prove that W can be so chosen that it becomes convex. The crucial point of the proof is the following proposition, which is interesting in its own right. As usual, we denote by $B_r(p)$ the interior of the region bounded by a geodesic circle $S_r(p)$ of radius r and center p.

PROPOSITION 3. *For each point* $p \in S$ *there exists a positive number* ϵ *with the following property: If a geodesic* $\gamma(t)$ *is tangent to the geodesic circle* $S_r(p)$, $r < \epsilon$, *at* $\gamma(0)$, *then, for* $t \neq 0$ *small,* $\gamma(t)$ *is outside* $B_r(p)$ (Fig. 4-45).

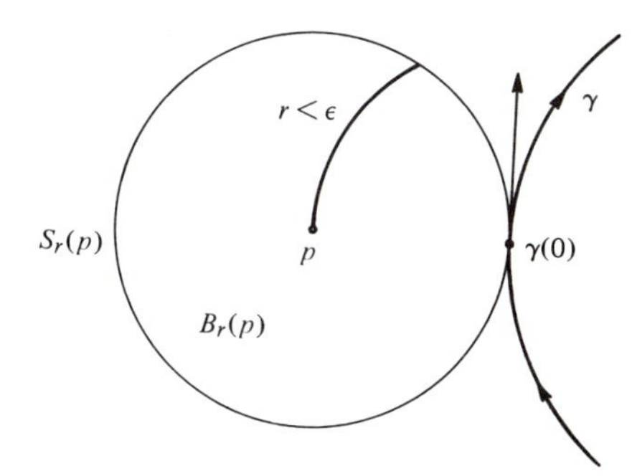

Figure 4-45

Proof. Let W be the neighborhood of p given by Prop. 1. For each pair (q, v), $q \in W$, $v \in T_p(S)$, $|v| = 1$, consider the geodesic $\gamma(t, q, v)$ and set, for a fixed pair (q, v) (Fig. 4-46),

$$\exp_p^{-1} \gamma(t, q, v) = u(t),$$
$$F(t, q, v) = |u(t)|^2 = F(t).$$

Thus, for a fixed (q, v), $F(t)$ is the square of the distance of the point

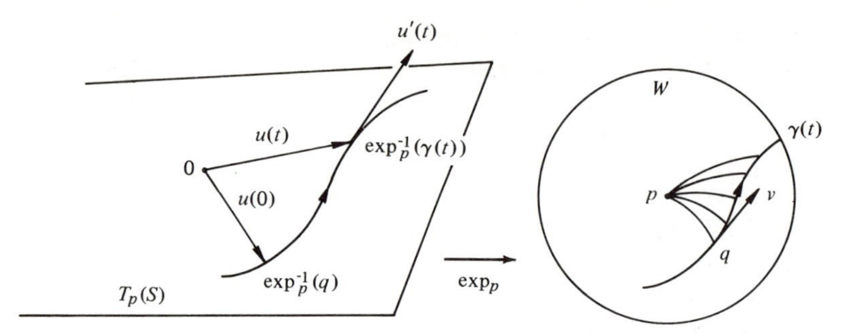

Figure 4-46

$\gamma(t, q, v)$ to p. Clearly, $F(t, q, v)$ is differentiable. Observe that $F(t, p, v) = |vt|^2$.

Now denote by $\mathfrak{U}^1$ the set

$$\mathfrak{U}^1 = \{(q, v), q \in W, v \in T_q(S), |v| = 1\},$$

and define a function $Q\colon \mathfrak{U}^1 \longrightarrow R$ by

$$Q(q, v) = \left.\frac{\partial^2 F}{\partial t^2}\right|_{t=0}.$$

Since F is differentiable, Q is continuous. Furthermore, since

$$\frac{\partial F}{\partial t} = 2\langle u(t), u'(t)\rangle,$$

$$\frac{\partial^2 F}{\partial t^2} = 2\langle u(t), u''(t)\rangle + 2\langle u'(t), u'(t)\rangle,$$

and at (p, v)

$$u'(t) = v, \qquad u''(t) = 0,$$

we obtain

$$Q(p, v) = 2|v|^2 = 2 > 0 \qquad \text{for all } v \in T_p(S), |v| = 1.$$

It follows, by continuity, that there exists a neighborhood $V \subset W$ such that $Q(q, v) > 0$ for all $q \in V$ and $v \in T_q(S)$ with $|v| = 1$. Let $\epsilon > 0$ be such that $B_\epsilon(p) \subset V$. We claim that this ϵ satisfies the statement of the proposition.

In fact, let $r < \epsilon$ and let $\gamma(t, q, v)$ be a geodesic tangent to $S_r(p)$ at $\gamma(0) = q$. By introducing geodesic polar coordinates around p, we see that $\langle u(0), u'(0)\rangle = 0$ (see Fig. 4-47). Thus, $\partial F/\partial t(0) = 0$. Since $F(0, q, v) = r^2$, and $(\partial^2 F/\partial t^2)(0) > 0$, we have that $F(t) > r^2$ for $t \neq 0$ small; hence, $\gamma(t)$ is outside $B_r(p)$. **Q.E.D.**

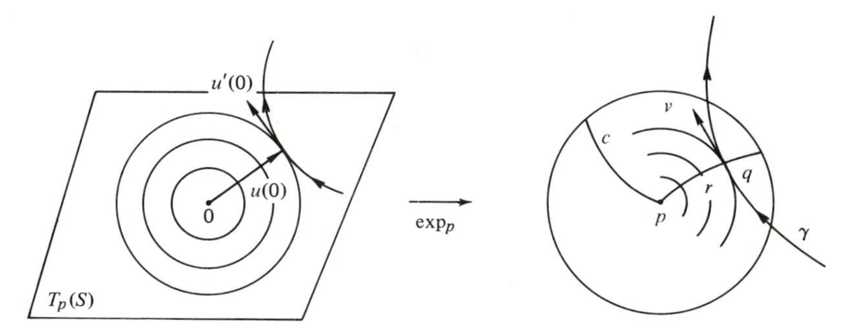

Figure 4-47

We can now prove

PROPOSITION 4 (Existence of Convex Neighborhoods). *For each point* p $\in$ S *there exists a number* c $>$ 0 *such that* $B_c(p)$ *is convex; that is, any two points of* $B_c(p)$ *can be joined by a unique minimal geodesic in* $B_c(p)$.

Proof. Let ϵ be given as in Prop. 3. Choose δ and W in Prop. 1 in such a way that $\delta < \epsilon/2$. Choose $c < \delta$ and such that $B_c(p) \subset W$. We shall prove that $B_c(p)$ is convex.

Let $q_1, q_2 \in B_c(p)$ and let $\gamma: I \longrightarrow S$ be the geodesic with length less that $\delta < \epsilon/2$ joining q_1 to q_2. $\gamma(I)$ is clearly contained in $B_\epsilon(p)$, and we want to prove that $\gamma(I)$ is contained in $B_c(p)$. Assume the contrary. Then there is a point $m \in B_\epsilon(p)$ where the maximum distance v of $\gamma(I)$ to p is attained (Fig. 4-48). In a neighborhood of m, the points of $\gamma(I)$ will be in $B_r(p)$. But this contadicts Prop. 3. **Q.E.D.**

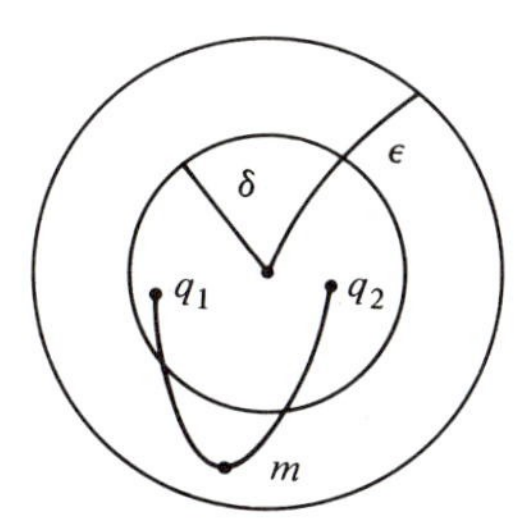

Figure 4-48

EXERCISES

***1.** Let y and w be differentiable vector fields on an open set $U \subset S$. Let $p \in S$ and let $\alpha: I \longrightarrow U$ be a curve such that $\alpha(0) = p$, $\alpha'(0) = y$. Denote by $P_{\alpha,t}$: $T_{\alpha(0)}(S) \longrightarrow T_{\alpha(t)}(S)$ the parallel transport along α from $\alpha(0)$ to $\alpha(t)$, $t \in I$. Prove that

$$(D_y w)(p) = \frac{d}{dt}(P^{-1}_{\alpha,t}(w(\alpha(t)))\bigg|_{t=0},$$

where the second member is the velocity vector of the curve $P^{-1}_{\alpha,t}(w(\alpha(t)))$ in $T_p(S)$ at $t = 0$. (Thus, the notion of covariant derivative can be derived from the notion of parallel transport.)

2. a. Show that the covariant derivative has the following properties. Let v, w, and y be differentiable vector fields in $U \subset S$, $f\colon U \longrightarrow R$ be a differentiable function in S, $y(f)$ be the derivative of f in the direction of y (cf. Exercise 7, Sec. 3-4), and λ, μ be real numbers. Then

1. $D_y(\lambda v + \mu w) = \lambda D_y(v) + \mu D_y(w)$; $D_{\lambda y + \mu v}(w) = \lambda D_y(w) + \mu D_v(w)$.
2. $D_y(fv) = y(f)v + fD_y(v)$; $D_{fy}(v) = fD_y(v)$.
3. $y(\langle v, w\rangle) = \langle D_y v, w\rangle + \langle v, D_y w\rangle$.
4. $D_{\mathbf{x}_v}\mathbf{x}_u = D_{\mathbf{x}_u}\mathbf{x}_v$, where $\mathbf{x}(u, v)$ is a parametrization of S.

***b.** Show that property 3 is equivalent to the fact that the parallel transport along a given piecewise regular parametrized curve $\alpha\colon I \longrightarrow S$ joining two points $p, q \in S$ is an isometry between $T_p(S)$ and $T_q(S)$. Show that property 4 is equivalent to the symmetry of the lower indices of the Christoffel symbols.

***c.** Let $\mathcal{V}(U)$ be the space of (differentiable) vector fields in $U \subset S$ and let $D\colon \mathcal{V} \times \mathcal{V} \longrightarrow \mathcal{V}$ (where we denote $D(y, v) = D_y(v)$) be a map satisfying properties 1–4. Verify that $D_y(v)$ coincides with the covariant derivative of the text. (In general, a D satisfying properties 1 and 2 is called a *connection* in U. The point of the exercise is to prove that on a surface with a given scalar product there exists a unique connection with the additional properties 3 and 4).

***3.** Let $\alpha\colon I = [0, l] \longrightarrow S$ be a simple, parametrized, regular curve. Consider a unit vector field $v(t)$ along α, with $\langle \alpha'(t), v(t)\rangle = 0$ and a mapping $\mathbf{x}\colon R \times I \longrightarrow S$ given by

$$\mathbf{x}(s, t) = \exp_{\alpha(t)}(sv(t)), \qquad s \in R,\ t \in I.$$

a. Show that $\mathbf{x}$ is differentiable in a neighborhood of I in $R \times I$ and that $d\mathbf{x}$ is nonsingular in $(0, t)$, $t \in I$.

b. Show that there exists $\epsilon > 0$ such that $\mathbf{x}$ is one-to-one in the rectangle $t \in I$, $|s| < \epsilon$.

c. Show that in the open set $t \in (0, l)$, $|s| < \epsilon$, $\mathbf{x}$ is a parametrization of S, the coordinate neighborhood of which contains $\alpha((0, l))$. The coordinates thus obtained are called *geodesic coordinates* (or Fermi's coordinates) of basis α. Show that in such a system $F = 0$, $E = 1$. Moreover, if α is a geodesic parametrized by the arc length, $G(0, t) = 1$ and $G_s(0, t) = 0$.

d. Establish the following analogue of the Gauss lemma (Remark 1 after Prop. 3, Sec. 4-6). Let $\alpha\colon I \longrightarrow S$ be a regular parametrized curve and let $\gamma_t(s)$, $t \in I$, be a family of geodesics parametrized by arc length s and given by; $\gamma_t(0) = \alpha(t)$, $\{\gamma_t'(0), \alpha'(t)\}$ is a positive orthogonal basis. Then, for a fixed $\bar{s}$, sufficiently small, the curve $t \longrightarrow \gamma_t(\bar{s})$, $t \in I$, intersects all γ_t orthogonally (such curves are called *geodesic parallels*).

4. The *energy* E of a curve $\alpha: [a, b] \longrightarrow S$ is defined by

$$E(\alpha) = \int_a^b |\alpha'(t)|^2 \, dt.$$

***a.** Show that $(l(\alpha))^2 \leq (b - a)E(\alpha)$ and that equality holds if and only if t is proportional to the arc length.

b. Conclude from part a that if $\gamma: [a, b] \longrightarrow S$ is a minimal geodesic with $\gamma(a) = p$, $\gamma(b) = q$, then for any curve $\alpha: [a, b] \longrightarrow S$, joining p to q, we have $E(\gamma) \leq E(\alpha)$ and equality holds if and only if α is a minimal geodesic.

5. Let $\gamma: [0, l] \longrightarrow S$ be a *simple* geodesic, parametrized by arc length, and denote by u and v the Fermi coordinates in a neighborhood of $\gamma([0, l])$ which is given as $u = 0$ (cf. Exercise 3). Let $u = \gamma(v, t)$ be a family of curves depending on a parameter t, $-\epsilon < t < \epsilon$, such that γ is differentiable and

$$\gamma(0, t) = \gamma(0) = p, \qquad \gamma(l, t) = \gamma(l) = q, \qquad \gamma(v, 0) = \gamma(v) \equiv 0.$$

Such a family is called a *variation* of γ keeping the end points p and q fixed. Let $E(t)$ be the energy of the curve $\gamma(v, t)$ (cf. Exercise 4); that is,

$$E(t) = \int_0^l \left(\frac{\partial \gamma}{\partial v}(v, t)\right)^2 dv.$$

***a.** Show that

$$E'(0) = 0,$$

$$\frac{1}{2}E''(0) = \int_0^l \left\{\left(\frac{d\eta}{dv}\right)^2 - K\eta^2\right\} dv,$$

where $\eta(v) = \partial\gamma/\partial t|_{t=0}$, $K = K(v)$ is the Gaussian curvature along γ, and $'$ denotes the derivative with respect to t (the above formulas are called *the first and second variations*, respectively, of the energy of γ; a more complete treatment of these formulas, including the case where γ is not simple, will be given in Sec. 5-4).

b. Conclude from part a that if $K \leq 0$, then any simple geodesic $\gamma: [0, l] \longrightarrow S$ is minimal relatively to the curves sufficiently close to γ and joining $\gamma(0)$ to $\gamma(l)$.

6. Let S be the cone $z = k\sqrt{x^2 + y^2}$, $k > 0$, $(x, y) \neq (0, 0)$, and let $V \subset R^2$ be the open set of R^2 given in polar coordinates by $0 < \rho < \infty$, $0 < \theta < 2\pi n \sin \beta$, where $\cot \beta = k$ and n is the largest integer such that $2\pi n \sin \beta < 2\pi$ (cf. Example 3, Sec. 4-2). Let $\varphi: V \longrightarrow S$ be the map

$$\varphi(\rho, \theta) = \left(\rho \sin \beta \cos\left(\frac{\theta}{\sin \beta}\right), \rho \sin \beta \sin\left(\frac{\theta}{\sin \beta}\right), \rho \cos \beta\right).$$

a. Prove that φ is a local isometry.

***b.** Let $q \in S$. Assume that $\beta < \pi/6$ and let k be the largest integer such that $2\pi k \sin \beta < \pi$. Prove that there exist at least k geodesics that leaving from q return to q. Show that these geodesics are broken at q and that, therefore, none of them is a closed geodesic (Fig. 4-49).

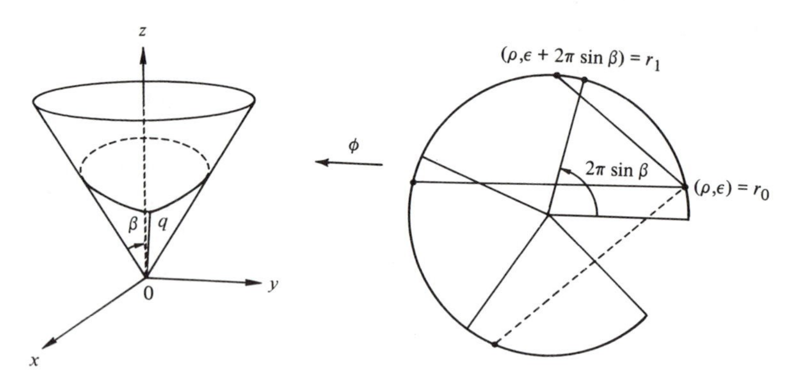

Figure 4-49

***c.** Under the conditions of part b, prove that there are exactly k such geodesics.

7. Let $\alpha: I \longrightarrow R^3$ be a parametrized regular curve. For each $t \in I$, let $P(t) \subset R^3$ be a plane through $\alpha(t)$ which contains $\alpha'(t)$. When the unit normal vector $N(t)$ of $P(t)$ is a differentiable function of t and $N'(t) \neq 0$, $t \in I$, we say that the map $t \longrightarrow \{\alpha(t), N(t)\}$ is a *differentiable family of tangent planes*. Given such a family, we determine a parametrized surface (cf. Def. 2, Sec. 2-3) by

$$\mathbf{x}(t, v) = \alpha(t) + v\frac{N(t) \wedge N'(t)}{|N'(t)|}.$$

The parametrized surface $\mathbf{x}$ is called the *envelope* of the family $\{\alpha(t), N(t)\}$ (cf. Example 4, Sec. 3-5).

a. Let S be an oriented surface and let $\gamma: I \longrightarrow S$ be a geodesic parametrized by arc length with $k(s) \neq 0$ and $\tau(s) \neq 0$, $s \in I$. Let $N(s)$ be the unit normal vector of S along γ. Prove that the envelope of the family of tangent planes $\{\gamma(s), N(s)\}$ is regular in a neighborhood of γ, has Gaussian curvature $K \equiv 0$, and is tangent to S along γ. (*Thus, we have obtained a surface locally isometric to the plane which contains γ as a geodesic.*)

b. Let $\alpha: I \longrightarrow R^3$ be a curve parametrized by arc length with $k(s) \neq 0$ and $\tau(s) \neq 0$, $s \in I$, and let $\{\alpha(s), n(s)\}$ be the family of its rectifying planes. Prove that the envelope of this family is regular in a neighborhood of α, has Gaussian curvature $K \equiv 0$, and contains α as a geodesic. (*Thus, every curve is a geodesic in the envelope of its rectifying planes; since this envelope is locally isometric to the plane, this justifies the name rectifying plane.*)

Appendix Proofs of the Fundamental Theorems of the Local Theory of Curves and Surfaces

In this appendix we shall show how the fundamental theorems of existence and uniqueness of curves and surfaces (Secs. 1-5 and 4-2) may be obtained from theorems on differential equations.

Proof of the Fundamental Theorem of the Local Theory of Curves (cf. statement in Sec. 1-5). The starting point is to observe that Frenet's equations

$$\begin{aligned}\frac{dt}{ds} &= kn,\\ \frac{dn}{ds} &= -kt - \tau b,\\ \frac{db}{ds} &= \tau n\end{aligned} \tag{1}$$

may be considered as a differential system in $I \times R^9$,

$$\left.\begin{aligned}\frac{d\xi_1}{ds} &= f_1(s, \xi_1, \dots, \xi_9)\\ &\vdots\\ \frac{d\xi_9}{ds} &= f_9(s, \xi_1, \dots, \xi_9)\end{aligned}\right\}, \qquad s \in I, \tag{1a}$$

where $(\xi_1, \xi_2, \xi_3) = t$, $(\xi_4, \xi_5, \xi_6) = n$, $(\xi_7, \xi_8, \xi_9) = b$, and f_i, $i = 1, \dots, 9$, are linear functions (with coefficients that depend on s) of the coordinates ξ_i.

In general, a differential system of type (1a) cannot be associated to a "steady" vector field (as in Sec. 3-4). At any rate, a theorem of existence and uniqueness holds in the following form:

Given initial conditions $s_0 \in I, (\xi_1)_0, \ldots, (\xi_9)_0$, *there exist an open interval* $J \subset I$ *containing* s_0 *and a unique differentiable mapping* $\alpha: J \to R^9$, *with*

$$\alpha(s_0) = ((\xi_1)_0, \ldots, (\xi_9)_0) \quad \textit{and} \quad \alpha'(s) = (f_1, \ldots, f_9),$$

where each f_i, $i = 1, \ldots, 9$, *is calculated in* $(s, \alpha(s)) \in J \times R^9$. *Furthermore, if the system is linear,* $J = I$ (cf. S. Lang, *Analysis* I, Addison-Wesley, Reading, Mass., 1968, pp. 383–386).

It follows that given an orthonormal, positively oriented trihedron $\{t_0, n_0, b_0\}$ in R^3 and a value $s_0 \in I$, there exists a family of trihedrons $\{t(s), n(s), b(s)\}$, $s \in I$, with $t(s_0) = t_0$, $n(s_0) = n_0$, $b(s_0) = b_0$.

We shall first show that the family $\{t(s), n(s), b(s)\}$ thus obtained remains orthonormal for every $s \in I$. In fact, by using the system (1) to express the derivatives relative to s of the six quantities

$$\langle t, n\rangle, \qquad \langle t, b\rangle, \qquad \langle n, b\rangle, \qquad \langle t, t\rangle, \qquad \langle n, n\rangle, \qquad \langle b, b\rangle$$

as functions of these same quantities, we obtain the system of differential equations

$$\frac{d}{ds}\langle t, n\rangle = k\langle n, n\rangle - k\langle t, t\rangle - \tau\langle t, b\rangle,$$

$$\frac{d}{ds}\langle t, b\rangle = k\langle n, b\rangle + \tau\langle t, n\rangle,$$

$$\frac{d}{ds}\langle n, b\rangle = -k\langle t, b\rangle - \tau\langle b, b\rangle + \tau\langle n, n\rangle,$$

$$\frac{d}{ds}\langle t, t\rangle = 2k\langle t, n\rangle,$$

$$\frac{d}{ds}\langle n, n\rangle = -2k\langle n, t\rangle - 2\tau\langle n, b\rangle,$$

$$\frac{d}{ds}\langle b, b\rangle = 2\tau\langle b, n\rangle.$$

It is easily checked that

$$\langle t, n\rangle \equiv 0, \qquad \langle b, b\rangle \equiv 0, \qquad \langle n, b\rangle \equiv 0,$$
$$t^2 \equiv 1, n^2 \equiv 1, b^2 \equiv 1,$$

is a solution of the above system with initial conditions 0, 0, 0, 1, 1, 1. By uniqueness, the family $\{t(s), n(s), b(s)\}$ is orthonormal for every $s \in I$, as we claimed.

From the family $\{t(s), n(s), b(s)\}$ it is possible to obtain a curve by setting

$$\alpha(s) = \int t(s)\, ds, \qquad s \in I,$$

where by the integral of a vector we understand the vector function obtained by integrating each component. It is clear that $\alpha'(s) = t(s)$ and that $\alpha''(s) = kn$. Therefore, $k(s)$ is the curvature of α at s. Moreover, since

$$\alpha'''(s) = k'n + kn' = k'n - k^2t - k\tau b,$$

the torsion of α will be given by (cf. Exercise 3, Sec. 1-5)

$$-\frac{\langle \alpha' \wedge \alpha'', \alpha''' \rangle}{k^2} = -\frac{\langle t \wedge kn, (-k^2t + k'n - k\tau b)\rangle}{k^2} = \tau;$$

α is, therefore, the required curve.

We still have to show that α is unique up to translations and rotations of R^3. Let $\bar{\alpha}\colon I \longrightarrow R^3$ be another curve with $\bar{k}(s) = k(s)$ and $\bar{\tau}(s) = \tau(s)$, $s \in I$, and let $\{\bar{t}_0, \bar{n}_0, \bar{b}_0\}$ be the Frenet trihedron of $\bar{\alpha}$ at s_0. It is clear that by a translation A and a rotation ρ it is possible to make the trihedron $\{\bar{t}_0, \bar{n}_0, \bar{b}_0\}$ coincide with the trihedron $\{t_0, n_0, b_0\}$ (both trihedrons are positive). By applying the uniqueness part of the above theorem on differential equations, we obtain the desired result. **Q.E.D.**

Proof of the Fundamental Theorem of the Local Theory of Surfaces (cf. statement in Sec. 4-3). The idea of the proof is the same as the one above; that is, we search for a family of trihedrons $\{\mathbf{x}_u, \mathbf{x}_v, N\}$, depending on u and v, which satisfies the system

$$\begin{aligned}
\mathbf{x}_{uu} &= \Gamma_{11}^1\mathbf{x}_u + \Gamma_{11}^2\mathbf{x}_v + eN,\\
\mathbf{x}_{uv} &= \Gamma_{12}^1\mathbf{x}_u + \Gamma_{12}^2\mathbf{x}_v + fN = \mathbf{x}_{vu},\\
\mathbf{x}_{vv} &= \Gamma_{22}^1\mathbf{x}_u + \Gamma_{22}^2\mathbf{x}_v + gN,\\
N_u &= a_{11}\mathbf{x}_u + a_{21}\mathbf{x}_v,\\
N_v &= a_{12}\mathbf{x}_u + a_{22}\mathbf{x}_v,
\end{aligned} \tag{2}$$

where the coefficients Γ_{ij}^k, a_{ij}, $i, j = 1, 2$, are obtained from E, F, G, e, f, g as if it were on a surface.

The above equations define a system of partial differential equations in $V \times R^9$,

$$\begin{aligned}
(\xi_1)_u &= f_1(u, v, \xi_1, \dots, \xi_9),\\
&\;\;\vdots\\
(\xi_9)_v &= f_{15}(u, v, \xi_1, \dots, \xi_9),
\end{aligned} \tag{2a}$$

where $\xi = (\xi_1, \xi_2, \xi_3) = \mathbf{x}_u$, $\eta = (\xi_4, \xi_5, \xi_6) = \mathbf{x}_v$, $\zeta = (\xi_7, \xi_8, \xi_9) = N$, and $f_i = 1, \ldots, 15$, are linear functions of the coordinates $\xi_j, j = 1, \ldots, 9$, with coefficients that depend on u and v.

In contrast to what happens with ordinary differential equations, a system of type (2a) is not integrable, in general. For the case in question, the conditions which guarantee the existence and uniqueness of a local solution, for given initial conditions, are

$$\xi_{uv} = \xi_{vu}, \qquad \eta_{uv} = \eta_{vu}, \qquad \zeta_{uv} = \zeta_{vu}.$$

A proof of this assertion is found in J. Stoker, *Differential Geometry*, Wiley-Interscience, New York, 1969, Appendix B.

As we have seen in Sec. 4-3, the conditions of integrability are equivalent to the equations of Gauss and Mainardi-Codazzi, which are, by hypothesis, satisfied. Therefore, the system (2a) is integrable.

Let $\{\xi, \eta, \zeta\}$ be a solution of (2a) defined in a neighborhood of $(u_0, v_0) \in V$, with the initial conditions $\xi(u_0, v_0) = \xi_0$, $\eta(u_0, v_0) = \eta_0$, $\zeta(u_0, v_0) = \zeta_0$. Clearly, it is possible to choose the initial conditions in such a way that

$$\begin{aligned} \xi_0^2 &= E(u_0, v_0), \\ \eta_0^2 &= G(u_0, v_0), \\ \langle \xi_0, \eta_0 \rangle &= F(u_0, v_0), \\ \zeta_0^2 &= 1, \\ \langle \xi_0, \zeta_0 \rangle &= \langle \eta_0, \zeta_0 \rangle = 0. \end{aligned} \tag{3}$$

With the given solution we form a new system,

$$\begin{aligned} \mathbf{x}_u &= \xi, \\ \mathbf{x}_v &= \eta, \end{aligned} \tag{4}$$

which is clearly integrable, since $\xi_v = \eta_u$. Let $\mathbf{x}\colon \bar{V} \to R^3$ be a solution of (4), defined in a neighborhood $\bar{V}$ of (u_0, v_0), with $\mathbf{x}(u_0, v_0) = p_0 \in R^3$. We shall show that by contracting $\bar{V}$ and interchanging v and u, if necessary, $\mathbf{x}(\bar{V})$ is the required surface.

We shall first show that the family $\{\xi, \eta, \zeta\}$, which is a solution of (2a), has the following property. For every (u, v) where the solution is defined, we have

$$\begin{aligned} \xi^2 &= E, \\ \eta^2 &= G, \\ \langle \xi, \eta \rangle &= F \\ \zeta^2 &= 1, \\ \langle \xi, \zeta \rangle &= \langle \eta, \zeta \rangle = 0. \end{aligned} \tag{5}$$

Indeed, by using (2) to express the partial derivatives of

$$\xi^2, \qquad \eta^2, \qquad \zeta^2, \qquad \langle \xi, \eta \rangle, \qquad \langle \xi, \zeta \rangle, \qquad \langle \eta, \zeta \rangle$$

as functions of these same 6 quantities, we obtain a system of 12 partial differential equations:

$$\begin{aligned}
(\xi^2)_u &= B_1(\xi^2, \eta^2, \ldots, \langle \eta, \zeta \rangle), \\
(\xi^2)_u &= B_2(\xi^2, \eta^2, \ldots, \langle \eta, \zeta \rangle), \\
&\vdots \\
\langle \eta, \zeta \rangle_v &= B_{12}(\xi^2, \eta^2, \ldots, \langle \eta, \zeta \rangle).
\end{aligned} \tag{6}$$

Since (6) was obtained from (2a), it is clear (and may be checked directly) that (6) is integrable and that

$$\begin{aligned}
\xi^2 &= E, \\
\eta^2 &= G, \\
\langle \eta, \xi \rangle &= F, \\
\zeta^2 &= 1, \\
(\xi, \zeta\rangle &= \langle \eta, \zeta \rangle = 0
\end{aligned}$$

is a solution of (6), with the initial conditions (3). By uniqueness, we obtain our claim.

It follows that

$$|\mathbf{x}_u \wedge \mathbf{x}_v|^2 = \mathbf{x}_u^2 \mathbf{x}_v^2 - \langle \mathbf{x}_u, \mathbf{x}_v \rangle^2 = EG - F^2 > 0.$$

Therefore, if $\mathbf{x}\colon \bar{V} \longrightarrow R^3$ is given by

$$\mathbf{x}(u, v) = (x(u, v), y(u, v), z(u, v)), \qquad (u, v) \in \bar{V},$$

one of the components of $\mathbf{x}_u \wedge \mathbf{x}_v$, say $\partial(x, y)/\partial(u, v)$, is different from zero in (u_0, v_0). Therefore, we may invert the system formed by the two first component functions of $\mathbf{x}$, in a neighborhood $U \subset \bar{V}$ of (u_0, v_0), to obtain a map $F(x, y) = (u, v)$. By restricting $\mathbf{x}$ to U, the mapping $\mathbf{x}\colon U \longrightarrow R^3$ is one-to-one, and its inverse $\mathbf{x}^{-1} = F \circ \pi$ (where π is the projection of R^3 on the xy plane) is continuous. Therefore, $\mathbf{x}\colon U \longrightarrow R^3$ is a differentiable homeomorphism with $\mathbf{x}_u \wedge \mathbf{x}_v \neq 0$; hence, $\mathbf{x}(U) \subset R^3$ is a regular surface.

From (5) it follows immediately that E, F, G are the coefficients of the first fundamental form of $\mathbf{x}(U)$ and that ζ is a unit vector normal to the surface. Interchanging v and u, if necessary, we obtain

$$\zeta = \frac{\mathbf{x}_u \wedge \mathbf{x}_v}{|\mathbf{x}_u \wedge \mathbf{x}_v|} = N.$$

From this, the coefficients of the second fundamental form of $\mathbf{x}(u, v)$ are computed by (2), yielding

$$\langle \zeta, \mathbf{x}_{uu} \rangle = e, \qquad \langle \zeta, \mathbf{x}_{uv} \rangle = f, \qquad \langle \zeta, \mathbf{x}_{vv} \rangle = g,$$

which shows that those coefficients are e, f, g and concludes the first part of the proof.

It remains to show that if U is connected, $\mathbf{x}$ is unique up to translations and rotations of R^3. To do this, let $\bar{\mathbf{x}}: U \longrightarrow R^3$ be another regular surface with $\bar{E} = E, \bar{F} = F, \bar{G} = G, \bar{e} = e, \bar{f} = f$, and $\bar{g} = g$. Since the first and second fundamental forms are equal, it is possible to bring the trihedron

$$\{\bar{\mathbf{x}}_u(u_0, v_0), \bar{\mathbf{x}}_v(u_0, v_0), \bar{N}(u_0, v_0)\}$$

into coincidence with the trihedron

$$\{\mathbf{x}_u(u_0, v_0), \mathbf{x}_v(u_0, v_0), N(u_0, v_0)\}$$

by means of a translation A and a rotation ρ.

The system (1a) is satisfied by the two solutions.

$$\begin{aligned} \xi &= \mathbf{x}_u, \qquad \eta = \mathbf{x}_v, \qquad \zeta = N; \\ \xi &= \bar{\mathbf{x}}_u, \qquad \eta = \bar{\mathbf{x}}_v, \qquad \zeta = \bar{N}. \end{aligned}$$

Since both solutions coincide in (u_0, v_0), we have by uniqueness that

$$\mathbf{x}_u = \bar{\mathbf{x}}_u, \qquad \mathbf{x}_v = \bar{\mathbf{x}}_v, \qquad N = \bar{N}, \tag{7}$$

in a neighborhood of (u_0, v_0). On the other hand, the subset of U where (7) holds is, by continuity, closed. Since U is connected, (7) holds for every $(u, v) \in U$.

From the first two equations of (7) and the fact that U is connected, we conclude that

$$\mathbf{x}(u, v) = \bar{\mathbf{x}}(u, v) + C,$$

where C is a constant vector. Since $\mathbf{x}(u_0, v_0) = \bar{\mathbf{x}}(u_0, v_0)$, we have that $C = 0$, which completes the proof of the theorem. **Q.E.D.**

5 Global Differential Geometry

5-1. Introduction

The goal of this chapter is to provide an introduction to global differential geometry. We have already met global theorems (the characterization of compact orientable surfaces in Sec. 2-7 and the Gauss-Bonnet theorem in Sec. 4-5 are some examples). However, they were more or less encountered in passing, our main task being to lay the foundations of the local theory of regular surfaces in R^3. Now, with that out of the way, we can start a more systematic study of global properties.

Global differential geometry deals with the relations between local and global (in general, topological) properties of curves and surfaces. We tried to minimize the requirements from topology by restricting ourselves to subsets of euclidean spaces. Only the most elementary properties of connected and compact subsets of euclidean spaces were used. For completeness, this material is presented with proofs in an appendix to Chap. 5.

In using this chapter, the reader can make a number of choices, and with this in mind, we shall now present a brief section-by-section description of the chapter. At the end of this introduction, a dependence table of the various sections will be given.

In Sec. 5-2 we shall prove that the sphere is rigid; that is, if a connected, compact, regular surface $S \subset R^3$ is isometric to a sphere, then S is a sphere. Except as a motivation for Sec. 5-3, this section is not used in the book.

In Sec. 5-3 we shall introduce the notion of a complete surface as a natural setting for global theorems. We shall prove the basic Hopf-Rinow theorem, which asserts the existence of a minimal geodesic joining any two points of a complete surface.

In Sec. 5-4 we shall derive the formulas for the first and second variations of arc length. As an application, we shall prove Bonnet's theorem: A complete surface with Gaussian curvature positive and bounded away from zero is compact.

In Sec. 5-5 we shall introduce the important notion of a Jacobi field along a geodesic γ which measures how rapidly the geodesics near γ pull away from γ. We shall prove that if the Gaussian curvature of a complete surface S is nonpositive, then $\exp_p: T_p(S) \longrightarrow S$ is a local diffeomorphism.

This raises the question of finding conditions for a local diffeomorphism to be a global diffeomorphism, which motivates the introduction of covering spaces in Sec. 5-6. Part A of Sec. 5-6 is entirely independent of the previous sections. In Part B we shall prove two theorems due to Hadamard: (1) If S is complete and simply connected and the Gaussian curvature of S is nonpositive, then S is diffeomorphic to a plane. (2) If S is compact and has positive Gaussian curvature, then the Gauss map $N: S \longrightarrow S^2$ is a diffeomorphism; in particular, S is diffeomorphic to a sphere.

In Sec. 5-7 we shall present some global theorems for curves. This section depends only on Part A of Sec. 5-6.

In Sec. 5-8 we shall prove that a complete surface in R^3 with vanishing Gaussian curvature is either a plane or a cylinder.

In Sec. 5-9 we shall prove the so-called Jacobi theorem: A geodesic arc is minimal relative to neighboring curves with the same end points if and only if such an arc contains no conjugate points.

In Sec. 5-10 we shall introduce the notion of abstract surface and extend to such surfaces the intrinsic geometry of Chap. 4. Except for the Exercises, this section is entirely independent of the previous sections. At the end of the section, we shall mention possible further generalizations, such as differentiable manifolds and Riemannian manifolds.

In Sec. 5-11 we shall prove Hilbert's theorem, which implies that there exists no complete regular surface in R^3 with constant negative Gaussian curvature.

In the accompanying diagram we present a dependence table of the sections of this chapter. For instance, for Sec. 5-11 one needs Secs. 5-3, 5-4, 5-5, 5-6, and 5-10; for Sec. 5-7, one needs Part A of Sec. 5-6; for Sec. 5-8 one needs Secs. 5-3, 5-4, and 5-5 and Part A of Sec. 5-6.

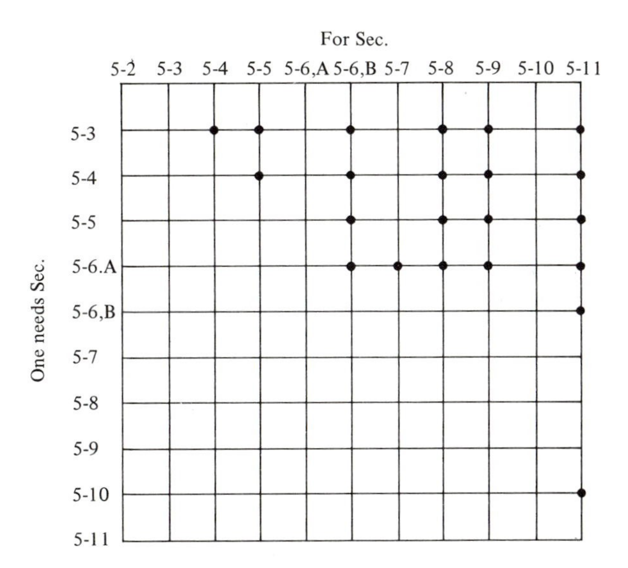

5-2. The Rigidity of the Sphere

It is perhaps convenient to begin with a typical, although simple, example of a global theorem. We choose the rigidity of the sphere.

We shall prove that the sphere is *rigid* in the following sense. Let $\varphi:\Sigma \longrightarrow S$ be an isometry of a sphere $\Sigma \subset R^3$ *onto* a regular surface $S = \varphi(\Sigma) \subset R^3$. Then S is a sphere. Intuitively, this means that it is not possible to deform a sphere made of a flexible but inelastic material.

Actually, we shall prove the following theorem.

THEOREM 1. *Let* S *be a compact, connected, regular surface with constant Gaussian curvature* K. *Then* S *is a sphere.*

The rigidity of the sphere follows immediately from Theorem 1. In fact, let $\varphi: \Sigma \longrightarrow S$ be an isometry of a sphere Σ onto S. Then $\varphi(\Sigma) = S$ has constant curvature, since the curvature is invariant under isometries. Furthermore, $\varphi(\Sigma) = S$ is compact and connected as a continuous image of the compact and connected set Σ (appendix to Chap. 5, Props. 6 and 12). It follows from Theorem 1 that S is a sphere.

The first proof of Theorem 1 is due to H. Liebmann (1899). The proof we shall present here is a modification by S. S. Chern of a proof given by D. Hilbert (S. S. Chern, "Some New Characterizations of the Euclidean

Sphere," *Duke Math. J.* 12 (1945), 270–290; and D. Hilbert, *Grundlagen der Geometrie*, 3rd ed., Leipzig, 1909, Appendix 5).

Remark 1. It should be noticed that there are surfaces homeomorphic to a sphere which are not rigid. An example is given in Fig. 5-1. We replace the plane region P of the surface S in Fig. 5-1 by a "bump" inwards so that the resulting surface S' is still regular. The surface S'' formed with the "symmetric bump" is isometric to S', but there is no linear orthogonal transformation that takes S' into S''. Thus, S' is not rigid.

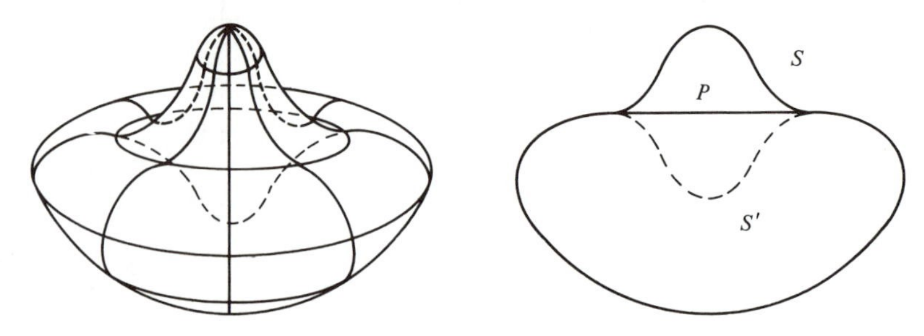

Figure 5-1

We recall the following convention. We choose the principal curvatures k_1 and k_2 so that $k_1(q) \geq k_2(q)$ for every $q \in S$. In this way we obtain k_1 and k_2 as continuous functions in S which are differentiable except, perhaps, at the umbilical points ($k_1 = k_2$) of S.

The proof of Theorem 1 is based on the following local lemma, for which we shall use the Mainardi-Codazzi equations (Sec. 4-3).

LEMMA 1. *Let* S *be a regular surface and* $\mathrm{p} \in \mathrm{S}$ *a point of* S *satisfying the following conditions:*

1. $\mathrm{K(p)} > 0$; *that is, the Gaussian curvature in* p *is positive.*
2. p *is simultaneously a point of local maximum for the function* k_1 *an a point of local minimum for the function* k_2 ($\mathrm{k}_1 \geq \mathrm{k}_2$).

Then p *is an umbilical point of* S.

Proof. Let us assume that p is not an umbilical point and obtain a contradiction.

If p is not an umbilical point of S, it is possible to parametrize a neighborhood of p by coordinates (u, v) so that the coordinate lines are lines of curvature (Sec. 3-4). In this situation, $F = f = 0$, and the principal curvatures are given by e/E, g/G. Since the point p is not umbilical, we may assume, by interchanging u and v if necessary, that in a neighborhood of p

$$k_1 = \frac{e}{E}, \qquad k_2 = \frac{g}{G}. \tag{1}$$

In the coordinate system thus obtained, the Mainardi-Codazzi equations are written as (Sec. 4-3, Eqs. (7) and (7a))

$$e_v = \frac{E_v}{2}(k_1 + k_2), \tag{2}$$

$$g_u = \frac{G_u}{2}(k_1 + k_2). \tag{3}$$

By differentiating the first equation of (1) with respect to v and using Eq. (2), we obtain

$$E(k_1)_v = \frac{E_v}{2}(-k_1 + k_2). \tag{4}$$

Similarly, by differentiating the second equation of (1) with respect to u and using Eq. (3),

$$G(k_2)_u = \frac{G_u}{2}(k_1 - k_2). \tag{5}$$

On the other hand, when $F = 0$, the Gauss formula for K reduces to (Sec. 4-3, Exercise 1)

$$K = -\frac{1}{2\sqrt{EG}}\left\{\left(\frac{E_v}{\sqrt{EG}}\right)_v + \left(\frac{G_u}{\sqrt{EG}}\right)_u\right\};$$

hence,

$$-2KEG = E_{vv} + G_{uu} + ME_v + NG_u, \tag{6}$$

where $M = M(u, v)$ and $N = N(u, v)$ are functions of (u, v), the expressions of which are immaterial for the proof. The same remark applies to $\bar{M}$, $\bar{N}$, $\tilde{M}$, and $\tilde{N}$, to be introduced below.

From Eqs. (4) and (5) we obtain expressions for E_v and G_u which, after being differentiated and introduced in Eq. (6), yield

$$-2KEG = -\frac{2E}{k_1 - k_2}(k_1)_{vv} + \frac{2G}{k_1 - k_2}(k_2)_{uu} + \bar{M}(k_1)_v + \bar{N}(k_2)_u;$$

hence,

$$-(k_1 - k_2)KEG = -2E(k_1)_{vv} + 2G(k_2)_{uu} + \tilde{M}(k_1)_v + \tilde{N}(k_2)_u. \tag{7}$$

Since $K > 0$ and $k_1 > k_2$ at p, the first member of Eq. (7) is strictly negative at p. Since k_1 reaches a local maximum at p and k_2 reaches a local minimum at p, we have

$$(k_1)_v = 0, \qquad (k_2)_u = 0, \qquad (k_1)_{vv} \leq 0, \qquad (k_2)_{uu} \geq 0$$

at p. However, this implies that the second member of Eq. (7) is positive or zero, which is a contradiction. This concludes the proof of Lemma 1. **Q.E.D.**

It should be observed that no contradiction arises in the proof if we assume that k_1 has a local *minimum* and k_2 has a local *maximum* at p. Actually, such a situation may happen on a surface of positive curvature without p being an umbilical point, as shown in the following example.

Example 1. Let S be a surface of revolution given by (cf. Sec. 3-3, Example 4)

$$x = \varphi(v) \cos u, \qquad y = \varphi(v) \sin u, \qquad z = \psi(v), \qquad 0 < u < 2\pi,$$

where

$$\varphi(v) = C \cos v, \qquad C > 1,$$
$$\psi(v) = \int \sqrt{1 - C^2 \sin^2 v}\, dv, \qquad \psi(0) = 0.$$

We take $|v| < \sin^{-1}(1/C)$, so that $\psi(v)$ is defined.

By using expressions already known (Sec. 3-3, Example 4), we obtain

$$E = C^2 \cos^2 v,$$
$$F = 0,$$
$$G = 1,$$
$$e = -C \cos v(\sqrt{1 - C^2 \sin^2 v}),$$
$$f = 0,$$
$$g = -\frac{C \cos v}{\sqrt{1 - C^2 \sin^2 v}};$$

hence,

$$k_1 = \frac{e}{E} = -\frac{\sqrt{1 - C^2 \sin^2 v}}{C \cos v}, \qquad k_2 = \frac{g}{G} = -\frac{C \cos v}{\sqrt{1 - C^2 \sin^2 v}}.$$

Therefore, S has curvature $K = k_1 k_2 = 1 > 0$, positive and constant (cf. Exercise 7, Sec. 3-3).

It is easily seen that $k_1 > k_2$ everywhere in S, since $C > 1$. Therefore, S has no umbilical points. Furthermore, since $k_1 = -(1/C)$ for $v = 0$, and

$$k_1 = -\frac{\sqrt{1 - C^2 \sin^2 v}}{C \cos v} > -\frac{1}{C} \qquad \text{for } v \neq 0,$$

we conclude that k_1 reaches a minimum (and therefore k_2 reaches a maximum, since $K = 1$) at the points of the parallel $v = 0$.

Incidentally, this example shows that the assumption of compactness in Theorem 1 is essential, since the surface S (see Fig. 5-2) has constant positive curvature but is not a sphere.

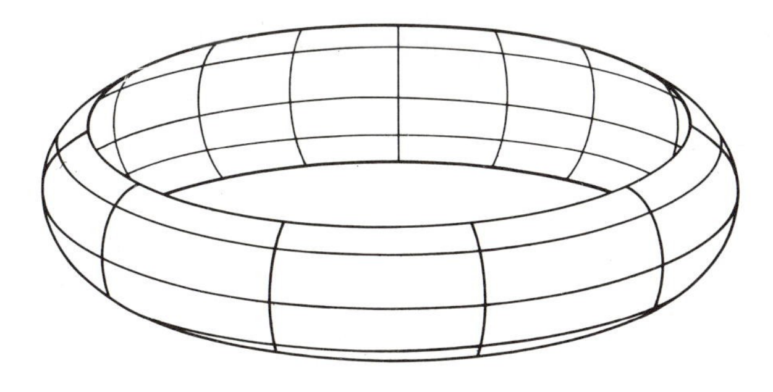

Figure 5-2

In the proof of Theorem 1 we shall use the following fact, which we establish as a lemma.

LEMMA 2. *A regular compact surface* $S \subset R^3$ *has at least one elliptic point.*

Proof. Since S is compact, S is bounded. Therefore, there are spheres of R^3, centered in a fixed point $O \in R^3$, such that S is contained in the interior of the region bounded by any of them. Consider the set of all such spheres. Let r be the infimum of their radii and let $\Sigma \subset R^3$ be a sphere of radius r centered in O. It is clear that Σ and S have at least one common point, say p. The tangent plane to Σ at p has only the common point p with S, in a neighborhood of p. Therefore, Σ and S are tangent at p. By observing the normal sections at p, it is easy to conclude that any normal curvature of S at p is greater than or equal to the corresponding curvature of Σ at p. Therefore, $K_S(p) \geq K_\Sigma(p) > 0$, and p is an elliptic point, as we wished. **Q.E.D.**

Proof of Theorem 1. Since S is compact, there is an elliptic point, by Lemma 2. Because K is constant, $K > 0$ in S.

By compactness, the continuous function k_1 on S reaches a maximum at a point $p \in S$ (appendix to Chap. 5, Prop. 13). Since $K = k_1 k_2$ is a positive constant, k_2 is a decreasing function of k_1, and, therefore, it reaches a minimum at p. It follows from Lemma 1 that p is an umbilical point; that is, $k_1(p) = k_2(p)$.

Now let q be any given point of S. Since we assumed $k_1(q) \geq k_2(q)$, we have that

$$k_1(p) \geq k_1(q) \geq k_2(q) \geq k_2(p) = k_1(p).$$

Therefore, $k_1(q) = k_2(q)$ for every $q \in S$.

It follows that all the points of S are umbilical points and, by Prop. 5 of Sec. 3-2, S is contained in a sphere or a plane. Since $K > 0$, S is contained in a sphere Σ. By compactness, S is closed in Σ, and since S is a regular surface, S is open in Σ. Since Σ is connected and S is open and closed in Σ, $S = \Sigma$ (appendix to Chap. 5, Prop. 5).

Therefore, the surface S is a sphere. **Q.E.D.**

Observe that in the proof of Theorem 1 the assumption that $K = k_1 k_2$ is constant is used only to guarantee that k_2 is a decreasing function of k_1. The same conclusion follows if we assume that the mean curvature $H = \frac{1}{2}(k_1 + k_2)$ is constant. This allows us to state

THEOREM 1a. *Let* S *be a regular, compact, and connected surface with Gaussian curvature* $K > 0$ *and mean curvature* H *constant. Then* S *is a sphere.*

The proof is entirely analogous to that of Theorem 1. Actually, the argument applies whenever $k_2 = f(k_1)$, where f is a decreasing function of k_1. More precisely, we have

THEOREM 1b. *Let* S *be a regular, compact, and connected surface of positive Gaussian curvature. If there exists a relation* $k_2 = f(k_1)$ *in* S, *where* f *is a decreasing function of* k_1, $k_1 \geq k_2$, *then* S *is a sphere.*

Remark 2. The compact, connected surfaces in R^3 for which the Gaussian curvature $K > 0$ are called *ovaloids*. Therefore Theorem 1a may be stated as follows: *An ovaloid of constant mean curvature is a sphere.*

On the other hand, it is a simple consequence of the Gauss-Bonnet theorem that an ovaloid is *homeomorphic* to a sphere (cf. Sec. 4-5, application 1). H. Hopf proved that Theorem 1a still holds with the following (stronger) statement: *A regular surface of constant mean curvature that is homemorphic to a sphere is a sphere.* A theorem due to A. Alexandroff extends this result further by replacing the condition of being homeomorphic to a sphere by compactness: *A regular, compact, and connected surface of constant mean curvature is a sphere.*

An exposition of the above-mentioned results can be found in Hopf [11]. (References are listed at the end of the book.)

Remark 3. The rigidity of the sphere may be obtained as a consequence of a general theorem of rigidity on ovaloids. This theorem, due to Cohn-Vossen, states the following: *Two isometric ovaloids differ by an orthogonal linear transformation of* R^3. A proof of this result may be found in Chern [10].

Theorem 1 is a typical result of global differential geometry, that is, information on local entities (in this case, the curvature) together with weak global hypotheses (in this case, compactness and connectedness) imply strong restrictions on the entire surface (in this case, being a sphere). Observe that the only effect of the connectedness is to prevent the occurrence of two or more spheres in the conclusion of Theorem 1. On the other hand, the hypothesis of compactness is essential in several ways, one of its functions being to ensure that we obtain an entire sphere and not a surface contained in a sphere.

EXERCISES

1. Let $S \subset R^3$ be a compact regular surface and fix a point $p_0 \in R^3$, $p_0 \notin S$. Let $d: S \longrightarrow R$ be the differentiable function defined by $d(q) = \frac{1}{2}|q - p_0|^2$, $q \in S$. Since S is compact, there exists $q_0 \in S$ such that $d(q_0) \geq d(q)$ for all $q \in S$. Prove that q_0 is an elliptic point of S (*this gives another proof of Lemma 1*).

2. Let $S \subset R^3$ be a regular surface with Gaussian curvature $K > 0$ and without umbilical points. Prove that there exists no point on S where H is a maximum and K is a minimum.

3. (*Kazdan-Warner's Remark.*) Let $S \subset R^3$ be an extended compact surface of revolution (cf. Remark 4, Sec. 2-3) obtained by rotating the curve

$$\alpha(s) = (0, \varphi(s), \psi(s)),$$

parametrized by arc length $s \in [0, l]$, about the z axis. Here $\varphi(0) = \varphi(l) = 0$ and $\varphi(s) > 0$ for all $s \in [0, l]$. The regularity of S at the poles implies further that $\varphi'(0) = 1$, $\varphi'(l) = -1$ (cf. Exercise 10, Sec. 2-3). We also know that the Gaussian curvature of S is given by $K = -\varphi''(s)/\varphi(s)$ (cf. Example 4, Sec. 3-3).

***a.** Prove that

$$\int_0^l K'\varphi^2\, ds = 0, \qquad K' = \frac{dK}{ds}.$$

b. Conclude from part a that *there exists no compact* (*extended*) *surface of revolution in* R^3 *with monotonic increasing curvature.*

The following exercise outlines a proof of Hopf's theorem: *A regular surface with constant mean curvature which is homeomorphic to a sphere is a sphere* (cf. Remark 2). Hopf's main idea has been used over and over again in recent work. The exercise requires some elementary facts on functions of complex variables.

4. Let $U \subset R^3$ be an open connected subset of R^2 and let $\mathbf{x}: U \longrightarrow S$ be an isothermal parametrization (i.e., $E = G$, $F = 0$; cf. Sec. 4-2) of a regular surface S. We identify R^2 with the complex plane $\mathbb{C}$ by setting $u + iv = \zeta$, $(u, v) \in R^2$, $\zeta \in \mathbb{C}$. ζ is called the *complex parameter* corresponding to $\mathbf{x}$. Let $\phi: \mathbf{x}(U) \longrightarrow \mathbb{C}$ be the complex-valued function given by

$$\phi(\zeta) = \phi(u, v) = \frac{e - g}{2} - if = \phi_1 + i\phi_2,$$

where e, f, g are the coefficients of the second fundamental form of S.

a. Show that the Mainardi-Codazzi equations (cf. Sec. 4-3) can be written, in the isothermal parametrization $\mathbf{x}$, as

$$\left(\frac{e - g}{2}\right)_u + f_v = EH_u, \qquad \left(\frac{e - g}{2}\right)_v - f_u = -EH_v$$

and conclude that the mean curvature H of $\mathbf{x}(U) \subset S$ is constant if and only if ϕ is an analytic function of ζ (i.e., $(\phi_1)_u = (\phi_2)_v$, $(\phi_1)_v = -(\phi_2)_u$).

b. Define the "complex derivative"

$$\frac{\partial}{\partial \zeta} = \frac{1}{2}\left(\frac{\partial}{\partial u} - i\frac{\partial}{\partial v}\right),$$

and prove that $\phi(\zeta) = -2\langle \mathbf{x}_\zeta, N_\zeta \rangle$, where by $\mathbf{x}_\zeta$, for instance, we mean the vector with complex coordinates

$$\mathbf{x}_\zeta = \left(\frac{\partial x}{\partial \zeta}, \frac{\partial y}{\partial \zeta}, \frac{\partial z}{\partial \zeta}\right).$$

c. Let $f: U \subset \mathbb{C} \longrightarrow V \subset \mathbb{C}$ be a one-to-one complex function given by $f(u + iv) = x + iy = \eta$. Show that (x, y) are isothermal parameters on S (i.e., η is a complex parameter on S) if and only if f is analytic and $f'(\zeta) \neq 0$, $\zeta \in U$. Let $\mathbf{y} = \mathbf{x} \circ f^{-1}$ be the corresponding parametrization and define $\psi(\eta) = -2\langle \mathbf{y}_\eta, N_\eta \rangle$. Show that on $\mathbf{x}(U) \cap \mathbf{y}(V)$,

$$\phi(\zeta) = \psi(\eta)\left(\frac{d\eta}{d\zeta}\right)^2. \qquad (*)$$

d. Let S^2 be the unit sphere of R^3. Use the stereographic projection (cf. Exercise 16, Sec. 2-2) from the poles $N = (0, 0, 1)$ and $S = (0, 0, -1)$ to cover S^2 by the coordinate neighborhoods of two (isothermal) complex parameters, ζ and η, with $\zeta(S) = 0$ and $\eta(N) = 0$, in such a way that in the intersection W of these coordinate neighborhoods (the sphere minus the two poles) $\eta = \zeta^{-1}$. Assume that there exists on each coordinate neighborhood analytic functions $\varphi(\zeta)$, $\psi(\eta)$ such that $(*)$ holds in W. Use Liouville's theorem to prove that $\varphi(\zeta) \equiv 0$ (hence, $\psi(\eta) \equiv 0$).

e. Let $S \subset R^3$ be a regular surface with constant mean curvature homeomorphic to a sphere. Assume that there exists a conformal diffeomorphism $\varphi: S \longrightarrow S^2$ of S onto the unit sphere S^2 (this is a consequence of the uniformization theorem for Riemann surfaces and will be assumed here). Let $\tilde{\zeta}$ and $\tilde{\eta}$ be the complex parameters corresponding under φ to the parameters ζ and η of S^2 given in part d. By part a, the function $\phi(\tilde{\zeta}) = ((e - g)/2) - if$ is analytic. The similar function $\psi(\tilde{\eta})$ is also analytic, and by part c they are related by $(*)$. Use part d to show that $\phi(\tilde{\zeta}) \equiv 0$ (hence, $\psi(\tilde{\eta}) \equiv 0$). Conclude that S is made up of umbilical points and hence is a sphere. This proves Hopf's theorem.

5-3. Complete Surfaces. Theorem of Hopf-Rinow

All the surfaces to be considered from now on will be regular and connected, except when otherwise stated.

The considerations at the end of Sec. 5-1 have shown that in order to obtain global theorems we require, besides the connectedness, some global hypothesis to ensure that the surface cannot be "extended" further as a regular surface. It is clear that the compactness serves this purpose. However, it would be useful to have a hypothesis weaker than compactness which could still have the same effect. That would allow us to expect global theorems in a more general situation than that of compactness.

A more precise formulation of the concept that a surface cannot be extended is given in the following definition.

DEFINITION 1. *A regular (connected) surface* S *is said to be* extendable *if there exists a regular (connected) surface* $\bar{S}$ *such that* $S \subset \bar{S}$ *as a proper subset. If there exists no such* $\bar{S}$, S *said to be* nonextendable.

Unfortunately, the class of nonextendable surfaces is much too large to allow interesting results. A more adequate hypothesis is given by

DEFINITION 2. *A regular surface* S *is said to be* complete *when for every point* $p \in S$, *any parametrized geodesic* $\gamma: [0, \epsilon) \longrightarrow S$ *of* S, *starting from* $p = \gamma(0)$, *may be extended into a parametrized geodesic* $\bar{\gamma}: R \longrightarrow S$, *defined on the entire line* R.

In other words, S *is complete when for every* $p \in S$ *the mapping* $\exp_p: T_p(S) \longrightarrow S$ *(Sec. 4-6) is defined for every* $v \in T_p(S)$.

We shall prove later (Prop. 1) that every complete surface is nonextendable and that there exist nonextendable surfaces which are not complete (Example 1). Therefore, the hypothesis of completeness is stronger than that of nonextendability. Furthermore, we shall prove (Prop. 5) that every closed surface in R^3 is complete; that is, the hypothesis of completeness is weaker than that of compactness.

The object of this section is to prove that given two points $p, q \in S$ of a complete surface S there exists a geodesic joining p to q which is minimal (that is, its length is smaller than or equal to that of any other curve joining p to q). This fundamental result was first proved by Hopf and Rinow (H. Hopf, W. Rinow, "Über den Begriff der vollständigen differentialgeometrischen Flächen," *Comm. Math. Helv.* 3 (1931), 209–225). This theorem is the main reason the complete surfaces are more adequate for differential geometry than the nonextendable ones.

Let us now look at some examples. The plane is clearly a complete surface. The cone minus the vertex is a noncomplete surface, since by extending a generator (which is a geodesic) sufficiently we reach the vertex, which does not belong to the surface. A sphere is a complete surface, since its parametrized geodesics (the traces of which are the great circles of the sphere) may be defined for every real value. The cylinder is also a complete surface since its geodesics are circles, lines, and helices, which are defined for all real values.

On the other hand, a surface $S - \{p\}$ obtained by removing a point p from a complete surface S is not complete. In fact, a geodesic γ of S should pass through p. By taking a point q, nearby p on γ (Fig. 5-3), there exists a parametrized geodesic of $S - \{p\}$ that starts from q and cannot be extended through p (this argument will be given in detail in Prop. 1). Thus, a sphere minus a point and a cylinder minus a point are not complete surfaces.

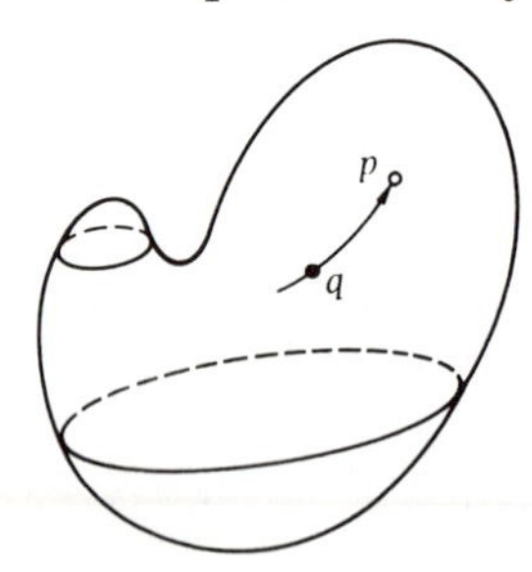

Figure 5-3

PROPOSITION 1. *A complete surface* S *is nonextendable.*

Proof. Let us assume that S is extendable and obtain a contradiction. To say that S is extendable means that there exists a regular (connected) surface $\bar{S}$ with $S \subset \bar{S}$. Since S is a regular surface, S is open in $\bar{S}$. The boundary (appendix to Chap. 5, Def. 4) Bd S of S in $\bar{S}$ is nonempty; otherwise $\bar{S} = S \cup (\bar{S} - S)$ would be the union of two disjoint open sets S and $\bar{S} - S$, which contradicts the connectedness of $\bar{S}$ (appendix to Chap. 5, Def. 10).

Therefore, there exists a point $p \in \text{Bd } S$, and since S is open in $\bar{S}$, $p \notin S$.

Let $\bar{V} \subset \bar{S}$ be a neighborhood of p in $\bar{S}$ such that every $q \in \bar{V}$ may be joined to p by a unique geodesic of $\bar{S}$ (Sec. 4-6, Prop. 2). Since $p \in \text{Bd } S$, some $q_0 \in \bar{V}$ belongs to S. Let $\bar{\gamma}: [0, 1] \longrightarrow \bar{S}$ be a geodesic of $\bar{S}$, with $\bar{\gamma}(0) = p$ and $\bar{\gamma}(1) = q_0$. It is clear that $\alpha: [0, \epsilon) \longrightarrow S$, given by $\alpha(t) = \bar{\gamma}(1 - t)$, is a geodesic of S, with $\alpha(0) = q_0$, the extension of which to the line R would pass through p for $t = 1$ (Fig. 5-4). Since $p \notin S$, this geodesic cannot be extended, which contradicts the hypothesis of completeness and concludes the proof. **Q.E.D.**

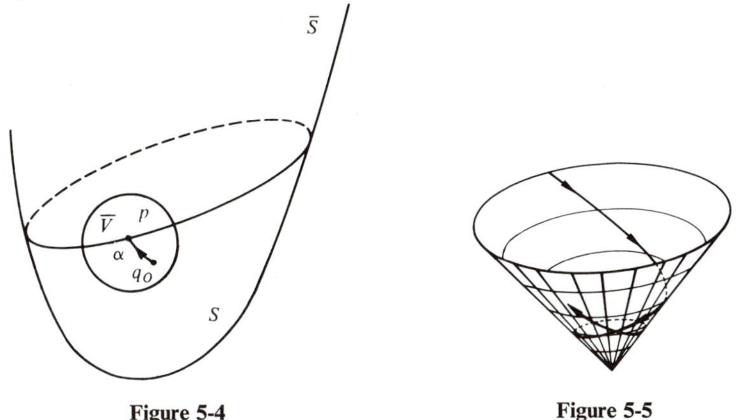

Figure 5-4 **Figure 5-5**

The converse of Prop. 1 is false, as shown in the following example.

Example 1. When we remove the vertex p_0 from the one-sheeted cone given by

$$z = \sqrt{x^2 + y^2}, \qquad (x, y) \in R^2,$$

we obtain a regular surface S. S is not complete since the generators cannot be extended for every value of the arc length without reaching the vertex.

Let us show that S is nonextendable by assuming that $S \subset \bar{S}$, where $\bar{S} \neq S$ is a regular surface, and by obtaining a contradiction. The argument consists of showing that the boundary of S in $\bar{S}$ reduces to the vertex p_0 and that there exists a neighborhood $\bar{W}$ of p_0 in $\bar{S}$ such that $\bar{W} - \{p_0\} \subset S$. But this contradicts the fact that the cone (vertex p_0 included) is not a regular surface in p_0 (Sec. 2-2, Example 5).

First, we observe that the only geodesic of S, starting from a point $p \in S$, that cannot be extended for every value of the parameter is the meridian (generator) that passes through p (see Fig. 5-5). This fact may easily be seen by using, for example, Clairaut's relation (Sec. 4-4, Example 5) and will be left as an exercise (Exercise 2).

Now let $p \in \text{Bd } S$, where Bd S denotes the boundary of S in $\bar{S}$ (as we have seen in Prop. 1, Bd $S \neq \phi$). Since S is an open set in $\bar{S}$, $p \notin S$. Let $\bar{V}$ be a neighborhood of p in $\bar{S}$ such that every point of $\bar{V}$ may be joined to p by a unique geodesic of $\bar{S}$ in $\bar{V}$. Since $p \in \text{Bd } S$, there exists $q \in \bar{V} \cap S$. Let $\bar{\gamma}$ be a geodesic of $\bar{S}$ joining p to q. Because S is an open set in $\bar{S}$, $\bar{\gamma}$ agrees with a geodesic γ of S in a neighborhood of q. Let p_0 be the first point of $\bar{\gamma}$ that does not belong to S. By the initial observation, $\bar{\gamma}$ is a meridian and p_0 is the vertex of S. Furthermore, $p_0 = p$; otherwise there would exist a neighborhood of p that does not contain p_0. By repeating the argument for that neighborhood, we obtain a vertex different from p_0, which is a contradiction. It follows that Bd S reduces to the vertex p_0.

Now let $\bar{W}$ be a neighborhood of p_0 in $\bar{S}$ such that any two points of $\bar{W}$ may be joined by a geodesic of $\bar{S}$ (Sec. 4-7, Prop. 1). We shall prove that $\bar{W} - \{p_0\} \subset S$. In fact, the points of γ belong to S. On the other hand, a point $r \in \bar{W}$ which does not belong to γ or to its extension may be joined to a point t of γ, $t \neq p_0$, $t \in \bar{W}$, by a geodesic α, different from γ (see Fig. 5-6). By the initial observation, every point of α, in particular r, belongs to S. Finally, the points of the extension of γ, except p_0, also belong to S; otherwise, they would belong to the boundary of S which we have proved to be made up only of p_0.

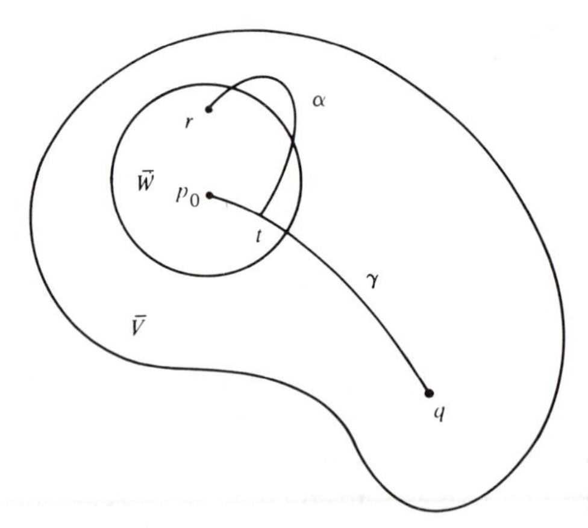

Figure 5-6

In this way, our assertions are completely proved. Thus, S is nonextendable and the desired example is obtained.

For what follows, it is convenient to introduce a notion of distance between two points of S which depends only on the intrinsic geometry of S and not on the way S is immersed in R^3 (cf. Remark 1, Sec. 4-2). Observe that, since $S \subset R^3$, it is possible to define a distance between two points of S as the distance between these two points in R^3. However, this distance depends on the second fundamental form, and, thus, it is not adequate for the purposes of this chapter.

We need some preliminaries.

A continuous mapping $\alpha: [a, b] \longrightarrow S$ of a closed interval $[a, b] \subset R$ of the line R onto the surface S is said to be a *parametrized, piecewise differentiable curve* joining $\alpha(a)$ to $\alpha(b)$ if there exists a partition of $[a, b]$ by points $a = t_0 < t_1 < t_2 < \cdots < t_k < t_{k+1} = b$ such that α is differentiable in $[t_i, t_{i+1}]$, $i = 0, \ldots, k$. The length $l(\alpha)$ of α is defined as

$$l(\alpha) = \sum_{i=0}^{k} \int_{t_i}^{t_{i+1}} |\alpha'(t)|\, dt.$$

PROPOSITION 2. *Given two points* p, q $\in$ S *of a regular* (*connected*) *surface* S, *there exists a parametried piecewise differentiable curve joining* p *to* q.

Proof: Since S is connected, there exists a continuous curve $\alpha: [a, b] \longrightarrow S$ with $\alpha(a) = p$, $\alpha(b) = q$. Let $t \in [a, b]$ and let I_t be an open interval in $[a, b]$, containing t, such that $\alpha(I_t)$ is contained in a coordinate neighborhood of $\alpha(t)$. The union $\cup I_t$, $t \in [a, b]$, covers $[a, b]$ and, by compactness, a finite number $I_1, \ldots, I_n$ still covers $[a, b]$. Therefore, it is possible to decompose I by points $a = t_0 < t_1 < \cdots < t_k < t_{k+1} = b$ in such a way that $[t_i, t_{i+1}]$ is contained in some I_j, $j = 1, \ldots, n$. Thus, $\alpha(t_i, t_{i+1})$ is contained in a coordinate neighborhood.

Since $p = \alpha(t_0)$ and $\alpha(t_1)$ lie in a same coordinate neighborhood $\mathbf{x}(U) \subset S$, it is possible to join them by a differentiable curve, namely, the image by $\mathbf{x}$ of a differentiable curve in $U \subset R^2$ joining $\mathbf{x}^{-1}(\alpha(t_0))$ to $\mathbf{x}^{-1}(\alpha(t_1))$. By this process, we join $\alpha(t_i)$ to $\alpha(t_{i+1})$, $i = 0, \ldots, k$, by a differentiable curve. This gives a piecewise differentiable curve, joining $p = \alpha(t_0)$ and $q = \alpha(t_{k+1})$, and concludes the proof of the proposition. **Q.E.D.**

Now let $p, q \in S$ be two points of a regular surface S. We denote by $\alpha_{p,q}$ a parametrized, piecewise differentiable curve joining p to q and by $l(\alpha_{p,q})$ its length. Proposition 2 shows that the set of all such $\alpha_{p,q}$ is not empty. Thus, we can set the following:

DEFINITION 3. *The* (intrinsic) distance d(p, q) *from the point* p $\in$ S *to the point* q $\in$ S *is the number*

$$\mathrm{d}(\mathrm{p}, \mathrm{q}) = \mathit{inf}\, \mathrm{l}(\alpha_{\mathrm{p},\mathrm{q}}),$$

where the inf *is taken over all piecewise differentiable curves joining* p *to* q.

PROPOSITION 3. *The distance* d *defined above has the following properties,*

1. $d(p, q) = d(q, p)$,
2. $d(p, q) + d(q, r) \geq d(p, r)$,
3. $d(p, q) \geq 0$,
4. $d(p, q) = 0$ *if and only if* $p = q$,

where p, q, r *are arbitrary points of* S.

Proof. Property 1 is immediate, since each parametrized curve

$$\alpha: [a, b] \longrightarrow S,$$

with $\alpha(a) = p$, $\alpha(b) = q$, gives rise to a parametrized curve $\tilde{\alpha}: [a, b] \longrightarrow S$, defined by $\tilde{\alpha}(t) = \alpha(a - t + b)$. It is clear that $\tilde{\alpha}(a) = q$, $\tilde{\alpha}(b) = p$, and $l(\alpha_{p,q}) = l(\tilde{\alpha}_{p,q})$.

Property 2 follows from the fact that when A and B are sets of real numbers and $A \subseteq B$ then $\inf A \geq \inf B$.

Property 3 follows from the fact that the infimum of positive numbers is positive or zero.

Let us now prove property 4. Let $p = q$. Then, by taking the constant curve $\alpha: [a, b] \longrightarrow S$, given by $\alpha(t) = p$, $t \in [a, b]$, we get $l(\alpha) = 0$; hence, $d(p, q) = 0$.

To prove that $d(p, q) = 0$ implies that $p = q$ we proceed as follows. Let us assume that $d(p, q) = \inf l(\alpha_{p,q}) = 0$ and $p \neq q$. Let V be a neighborhood of p in S, with $q \notin V$, and such that every point of V may be joined to p by a unique geodesic in V. Let $B_r(p) \subset V$ be the region bounded by a geodesic circle of radius r, centered in p, and contained in V. By definition of infimum, given $\epsilon > 0$, $0 < \epsilon < r$, there exists a parametrized, piecewise differentiable curve $\alpha: [a, b] \longrightarrow S$ joining p to q and with $l(\alpha) < \epsilon$. Since $\alpha([a, b])$ is connected and $q \notin B_r$, there exists a point $t_0 \in [a, b]$ such that $\alpha(t_0)$ belongs to the boundary of $B_r(p)$. It follows that $l(\alpha) \geq r > \epsilon$, which is a contradiction. Therefore, $p = q$, and this concludes the proof of the proposition. **Q.E.D.**

COROLLARY. $|d(p, r) - d(r, q)| \leq d(p, q)$.

It suffices to observe that

$$d(p, r) \leq d(p, q) + d(q, r),$$
$$d(r, q) \leq d(r, p) + d(p, q);$$

hence,

$$-d(p, q) \leq d(p, r) - d(r, q) \leq d(p, q).$$

PROPOSITION 4. *If we let* $p_0 \in S$ *be a point of* S, *then the function* $f: S \longrightarrow R$ *given by* $f(p) = d(p_0, p)$, $p \in S$, *is continuous on* S.

Proof. We have to show that for each $p \in S$, given $\epsilon > 0$, there exists $\delta > 0$ such that if $q \in B_\delta(p) \cap S$, where $B_\delta(p) \subset R^3$ is an open ball of R^3 centered at p and of radius δ, then $|f(p) - f(q)| = |d(p_0, p) - d(p_0, q)| < \epsilon$.

Let $\epsilon' < \epsilon$ be such that the exponential map $\exp_p = T_p(S) \to S$ is a diffeomorphism in the disk $B_{\epsilon'}(0) \subset T_p(S)$, where 0 is the origin of $T_p(S)$, and set $\exp(B_{\epsilon'}(0)) = V$. Clearly, V is an open set in S; hence, there exists an open ball $B_\delta(p)$ in R^3 such that $B_\delta(p) \cap S \subset V$. Thus, if $q \in B_\delta(p) \cap S$,

$$|d(p_0, p) - d(p_0, q)| \leq d(p, q) < \epsilon' < \epsilon,$$

which completes the proof. **Q.E.D.**

Remark 1. The readers with an elementary knowledge of topology will notice that Prop. 3 shows that the function $d: S \times S \to R$ gives S the structure of a metric space. On the other hand, as a subset of a metric space, $S \subset R^3$ has an induced metric $\bar{d}$. It is an important fact that these two metrics determine the same topology, that is, the same family of open sets in S. This follows from the fact that $\exp_p: U \subset T_p(S) \to S$ is a local diffeomorphism, and its proof is analogous to that of Prop. 4.

Having finished the preliminaries, we may now make the following observation.

PROPOSITION 5. *A closed surface* $S \subset R^3$ *is complete.*

Proof. Let $\gamma: [0, \epsilon) \to S$ be a parametrized geodesic of S, $\gamma(0) = p \in S$, which we may assume, without loss of generality, to be parametrized by arc length. We need to show that it is possible to extend γ to a geodesic $\bar{\gamma}: R \to S$, defined on the entire line R. Observe first that when $\bar{\gamma}(s_0)$, $s_0 \in R$, is defined, then, by the theorem of existence and uniqueness of geodesics (Sec. 4-4, Prop. 5), it is possible to extend $\bar{\gamma}$ to a neighborhood of s_0 in R. Therefore, the set of all $s \in R$ where $\bar{\gamma}$ is defined is open in R. If we can prove that this set is closed in R (which is connected), it will be possible to define $\bar{\gamma}$ for all of R, and the proof will be completed.

Let us assume that $\bar{\gamma}$ is defined for $s < s_0$ and let us show that $\bar{\gamma}$ is defined for $s = s_0$. Consider a sequence $\{s_n\} \to s_0$, with $s_n < s_0$, $n = 1, 2, \ldots$.

We shall first prove that the sequence $\{\bar{\gamma}(s_n)\}$ converges in S. In fact, given $\epsilon > 0$, there exists n_0 such that if n, $m > n_0$, then $|s_n - s_m| < \epsilon$. Denote by $\bar{d}$ the distance in R^3, and observe that if p, $q \in S$, then $\bar{d}(p, q) \leq d(p, q)$. Thus,

$$\bar{d}(\bar{\gamma}(s_n), \bar{\gamma}(s_m)) \leq d(\bar{\gamma}(s_n), \bar{\gamma}(s_m)) \leq |s_n - s_m| < \epsilon,$$

where the second inequality comes from the definition of d and the fact that $|s_n - s_m|$ is equal to the arc length of the curve $\bar{\gamma}$ between s_n and s_m. It follows that $\{\bar{\gamma}(s_n)\}$ is a Cauchy sequence in R^3; hence, it converges to a point $q \in R^3$ (appendix to Chap. 5, Prop. 4). Since q is a limit point of $\{\bar{\gamma}(s_n)\}$ and S is closed, $q \in S$, which proves our assertion.

Now let W and δ be the neighborhood of q and the number given by Prop. 1 of Sec. 4-7. Let $\bar{\gamma}(s_n), \bar{\gamma}(s_m) \in W$ be points such that $|s_n - s_m| < \delta$, and let γ be the unique geodesic with $l(\gamma) < \delta$ joining $\bar{\gamma}(s_n)$ to $\bar{\gamma}(s_m)$. Clearly, $\bar{\gamma}$ agrees with γ. Since $\exp_{\bar{\gamma}(s_n)}$ is a diffeomorphism in $B_\delta(0)$ and $\exp_{\bar{\gamma}(s_n)}(B_\delta(0)) \supset W$, γ extends $\bar{\gamma}$ beyond q. Thus, $\bar{\gamma}$ is defined at $s = s_0$, which completes the proof. **Q.E.D.**

COROLLARY. *A compact surface is complete.*

Remark 2. The converse of Prop. 5 does not hold. For instance, a right cylinder erected over a plane curve that is asymptotic to a circle is easily seen to be complete but not closed (Fig. 5-7).

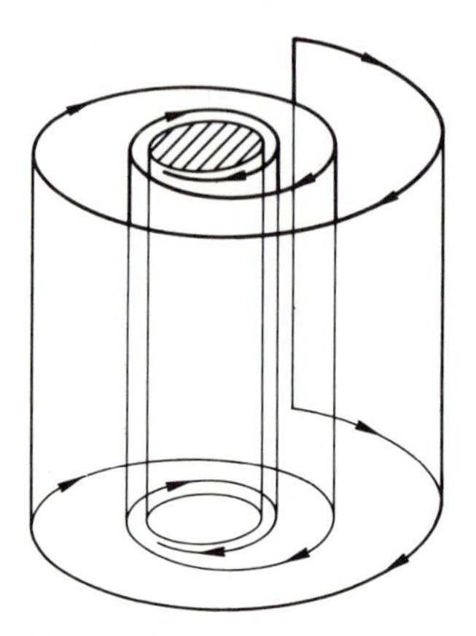

Figure 5-7. A closed noncomplete surface.

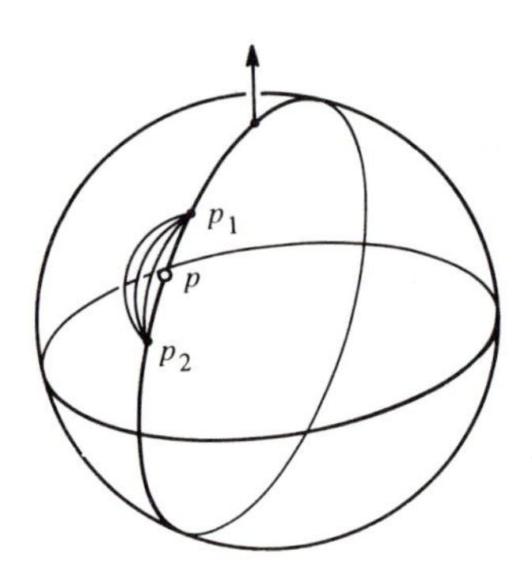

Figure 5-8

We say that a geodesic γ joining two points $p, q \in S$ is *minimal* if its length $l(\gamma)$ is smaller than or equal to the length of any piecewise regular curve joining p to q (cf. Sec. 4-7). This is equivalent to saying that $l(\gamma) = d(p, q)$, since, given a piecewise differentiable curve α joining p to q, we can find a piecewise regular curve joining p to q that is shorter (or at least not longer) than α. The proof of the last assertion is left as an exercise.

Observe that a minimal geodesic may not exist, as shown in the following example.

Let $S^2 - \{p\}$ be the surface formed by the sphere S^2 minus the point $p \in S^2$. By taking, on the meridian that passes through p, two points p_1 and p_2, symmetric relative to p and sufficiently near to p, we see that there exists no minimal geodesic joining p_1 to p_2 in the surface $S^2 - \{p\}$ (see Fig. 5-8).

On the other hand, there may exist an infinite number of minimal geodesics joining two points of a surface, as happens, for example, with two antipodal points of a sphere; all the meridians that join these antipodal points are minimal geodesics.

The main result of this section is that in a complete surface there always exists a minimal geodesic joining two given points.

THEOREM (Hopf-Rinow). *Let* S *be a complete surface. Given two points* p, q ∈ S, *there exists a minimal geodesic joining* p *to* q.

Proof. Let $r = d(p, q)$ be the distance between the points p and q. Let $B_\delta(0) \in T_p(S)$ be a disk of radius δ, centered in the origin 0 of the tangent plane $T_p(S)$ and contained in a neighborhood $U \subset T_p(S)$ of 0, where $\exp_p$ is a diffeomorphism. Let $B_\delta(p) = \exp_p(B_\delta(0))$. Observe that the boundary Bd $B_\delta(p) = \Sigma$ is compact since it is the continuous image of the compact set Bd $B_\delta(0) \subset T_p(S)$.

If $x \in \Sigma$, the continuous function $d(x, q)$ reaches a minimum at a point x_0 of the compact set Σ. The point x_0 may be written as

$$x_0 = \exp_p(\delta v), \qquad |v| = 1, v \in T_p(S).$$

Let γ be the geodesic parametrized by arc length, given by (see Fig. 5-9)

$$\gamma(s) = \exp_p(sv).$$

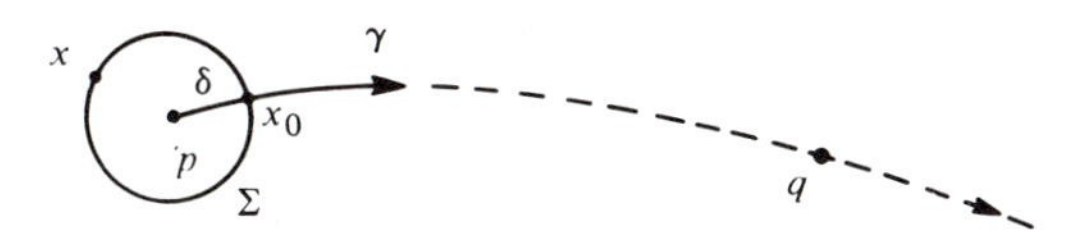

Figure 5-9

Since S is complete, γ is defined for every $s \in R$. In particular, γ is defined in the interval $[0, r]$. If we show that $\gamma(r) = q$, then γ must be a geodesic joining p to q which is minimal, since $l(\gamma) = r = d(p, q)$, and this will conclude the proof.

To prove this, we shall show that if $s \in [\delta, r]$, then

$$d(\gamma(s), q) = r - s. \tag{1}$$

Equation (1) implies, for $s = r$, that $\gamma(r) = q$, as desired.

To prove Eq. (1), we shall first show that it holds for $s = \delta$. Now the set $A = \{s \in [\delta, r];$ where Eq. (1) holds$\}$ is clearly closed in $[0, r]$. Next we show that if $s_0 \in A$ and $s_0 < r$, then Eq. (1) holds for $s_0 + \delta'$, where $\delta' > 0$ and δ' is sufficiently small. It follows that $A = [\delta, r]$ and that Eq. (1) will be proved.

We shall now show that Eq. (1) holds for $s = \delta$. In fact, since every curve joining p to q intersects Σ, we have, denoting by x an arbitrary point of Σ,

$$\begin{aligned} d(p, q) &= \inf_{\alpha} l(\alpha_{p,q}) = \inf_{x \in \Sigma} \{\inf_{\alpha} l(\alpha_{p,x}) + \inf_{\alpha} l(\alpha_{x,q})\} \\ &= \inf_{x \in \Sigma} (d(p, x) + d(x, q)) = \inf_{x \in \Sigma} (\delta + d(x, q)) \\ &= \delta + d(x_0, q). \end{aligned}$$

Hence,

$$d(\gamma(\delta), q) = r - \delta,$$

which is Eq. (1) for $s = \delta$.

Now we shall show that if Eq. (1) holds for $s_0 \in [\delta, r]$, then, for $\delta' > 0$ and sufficiently small, it holds for $s_0 + \delta'$.

Let $B_{\delta'}(0)$ be a disk in the tangent plane $T_{\gamma(s_0)}(S)$, centered in the origin 0 of this tangent plane and contained in a neighborhood U', where $\exp_{\gamma(s_0)}$ is a diffeormophism. Let $B_{\delta'}(\gamma(s_0)) = \exp_{\gamma(s_0)} B_{\delta'}(0)$ and $\Sigma' = \mathrm{Bd}(B_{\delta'}(\gamma(s_0)))$. If $x' \in E'$, the continuous function $d(x', q)$ reaches a minimum at $x'_0 \in \Sigma'$ (see Fig. 5-10). Then, as previously,

$$\begin{aligned} d(\gamma(s_0), q) &= \inf_{x' \in \Sigma'} \{d(\gamma(s_0), x') + d(x', q)\} \\ &= \delta' + d(x'_0, q). \end{aligned}$$

Since Eq. (1) holds in s_0, we have that $d(\gamma(s_0), q) = r - s_0$. Therefore,

$$d(x'_0, q) = r - s_0 - \delta'. \qquad \textbf{(2)}$$

Furthermore, since

$$d(p, x'_0) \geq d(p, q) - d(q, x'_0),$$

we obtain from Eq. (2)

$$d(p, x'_0) \geq r - (r - s_0) + \delta' = s_0 + \delta'.$$

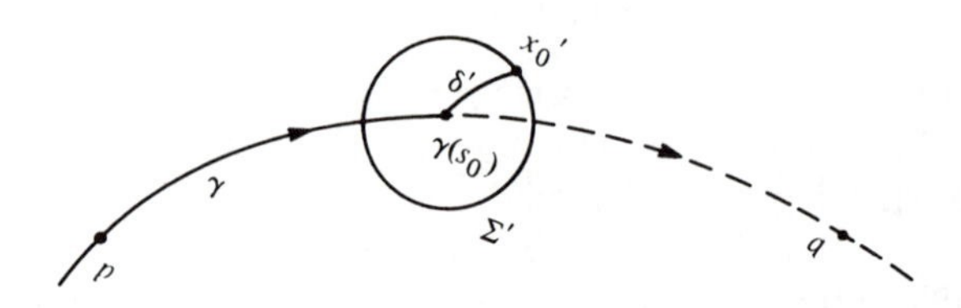

Figure 5-10

Observe now that the curve that goes from p to $\gamma(s_0)$ through γ and from $\gamma(s_0)$ to x_0' through a geodesic radius of $B_{\delta'}(\gamma(s_0))$ has length exactly equal to $s_0 + \delta'$. Since $d(p, x_0') \geq s_0 + \delta'$, this curve, which joins p to x_0', has minimal length. It follows (Sec. 4-6, Prop. 2) that it is a geodesic, and hence regular in all its points. Therefore, it should coincide with γ; hence, $x_0' = \gamma(s + \delta')$. Thus, Eq. (2) may be written as

$$d(\gamma(s_0 + \delta'), q) = r - (s_0 + \delta'),$$

which is Eq. (1) for $s = s_0 + \delta'$.

This proves our assertion and concludes the proof. **Q.E.D.**

COROLLARY 1. *Let* S *be complete. Then for every point* $p \in S$ *the map* $exp_p\colon T_p(S) \longrightarrow S$ *is onto* S.

This is true because if $q \in S$ and $d(p, q) = r$, then $q = \exp_p rv$, where $v = \gamma'(0)$ is the tangent vector of a minimal geodesic γ parametrized by the arc length and joining p to q.

COROLLARY 2. *Let* S *be complete and bounded in the metric* d (*that is, there exists* $r > 0$ *such that* $d(p, q) < r$ *for every pair* $p, q \in S$). *Then* S *is compact.*

Proof. By fixing $p \in S$, the fact that S is bounded implies the existence of a closed ball $B \subset T_p(S)$ of radius r, centered at the origin 0 of the tangent plane $T_p(S)$, such that $\exp_p(B) = \exp_p(T_p(S))$. By the fact that $\exp_p$ is onto, we have $S = \exp_p(T_p(S)) = \exp_p(B)$. Since B is compact and $\exp_p$ is continuous, we conclude that S is compact. **Q.E.D.**

From now on, the metric notions to be used will refer, except when otherwise stated, to the distance d in Def. 3. For instance, the diameter $\rho(S)$ of a surface S is, by definition,

$$\rho(s) = \sup_{p,q \in S} d(p, q).$$

With this definition, the diameter of a unit sphere S^2 is $\rho(S^2) = \pi$.

EXERCISES

1. Let $S \subset R^3$ be a complete surface and let $F \subset S$ be a nonempty, closed subset of S such that the complement $S - F$ is connected. Show that $S - F$ is a noncomplete regular surface.

2. Let S be the one-sheeted cone of Example 1. Show that, given $p \in S$, the only geodesic of S that passes through p and cannot be extended for every value of the parameter is the meridian of S through p.

3. Let S be the one-sheeted cone of Example 1. Use the isometry of Example 3 of Sec. 4-2 to show that any two points $p, q \in S$ (see Fig. 5-11) can be joined by a minimal geodesic on S.

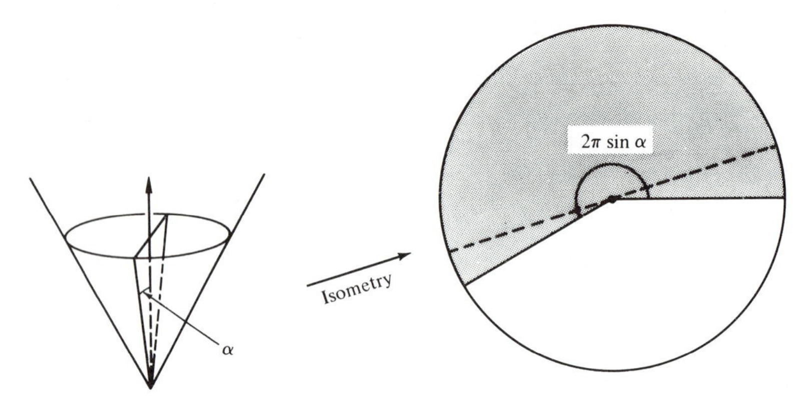

Figure 5-11

4. We say that a sequence $\{p_n\}$ of points on a regular surface $S \subset R^3$ *converges* to a point $p_0 \in S$ *in the (intrinsic) distance d* if given $\epsilon > 0$ there exists an index n_0 such that $n \geq n_0$ implies that $d(p_n, p_0) < \epsilon$. Prove that a sequence $\{p_n\}$ of points in S converges in d to $p_0 \in S$ if and only if $\{p_n\}$ converges to p_0 as a sequence of points in R^3 (i.e., in the euclidean distance).

***5.** Let $S \subset R^3$ be a regular surface. A sequence $\{p_n\}$ of points on S is a *Cauchy sequence in the (intrinsic) distance d* if given $\epsilon > 0$ there exists an index n_0 such that when $n, m \geq n_0$ then $d(p_n, p_m) < \epsilon$. Prove that S is complete if and only if every Cauchy sequence on S converges to a point in S.

***6.** A geodesic $\gamma\colon [0, \infty) \longrightarrow S$ on a surface S is a *ray issuing from* $\gamma(0)$ if it realizes the (intrinsic) distance between $\gamma(0)$ and $\gamma(s)$ for all $s \in [0, \infty)$. Let p be a point on a complete, noncompact surface S. Prove that S contains a ray issuing from p.

7. A *divergent curve* on S is a differentiable map $\alpha\colon [0, \infty) \longrightarrow S$ such that for every compact subset $K \subset S$ there exists a $t_0 \in (0, \infty)$ with $\alpha(t) \notin K$ for $t > t_0$ (i.e., α "leaves" every compact subset of S). The *length of a divergent curve* is defined as

$$\lim_{t \to \infty} \int_0^t |\alpha'(t)|\, dt.$$

Prove that $S \subset R^3$ is complete if and only if the length of every divergent curve is unbounded.

***8.** Let S and $\bar{S}$ be regular surfaces and let $\varphi: S \longrightarrow \bar{S}$ be a diffeomorphism. Assume that $\bar{S}$ is complete and that a constant $c > 0$ exists such that

$$I_p(v) \geq c\bar{I}_{\varphi(p)}(d\varphi_p(v))$$

for all $p \in S$ and all $v \in T_p(S)$, where I and $\bar{I}$ denote the first fundamental forms of S and $\bar{S}$, respectively. Prove that S is complete.

***9.** Let $S_1 \subset R^3$ be a (connected) complete surface and $S_2 \subset R^3$ be a connected surface such that any two points of S_2 can be joined by a *unique* geodesic. Let $\varphi: S_1 \longrightarrow S_2$ be a local isometry. Prove that φ is a global isometry.

***10.** Let $S \subset R^3$ be a complete surface. Fix a unit vector $v \in R$, and let $h: S \longrightarrow R$ be the height function $h(p) = \langle p, v \rangle$, $p \in S$. We recall that the gradient of h is the (tangent) vector field grad h on S defined by

$$\langle \text{grad } h(p), w \rangle_p = dh_p(w) \qquad \text{for all } w \in T_p(S)$$

(cf. Exercise 14, Sec. 2-5). Let $\alpha(t)$ be a trajectory of grad h; i.e., $\alpha(t)$ is a curve on S such that $\alpha'(t) = \text{grad } h(\alpha(t))$. Prove that

a. $|\text{grad } h(p)| \leq 1$ for all $p \in S$.

b. A trajectory $\alpha(t)$ of grad h is defined for all $t \in R$.

The following exercise presumes the material of Sec. 3-5, part, B and an elementary knowledge of functions of complex variables.

11. (*Osserman's Lemma.*) Let $D_1 = \{\zeta \in \mathbb{C}; |\zeta| \leq 1\}$ be the unit disk in the complex plane $\mathbb{C}$. As usual, we identify $\mathbb{C} \approx R^2$ by $\zeta = u + iv$. Let $\mathbf{x}: D_1 \longrightarrow R^3$ be an isothermal parametrization of a minimal surface $\mathbf{x}(D_1) \subset R^3$. This means (cf. Sec. 3-5, Part B) that

$$\langle \mathbf{x}_u, \mathbf{x}_u \rangle = \langle \mathbf{x}_v, \mathbf{x}_v \rangle, \qquad \langle \mathbf{x}_u, \mathbf{x}_v \rangle = 0$$

and (the minimality condition) that

$$\mathbf{x}_{uu} + \mathbf{x}_{vv} = 0.$$

Assume that the unit normal vectors of $\mathbf{x}(D_1)$ omit a neighborhood of a unit sphere. More precisely, assume that for some vector $w \in R^3$, $|w| = 1$, there exists an $\epsilon > 0$ such that

$$\frac{\langle \mathbf{x}_u, w \rangle^2}{|\mathbf{x}_u|^2} \geq \epsilon^2 \quad \text{and} \quad \frac{\langle \mathbf{x}_v, w \rangle^2}{|\mathbf{x}_v|^2} \geq \epsilon^2. \qquad (*)$$

The goal of the exercise is to *prove that* $\mathbf{x}(D)$ *is not a complete surface*. (This is the crucial step in the proof of Osserman's theorem quoted at the end of Sec. 3-5.) Proceed as follows:

a. Define $\varphi\colon D_1 \longrightarrow \mathbb{C}$ by

$$\varphi(u, v) = \varphi(\zeta) = \langle \mathbf{x}_u, w \rangle + i\langle \mathbf{x}_v, w \rangle.$$

Show that the minimality condition implies that φ is analytic.

b. Define $\theta\colon D_1 \longrightarrow \mathbb{C}$ by

$$\theta(\zeta) = \int_0^\zeta \varphi(\zeta)\, d\zeta = \eta.$$

By part a, θ is an analytic function. Show that $\theta(0) = 0$ and that the condition $(*)$ implies that $\theta'(\zeta) \neq 0$. Thus, in a neighborhood of 0, θ has an analytic inverse θ^{-1}. Use Liouville's theorem to show that θ^{-1} cannot be analytically extended to all of $\mathbb{C}$.

c. By part b there is a disk

$$D_R = \{\eta \in \mathbb{C};\, |\eta| \leq R\}$$

and a point η_0, with $|\eta_0| = R$, such that θ^{-1} is analytic in D and cannot be analytically extended to a neighborhood of η_0 (Fig. 5-12). Let L be the segment of D_R that joins η_0 to 0; i.e., $L = \{t\eta_0 \in \mathbb{C};\, 0 \leq t \leq 1\}$. Set $\alpha = \theta^{-1}(L)$ and show that the arc length l of $\mathbf{x}(\alpha)$ is

$$\begin{aligned} l &= \int_\alpha \sqrt{2\langle \mathbf{x}_u, \mathbf{x}_u \rangle \left\{ \left(\frac{du}{dt}\right)^2 + \left(\frac{dv}{dt}\right)^2 \right\}}\, dt \\ &\leq \frac{1}{\epsilon} \int_\alpha \sqrt{\langle \mathbf{x}_u, w \rangle^2 + \langle \mathbf{x}_v, w \rangle^2}\, |d\zeta| = \frac{1}{\epsilon} \int_\alpha |\varphi(\zeta)||d\zeta| \\ &= \frac{R}{\epsilon} < +\infty. \end{aligned}$$

Use Exercise 7 to conclude that $\mathbf{x}(D)$ is not complete.

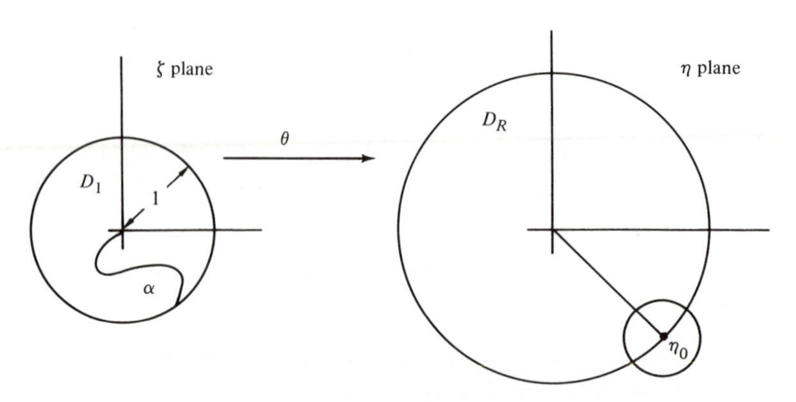

Figure 5-12

5-4. First and Second Variations of Arc Length; Bonnet's Theorem

The goal of this section is to prove that a complete surface S with Gaussian curvature $K \geq \delta > 0$ is compact (Bonnet's theorem).

The crucial point of the proof is to show that if $K \geq \delta > 0$, a geodesic γ joining two arbitrary points $p, q \in S$ and having length $l(\gamma) > \pi/\sqrt{\delta}$ is no longer minimal; that is, there exists a parametrized curve joining p and q, the length of which is smaller than $l(\gamma)$.

Once this is proved, it follows that every minimal geodesic has length $l \leq \pi/\sqrt{\delta}$; thus, S is bounded in the distance d. Since S is complete, S is compact (Corollary 2, Sec. 5-3). We remark that, in addition, we obtain an estimate for the diameter of S, namely, $\rho(S) \leq \pi/\sqrt{\delta}$.

To prove the above point, we need to compare the arc length of a parametrized curve with the arc length of "neighboring curves." For this, we shall introduce a number of ideas which are useful in other problems of differential geometry. Actually, these ideas are adaptations to the purposes of differential geometry of more general concepts found in calculus of variations. No knowledge of calculus of variations will be assumed.

In this section, S will denote a regular (not necessarily complete) surface.

We shall begin by making precise the idea of neighboring curves of a given curve.

DEFINITION 1. *Let* $\alpha: [0, l] \rightarrow S$ *be a regular parametrized curve, where the parameter* $s \in [0, l]$ *is the arc length. A* variation *of* α *is a differentiable map* $h: [0, l] \times (-\epsilon, \epsilon) \subset R^2 \rightarrow S$ *such that*

$$h(s, 0) = \alpha(s), \qquad s \in [0, l].$$

For each $t \in (-\epsilon, \epsilon)$, *the curve* $h_t: [0, l] \rightarrow S$, *given by* $h_t(s) = h(s, t)$, *is called a* curve of the variation h. *A variation* h *is said to be* proper *if*

$$h(0, t) = \alpha(0), \qquad h(l, t) = \alpha(l), \qquad t \in (-\epsilon, \epsilon).$$

Intuitively, a variation of α is a family h_t of curves depending differentiably on a parameter $t \in (-\epsilon, \epsilon)$ and such that h_0 agrees with α (Fig. 5-13). The condition of being proper means that all curves h_t have the same initial point $\alpha(0)$ and the same end point $\alpha(l)$.

It is convenient to adopt the following notation. The parametrized curves in R^2 given by

$$s \longrightarrow (s, t_0),$$
$$t \longrightarrow (s_0, t),$$

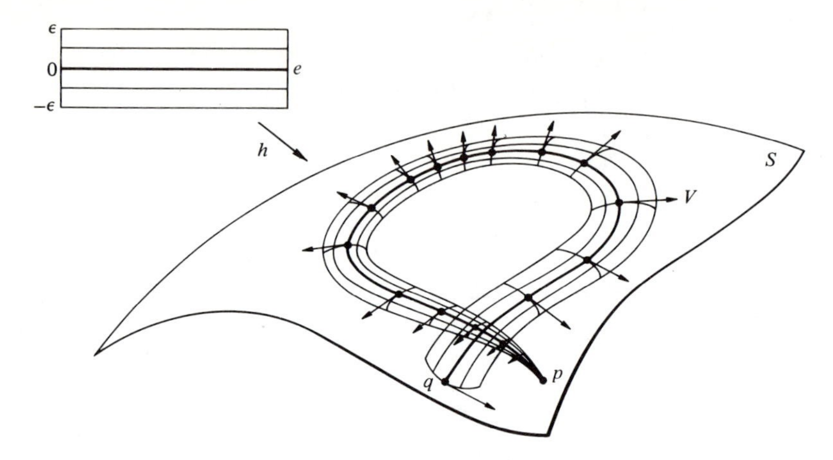

Figure 5-13

pass through the point $p_0 = (s_0, t_0) \in R^2$ and have $(1, 0)$ and $(0, 1)$ as tangent vectors at (s_0, t_0). Let $h: [0, l] \times (-\epsilon, \epsilon) \subset R^2 \longrightarrow S$ be a differentiable map and let $p_0 \in [0, l] \times (-\epsilon, \epsilon)$. Then $dh_{p_0}(1, 0)$ is the tangent vector to the curve $s \longrightarrow h(s, t_0)$ at $h(p_0)$, and $dh_{p_0}(0, 1)$ is the tangent vector to the curve $t \longrightarrow h(s_0, t)$ at $h(p_0)$. We shall denote

$$dh_{p_0}(1, 0) = \frac{\partial h}{\partial s}(p_0),$$

$$dh_{p_0}(0, 1) = \frac{\partial h}{\partial t}(p_0).$$

We recall (cf. Sec. 4-4, Def. 2) that a vector field w along a curve $\alpha: I \longrightarrow S$ is a correspondence that assigns to each $t \in I$ a vector $w(t)$ *tangent to the surface* S at $\alpha(t)$. Thus, $\partial h/\partial s$ and $\partial h/\partial t$ are differentiable tangent vector fields along α.

It follows that a variation h of α determines a differentiable vector field $V(s)$ along α by

$$V(s) = \frac{\partial h}{\partial s}(s, 0), \qquad s \in [0, l].$$

V is called the *variational vector field* of h; we remark that if h is proper, then

$$V(0) = V(l) = 0.$$

This terminology is justified by the following proposition.

PROPOSITION 1. *If we let* V(s) *be a differentiable vector field along a parametrized regular curve* $\alpha: [0, l] \longrightarrow$ S *then there exists a variation*

$h: [0, l] \times (-\epsilon, \epsilon) \to S$ *of* α *such that* $V(s)$ *is the variational vector field of* h. *Furthermore, if* $V(0) = V(l) = 0$, *then* h *can be chosen to be proper.*

Proof. We first show that there exists a $\delta > 0$ such that if $|v| < \delta$, $v \in T_{\alpha(s)}(S)$, then $\exp_{\alpha(s)} v$ is well defined for all $s \in [0, l]$. In fact, for each $p \in \alpha([0, l]) \subset S$ consider the neighborhood W_p (a normal neighborhood of all of its points) and the number $\delta_p > 0$ given by Prop. 1 of Sec. 4-7. The union $\bigcup_p W_p$ covers $\alpha([0, l])$ and, by compactness, a finite number of them, say, $W_1, \ldots, W_n$, still covers $\alpha([0, l])$. Set $\delta = \min(\delta_1, \ldots, \delta_n)$, where δ_i is the number corresponding to the neighborhood W_i, $i = 1, \ldots, n$. It is easily seen that δ satisfies the above condition.

Now let $M = \max_{s \in [0,l]} |V(s)|$, $\epsilon < \delta/M$, and define

$$h(s, t) = \exp_{\alpha(s)} tV(s), \qquad s \in [0, l], \quad t \in (-\epsilon, \epsilon).$$

h is clearly well defined. Furthermore, since

$$\exp_{\alpha(s)} tV(s) = \gamma(1, \alpha(s), tV(s)),$$

where γ is the (differentiable) map of Theorem 1 of Sec. 4-7 (i.e., for $t \neq 0$, and $V(s) \neq 0$, $\gamma(1, \alpha(s), tV(s))$ is the geodesic γ with initial conditions $\gamma(0) = \alpha(s)$, $\gamma'(0) = V(s)$), h is differentiable. It is immediately checked that $h(s, 0) = \alpha(s)$. Finally, the variational vector field of h is given by

$$\begin{aligned}\frac{\partial h}{\partial t}(s, 0) &= dh_{(s,0)}(0, 1) = \frac{d}{dt}(\exp_{\alpha(s)} tV(s))\Big|_{t=0} \\ &= \frac{d}{dt}\gamma(1, \alpha(s), tV(s))\Big|_{t=0} = \frac{d}{dt}\gamma(t, \alpha(s), V(s))\Big|_{t=0} = V(s),\end{aligned}$$

and it is clear, by the definition of h, that if $V(0) = V(l) = 0$, then h is proper.

Q.E.D.

We want to compare the arc length of α $(= h_0)$ with the arc length of h_t. Thus, we define a function $L: (-\epsilon, \epsilon) \to R$ by

$$L(t) = \int_0^l \left|\frac{\partial h}{\partial s}(s, t)\right| ds, \qquad t \in (-\epsilon, \epsilon). \tag{1}$$

The study of L in a neighborhood of $t = 0$ will inform us of the "arc length behavior" of curves neighboring α.

We need some preliminary lemmas.

LEMMA 1. *The function* L *defined by Eq. (1) is differentiable in a neighborhood of* $t = 0$; *in such a neighborhood, the derivative of* L *may be obtained by differentiation under the integral sign.*

Proof. Since $\alpha: [0, l] \longrightarrow S$ is parametrized by arc length,

$$\left|\frac{\partial h}{\partial s}\right| = \left|\frac{\partial h}{\partial s}(s, 0)\right| = 1.$$

It follows, by compactness of $[0, l]$, that there exists a $\delta > 0$, $\delta \leq \epsilon$, such that

$$\left|\frac{\partial h}{\partial s}(s, t)\right| \neq 0, \qquad s \in [0, l], \quad |t| < \delta.$$

Since the absolute value of a nonzero differentiable function is differentiable, the integrand of Eq. (1) is differentiable for $|t| < \delta$. By a classical theorem of calculus (see R. C. Buck, *Advanced Calculus*, 1965, p. 120), we conclude that L is differentiable for $|t| < \delta$ and that

$$L'(t) = \int_0^l \frac{\partial}{\partial t}\left|\frac{\partial h}{\partial s}(s, t)\right| ds. \qquad \textbf{Q.E.D.}$$

Lemmas 2, 3, and 4 below have some independent interest.

LEMMA 2. *Let* w(t) *be a differentiable vector field along the parametrized curve* α: [a, b] $\longrightarrow$ S *and let* f: [a, b] $\longrightarrow$ R *be a differentiable function. Then*

$$\frac{\mathrm{D}}{\mathrm{dt}}(\mathrm{f(t)w(t)}) = \mathrm{f(t)}\frac{\mathrm{Dw}}{\mathrm{dt}} + \frac{\mathrm{df}}{\mathrm{dt}}\mathrm{w(t)}.$$

Proof. It suffices to use the fact that the covariant derivative is the tangential component of the usual derivative to conclude that (here $(\ )_T$ denotes the tangential component of ( ))

$$\frac{D}{dt}(fw) = \left(\frac{df}{dt}w + f\frac{dw}{dt}\right)_T = \frac{df}{dt}w + f\left(\frac{dw}{dt}\right)_T$$
$$= \frac{df}{dt}w + f\frac{Dw}{dt}. \qquad \textbf{Q.E.D.}$$

LEMMA 3. *Let* v(t) *and* w(t) *be differentiable vector fields along the parametrized curve* α: [a, b] $\longrightarrow$ S. *Then*

$$\frac{\mathrm{d}}{\mathrm{dt}}\langle \mathrm{v(t), w(t)}\rangle = \left\langle\frac{\mathrm{Dv}}{\mathrm{dt}}, \mathrm{w(t)}\right\rangle + \left\langle \mathrm{v(t)}, \frac{\mathrm{Dw}}{\mathrm{dt}}\right\rangle.$$

Proof. Using the remarks of the above proof, we obtain

$$\frac{d}{dt}\langle v, w\rangle = \left\langle\frac{dv}{dt}, w\right\rangle + \left\langle v, \frac{dw}{dt}\right\rangle = \left\langle\left(\frac{dv}{dt}\right)_T, w\right\rangle + \left\langle v, \left(\frac{dw}{dt}\right)_T\right\rangle$$
$$= \left\langle\frac{Dv}{dt}, w\right\rangle + \left\langle v, \frac{Dw}{dt}\right\rangle. \qquad \textbf{Q.E.D.}$$

Before stating the next lemma it is convenient to introduce the following terminology. Let $h: [0, l] \times (-\epsilon, \epsilon) \rightarrow S$ be a differentiable map. *A differentiable vector field along h* is a differentiable map

$$V: [0, l] \times (-\epsilon, \epsilon) \longrightarrow S \subset R^3$$

such that $V(s, t) \in T_{h(s,t)}(S)$ for each $(s, t) \in [0, l] \times (-\epsilon, \epsilon)$. This generalizes the definition of a differentiable vector field along a parametrized curve (Sec. 4-4, Def. 2).

For instance, the vector fields $(\partial h/\partial s)(s, t)$ and, $(\partial h/\partial t)(s, t)$, introduced above, are vector fields along h.

If we restrict $V(s, t)$ to the curves $s =$ const., $t =$ const., we obtain vector fields along curves. In this context, the notation $(DV/\partial t)(s, t)$ means the covariant derivative, at the point (s, t), of the restriction of $V(s, t)$ to the curve $s =$ const.

LEMMA 4. *Let* h: $[0, l] \times (-\epsilon, \epsilon) \subset R^2 \rightarrow S$ *be a differentiable mapping. Then*

$$\frac{D}{\partial s}\frac{\partial h}{\partial t}(s, t) = \frac{D}{\partial t}\frac{\partial h}{\partial s}(s, t).$$

Proof. Let $\mathbf{x}: U \rightarrow S$ be a parametrization of S at the point $h(s, t)$, with parameters u, v, and let

$$u = h_1(s, t), \qquad v = h_2(s, t)$$

be the expression of h in this parametrization. Under these conditions, when $(s, t) \in h^{-1}(\mathbf{x}(U)) = W$, the curve $h(s, t_0)$ may be expressed by

$$u = h_1(s, t_0), \qquad v = h_2(s, t_0).$$

Since $(\partial h/\partial s)(s_0, t_0)$ is tangent to the curve $h(s, t_0)$ at $s = s_0$, we have that

$$\frac{\partial h}{\partial s}(s_0, t_0) = \frac{\partial h_1}{\partial s}(s_0, t_0)\mathbf{x}_u + \frac{\partial h_2}{\partial s}(s_0, t_0)\mathbf{x}_v.$$

By the arbitrariness of $(s_0, t_0) \in W$, we conclude that

$$\frac{\partial h}{\partial s} = \frac{\partial h_1}{\partial s}\mathbf{x}_u + \frac{\partial h_2}{\partial s}\mathbf{x}_v,$$

where we omit the indication of the point (s, t) for simplicity of notation.

Similarly,

$$\frac{\partial h}{\partial t} = \frac{\partial h_1}{\partial t}\mathbf{x}_u + \frac{\partial h_2}{\partial t}\mathbf{x}_v.$$

We shall now compute the covariant derivatives $(D/\partial s)(\partial h/\partial t)$ and $(D/\partial t)/(\partial h/\partial s)$ using the expression of the covariant derivative in terms of the Christoffel symbols Γ_{ij}^k (Sec. 4-4, Eq. (1)) and obtain the asserted equality. For instance, the coefficient of $\mathbf{x}_u$ in both derivatives is given by

$$\frac{\partial^2 h_1}{\partial s \partial t} + \Gamma_{11}^1 \frac{\partial h_1}{\partial t}\frac{\partial h_1}{\partial s} + \Gamma_{12}^1 \frac{\partial h_1}{\partial t}\frac{\partial h_2}{\partial s} + \Gamma_{12}^1 \frac{\partial h_2}{\partial t}\frac{\partial h_1}{\partial s} + \Gamma_{22}^1 \frac{\partial h_2}{\partial t}\frac{\partial h_2}{\partial s}.$$

The equality of the coefficientes of $\mathbf{x}_v$ may be shown in the same way, thus concluding the proof. **Q.E.D.**

We are now in a position to compute the first derivative of L at $t = 0$ and obtain

PROPOSITION 2. *Let* $h: [0, l] \times (-\epsilon, \epsilon)$ *be a proper variation of the curve* $\alpha: [0, l] \to S$ *and let* $V(s) = (\partial h/\partial t)(s, 0)$, $s \in [0, l]$, *be the variational vector field of* h. *Then*

$$L'(0) = -\int_0^l \langle A(s), V(s) \rangle \, ds, \tag{2}$$

where $A(s) = (D/\partial s)(\partial h/\partial s)(s, 0)$.

Proof. If t belongs to the interval $(-\delta, \delta)$ given by Lemma 1, then

$$L'(t) = \int_0^l \left\{ \frac{d}{dt}\left\langle \frac{\partial h}{\partial s}, \frac{\partial h}{\partial s} \right\rangle^{1/2} \right\} ds.$$

By applying Lemmas 3 and 4, we obtain

$$L'(t) = \int_l^0 \frac{\left\langle \dfrac{D}{\partial t}\dfrac{\partial h}{\partial s}, \dfrac{\partial h}{\partial s} \right\rangle}{\left| \dfrac{\partial h}{\partial s} \right|} ds = \int_0^l \frac{\left\langle \dfrac{D}{\partial s}\dfrac{\partial h}{\partial t}, \dfrac{\partial h}{\partial s} \right\rangle}{\left| \dfrac{\partial h}{\partial s} \right|} ds.$$

Since $|(\partial h/\partial s)(s, 0)| = 1$, we have that

$$L'(0) = \int_0^l \left\langle \frac{D}{\partial s}\frac{\partial h}{\partial t}, \frac{\partial h}{\partial s} \right\rangle ds,$$

where the integrand is calculated at $(s, 0)$, which is omitted for simplicity of notation.

According to Lemma 3,

$$\frac{\partial}{\partial s}\left\langle \frac{\partial h}{ds}, \frac{\partial h}{\partial t} \right\rangle = \left\langle \frac{D}{\partial s}\frac{\partial h}{\partial s}, \frac{\partial h}{\partial t} \right\rangle + \left\langle \frac{\partial h}{\partial s}, \frac{D}{\partial s}\frac{\partial h}{\partial t} \right\rangle.$$

Therefore,

$$L'(0) = \int_0^l \frac{\partial}{\partial s}\left\langle \frac{\partial h}{\partial s}, \frac{\partial h}{\partial t}\right\rangle ds - \int_0^l \left\langle \frac{D}{\partial s}\frac{\partial h}{\partial s}, \frac{\partial h}{\partial t}\right\rangle ds$$

$$= -\int_0^l \left\langle \frac{D}{\partial s}\frac{\partial h}{\partial s}, \frac{\partial h}{\partial t}\right\rangle ds,$$

since $(\partial h/\partial t)(0, 0) = (\partial h/\partial t)\,(l, 0) = 0$, due to the fact that the variation is proper. By recalling the definitions of $A(s)$ and $V(s)$, we may write the last expression in the form

$$L'(0) = -\int_0^l \langle A(s), V(s)\rangle\, ds. \qquad \textbf{Q.E.D.}$$

Remark 1. The vector $A(s)$ is called the *acceleration vector* of the curve α, and its norm is nothing but the absolute value of the geodesic curvature of α. Observe that $L'(0)$ depends only on the variational field $V(s)$ and not on the variation h itself. Expression (2) is usually called the *formula for the first variation* of the arc length of the curve α.

Remark 2. The condition that h is proper was only used at the end of the proof in order to eliminate the terms

$$\left\langle \frac{\partial h}{\partial s}, \frac{\partial h}{\partial t}\right\rangle(l, 0) - \left\langle \frac{\partial h}{\partial s}, \frac{\partial h}{\partial t}\right\rangle(0, 0).$$

Therefore, if h is not proper, we obtain a formula which is similar to Eq. (2) and contains these additional boundary terms.

An interesting consequence of Prop. 2 is a characterization of the geodesics as solutions of a "variational problem." More precisely,

PROPOSITION 3. *A regular parametrized curve* $\alpha\colon [0, l] \longrightarrow S$, *where the parameter* $s \in [0, l]$ *is the arc length of* α, *is a geodesic if and only if, for every proper variation* $h\colon [0, l] \times (-\epsilon, \epsilon) \longrightarrow S$ *of* α, $L'(0) = 0$.

Proof. The necessity is trivial since the acceleration vector $A(s) = (D/\partial s)(\partial\alpha/\partial s)$ of a geodesic α is identically zero. Therefore, $L'(0) = 0$ for every proper variation.

Suppose now that $L'(0) = 0$ for every proper variation of α and consider a vector field $V(s) = f(s)A(s)$, where $f\colon [0, l] \longrightarrow R$ is a real differentiable function, with $f(s) \geq 0$, $f(0) = f(l) = 0$, and $A(s)$ is the acceleration vector of α. By constructing a variation corresponding to $V(s)$, we have

$$L'(0) = -\int_0^l \langle f(s)A(s), A(s)\rangle\, ds$$
$$= -\int_0^l f(s)|A(s)|^2\, ds = 0.$$

Therefore, since $f(s)|A(s)|^2 \geq 0$, we obtain

$$f(s)|A(s)|^2 \equiv 0.$$

We shall prove that the above relation implies that $A(s) = 0$, $s \in [0, l]$. In fact, if $|A(s_0)| \neq 0$, $s_0 \in (0, l)$, there exists an interval $I = (s_0 - \epsilon, s_0 + \epsilon)$ such that $|A(s)| \neq 0$ for $s \in I$. By choosing f such that $f(s_0) > 0$, we contradict $f(s_0)|A(s_0)| = 0$. Therefore, $|A(s)| = 0$ when $s \in (0, l)$. By continuity, $A(0) = A(l) = 0$ as asserted.

Since the acceleration vector of α is identically zero, α is geodesic.

Q.E.D.

From now on, we shall only consider proper variations of geodesics $\gamma: [0, l] \to S$, parametrized by arc length; that is, we assume $L'(0) = 0$. To simplify the computations, we shall restrict ourselves to *orthogonal variations*; that is, we shall assume that the variational field $V(s)$ satisfies the condition $\langle V(s), \gamma'(s)\rangle = 0$, $s \in [0, l]$. To study the behavior of the function L in a neighborhood of 0 we shall compute $L''(0)$.

For this computation, we need some lemmas that relate the Gaussian curvature to the covariant derivative.

LEMMA 5. *Let* $\mathbf{x}: U \to S$ *be a parametrization at a point* $p \in S$ *of a regular surface* S, *with parameters* u, v, *and let* K *be the Gaussian curvature of* S. *Then*

$$\frac{D}{\partial v}\frac{D}{\partial u}\mathbf{x}_u - \frac{D}{\partial u}\frac{D}{\partial v}\mathbf{x}_u = K(\mathbf{x}_u \wedge \mathbf{x}_v) \wedge \mathbf{x}_u.$$

Proof. By observing that the covariant derivative is the component of the usual derivative in the tangent plane, we have that (Sec. 4-3)

$$\frac{D}{\partial u}\mathbf{x}_u = \Gamma_{11}^1\mathbf{x}_u + \Gamma_{11}^2\mathbf{x}_v.$$

By applying to the above expression the formula for the covariant derivative (Sec. 4-4, Eq. (1)), we obtain

$$\frac{D}{\partial v}\left(\frac{D}{\partial u}\mathbf{x}_u\right) = \{(\Gamma_{11}^1)_v + \Gamma_{12}^1\Gamma_{11}^1 + \Gamma_{22}^1\Gamma_{11}^2\}\mathbf{x}_u$$
$$+ \{(\Gamma_{11}^2)_v + \Gamma_{12}^2\Gamma_{11}^1 + \Gamma_{22}^2\Gamma_{11}^2\}\mathbf{x}_v.$$

We verify, by means of a similar computation, that

$$\frac{D}{\partial u}\left(\frac{D}{\partial v}\mathbf{x}_u\right) = \{(\Gamma_{12}^1)_u + \Gamma_{12}^1\Gamma_{11}^1 + \Gamma_{12}^1\Gamma_{12}^2\}\mathbf{x}_u$$
$$+ \{(\Gamma_{12}^2)_u + \Gamma_{11}^2\Gamma_{12}^1 + \Gamma_{12}^2\Gamma_{12}^2\}\mathbf{x}_v.$$

Therefore,

$$\frac{D}{\partial v}\frac{D}{\partial u}\mathbf{x}_u - \frac{D}{\partial u}\frac{D}{\partial v}\mathbf{x}_u = \{(\Gamma_{11}^1)_v - (\Gamma_{12}^1)_u + \Gamma_{22}^1\Gamma_{11}^2 - \Gamma_{12}^1\Gamma_{12}^2\}\mathbf{x}_u$$
$$+ \{(\Gamma_{11}^2)_v - (\Gamma_{12}^2)_u + \Gamma_{12}^2\Gamma_{11}^1 + \Gamma_{22}^2\Gamma_{11}^2$$
$$- \Gamma_{11}^2\Gamma_{12}^1 - \Gamma_{12}^2\Gamma_{12}^2\}\mathbf{x}_v.$$

We now use the expressions of the curvature in terms of Christoffel symbols (Sec. 4-3, Eqs. (5) and (5a) and conclude that

$$\frac{D}{\partial v}\frac{D}{\partial u}\mathbf{x}_u - \frac{D}{\partial u}\frac{D}{\partial v}\mathbf{x}_u = -FK\mathbf{x}_u + EK\mathbf{x}_v$$
$$= K\{\langle \mathbf{x}_u, \mathbf{x}_u\rangle\mathbf{x}_v - \langle \mathbf{x}_u, \mathbf{x}_v\rangle\mathbf{x}_u\}$$
$$= K(\mathbf{x}_u \wedge \mathbf{x}_v) \wedge \mathbf{x}_u. \qquad \textbf{Q.E.D.}$$

LEMMA 6. *Let* $h\colon [0, l] \times (-\epsilon, \epsilon) \to S$ *be a differentiable mapping and let* $V(s, t)$, $(s, t) \in [0, l] \times (-\epsilon, \epsilon)$, *be a differentiable vector field along* h. *Then*

$$\frac{D}{\partial t}\frac{D}{\partial s}V - \frac{D}{\partial s}\frac{D}{\partial t}V = K(s, t)\left(\frac{\partial h}{\partial s} \wedge \frac{\partial h}{\partial t}\right) \wedge V,$$

where $K(s, t)$ *is the curvature of* S *at the point* $h(s, t)$.

Proof. Let $\mathbf{x}(u, v)$ be a system of coordinates of S around $h(s, t)$ and let

$$V(s, t) = a(s, t)\mathbf{x}_u + b(s, t)\mathbf{x}_v$$

be the expression of $V(s, t) = V$ in this system of coordinates. By Lemma 2, we have

$$\frac{D}{\partial s}V = \frac{D}{\partial s}(a\mathbf{x}_u + b\mathbf{x}_v)$$
$$= a\frac{D}{\partial s}\mathbf{x}_u + b\frac{D}{\partial s}\mathbf{x}_v + \frac{\partial a}{\partial s}\mathbf{x}_u + \frac{\partial b}{\partial s}\mathbf{x}_v.$$

Therefore,

$$\frac{D}{\partial t}\frac{D}{\partial s}V = a\frac{D}{\partial t}\frac{D}{\partial s}\mathbf{x}_u + b\frac{D}{\partial t}\frac{D}{\partial s}\mathbf{x}_v + \frac{\partial a}{\partial s}\frac{D}{\partial t}\mathbf{x}_u$$
$$+ \frac{\partial b}{\partial s}\frac{D}{\partial t}\mathbf{x}_v + \frac{\partial a}{\partial t}\frac{D}{\partial s}\mathbf{x}_u + \frac{\partial b}{\partial t}\frac{D}{\partial s}\mathbf{x}_v + \frac{\partial^2 a}{\partial t\,\partial s}\mathbf{x}_u + \frac{\partial^2 b}{\partial t\,\partial s}\mathbf{x}_v.$$

By a similar computation, we obtain a formula for $(D/\partial s)(D/\partial t)V$, which is given by interchanging s and t in the last expression. It follows that

$$\begin{aligned}\frac{D}{\partial t}\frac{D}{\partial s}V - \frac{D}{\partial s}\frac{D}{\partial t}V = a\left(\frac{D}{\partial t}\frac{D}{\partial s}\mathbf{x}_u - \frac{D}{\partial s}\frac{D}{\partial t}\mathbf{x}_u\right)\\ + b\left(\frac{D}{\partial t}\frac{D}{\partial s}\mathbf{x}_v - \frac{D}{\partial s}\frac{D}{\partial t}\mathbf{x}_v\right).\end{aligned} \tag{3}$$

To compute $(D/\partial t)(D/\partial s)\mathbf{x}_u$, we shall take the expression of h,

$$u = h_1(s, t), \qquad v = h_2(s, t),$$

in the parametrization $\mathbf{x}(u, v)$ and write

$$\mathbf{x}_u(u, v) = \mathbf{x}_u(h_1(s, t), h_2(s, t)) = \mathbf{x}_u.$$

Since the covariant derivative $(D/\partial s)\mathbf{x}_u$ is the projection onto the tangent plane of the usual derivative $(d/ds)\mathbf{x}_u$, we have

$$\begin{aligned}\frac{D}{\partial s}\mathbf{x}_u &= \left\{\frac{d}{ds}\mathbf{x}_u\right\}_T = \left\{\mathbf{x}_{uu}\frac{\partial h_1}{\partial s} + \mathbf{x}_{uv}\frac{\partial h_2}{\partial s}\right\}_T\\ &= \frac{\partial h_1}{\partial s}\{\mathbf{x}_{uu}\}_T + \frac{\partial h_2}{\partial s}\{\mathbf{x}_{uv}\}_T\\ &= \frac{\partial h_1}{\partial s}\frac{D}{\partial u}\mathbf{x}_u + \frac{\partial h_2}{\partial s}\frac{D}{\partial v}\mathbf{x}_u,\end{aligned}$$

where T denotes the projection of a vector onto the tangent plane.

With the same notation, we obtain

$$\begin{aligned}\frac{D}{\partial t}\frac{D}{\partial s}\mathbf{x}_u &= \left\{\frac{d}{dt}\left(\frac{\partial h_1}{\partial s}\frac{D}{\partial u}\mathbf{x}_u + \frac{\partial h_2}{\partial s}\frac{D}{\partial v}\mathbf{x}_u\right)\right\}_T = \frac{\partial^2 h_1}{\partial t\,\partial s}\frac{D}{\partial u}\mathbf{x}_u\\ &+ \frac{\partial^2 h_2}{\partial t\,\partial s}\frac{D}{\partial v}\mathbf{x}_u + \frac{\partial h_1}{\partial s}\left(\frac{\partial h_1}{\partial t}\frac{D}{\partial u}\frac{D}{\partial u}\mathbf{x}_u + \frac{\partial h_2}{\partial t}\frac{D}{\partial v}\frac{D}{\partial u}\mathbf{x}_u\right)\\ &+ \frac{\partial h_2}{\partial s}\left(\frac{\partial h_1}{\partial t}\frac{D}{\partial u}\frac{D}{\partial v}\mathbf{x}_u + \frac{\partial h_2}{\partial t}\frac{D}{\partial v}\frac{D}{\partial u}\mathbf{x}_u\right).\end{aligned}$$

In a similar way, we obtain $(D/\partial s)(D/\partial t)\mathbf{x}_u$, which is given by interchanging s and t in the above expression. It follows that

$$\begin{aligned}\frac{D}{\partial t}\frac{D}{\partial s}\mathbf{x}_u - \frac{D}{\partial s}\frac{D}{\partial t}\mathbf{x}_u &= \frac{\partial h_2}{\partial s}\frac{\partial h_1}{\partial t}\left(\frac{D}{\partial u}\frac{D}{\partial v}\mathbf{x}_u - \frac{D}{\partial v}\frac{D}{\partial u}\mathbf{x}_u\right)\\ &+ \frac{\partial h_1}{\partial s}\frac{\partial h_2}{\partial t}\left(\frac{D}{\partial v}\frac{D}{\partial u}\mathbf{x}_u - \frac{D}{\partial u}\frac{D}{\partial v}\mathbf{x}_u\right)\\ &= \Delta\left(\frac{D}{\partial v}\frac{D}{\partial u}\mathbf{x}_u - \frac{D}{\partial u}\frac{D}{\partial v}\mathbf{x}_u\right),\end{aligned}$$

where

$$\Delta = \left(\frac{\partial h_1}{\partial s}\frac{\partial h_2}{\partial t} - \frac{\partial h_2}{\partial s}\frac{\partial h_1}{\partial t}\right).$$

By replacing $\mathbf{x}_u$ for $\mathbf{x}_v$ in the last expression, we obtain

$$\frac{D}{\partial t}\frac{D}{\partial s}\mathbf{x}_v - \frac{D}{\partial s}\frac{D}{\partial t}\mathbf{x}_v = \Delta\left(\frac{D}{\partial v}\frac{D}{\partial u}\mathbf{x}_v - \frac{D}{\partial u}\frac{D}{\partial v}\mathbf{x}_v\right).$$

By introducing the above expression in Eq. (3) and using Lemma 5, we conclude that

$$\begin{aligned}\frac{D}{\partial t}\frac{D}{\partial s}V - \frac{D}{\partial s}\frac{D}{\partial t}V &= a\,\Delta K(\mathbf{x}_u \wedge \mathbf{x}_v) \wedge \mathbf{x}_u + b\,\Delta K(\mathbf{x}_u \wedge \mathbf{x}_v) \wedge \mathbf{x}_v \\ &= K(\Delta \mathbf{x}_u \wedge \mathbf{x}_v) \wedge (a\mathbf{x}_u + b\mathbf{x}_v).\end{aligned}$$

On the other hand, as we saw in the proof of Lemma 4,

$$\frac{\partial h}{\partial s} = \frac{\partial h_1}{\partial s}\mathbf{x}_u + \frac{\partial h_2}{\partial s}\mathbf{x}_v, \qquad \frac{\partial h}{\partial t} = \frac{\partial h_1}{\partial t}\mathbf{x}_u + \frac{\partial h_2}{\partial t}\mathbf{x}_v;$$

hence,

$$\frac{\partial h}{\partial s} \wedge \frac{\partial h}{\partial t} = \Delta \mathbf{x}_u \wedge \mathbf{x}_v.$$

Therefore,

$$\frac{D}{\partial t}\frac{D}{\partial s}V - \frac{D}{\partial s}\frac{D}{\partial t}V = K\left(\frac{\partial h}{\partial s} \wedge \frac{\partial h}{\partial t}\right) \wedge V. \qquad \textbf{Q.E.D.}$$

We are now in a position to compute $L''(0)$.

PROPOSITION 4. *Let* $h\colon [0, l] \times (-\epsilon, \epsilon) \to S$ *be a proper orthogonal variation of a geodesic* $\gamma\colon [0, l] \to S$ *parametrized by the arc length* $s \in [0, l]$. *Let* $V(s) = (\partial h/\partial t)(s, 0)$ *be the variational vector field of* h. *Then*

$$L''(0) = \int_0^l \left(\left|\frac{D}{\partial s}V(s)\right|^2 - K(s)\,|V(s)|^2\right) ds, \tag{4}$$

where $K(s) = K(s, 0)$ *is the Gaussian curvature of* S *at* $\gamma(s) = h(s, 0)$.

Proof. As we saw in the proof of Prop. 2,

$$L'(t) = \int_0^l \frac{\left\langle \dfrac{D}{\partial s}\dfrac{\partial h}{\partial t}, \dfrac{\partial h}{\partial s}\right\rangle}{\left\langle \dfrac{\partial h}{\partial s}, \dfrac{\partial h}{\partial s}\right\rangle^{1/2}}\, ds$$

for t belonging to the interval $(-\delta, \delta)$ given by Lemma 1. By differentiating the above expression, we obtain

$$L''(t) = \int_0^l \frac{\left(\dfrac{d}{dt}\left\langle \dfrac{D}{\partial s}\dfrac{\partial h}{\partial t}, \dfrac{\partial h}{\partial s}\right\rangle\right)\left\langle \dfrac{\partial h}{\partial s}, \dfrac{\partial h}{\partial s}\right\rangle^{1/2}}{\left\langle \dfrac{\partial h}{\partial s}, \dfrac{\partial h}{\partial s}\right\rangle} ds$$

$$- \int_0^l \frac{\left(\left\langle \dfrac{D}{\partial s}\dfrac{\partial h}{\partial t}, \dfrac{\partial h}{\partial s}\right\rangle\right)^2}{\left|\dfrac{\partial h}{\partial s}\right|^{3/2}} ds.$$

Observe now that for $t = 0$, $|(\partial h/\partial s)(s, 0)| = 1$. Furthermore,

$$\frac{d}{ds}\left\langle \frac{\partial h}{\partial s}, \frac{\partial h}{\partial t}\right\rangle = \left\langle \frac{D}{\partial s}\frac{\partial h}{\partial s}, \frac{\partial h}{\partial t}\right\rangle + \left\langle \frac{\partial h}{\partial s}, \frac{D}{\partial s}\frac{\partial h}{\partial t}\right\rangle.$$

Since γ is a geodesic, $(D/\partial s)(\partial h/\partial s) = 0$ for $t = 0$, and since the variation is orthogonal,

$$\left\langle \frac{\partial h}{\partial s}, \frac{\partial h}{\partial t}\right\rangle = 0 \qquad \text{for } t = 0.$$

It follows that

$$L''(0) = \int_0^l \frac{d}{dt}\left\langle \frac{D}{\partial s}\frac{\partial h}{\partial t}, \frac{\partial h}{\partial s}\right\rangle ds, \tag{5}$$

where the integrand is calculated at $(s, 0)$.

Let us now transform the integrand of Eq. (5) into a more convenient expression. Observe first that

$$\begin{aligned}\frac{d}{dt}\left\langle \frac{D}{\partial s}\frac{\partial h}{\partial t}, \frac{\partial h}{\partial s}\right\rangle &= \left\langle \frac{D}{\partial t}\frac{D}{\partial s}\frac{\partial h}{\partial t}, \frac{\partial h}{\partial s}\right\rangle + \left\langle \frac{D}{\partial s}\frac{\partial h}{\partial t}, \frac{D}{\partial t}\frac{\partial h}{\partial s}\right\rangle \\ &= \left\langle \frac{D}{\partial t}\frac{D}{\partial s}\frac{\partial h}{\partial t}, \frac{\partial h}{\partial s}\right\rangle - \left\langle \frac{D}{\partial s}\frac{D}{\partial t}\frac{\partial h}{\partial t}, \frac{\partial h}{\partial s}\right\rangle \\ &\quad + \left\langle \frac{D}{\partial s}\frac{D}{\partial t}\frac{\partial h}{\partial t}, \frac{\partial h}{\partial s}\right\rangle + \left|\frac{D}{\partial s}\frac{\partial h}{\partial t}\right|^2.\end{aligned}$$

On the other hand, for $t = 0$,

$$\frac{d}{ds}\left\langle \frac{D}{\partial t}\frac{\partial h}{\partial t}, \frac{\partial h}{\partial s}\right\rangle = \left\langle \frac{D}{\partial s}\frac{D}{\partial t}\frac{\partial h}{\partial t}, \frac{\partial h}{\partial s}\right\rangle,$$

since $(D/\partial s)(\partial h/\partial s)(s, 0) = 0$, owing to the fact that γ is a geodesic. Moreover, by using Lemma 6 plus the fact that the variation is orthogonal, we obtain (for $t = 0$)

$$\left\langle \frac{D}{\partial t}\frac{D}{\partial s}\frac{\partial h}{\partial t}, \frac{\partial h}{\partial s}\right\rangle - \left\langle \frac{D}{\partial s}\frac{D}{\partial t}\frac{\partial h}{\partial t}, \frac{\partial h}{\partial s}\right\rangle = K(s)\left\langle\left(\frac{\partial h}{\partial s} \wedge \frac{\partial h}{\partial t}\right) \wedge \frac{\partial h}{\partial t}, \frac{\partial h}{\partial s}\right\rangle$$

$$= -K(s)\left\langle | V(s)|^2 \frac{\partial h}{\partial s}, \frac{\partial h}{\partial s}\right\rangle$$

$$= -K| V(s)|^2.$$

By introducing the above values in Eq. (5), we have

$$L''(0) = \int_0^l \left(-K(s)| V(s)|^2 + \left|\frac{D}{\partial s}V(s)\right|^2\right) ds$$
$$+ \left\langle \frac{D}{\partial t}\frac{\partial h}{\partial t}, \frac{\partial h}{\partial s}\right\rangle(l, 0) - \left\langle \frac{D}{\partial t}\frac{\partial h}{\partial t}, \frac{\partial h}{\partial s}\right\rangle(0, 0).$$

Finally, since the variation is proper, $(\partial h/\partial t)(0, t) = (\partial h/\partial t)(l, t) = 0$, $t \in (-\delta, \delta)$. Thus,

$$L''(0) = \int_0^l \left(\left|\frac{D}{\partial s}V(s)\right|^2 - K| V(s)|^2\right) ds. \qquad \textbf{Q.E.D.}$$

Remark 3. Expression (4) is called the *formula for the second variation of the arc length of* γ. Observe that it depends only on the variational field of h and not on the variation h itself. Sometimes it is convenient to indicate this dependence by writing $L''_V(0)$.

Remark 4. It is often convenient to have the formula (4) for the second variation written as follows:

$$L''(0) = = \int_0^l \left\langle \frac{D^2 V}{ds^2} + KV, V\right\rangle ds. \qquad \textbf{(4a)}$$

Equation (4a) comes from Eq. (4), by noticing that $V(0) = V(l) = 0$ and that

$$\frac{d}{ds}\left\langle V, \frac{DV}{ds}\right\rangle = \left\langle \frac{DV}{ds}, \frac{DV}{ds}\right\rangle + \left\langle V, \frac{D^2 V}{ds^2}\right\rangle.$$

Thus,

$$\int_0^l \left(\left\langle \frac{DV}{ds}, \frac{DV}{ds}\right\rangle - K\langle V, V\rangle\right) ds = \left[\left\langle V, \frac{DV}{ds}\right\rangle\right]_0^l$$
$$- \int_0^l \left\langle \frac{D^2 V}{ds^2} + KV, V\right\rangle ds$$
$$= -\int_0^l \left\langle \frac{D^2 V}{ds^2} + KV, V\right\rangle ds.$$

The second variation $L''(0)$ of the arc length is the tool that we need to prove the crucial step in Bonnet's theorem, which was mentioned in the beginning of this section. We may now prove

THEOREM (Bonnet). *Let the Gaussian curvature* K *of a complete surface* S *satisfy the condition*

$$K \geq \delta > 0.$$

Then S *is compact and the diameter* ρ *of* S *satisfies the inequality*

$$\rho \leq \frac{\pi}{\sqrt{\delta}}.$$

Proof. Since S is complete, given two points $p, q \in S$, there exists, by the Hopf-Rinow theorem, a minimal geodesic γ of S joining p to q. We shall prove that the length $l = d(p, q)$ of this geodesic satisfies the inequality

$$l \leq \frac{\pi}{\sqrt{\delta}}.$$

We shall assume that $l > \pi/\sqrt{\delta}$ and consider a variation of the geodesic $\gamma: [0, l] \to S$, defined as follows. Let w_0 be a unit vector of $T_{\gamma(0)}(S)$ such that $\langle w_0, \gamma'(0)\rangle = 0$ and let $w(s)$, $s \in [0, l]$, be the parallel transport of w_0 along γ. It is clear that $|w(s)| = 1$ and that $\langle w(s), \gamma'(s)\rangle = 0$, $s \in [0, l]$. Consider the vector field $V(s)$ defined by

$$V(s) = w(s) \sin \frac{\pi}{l} s, \qquad s \in [0, l].$$

Since $V(0) = V(l) = 0$ and $\langle V(s), \gamma'(s)\rangle = 0$, the vector field $V(s)$ determines a proper, orthogonal variation of γ. By Prop. 4,

$$L''_V(0) = \int_0^l \left(\left| \frac{D}{\partial s} V(s) \right|^2 - K(s) | V(s) |^2 \right) ds.$$

Since $w(s)$ is a parallel vector field,

$$\frac{D}{\partial s} V(s) = \left(\frac{\pi}{l} \cos \frac{\pi}{l} s \right) w(s).$$

Thus, since $l > \pi/\sqrt{\delta}$, so that $K \geq \delta > \pi^2/l^2$, we obtain

$$L''_V(0) = \int_0^l \left(\frac{\pi^2}{l^2} \cos^2 \frac{\pi}{l} s - K \sin^2 \frac{\pi}{l} s \right) ds$$

$$< \int_0^l \frac{\pi^2}{l^2}\left(\cos^2 \frac{\pi}{l}s - \sin^2 \frac{\pi}{l}s\right) ds$$

$$= \frac{\pi^2}{l^2} \int_0^l \cos \frac{2\pi}{l} s\, ds = 0.$$

Therefore, there exists a variation of γ for which $L''(0) < 0$. However, since γ is a minimal geodesic, its length is smaller than or equal to that of any curve joining p to q. Thus, for every variation of γ we should have $L'(0) = 0$ and $L''(0) \geq 0$. We obtained therefore a contradiction, which shows that $l = d(p, q) \leq \pi/\sqrt{\delta}$, as we asserted.

Since $d(p, q) \leq \pi/\sqrt{\delta}$ for any two given points of S, we have that S is bounded and that its diameter $\rho \leq \pi/\sqrt{\delta}$. Moreover, since S is complete and bounded, S is compact. **Q.E.D.**

Remark 5. The choice of the variation $V(s) = w(s) \sin (\pi/l)s$ in the above proof may be better understood if we look at the second variation in the form (4a) of Remark 4. Since $K > l^2/\pi^2$, we can write

$$L''_V(0) = -\int_0^l \left\langle V, \frac{D^2V}{ds^2} + \frac{\pi^2}{l^2}V\right\rangle ds - \int_0^l \left(K - \frac{\pi^2}{l^2}\right) |V|^2\, ds$$

$$< -\int_0^l \left\langle V, \frac{D^2V}{ds^2} + \frac{\pi^2}{l^2}V\right\rangle ds.$$

Now it is easy to guess that the above $V(s)$ makes the last integrand equal to zero; hence, $L''_V(0) < 0$.

Remark 6. The hypothesis $K \geq \delta > 0$ may not be weakened to $K > 0$. In fact, the paraboloid

$$\{(x, y, z) \in R^3; z = x^2 + y^2\}$$

has Gaussian curvature $K > 0$, is complete, and is not compact. Observe that the curvature of the paraboloid tends toward zero when the distance of the point $(x, y) \in R^2$ to the origin $(0, 0)$ becomes arbitrarily large (cf. Remark 8 below).

Remark 7. The estimate of the diameter $\rho \leq \pi/\sqrt{\delta}$ given by Bonnet's theorem is the best possible, as shown by the example of the unit sphere: $K \equiv 1$ and $\rho = \pi$.

Remark 8. The first proof of the above theorem was obtained by O. Bonnet, "Sur quelquer propriétés des lignes géodésiques," *C.R.Ac. Sc. Paris* XL (1850), 1331, and "Note sur les lignes géodésiques," *ibid.* XLI (1851),

32. A formulation of the theorem in terms of complete surfaces is found in the article of Hopf-Rinow quoted in the previous section. Actually, it is not necessary that K be bounded away from zero but only that it not approach zero too fast. See E. Calabi, "On Ricci Curvature and Geodesics," *Duke Math. J.* 34 (1967), 667–676; or R. Schneider, "Konvexe Flächen mit langsam abnehmender Krümmung," *Archiv der Math.* 23 (1972), 650–654 (cf. also Exercise 2 below).

EXERCISES

1. Is the converse of Bonnet's theorem true; i.e., if S is compact and has diameter $\rho \leq \pi/\sqrt{\delta}$, is $K \geq \delta$?

***2.** (*Kazdan-Warner's Remark.* cf. Exercise 10, Sec. 5-10.) Let $S = \{z = f(x, y); (x, y) \in R^2\}$ be a complete noncompact regular surface. Show that

$$\lim_{r\to\infty} \left(\inf_{x^2+y^2 \geq r} K(x, y) \right) \leq 0.$$

3. **a.** Derive a formula for the first variation of arc length without assuming that the variation is proper.

b. Let S be a complete surface. Let $\gamma(s)$, $s \in R$, be a geodesic on S and let $d(s)$ be the distance $d(\gamma(s), p)$ from $\gamma(s)$ to a point $p \in S$ not in the trace of γ. Show that there exists a point $s_0 \in R$ such that $d(s_0) \leq d(s)$ for all $s \in R$ and that the geodesic Γ joining p to $\gamma(s_0)$ is perpendicular to γ (Fig. 5-14).

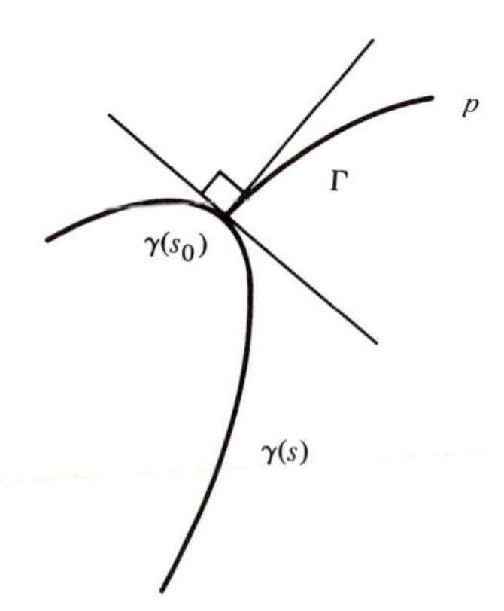

Figure 5-14

c. Assume further that S is homeomorphic to a plane and has Gaussian curvature $K \leq 0$. Prove that s_0 (hence, Γ) is unique.

4. (*Calculus of Variations.*) Geodesics are particular cases of solutions to variational problems. In this exercise, we shall discuss some points of a simple, although quite representative, variational problem. In the next exercise we shall make some applications of the ideas presented here.

Let $y = y(x)$, $x \in [x_1, x_2]$ be a differentiable curve in the xy plane and let a variation of y be given by a differentiable map $y = y(x, t)$, $t \in (-\epsilon, \epsilon)$. Here $y(x, 0) = y(x)$ for all $x \in [x_1, x_2]$, and $y(x_1, t) = y(x_1)$, $y(x_2, t) = y(x_2)$ for all $t \in (-\epsilon, \epsilon)$ (i.e., the end points of the variation are fixed). Consider the integral

$$I(t) = \int_{x_1}^{x_2} F(x, y(x, t), y'(x, t))\, dx, \qquad t \in (-\epsilon, \epsilon),$$

where $F(x, y, y')$ is a differentiable function of three variables and $y' = \partial y/\partial x$. The problem of finding the critical points of $I(t)$ is called a *variational problem with integrand F.*

a. Assume that the curve $y = y(x)$ is a critical point of $I(t)$ (i.e., $dI/dt = 0$ for $t = 0$). Use integration by parts to conclude that ($\dot{I} = dI/dt$)

$$\begin{aligned}\dot{I}(t) &= \int_{x_1}^{x_2} \left(F_y \frac{\partial y}{\partial t} + F_{y'} \frac{\partial y'}{\partial t}\right) dx \\ &= \left[\frac{\partial y}{\partial t} F_{y'}\right]_{x_1}^{x_2} + \int_{x_1}^{x_2} \frac{\partial y}{\partial t}\left(F_y - \frac{d}{dx} F_{y'}\right) dx.\end{aligned}$$

Then, by using the boundary conditions, obtain

$$0 = \dot{I}(0) = \int_{x_1}^{x_2} \left\{\eta\left(F_y - \frac{d}{dx} F_{y'}\right)\right\} dx, \tag{$*$}$$

where $\eta = (\partial y/\partial t)(x, 0)$. (The function η corresponds to the variational vector field of $y(x, t)$.)

b. Prove that if $\dot{I}(0) = 0$ for all variations with fixed end points (i.e., for all η in ($*$) with $\eta(x_1) = \eta(x_2) = 0$), then

$$F_y - \frac{d}{dx} F_{y'} = 0. \tag{$**$}$$

Equation ($**$) is called the *Euler-Lagrange equation* for the variational problem with integrand F.

c. Show that if F does not involve explicitly the variable x, i.e., $F = F(y, y')$, then, by differentiating $y'F_{y'} - F$, and using ($**$) we obtain that

$$y'F_{y'} - F = \text{const.}$$

5. (*Calculus of Variations; Applications.*)

a. (*Surfaces of Revolution of Least Area.*) Let S be a surface of revolution obtained by rotating the curve $y = f(x)$, $x \in [x_1, x_2]$, about the x axis. Suppose that S has least area among all surfaces of revolution generated by curves joining $(x_1, f(x_1))$ to $(x_2, f(x_2))$. Thus, $y = f(x)$ minimizes the integral (cf. Exercise 11, Sec. 2-5)

$$I(t) = \int_{x_1}^{x_2} y\sqrt{1 + (y')^2}\, dx$$

for all variations $y(x, t)$ of y with fixed end points $y(x_1), y(x_2)$. By part b of Exercise 4, $F(y, y') = y\sqrt{1 + (y')^2}$ satisfies the Euler-Lagrange equation (**). Use part c of Exercise 4 to obtain that

$$y'F_{y'} - F = -\frac{y}{\sqrt{1 + (y')^2}} = -\frac{1}{c}, \qquad c = \text{const.};$$

hence,

$$y = \frac{1}{c}\cosh(cx + c_1), \qquad c_1 = \text{const.}$$

Conclude that *if there exists a regular surface of revolution of least area connecting two given parallel circles, this surface is the catenoid which contains the two given circles as parallels.*

b. (*Geodesics of Surfaces of Revolution.*) Let

$$\mathbf{x}(u, v) = (f(v)\cos u, f(v)\sin u, g(v))$$

be a parametrization of a surface of revolution S. Let $u = u(v)$ be the equation of a geodesic of S which is neither a parallel nor a meridian. Then $u = u(v)$ is a critical point for the arc length integral ($F = 0$)

$$\int \sqrt{E(u')^2 + G}\, dv, \qquad u' = \frac{du}{dv}.$$

Since $E = f^2$, $G = (f')^2 + (g')^2$, we see that the Euler-Lagrange equation for this variational problem is

$$F_u - \frac{d}{dv}F_{u'} = 0, \qquad F = \sqrt{f^2(u')^2 + (f')^2 + (g')^2}.$$

Notice that F does not depend on u. Thus, $(d/dv)F_{u'} = 0$, and

$$c = \text{const.} = F_{u'} = \frac{u'f^2}{\sqrt{f^2(u')^2 + (f')^2 + (g')^2}}.$$

From this, obtain the following equation for the geodesic $u = u(v)$ (cf. Example 5, Sec. 4-4):

$$u = c\int \frac{1}{f}\frac{\sqrt{(f')^2 + (g')^2}}{f^2 - c^2}\, dv + \text{const.}$$

5-5. Jacobi Fields and Conjugate Points

In this section we shall explore some details of the variational techniques which were used to prove Bonnet's theorem.

We are interested in obtaining information on the behavior of geodesics neighboring a given geodesic γ. The natural way to proceed is to consider variations of γ which satisfy the further condition that the curves of the variation are themselves geodesics. The variational field of such a variation gives an idea of how densely the geodesics are distributed in a neighborhood of γ.

To simplify the exposition we shall assume that the surfaces are complete, although this assumption may be dropped with further work. The notation $\gamma: [0, l] \to S$ will denote a geodesic parametrized by arc length on the complete surface S.

DEFINITION 1. *Let* $\gamma: [0, l] \to \mathrm{S}$ *be a parametrized geodesic on* S *and let* $\mathrm{h}: [0, l] \times (-\epsilon, \epsilon) \to \mathrm{S}$ *be a variation of* γ *such that for every* $\mathrm{t} \in (-\epsilon, \epsilon)$ *the curve* $\mathrm{h_t(s) = h(s, t)}$, $\mathrm{s} \in [0, l]$, *is a parametrized geodesic (not necessarily parametrized by arc length). The variational field* $(\partial \mathrm{h}/\partial \mathrm{t})(\mathrm{s}, 0) = \mathrm{J(s)}$ *is called a* Jacobi field *along* γ.

A trivial example of a Jacobi field is given by the field $\gamma'(s)$, $s \in [0, l]$, of tangent vectors to the geodesic γ. In fact, by taking $h(s, t) = \gamma(s + t)$, we have

$$J(s) = \frac{\partial h}{\partial t}(s, 0) = \frac{d\gamma}{ds}.$$

We are particularly interested in studying the behavior of the geodesics neighboring $\gamma: [0, l] \to S$, which start from $\gamma(0)$. Thus, we shall consider variations $h: [0, l] \times (-\epsilon, \epsilon) \to S$ that satisfy the condition $h(0, t) = \gamma(0)$, $t \in (-\epsilon, \epsilon)$. Therefore, the corresponding Jacobi field satisfies the condition $J(0) = 0$ (see Fig. 5-15).

Before presenting a nontrivial example of a Jacobi field, we shall prove that such a field may be characterized by an analytical condition.

PROPOSITION 1. *Let* J(s) *be a Jacobi field along* $\gamma: [0, l] \to \mathrm{S}$, $\mathrm{s} \in [0, l]$. *Then* J *satisfies the so-called* Jacobi equation

$$\frac{\mathrm{D}}{\mathrm{ds}}\frac{\mathrm{D}}{\mathrm{ds}}\mathrm{J(s)} + \mathrm{K(s)}(\gamma'(\mathrm{s}) \wedge \mathrm{J(s)}) \wedge \gamma'(\mathrm{s}) = 0, \qquad \textbf{(1)}$$

where K(s) *is the Gaussian curvature of* S *at* $\gamma(\mathrm{s})$.

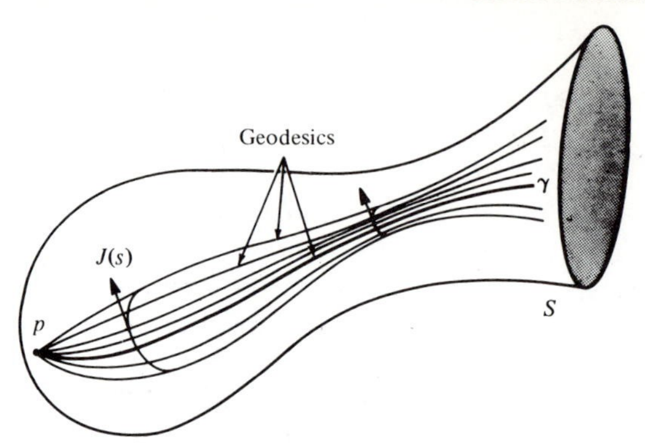

Figure 5-15

Proof. By the definition of $J(s)$, there exists a variation

$$h\colon [0, l] \times (-\epsilon, \epsilon) \longrightarrow S$$

of γ such that $(\partial h/\partial t)(s, 0) = J(s)$ and $h_t(s)$ is a geodesic, $t \in (-\epsilon, \epsilon)$. It follows that $(D/\partial s)(\partial h/\partial s)(s, t) = 0$. Therefore,

$$\frac{D}{\partial t}\frac{D}{\partial s}\frac{\partial h}{\partial s}(s, t) = 0, \qquad (s, t) \in [0, l] \times (-\epsilon, \epsilon).$$

On the other hand, by using Lemma 6 of Sec. 5-4 we have

$$\frac{D}{\partial t}\frac{D}{\partial s}\frac{\partial h}{\partial s} = \frac{D}{\partial s}\frac{D}{\partial t}\frac{\partial h}{\partial s} + K(s, t)\left(\frac{\partial h}{\partial s} \wedge \frac{\partial h}{\partial t}\right) \wedge \frac{\partial h}{\partial s} = 0.$$

Since $(D/\partial t)(\partial h/\partial s) = (D/\partial s)(\partial h/\partial t)$, we have, for $t = 0$,

$$\frac{D}{\partial s}\frac{D}{\partial s}J(s) + K(s)(\gamma'(s) \wedge J(s)) \wedge \gamma'(s) = 0. \qquad \textbf{Q.E.D.}$$

To draw some consequences from Prop. 1, it is convenient to put the Jacobi equation (1) in a more familiar form. For that, let $e_1(0)$ and $e_2(0)$ be unit orthogonal vectors in the tangent plane $T_{\gamma(0)}(S)$ and let $e_1(s)$ and $e_2(s)$ be the parallel transport of $e_1(0)$ and $e_2(0)$, respectively, along $\gamma(s)$.

Assume that

$$J(s) = a_1(s)e_1(s) + a_2(s)e_2(s)$$

for some functions $a_1 = a_1(s)$, $a_2 = a_2(s)$. Then, by using Lemma 2 of the last section and omitting s for notational simplicity, we obtain

$$\frac{D}{\partial s}J = a_1'e_1 + a_2'e_2,$$

$$\frac{D}{\partial s}\frac{D}{\partial s}J = a_1''e_1 + a_2''e_2.$$

On the other hand, if we write

$$(\gamma' \wedge J) \wedge \gamma' = \lambda_1 e_1 + \lambda_2 e_2,$$

we have

$$\begin{aligned}\lambda_1 e_1 + \lambda_2 e_2 &= (\gamma' \wedge (a_1 e_1 + a_2 e_2)) \wedge \gamma' \\ &= a_1(\gamma' \wedge e_1) \wedge \gamma' + a_2(\gamma' \wedge e_2) \wedge \gamma'.\end{aligned}$$

Therefore, by setting $\langle(\gamma' \wedge e_i) \wedge \gamma', e_j\rangle = \alpha_{ij}$, $i, j = 1, 2$, we obtain

$$\lambda_1 = a_1\alpha_{11} + a_2\alpha_{21}, \qquad \lambda_2 = a_1\alpha_{12} + a_2\alpha_{22}.$$

It follows that Eq. (1) may be written

$$\begin{aligned}a_1'' + K(\alpha_{11}a_1 + \alpha_{21}a_2) &= 0,\\ a_2'' + K(\alpha_{12}a_1 + \alpha_{22}a_2) &= 0,\end{aligned} \tag{1a}$$

where all the elements are functions of s. Note that (1a) is a system of linear, second-order differential equations. The solutions $(a_1(s), a_2(s)) = J(s)$ of such a system are defined for every $s \in [0, l]$ and constitute a vector space. Moreover, a solution $J(s)$ of (1a) (or (1)) is completely determined by the initial conditions $J(0)$, $(DJ/\partial s)(0)$, and the space of the solutions has $2 \times 2 = 4$ dimensions.

One can show that every vector field $J(s)$ along a geodesic $\gamma: [0, l] \longrightarrow S$ which satisfies Eq. (1) is, in fact, a Jacobi field. Since we are interested only in Jacobi fields $J(s)$ which satisfy the condition $J(0) = 0$, we shall prove the proposition only for this particular case.

We shall use the following notation. Let $T_p(S)$, $p \in S$, be the tangent plane to S at point p, and denote by $(T_p(S))_v$ the tangent space at v of $T_p(S)$ considered as a surface in R^3. Since $\exp_p: T_p(S) \longrightarrow S$,

$$d(\exp_p)_v: (T_p(S))_v \longrightarrow T_{\exp_p(v)}(S).$$

We shall frequently make the following notational abuse: If $v, w \in T_p(S)$, then w denotes also the vector of $(T_p(S))_v$ obtained from w by a translation of vector v (see Fig. 5-16). This is equivalent to identifying the spaces $T_p(S)$ and $(T_p(S))_v$ by the translation of vector v.

LEMMA 1. *Let* $p \in S$ *and choose* $v, w \in T_p(S)$, *with* $|v| = l$. *Let* $\gamma: [0, l] \longrightarrow S$ *be the geodesic on* S *given by*

$$\gamma(s) = \exp_p(sv), \qquad s \in [0, l].$$

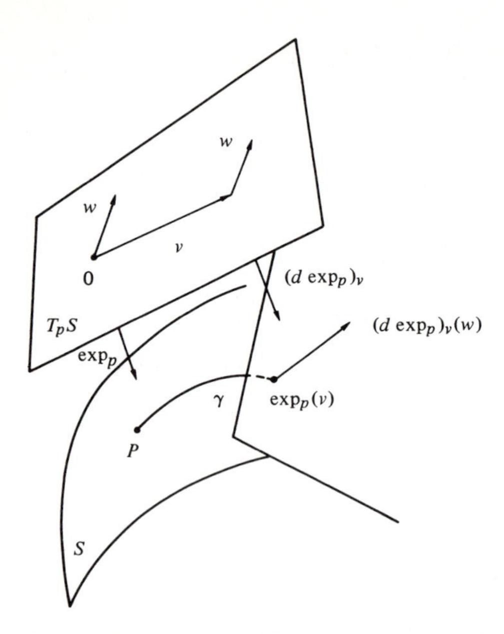

Figure 5-16

Then, the vector field J(s) *along* γ *given by*

$$\mathrm{J(s)} = \mathrm{s(d\,exp_p)_{sv}(w)}, \qquad \mathrm{s} \in [0, l],$$

is a Jacobi field. Furthermore, $\mathrm{J(0)} = 0$, $\mathrm{(DJ/ds)(0)} = \mathrm{w}$.

Proof. Let $t \longrightarrow v(t)$, $t \in (-\epsilon, \epsilon)$, be a parametrized curve in $T_p(S)$ such that $v(0) = v$ and $(dv/dt)(0) = w$. (Observe that we are making the notational abuse mentioned above.) Define (see Fig. 5-17)

$$h(s, t) = \exp_p(sv(t)), \qquad t \in (-\epsilon, \epsilon), s \in [0, l].$$

The mapping h is obviously differentiable, and the curves $s \longrightarrow h_t(s) = h(s, t)$ are the geodesics $s \longrightarrow \exp_p(sv(t))$. Therefore, the variational field of h is a Jacobi field along γ.

To compute the variational field $(\partial h/\partial t)(s, 0)$, observe that the curve of $T_p(S)$, $s = s_0$, $t = t$, is given by $t \longrightarrow s_0 v(t)$ and that the tangent vector to this curve at the point $t = 0$ is

$$s_0 \frac{dv}{dt}(0) = s_0 w.$$

It follows that

$$\frac{\partial h}{\partial t}(s, 0) = (d\exp_p)_{sv}(sw) = s(d\exp_p)_{sv}(w).$$

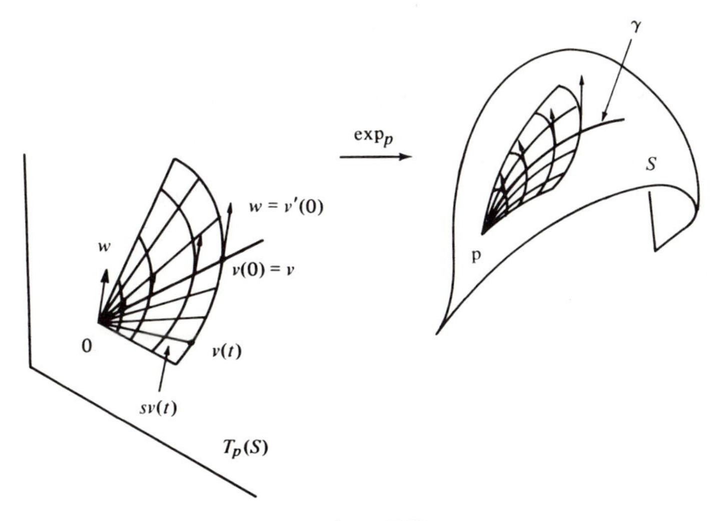

Figure 5-17

The vector field $J(s) = s(d\exp_p)_{sv}(w)$ is, therefore, a Jacobi field. It is immediate to check that $J(0) = 0$. To verify the last assertion of the lemma, we compute the covariant derivative of the above expression (cf. Lemma 2, Sec. 5-4), obtaining

$$\frac{D}{\partial s}s(d\exp_p)_{sv}(w) = (d\exp_p)_{sv}(w) + s\frac{D}{\partial s}(d\exp_p)_{sv}(w).$$

Hence, at $s = 0$,

$$\frac{DJ}{\partial s}(0) = (d\exp_p)_0(w) = w. \qquad \textbf{Q.E.D.}$$

PROPOSITION 2. *If we let* J(s) *be a differentiable vector field along* $\gamma: [0, l] \to S$, $s \in [0, l]$, *satisfying the Jacobi equation* (1), *with* $J(0) = 0$, *then* J(s) *is a Jacobi field along* γ.

Proof. Let $w = (DJ/ds)(0)$ and $v = \gamma'(0)$. By Lemma 1, there exists a Jacobi field $s(d\exp_p)_{sv}(w) = \bar{J}(s)$, $s \in [0, l]$, satisfying

$$\bar{J}(0) = 0, \quad \left(\frac{D\bar{J}}{ds}\right)(0) = w.$$

Then, J and $\bar{J}$ are two vector fields satisfying the system (1) with the same initial conditions. By uniqueness, $J(s) = \bar{J}(s)$, $s \in [0, l]$; hence, J is a Jacobi field. **Q.E.D.**

We are now in a position to present a nontrivial example of a Jacobi field.

Example. Let $S^2 = \{(x, y, z) \in R^3; x^2 + y^2 + z^2 = 1\}$ be the unit sphere and $\mathbf{x}(\theta, \varphi)$ be a parametrization at $p \in S$, by the colatitude θ and the longitude φ (Sec. 2-2, Example 1). Consider on the parallel $\theta = \pi/2$ the segment between $\varphi_0 = \pi/2$ and $\varphi_1 = 3\pi/2$. This segment is a geodesic γ, which we assume to be parametrized by $\varphi - \varphi_0 = s$. Let $w(s)$ be the parallel transport along γ of a vector $w(0) \in T_{\gamma(0)}(S)$, with $|w(0)| = 1$ and $\langle w(0), \gamma'(0)\rangle = 0$. We shall prove that the vector field (see Fig. 5-18)

$$J(t) = (\sin s)w(s), \qquad s \in [0, \pi],$$

is a Jacobi field along γ.

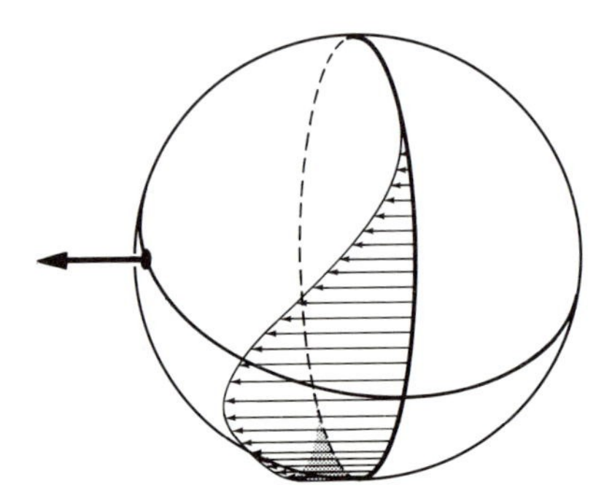

Figure 5-18. A Jacobi field on a sphere.

In fact, since $J(0) = 0$, it suffices to verify that J satisfies Eq. (1). By using the fact that $K = 1$ and w is a parallel field we obtain, sucessively,

$$\frac{DJ}{ds} = (\cos s)w(s),$$

$$\frac{D}{ds}\frac{DJ}{ds} = (-\sin s)w(s),$$

$$\frac{D}{ds}\frac{DJ}{ds} + K(\gamma' \wedge J) \wedge \gamma' = (-\sin s)w(s) + (\sin s)w(s) = 0,$$

which shows that J is a Jacobi field. Observe that $J(\pi) = 0$.

DEFINITION 2. *Let* $\gamma: [0, l] \to \mathrm{S}$ *be a geodesic of* S *with* $\gamma(0) = \mathrm{p}$. *We say that the point* $\mathrm{q} = \gamma(\mathrm{s}_0)$, $\mathrm{s}_0 \in [0, l]$, *is* conjugate *to* p *relative to the geodesic* γ *if there exists a Jacobi field* J(s) *which is not identically zero along* γ *with* $\mathrm{J}(0) = \mathrm{J}(\mathrm{s}_0) = 0$.

As we saw in the previous example, given a point $p \in S^2$ of a unit sphere S^2, its antipodal point is conjugate to p along any geodesic that starts from p.

However, the example of the sphere is not typical. In general, given a point p of a surface S, the "first" conjugate point q to p varies as we change the direction of the geodesic passing through p and describes a parametrized curve. The trace of such a curve is called the *conjugate locus* to p and is denoted by $C(p)$.

Figure 5-19 shows the situation for the ellipsoid, which is typical. The geodesics starting from a point p are tangent to the curve $C(p)$ in such a way that when a geodesic $\bar{\gamma}$ near γ approaches γ, then the intersection point of $\bar{\gamma}$ and γ approaches the conjugate point q of p relative to γ. This situation was expressed in classical terminology by saying that the conjugate point is the point of intersection of two "infinitely close" geodesics.

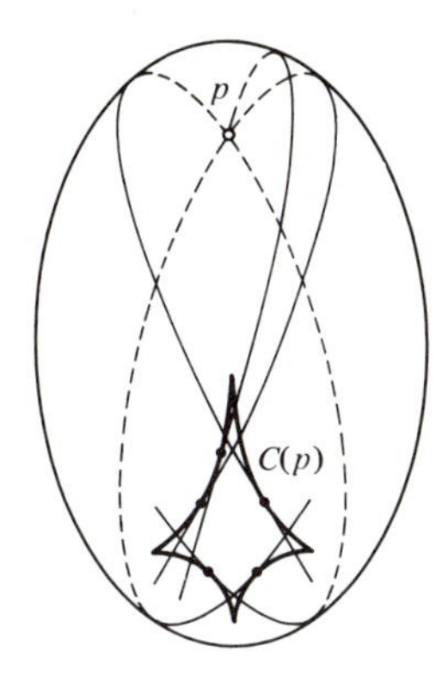

Figure 5-19. The conjugate locus of an ellipsoid.

Remark 1. The fact that, in the sphere S^2, the conjugate locus of each point $p \in S^2$ reduces to a single point (the antipodal point of p) is an exceptional situation. In fact, it can be proved that the sphere is the only such surface (cf. L. Green, "Aufwiedersehenfläche," *Ann. Math.* 78 (1963), 289–300).

Remark 2. The conjugate locus of the general ellipsoid was determined by A. Braunmühl, "Geodätische Linien auf dreiachsigen Flächen zweiten Grades," *Math. Ann.* 20 (1882), 557–586. Compare also H. Mangoldt, "Geodätische Linien auf positiv gekrümmten Flächen," *Crelles Journ.* 91 (1881), 23–52.

A useful property of Jacobi fields J along $\gamma\colon [0, l] \to S$ is the fact that when $J(0) = J(l) = 0$, then

$$\langle J(s), \gamma'(s) \rangle = 0$$

for every $s \in [0, l]$. Actually, this is a consequence of the following properties of Jacobi fields.

PROPOSITION 3. *Let* $J_1(s)$ *and* $J_2(s)$ *be Jacobi fields along* $\gamma\colon [0, l] \to S$, $s \in [0, l]$. *Then*

$$\left\langle \frac{DJ_1}{ds}, J_2(s) \right\rangle - \left\langle J_1(s), \frac{DJ_2}{ds} \right\rangle = const.$$

Proof. It suffices to differentiate the expression of the statement and apply Prop. 1 (s is omitted for notational convenience):

$$\frac{d}{ds}\left\{\left\langle \frac{DJ_1}{ds}, J_2 \right\rangle - \left\langle J_1, \frac{DJ_2}{ds} \right\rangle\right\}$$

$$= \left\langle \frac{D}{ds}\frac{DJ_1}{ds}, J_2 \right\rangle - \left\langle J_1, \frac{D}{ds}\frac{DJ_2}{ds} \right\rangle + \left\langle \frac{DJ_1}{ds}, \frac{DJ_2}{ds} \right\rangle - \left\langle \frac{DJ_1}{ds}, \frac{DJ_2}{ds} \right\rangle$$

$$= -K\{\langle \gamma' \wedge J_1) \wedge \gamma', J_2\rangle - \langle(\gamma' \wedge J_2) \wedge \gamma', J_1\rangle\} = 0.$$

Q.E.D.

PROPOSITION 4. *Let* J(s) *be a Jacobi field along* $\gamma\colon [0, l] \longrightarrow$ S, *with*

$$\langle J(s), \gamma'(s_1)\rangle = \langle J(s_2), \gamma'(s_2)\rangle = 0, \qquad s_1, s_2 \in [0, l], s_1 \neq s_2.$$

Then

$$\langle J(s), \gamma'(s)\rangle = 0, \qquad s \in [0, l].$$

Proof. We set $J_1(s) = J(s)$ and $J_2(s) = \gamma'(s)$ (which is a Jacobi field) in the previous proposition and obtain

$$\left\langle \frac{DJ}{ds}, \gamma'(s) \right\rangle = \text{const.} = A.$$

Therefore,

$$\frac{d}{ds}\langle J(s), \gamma'(s)\rangle = \left\langle \frac{DJ}{ds}, \gamma'(s) \right\rangle = A;$$

hence,

$$\langle J(s), \gamma'(s)\rangle = As + B,$$

where B is a constant. Since the linear expression $As + B$ is zero for $s_1, s_2 \in [0, l]$, $s_1 \neq s_2$, it is identically zero. **Q.E.D.**

COROLLARY. *Let* J(s) *be a Jacobi field along* $\gamma\colon [0, l] \longrightarrow$ S, *with* J(0) = J(l) = 0. *Then* $\langle J(s), \gamma'(s)\rangle = 0$, s $\in [0, l]$.

We shall now show that the conjugate points may be characterized by the behavior of the exponential map. Recall that when $\varphi\colon S_1 \longrightarrow S_2$ is a differentiable mapping of the regular surface S_1 into the regular surface S_2, a point $p \in S_1$ is said to be a *critical* point of φ if the linear map

$$d\varphi_p\colon T_p(S_1) \longrightarrow T_{\varphi(p)}(S_2)$$

is singular, that is, if there exists $v \in T_p(S_1)$, $v \neq 0$, with $d\varphi_p(v) = 0$.

PROPOSITION 5. *Let* p, q $\in$ S *be two points of* S *and let* $\gamma\colon [0, l] \longrightarrow$ S *be a geodesic joining* p $= \gamma(0)$ *to* q $= exp_p(l\gamma'(0))$. *Then* q *is conjugate to* p *relative to* γ *if and only if* v $= l\gamma'(0)$ *is a critical point of* $exp_p\colon T_p(S) \longrightarrow S$.

Proof. As we saw in Lemma 1, for every $w \in T_p(S)$ (which we identify with $(T_p(S))_v$) there exists a Jacobi field $J(s)$ along γ with

$$J(0) = 0,$$

$$\frac{DJ}{ds}(0) = w$$

and
$$J(l) = l\{(d \exp_p)_v(w)\}.$$

If $v \in T_p(S)$ is a critical point of $\exp_p$, there exists $w \in T_p(S))_v$, $w \neq 0$, with $(d \exp_p)_v(w) = 0$. This implies that the above vector field $J(s)$ is not identically zero and that $J(0) = J(l) = 0$; that is, $\gamma(l)$ is conjugate to $\gamma(0)$ relative to γ.

Conversely, if $q = \gamma(l)$ is conjugate to $p = \gamma(0)$ relative to γ, there exists a Jacobi field $\bar{J}(s)$, not identically zero, with $\bar{J}(0) = \bar{J}(l) = 0$. Let $(D\bar{J}/ds)(0) = w \neq 0$. By constructing a Jacobi field $J(s)$ as above, we obtain, by uniqueness, $\bar{J}(s) = J(s)$. Since

$$J(l) = l\{(d \exp_p)_v(w)\} = \bar{J}(l) = 0,$$

we conclude that $(d \exp_p)_v(w) = 0$, with $w \neq 0$. Therefore v is a critical point of $\exp_p$. **Q.E.D.**

The fact that Eq. (1) of Jacobi fields involves the Gaussian curvature K of S is an indication that the "spreading out" of the geodesics which start from a point $p \in S$ is closely related to the distribution of the curvature in S (cf. Remark 2, Sec. 4-6). It is an elementary fact that two neighboring geodesics starting from a point $p \in S$ initially pull apart. In the case of a sphere or an ellipsoid ($K > \delta > 0$) they reapproach each other and become tangent to the conjugate locus $C(p)$. In the case of a plane they never get closer again. The following theorem shows that an "infinitesimal version" of the situation for the plane occurs in surfaces of negative or zero curvature. (See Remark 3 after the proof of the theorem.)

THEOREM. *Assume that the Gaussian curvature* K *of a surface* S *satisfies the condition* $K \leq 0$. *Then, for every* p $\in$ S, *the conjugate locus of* p *is empty. In short, a surface of curvature* $K \leq 0$ *does not have conjugate points.*

Proof. Let $p \in S$ and let $\gamma\colon [0, l] \longrightarrow S$ be a geodesic of S with $\gamma(0) = p$. Assume that there exists a nonvanishing Jacobi field $J(s)$, with $J(0) = J(l) = 0$. We shall prove that this gives a contradiction.

In fact, since $J(s)$ is a Jacobi field and $J(0) = J(l) = 0$, we have, by the corollary of Prop. 4, that $\langle J(s), \gamma'(s)\rangle = 0$, $s \in [0, l]$. Therefore,

$$\frac{D}{ds}\frac{DJ}{ds} + KJ = 0,$$

$$\left\langle \frac{D}{ds}\frac{DJ}{ds}, J\right\rangle = -K\langle J, J\rangle \geq 0,$$

since $K \leq 0$.

It follows that

$$\frac{d}{ds}\left\langle \frac{DJ}{ds}, J\right\rangle = \left\langle \frac{D}{ds}\frac{DJ}{ds}, J\right\rangle + \left\langle \frac{DJ}{ds}, \frac{DJ}{ds}\right\rangle \geq 0.$$

Therefore, the function $\langle DJ/ds, J\rangle$ does not decrease in the interval $[0, l]$. Since this function is zero for $s = 0$ and $s = l$, we conclude that

$$\left\langle \frac{DJ}{ds}, J(s)\right\rangle = 0, \quad s \in [0, l].$$

Finally, by observing that

$$\frac{d}{ds}\langle J, J\rangle = 2\left\langle \frac{DJ}{ds}, J\right\rangle = 0,$$

we have $|J|^2 = \text{const}$. Since $J(0) = 0$, we conclude that $|J(s)| = 0, s \in [0, l]$; that is, J is identically zero in $[0, l]$. This is a contradiction. **Q.E.D.**

Remark 3. The theorem does not assert that two geodesics starting from a given point will never meet again. Actually, this is false, as shown by the closed geodesics of a cylinder, the curvature of which is zero. The assertion is not true even if we consider geodesics that start from a given point with "nearby directions." It suffices to consider a meridian of the cylinder and to observe that the helices that follow directions nearby that of the meridian meet this meridian. What the proposition asserts is that the intersection point of two "neighboring" geodesics goes to "infinity" as these geodesics approach each other (this is precisely what occurs in the cylinder). In a classical terminology we can say that two "infinitely close" geodesics never meet. In this sense, the theorem is an infinitesimal version of the situation for the plane.

An immediate consequence of Prop. 5, the above theorem, and the inverse function theorem is the following corollary.

COROLLARY. *Assume the Gaussian curvature* K *of* S *to be negative or zero. Then for every* $p \in S$, *the mapping*

$$exp_p \colon T_p(S) \longrightarrow S$$

is a local diffeomorphism.

We shall use later the following lemma, which generalizes the fact that, in a normal neighborhood of p, the geodesic circles are orthogonal to the radial geodesics (Sec. 4-6, Prop. 3 and Remark 1).

LEMMA 2 (Gauss). *Let* $p \in S$ *be a point of a* (complete) *surface* S *and let* $u \in T_p(S)$ *and* $w \in (T_p(S))_u$. *Then*

$$\langle u, w \rangle = \langle (d\, exp_p)_u(u), (d\, exp_p)_u(w) \rangle,$$

where the identification $T_p(S) \approx (T_p(S))_u$ *is being used.*

Proof. Let $l = |u|$, $v = u/|u|$ and let $\gamma \colon [0, l] \longrightarrow S$ be a geodesic of S given by

$$\gamma(s) = \exp_p(sv), \qquad s \in [0, l].$$

Then $\gamma'(0) = v$. Furthermore, if we consider the curve $s \longrightarrow sv$ in $T_p(S)$ which passes through u for $s = l$ with tangent vector v (see Fig. 5-20), we obtain

$$\gamma'(l) = \frac{d}{ds}(\exp_p sv)\bigg|_{s=l} = (d \exp_p)_u(v).$$

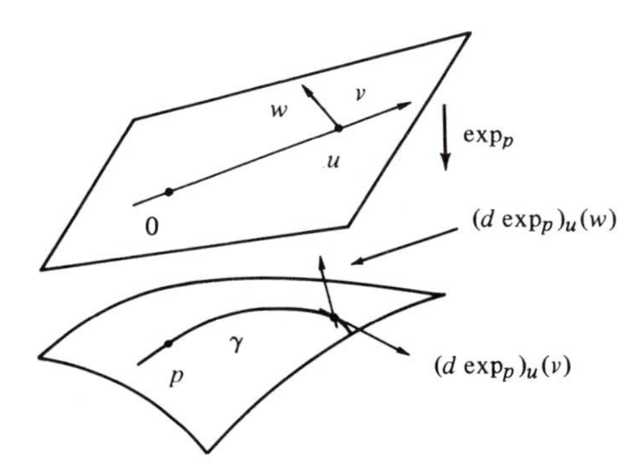

Figure 5-20

Consider now a Jacobi field J along γ, given by $J(0) = 0$, $(DJ/ds)(0) = w$ (cf. Lemma 1). Then, since $\gamma(s)$ is a geodesic,

$$\frac{d}{ds}\langle \gamma'(s), J(s) \rangle = \left\langle \gamma'(s), \frac{DJ}{ds} \right\rangle,$$

and since J is a Jacobi field,

$$\frac{d}{ds}\left\langle \gamma'(s), \frac{DJ}{ds} \right\rangle = \left\langle \gamma'(s), \frac{D^2J}{ds^2} \right\rangle = 0.$$

It follows that

$$\frac{d}{ds}\langle \gamma'(s), J(s) \rangle = \left\langle \gamma'(s), \frac{DJ}{ds} \right\rangle = \text{const.} = C; \tag{2}$$

hence (since $J(0) = 0$)

$$\langle \gamma'(s), J(s) \rangle = Cs. \tag{3}$$

To compute the constant C, set s equal to l in Eq. (3). By Lemma 1,

$$J(l) = l(d\exp_p)_u(w).$$

Therefore,

$$Cl = \langle \gamma'(l), J(l) \rangle = \langle (d\exp_p)_u(v), l(d\exp_p)_u(w) \rangle.$$

From Eq. (2) we conclude that

$$\left\langle \gamma'(l), \frac{DJ}{ds}(l) \right\rangle = C = \left\langle \gamma'(0), \frac{DJ}{ds}(0) \right\rangle = \langle v, w \rangle.$$

By using the value of C, we obtain from the above expression

$$\langle u, w \rangle = \langle (d\exp_p)_u(u), (d\exp_p)_u(w) \rangle. \qquad \textbf{Q.E.D.}$$

EXERCISES

1. **a.** Let $\gamma\colon [0, l] \longrightarrow S$ be a geodesic parametrized by arc length on a surface S and let $J(s)$ be a Jacobi field along γ with $J(0) = 0$, $\langle J'(0), \gamma'(0) \rangle = 0$. Prove that $\langle J(s), \gamma'(s) \rangle = 0$ for all $s \in [0, l]$.

   b. Assume further that $|J'(0)| = 1$. Take the parallel transport of $e_1(0) = \gamma'(0)$ and of $e_2(0) = J'(0)$ along γ and obtain orthonormal bases $\{e_1(s), e_2(s)\}$ for all $T_{\gamma(s)}(S)$, $s \in [0, l]$. By part a, $J(s) = u(s)e_2(s)$ for some function $u = u(s)$. Show that the Jacobi equation for J can be written as

   $$u''(s) + K(s)u(s) = 0,$$

   with initial conditions $u(0) = 0$, $u'(0) = 1$.

2. Show that the point $p = (0, 0, 0)$ of the paraboloid $z = x^2 + y^2$ has no conjugate point relative to a geodesic $\gamma(s)$ with $\gamma(0) = p$.

3. (*The Comparison Theorems.*) Let S and $\tilde{S}$ be complete surfaces. Let $p \in S$, $\tilde{p} \in \tilde{S}$ and choose a linear isometry $i: T_p(S) \longrightarrow T_{\tilde{p}}(\tilde{S})$. Let $\gamma: [0, \infty) \longrightarrow S$ be a geodesic on S with $\gamma(0) = p$, $|\gamma'(0)| = 1$, and let $J(s)$ be a Jacobi field along γ with $J(0) = 0$, $\langle J'(0), \gamma'(0)\rangle = 0$, $|J'(0)| = 1$. By using the linear isometry i, construct a geodesic $\tilde{\gamma}: [0, \infty) \longrightarrow \tilde{S}$ with $\tilde{\gamma}(0) = \tilde{p}$, $\tilde{\gamma}'(0) = i(\gamma'(0))$, and a Jacobi field $\tilde{J}$ along $\tilde{\gamma}$ with $\tilde{J}(0) = 0$, $\tilde{J}'(0) = i(J'(0))$ (Fig. 5-21). Below we shall describe two theorems (which are essentially geometric interpretations of the classical Sturm comparison theorems) that allow us to compare the Jacobi fields J and $\tilde{J}$ from a "comparison hypothesis" on the curvatures of S and $\tilde{S}$.

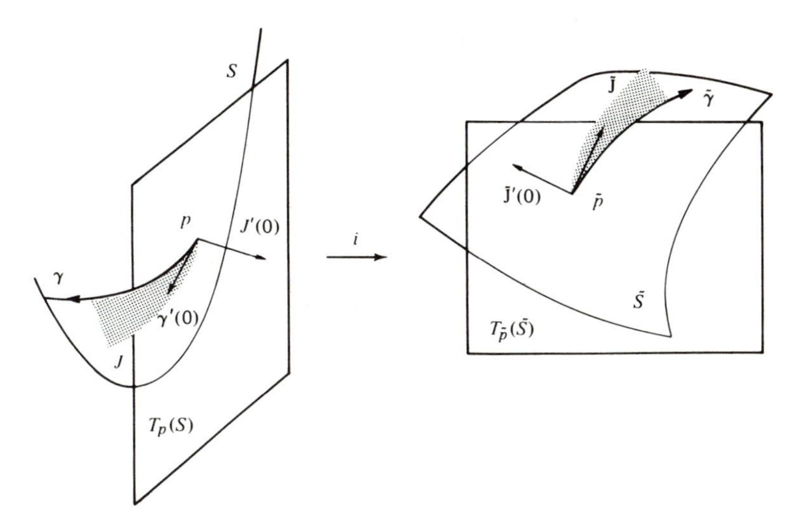

Figure 5-21

a. Use Exercise 1 to show that $J(s) = v(s)e_2(s)$, $\tilde{J}(s) = u(s)\tilde{e}_2(s)$, where $u = u(s)$, $v = v(s)$ are differentiable functions, and $e_2(s)$ (respectively, $\tilde{e}_2(s)$) is the parallel transport along γ (respectively, $\tilde{\gamma}$) of $J'(0)$ (respectively, $\tilde{J}'(0)$). Conclude that the Jacobi equations for J and $\tilde{J}$ are

$$v''(s) + K(s)v(s) = 0, \qquad v(0) = 0,\ v'(0) = 1,$$
$$u''(s) + \tilde{K}(s)u(s) = 0, \qquad u(0) = 0,\ u'(0) = 1,$$

respectively, where K and $\tilde{K}$ denote the Gaussian curvatures of S and $\tilde{S}$.

***b.** Assume that $K(s) \leq \tilde{K}(s)$, $s \in [0, \infty]$. Show that

$$\begin{aligned} 0 &= \int_0^s \{u(v'' + Kv) - v(u'' + \tilde{K}u)\}ds \\ &= [uv' - vu']_0^s + \int_0^s (K - \tilde{K})uv\, ds. \end{aligned} \tag{$*$}$$

Conclude that if a is the first zero of u in $(0, \infty)$ (i.e., $u(a) = 0$ and $u(s) > 0$ in

$(0, a)$) and b is the first zero of v in $(0, \infty)$, then $b \geq a$. *Thus, if* $\mathrm{K(s)} \leq \tilde{\mathrm{K}}(\mathrm{s})$ *for all* s, *the first conjugate point of* p *relative to* γ *does not occur before the first conjugate point of* $\tilde{\mathrm{p}}$ *relative to* $\tilde{\gamma}$. This is called the *first comparison theorem.*

***c.** Assume that $K(s) \leq \tilde{K}(s)$, $s \in [0, a)$. Use $(*)$ and the fact that u and v are positive in $(0, a)$ to obtain that $[uv' - vu']_0^s \geq 0$. Use this inequality to show that $v(s) \geq u(s)$ *for all* $s \in (0, a)$. *Thus, if* $\mathrm{K(s)} \leq \tilde{\mathrm{K}}(\mathrm{s})$ *for all* s *before the first conjugate point of* $\tilde{\gamma}$, *then* $|\mathrm{J(s)}| \geq |\tilde{\mathrm{J}}(\mathrm{s})|$ *for all such* s. This is called the *second comparison theorem* (of course, this includes the first one as a particular case; we have separated the first case because it is easier and because it is the one that we use more often).

d. Prove that in part c the equality $v(s) = u(s)$ holds for all $s \in [0, a)$ if and only if $K(s) = \tilde{K}(s)$, $s \in [0, a)$.

4. Let S be a complete surface with Gaussian curvature $K \leq K_0$, where K_0 is a positive constant. Compare S with a sphere $S^2(K_0)$ with curvature K_0 (that is, set, in Exercise 3, $\tilde{S} = S^2(K_0)$ and use the first comparison theorem, Exercise 3, part b) to conclude that any geodesic $\gamma\colon [0, \infty) \longrightarrow S$ on S has no point conjugate to $\gamma(0)$ in the interval $(0, \pi/\sqrt{K_0})$.

5. Let S be a complete surface with $K \geq K_1 > 0$, where K is the Gaussian curvature of S and K_1 is a constant. Prove that every geodesic $\gamma\colon [0, \infty) \longrightarrow S$ has a point conjugate to $\gamma(0)$ in the interval $(0, \pi/\sqrt{K_1}]$.

***6.** (*Sturm's Oscillation Theorem.*) The following slight generalization of the first comparison theorem (Exercise 3, part b) is often useful. Let S be a complete surface and $\gamma\colon [0, \infty) \longrightarrow S$ be a geodesic in S. Let $J(s)$ be a Jacobi field along γ with $J(0) = J(s_0) = 0$, $s_0 \in (0, \infty)$ and $J(s) \neq 0$, for $s \in (0, s_0)$. Thus, $J(s)$ is a normal field (corollary of Prop. 4). It follows that $J(s) = v(s)e_2(s)$, where $v(s)$ is a solution of

$$v''(s) + K(s)v(s) = 0, \qquad s \in [0, \infty),$$

and $e_2(s)$ is the parallel transport of a unit vector at $T_{\gamma(0)}(S)$ normal to $\gamma'(0)$. Assume that the Gaussian curvature $K(s)$ of S satisfies $K(s) \leq L(s)$, where L is a differentiable function on $[0, \infty)$. Prove that any solution of

$$u''(s) + L(s)u(s) = 0, \qquad s \in [0, \infty),$$

has a zero in the interval $(0, s_0]$ (i.e., there exists $s_1 \in (0, s_0]$ with $u(s_1) = 0$).

7. (*Kneser Criterion for Conjugate Points.*) Let S be a complete surface and let $\gamma\colon [0, \infty) \longrightarrow S$ be a geodesic on S with $\gamma(0) = p$. Let $K(s)$ be the Gaussian curvature of S along γ. Assume that

$$\int_t^\infty K(s)\,ds \leq \frac{1}{4(t+1)} \qquad \text{for all } t \geq 0 \tag{$*$}$$

in the sense that the integral converges and is bounded as indicated.

a. Define

$$w(t) = \int_t^\infty K(s)\,ds + \frac{1}{4(t+1)}, \qquad t \geq 0,$$

and show that $w'(t) + (w(t))^2 \leq -K(t)$.

b. Set, for $t \geq 0$, $w'(t) + (w(t))^2 = -L(t)$ (so that $L(t) \geq K(t)$) and define

$$v(t) = \exp\left(\int_0^t w(s)\,ds\right), \qquad t \geq 0.$$

Show that $v''(t) + L(t)v(t) = 0$, $v(0) = 1$, $v'(0) = 0$.

c. Notice that $v(t) > 0$ and use the Sturm oscillation theorem (Exercise 6) to show that there is no Jacobi field $J(s)$ along $\gamma(s)$ with $J(0) = 0$ and $J(s_0) = 0$, $s_0 \in (0, \infty)$. *Thus, if* $(*)$ *holds, there is no point conjugate to* p *along* γ.

***8.** Let $\gamma\colon [0, l] \longrightarrow S$ be a geodesic on a complete surface S, and assume that $\gamma(l)$ is not conjugate to $\gamma(0)$. Let $w_0 \in T_{\gamma(0)}(S)$ and $w_1 \in T_{\gamma(l)}(S)$. Prove that there exists a unique Jacobi field $J(s)$ along γ with $J(0) = w_0$, $J(l) = w_1$.

9. Let $J(s)$ be a Jacobi field along a geodesic $\gamma\colon [0, l] \longrightarrow S$ such that $\langle J(0), \gamma'(0)\rangle = 0$ and $J'(0) = 0$. Prove that $\langle J(s), \gamma'(s)\rangle = 0$ for all $s \in [0, l]$.

5-6. Covering Spaces; The Theorems of Hadamard

We saw in the last section that when the curvature K of a complete surface S satisfies the condition $K \leq 0$ then the mapping $\exp_p\colon T_p(S) \longrightarrow S$, $p \in S$, is a local diffeomorphism. It is natural to ask when this local diffeomorphism is a global diffeomorphism. It is convenient to put this question in a more general setting for which we need the notion of covering space.

A. Covering Spaces

DEFINITION 1. *Let* $\tilde{B}$ *and* B *be subsets of* R^3. *We say that* $\pi\colon \tilde{B} \longrightarrow B$ *is a* covering map *if*

1. π *is continuous and* $\pi(\tilde{B}) = B$.
2. *Each point* p $\in$ B *has a neighborhood* U *in* B (*to be called a* distinguished neighborhood *of* p) *such that*

$$\pi^{-1}(U) = \bigcup_\alpha V_\alpha,$$

where the V_α*'s are pairwise disjoint open sets such that the restriction of* π *to* V_α *is a homeomorphism of* V_α *onto* U.

$\tilde{B}$ *is then called a* covering space *of* B.

Example 1. Let $P \subset R^3$ be a plane of R^3. By fixing a point $q_0 \in P$ and two orthogonal unit vectors $e_1, e_2 \in P$, with origin in q_0, every point $q \in P$ is characterized by coordinates $(u, v) = q$ given by

$$q - u_0 = ue_1 + ve_2.$$

Now let $S = \{(x, y, z) \in R^3;\ x^2 + y^2 = 1\}$ be the right circular cylinder whose axis is the z axis, and let $\pi: P \longrightarrow S$ be the map defined by

$$\pi(u, v) = (\cos u, \sin u, v)$$

(the geometric meaning of this map is to wrap the plane P around the cylinder S an infinite number of times; see Fig. 5-22).

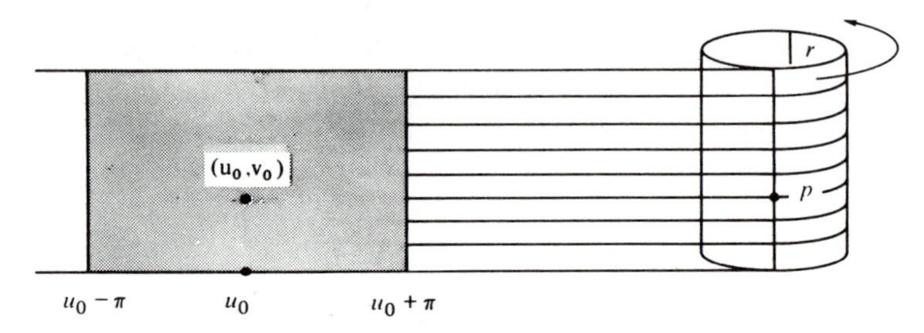

Figure 5-22

We shall prove that π is a covering map. We first observe that when $(u_0, v_0) \in P$, the mapping π restricted to the band

$$R = \{(u, v) \in P; u_0 - \pi \leq u \leq u_0 + \pi\}$$

covers S entirely. Actually, π restricted to the interior of R is a parametrization of S, the coordinate neighborhood of which covers S minus a generator. It follows that π is continuous (actually, differentiable) and that $\pi(P) = S$, thus verifying condition 1.

To verify condition 2, let $p \in S$ and $U = S - r$, where r is the generator opposite to the generator passing through p. We shall prove that U is a distinguished neighborhood of p.

Let $(u_0, v_0) \in P$ be such that $\pi(u_0, v_0) = p$ and choose for V_n the band given by

$$V_n = \{(u, v) \in P; u_0 + (2n - 1)\pi < u < u_0 + (2n + 1)\pi\},$$
$$n = 0, \pm 1, \pm 2, \ldots.$$

It is immediate to verify that if $n \neq m$, then $V_n \cap V_m = \phi$ and that $\bigcup_n V_n$

$= \pi^{-1}(U)$. Moreover, by the initial observation, π restricted to any V_n is a homeomorphism onto U. It follows that U is a distinguished neighborhood of p. This verifies condition 2 and shows that the plane P is a covering space of the cylinder S.

Example 2. Let H be the helix

$$H = \{(x, y, z) \in R^3; x = \cos t, y = \sin t; z = bt, t \in R\}$$

and let

$$S^1 = \{(x, y, 0) \in R^3; x^2 + y^2 = 1\}$$

be a unit circle. Let $\pi: H \longrightarrow S^1$ be defined by

$$\pi(x, y, z) = (x, y, 0).$$

We shall prove that π is a covering map (see Fig. 5-23).

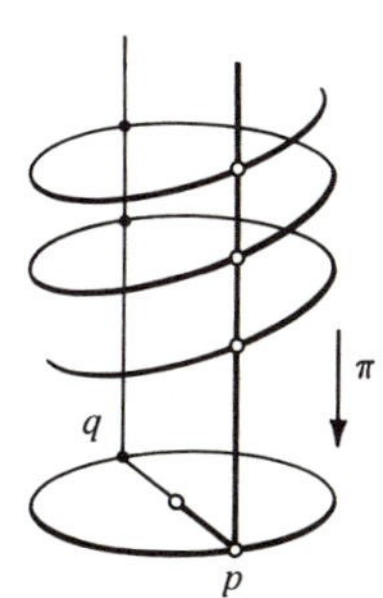

Figure 5-23

It is clear that π is continuous and that $\pi(H) = S^1$. This verifies condition 1.

To verify condition 2, let $p \in S^1$. We shall prove that $U = S^1 - \{q\}$, where $q \in S^1$ is the point symmetric to p, is a distinguished neighborhood of p. In fact, let $t_0 \in R$ be such that

$$\pi(\cos t_0, \sin t_0, bt_0) = p.$$

Let us take for V_n the arc of the helix corresponding to the interval

$$(t_0 + (2n - 1)\pi, t_0 + (2n + 1)\pi) \subset R, \qquad n = 0, \pm 1, \pm 2, \ldots.$$

Then it is easy to show that $\pi^{-1}(U) = \bigcup_n V_n$, that the V_n's are pairwise disjoint, and that π restricted to V_n is a homeomorphism onto U. This verifies condition 2 and concludes the example.

Now, let $\pi: \tilde{B} \longrightarrow B$ be a covering map. Since $\pi(\tilde{B}) = B$, each point $\tilde{p} \in \tilde{B}$ is such that $\tilde{p} \in \pi^{-1}(p)$ for some $p \in B$. Therefore, there exists a neighborhood V_α of $\tilde{p}$ such that π restricted to V_α is a homeomorphism. It follows that π is a local homeomorphism. The following example shows, however, that there exist local homeomorphisms which are not covering maps.

Before presenting the example it should be observed that if U is a distinguished neighborhood of p, then every neighborhood $\bar{U}$ of p such that $\bar{U} \subset U$ is again a distinguished neighborhood of p. Since $\pi^{-1}(\bar{U}) \subset \bigcup_\alpha V_\alpha$ and the V_α are pairwise disjoint, we obtain

$$\pi^{-1}(\bar{U}) = \bigcup_\alpha W_\alpha,$$

where the sets $W_\alpha = \pi^{-1}(\bar{U}) \cap V_\alpha$ still satisfy the disjointness condition 2 of Def. 1. In this way, when dealing with distinguished neighborhoods, we may restrict ourselves to "small" neighborhoods.

Example 3. Consider in Example 2 a segment $\tilde{H}$ of the helix H corresponding to the interval $(\pi, 4\pi) \subset R$. It is clear that the restriction $\tilde{\pi}$ of π to this open segment of helix is still a local homeomorphism and that $\tilde{\pi}(\tilde{H}) = S^1$. However, no neighborhood of

$$\pi(\cos 3\pi, \sin 3\pi, b3\pi) = (-1, 0, 0) = p \in S^1$$

can be a distinguished neighborhood. In fact, by taking U sufficiently small, $\tilde{\pi}^{-1}(U) = V_1 \cup V_2$, where V_1 is the segment of helix corresponding to $t \in (\pi, \pi + \epsilon)$ and V_2 is the segment corresponding to $t \in (3\pi - \epsilon, 3\pi + \epsilon)$. Now $\tilde{\pi}$ restricted to V_1 is not a homeomorphism onto U since $\tilde{\pi}(V_1)$ does not even contain p. It follows that $\tilde{\pi}\colon \tilde{H} \to S$ is a local homeomorphism onto S^1 but not a covering map.

We may now rephrase the question we posed in the beginning of this chapter in the following more general form: Under what conditions is a local homeomorphism a global homeomorphism?

The notion of covering space allows us to break up this question into two questions as follows:

1. Under what conditions is a local homeomorphism a covering map?
2. Under what conditions is a covering map a global homeomorphism?

A simple answer to question 1 is given by the following proposition.

PROPOSITION 1. *Let* $\pi\colon \tilde{\mathrm{B}} \to \mathrm{B}$ *be a local homeomorphism,* $\tilde{\mathrm{B}}$ *compact and* B *connected. Then* π *is a covering map.*

Proof. Since π is a local homeomorphism $\pi(\tilde{B}) \subset B$ is open in B. Moreover, by the continuity of π, $\pi(\tilde{B})$ is compact, and hence closed in B. Since $\pi(\tilde{B}) \subset B$ is open and closed in the connected set B, $\pi(\tilde{B}) = B$. Thus condition 1 of Def. 1 is verified.

To verify condition 2, let $b \in B$. Then $\pi^{-1}(B) \subset \tilde{B}$ is finite. Otherwise, it would have a limit point $\tilde{q} \in \tilde{B}$ which would contradict the fact that $\pi: \tilde{B} \longrightarrow B$ is a local homeomorphism. Therefore, we may write $\pi^{-1}(b) = \{\tilde{b}_1, \ldots, \tilde{b}_k\}$.

Let W_i be a neighborhood of $\tilde{b}_i$, $i = 1, \ldots, k$, such that the restriction of π to W_i is a homeomorphism (π is a local homeomorphism). Since $\pi^{-1}(b)$ is finite, it is possible to choose the W_i's sufficiently small so that they are pairwise disjoint. Clearly there exists a neighborhood U of b such that $U \subset \bigcap(\pi(W_i))$ (see Fig. 5-24). By setting $V_i = \pi^{-1}(U) \cap W_i$ we have that

$$\pi^{-1}(U) = \bigcup_i V_i$$

and that the V_i's are pairwise disjoint. Moreover, the restriction of π to V_i is clearly a homeomorphism onto U. It follows that U is a distinguished neighborhood of p. This verifies condition 2 and concludes the proof. **Q.E.D.**

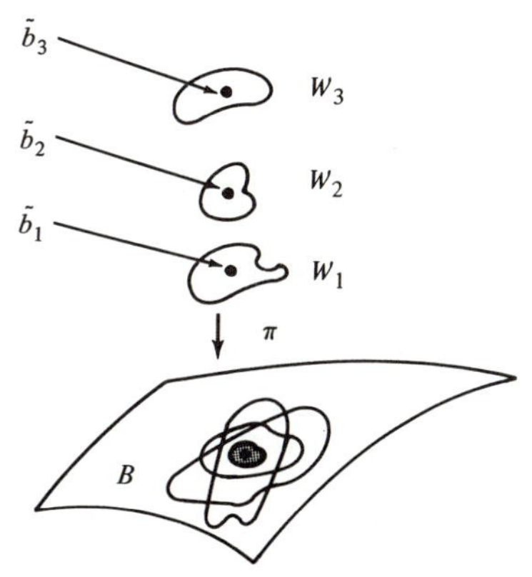

Figure 5-24

When $\tilde{B}$ is not compact there are few useful criteria for asserting that a local homeomorphism is a covering map. A special case will be treated later. For this special case as well as for a treatment of question 2 we need to return to covering spaces.

The most important property of a covering map is the possibility of "lifting" into $\tilde{B}$ continuous curves of B. To be more precise we shall introduce the following terminology.

Let $B \subset R^3$. Recall that a continuous mapping $\alpha: [0, l] \longrightarrow B$, $[0, l] \subset R$, is called an arc of B (see the appendix to Chap. 5, Def. 8). Now, let $\tilde{B}$ and B

be subsets of R^3. Let $\pi: \tilde{B} \longrightarrow B$ be a continuous map and $\alpha: [0, l] \longrightarrow B$ be an arc of B. If there exists an arc of $\tilde{B}$,

$$\tilde{\alpha}: [0, l] \longrightarrow \tilde{B},$$

with $\pi \circ \tilde{\alpha} = \alpha$, $\tilde{\alpha}$ is said to be a *lifting* of α with origin in $\tilde{\alpha}(0) \in \tilde{B}$. The situation is described in the accompanying diagram.

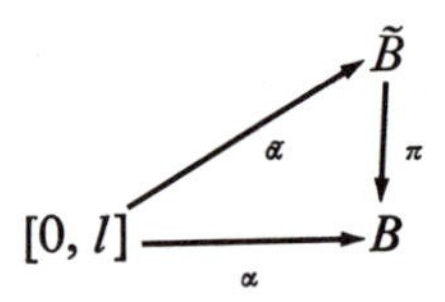

With the above terminology a fundamental property of covering spaces is expressed by the following proposition of existence and uniqueness.

PROPOSITION 2. *Let* $\pi: \tilde{B} \longrightarrow B$ *be a covering map,* $\alpha: [0, l] \longrightarrow B$ *an arc in* B, *and* $\tilde{p}_0 \in \tilde{B}$ *a point of* $\tilde{B}$ *such that* $\pi(\tilde{p}_0) = \alpha(0) = p_0$. *Then there exists a unique lifting* $\tilde{\alpha}: [0, l] \longrightarrow \tilde{B}$ *of* α *with origin at* $\tilde{p}_0$, *that is, with* $\tilde{\alpha}(0) = \tilde{p}_0$.

Proof. We first prove the uniqueness. Let $\tilde{\alpha}, \tilde{\beta}: [0, l] \longrightarrow \tilde{B}$ be two liftings of α with origin at $\tilde{p}_0$. Let $A \subset [0, l]$ be the set of points $t \in [0, l]$ such that $\tilde{\alpha}(t) = \tilde{\beta}(t)$. A is nonempty and clearly closed in $[0, l]$.

We shall prove that A is open in $[0, l]$. Suppose that $\tilde{\alpha}(t) = \tilde{\beta}(t) = \tilde{p}$. Consider a neighborhood V of $\tilde{p}$ in which π is a homeomorphism. Since $\tilde{\alpha}$ and $\tilde{\beta}$ are continuous maps, there exists an open interval $I_t \subset [0, l]$ containing t such that $\tilde{\alpha}(I_t) \subset V$ and $\tilde{\beta}(I_t) \subset V$. Since $\pi \circ \tilde{\alpha} = \pi \circ \tilde{\beta}$ and π is a homeomorphism in V, $\tilde{\alpha} = \tilde{\beta}$ in I_t, and thus A is open. It follows that $A = [0, l]$, and the two liftings coincide for every $t \in [0, l]$.

We shall now prove the existence. Since α is continuous, for every $\alpha(t) \in B$ there exists an interval $I_t \subset [0, l]$ containing t such that $\alpha(I_t)$ is contained in a distinguished neighborhood of $\alpha(t)$. The family I_t, $t \in [0, l]$, is an open covering of $[0, l]$ that, by compactness of $[0, l]$, admits a finite subcovering, say, $I_0, \ldots, I_n$.

Assume that $0 \in I_0$. (If it did not, we would change the enumeration of the intervals.) Since $\alpha(I_0)$ is contained in a distinguished neighborhood U_0 of p, there exists a neighborhood V_0 of $\tilde{p}_0$ such that the restriction π_0 of π to V_0 is a homeomorphism onto U_0. We define, for $t \in I_0$ (see Fig. 5-25),

$$\tilde{\alpha}(t) = \pi_0^{-1} \circ \alpha(t),$$

where π_0^{-1} is the inverse map in U_0 of the homeomorphism π_0. It is clear that

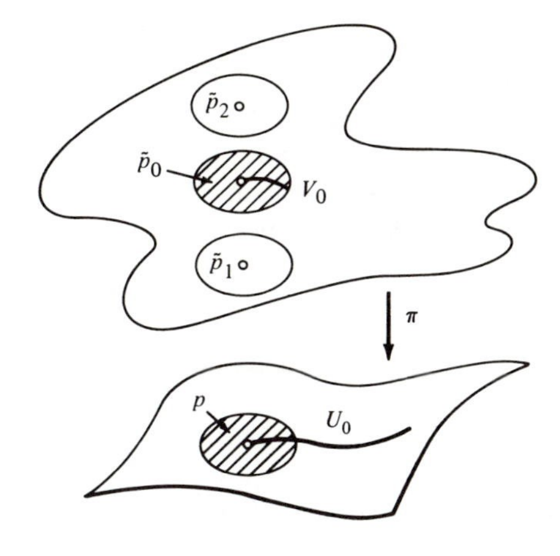

Figure 5-25

$$\tilde{\alpha}(0) = \tilde{p}_0,$$
$$\pi \circ \tilde{\alpha}(t) = \alpha(t), \qquad t \in I_0.$$

Suppose now that $I_1 \cap I_0 \neq \phi$ (otherwise we would change the order of the intervals). Let $t_1 \in I_1 \cap I_0$. Since $\alpha(I_1)$ is contained in a distinguished neighborhood U_1 of $\alpha(t_1)$, we may define a lifting of α in I_1 with origin at $\tilde{\alpha}(t_1)$. By uniqueness, this arc agrees with $\tilde{\alpha}$ in $I_1 \cap I_0$, and, therefore, it is an extension of $\tilde{\alpha}$ to $I_0 \cup I_1$. Proceeding in this manner, we build an arc $\tilde{\alpha}: [0, l] \longrightarrow \tilde{B}$ such that $\tilde{\alpha}(0) = \tilde{p}_0$ and $\pi \circ \tilde{\alpha}(t) = \alpha(t)$, $t \in [0, l]$. **Q.E.D.**

An interesting consequence of the arc lifting property of a covering map $\pi: \tilde{B} \longrightarrow B$ is the fact that when B is arcwise connected there exists a one-to-one correspondence between the sets $\pi^{-1}(p)$ and $\pi^{-1}(q)$, where p and q are two arbitrary points of B. In fact, if B is arcwise connected, there exists an arc $\alpha: [0, l] \longrightarrow B$, with $\alpha(0) = p$ and $\alpha(l) = q$. For every $\tilde{p} \in \pi^{-1}(p)$, there is a lifting $\tilde{\alpha}_p: [0, l] \longrightarrow \tilde{B}$, with $\tilde{\alpha}_p(0) = \tilde{p}$. Now define $\varphi: \pi^{-1}(p) \longrightarrow \pi^{-1}(q)$ by $\varphi(\tilde{p}) = \tilde{\alpha}_p(l)$; that is, let $\varphi(\tilde{p})$ be the extremity of the lifting of α with origin $\tilde{p}$. By the uniqueness of the lifting, φ is a one-to-one correspondence as asserted.

It follows that the "number" of points of $\pi^{-1}(p)$, $p \in B$, does not depend on p when B is arcwise connected. If this number is finite, it is called the *number of sheets* of the covering. If $\pi^{-1}(p)$ is not finite, we say that the covering is infinite. Examples 1 and 2 are infinite coverings. Observe that when $\tilde{B}$ is compact the covering is always finite.

Example 4. Let

$$S^1 = \{(x, y) \in R^2; x = \cos t, y = \sin t, t \in R\}$$

be the unit circle and define a map $\pi: S^1 \longrightarrow S^1$ by

$$\pi(\cos t, \sin t) = (\cos kt, \sin kt),$$

where k is a positive integer and $t \in R$. By the inverse function theorem, π is a local diffeomorphism, and hence a local homeomorphism. Since S^1 is compact, Prop. 1 can be applied. Thus, $\pi: S^1 \longrightarrow S^1$ is a covering map.

Geometrically, π wraps the first S^1 k times onto the second S^1. Notice that the inverse image of a point $p \in S^1$ contains exactly k points. Thus, π is a k-sheeted covering of S^1.

For the treatment of question 2 we also need to make precise some intuitive ideas which arise from the following considerations. In order that a covering map $\pi: \tilde{B} \longrightarrow B$ be a homeomorphism it suffices that it is a one-to-one map. Therefore, we shall have to find a condition which ensures that when two points $\tilde{p}_1, \tilde{p}_2$ of $\tilde{B}$ project by π onto the same point

$$p = \pi(\tilde{p}_1) = \pi(\tilde{p}_2)$$

of B, this implies that $\tilde{p}_1 = \tilde{p}_2$. We shall assume $\tilde{B}$ to be arcwise connected and project an arc $\tilde{\alpha}$ of $\tilde{B}$, which joins $\tilde{p}_1$ to $\tilde{p}_2$, onto the closed arc α of B, which joins p to p (see Fig. 5-26). If B does not have "holes" (in a sense to be made precise), it is possible to "deform α continuously to the point p." That is, there exists a family of arcs α_t, continuous in t, $t \in [0, 1]$, with $\alpha_0 = \alpha$ and α_1 equal to the constant arc p. Since $\tilde{\alpha}$ is a lifting of α, it is natural to expect that the arcs α_t may also be lifted in a family $\tilde{\alpha}_t$, continuous in t, $t \in [0, l]$, with $\alpha_0 = \tilde{\alpha}$. It follows that $\tilde{\alpha}_1$ is a lifting of the constant arc p and, therefore, reduces to a single point. On the other hand, $\tilde{\alpha}_1$ joins $\tilde{p}_1$ to $\tilde{p}_2$ and hence we conclude that $\tilde{p}_1 = \tilde{p}_2$.

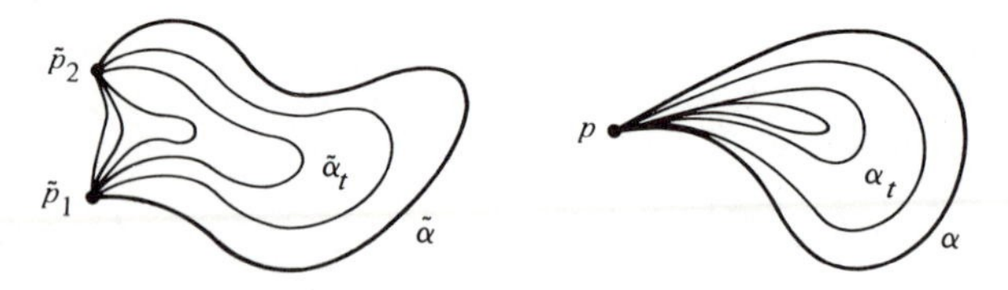

Figure 5-26

To make the above heuristic argument rigorous we have to define a "continuous family of arcs joining two given arcs" and to show that such a family may be "lifted."

DEFINITION 2. *Let* $B \subset R^3$ *and let* $\alpha_0: [0, l] \longrightarrow B$, $\alpha_1: [0, l] \longrightarrow B$ *be two arcs of* B, *joining the points*

$$p = \alpha_0(0) = \alpha_1(0) \quad \textit{and} \quad q = \alpha_0(l) = \alpha_1(l).$$

We say that α_0 *and* α_1 *are* homotopic *if there exists a continuous map* $H: [0, l] \times [0, 1] \longrightarrow B$ *such that*

1. $H(s, 0) = \alpha_0(s)$, $H(s, 1) = \alpha_1(s)$, $\quad s \in [0, l]$.
2. $H(0, t) = p$, $H(l, t) = q$, $\quad t \in [0, 1]$.

The map H *is called a* homotopy *between* α_0 *and* α_1.

For every $t \in [0, 1]$, the arc $\alpha_t: [0, l] \longrightarrow B$ given by $\alpha_t(s) = H(s, t)$ is called an arc of the homotopy H. Therefore, the homotopy is a family of arcs α_t, $t \in [0, 1]$, which constitutes a continuous deformation of α_0 into α_1 (see Fig. 5-27) in such a way that the extremities p and q of the arcs α_t remain fixed during the deformation (condition 2).

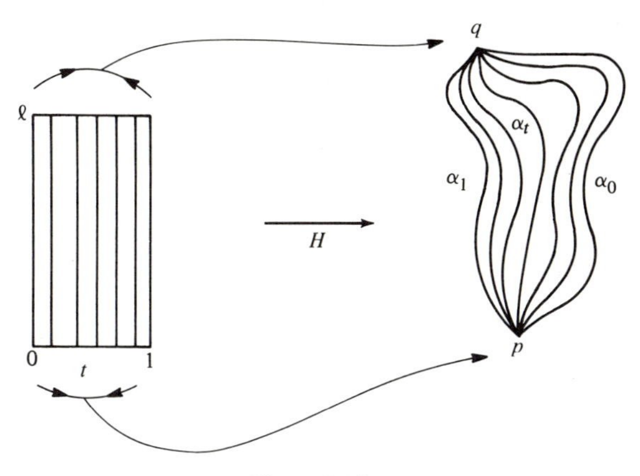

Figure 5-27

The notion of lifting of homotopies is entirely analogous to that of lifting of arcs. Let $\pi: \tilde{B} \longrightarrow B$ be a continuous map and let $\alpha_0, \alpha_1: [0, l] \longrightarrow B$ be two arcs of B joining the points p and q. Let $H: [0, l] \times [0, 1] \longrightarrow B$ be a homotopy between α_0 and α_1. If there exists a continuous map

$$\tilde{H}: [0, l] \times [0, 1] \longrightarrow \tilde{B}$$

such that $\pi \circ \tilde{H} = H$, we say that $\tilde{H}$ is a *lifting of the homotopy H*, with origin at $\tilde{H}(0, 0) = \tilde{p} \in \tilde{B}$.

We shall now show that a covering map has the property of lifting homotopies. Actually, we shall prove a more general proposition. Observe that a covering map $\pi: \tilde{B} \longrightarrow B$ is a local homeomorphism and, furthermore, that every arc of B may be lifted into an arc of $\tilde{B}$. For the proofs of Props. 3, 4, and 5 below we shall use only these two properties of covering maps, and so, for future use, we shall state these propositions in this generality.

Thus, we shall say that a continuous map $\pi: \tilde{B} \to B$ has the *property of lifting arcs* when every arc of B may be lifted. Notice that this implies that π maps $\tilde{B}$ onto B.

PROPOSITION 3. *Let $\pi: \tilde{B} \to B$ be a local homeomorphism with the property of lifting arcs. Let $\alpha_0, \alpha_1: [0, l] \to B$ be two arcs of B joining the points p and q, let*

$$H: [0, l] \times [0, 1] \longrightarrow B$$

be a homotopy between α_0 and α_1, and let $\tilde{p} \in \tilde{B}$ be a point of $\tilde{B}$ such that $\pi(\tilde{p}) = p$. Then there exists a unique lifting $\tilde{H}$ of H with origin at $\tilde{p}$.

Proof. The proof of the uniqueness is entirely analogous to that of the lifting of arcs. Let $\tilde{H}_1$ and $\tilde{H}_2$ be two liftings of H with $\tilde{H}_1(0, 0) = \tilde{H}_2(0, 0) = \tilde{p}$. Then the set A of points $(s, t) \in [0, l] \times [0, 1] = Q$ such that $\tilde{H}_1(s, t) = \tilde{H}_2(s, t)$ is nonempty and closed in Q. Since $\tilde{H}_1$ and $\tilde{H}_2$ are continuous and π is a local homeomorphism, A is open in Q. By connectedness of Q, $A = Q$; hence, $\tilde{H}_1 = \tilde{H}_2$.

To prove the existence, let $\alpha_t(s) = H(s, t)$ be an arc of the homotopy H. Define $\tilde{H}$ by

$$\tilde{H}(s, t) = \tilde{\alpha}_t(s), \qquad s \in [0, l], t \in [0, 1],$$

where $\tilde{\alpha}_t$ is the lifting of α_t, with origin at $\tilde{p}$. It is clear that

$$\pi \circ \tilde{H}(s, t) = \alpha_t(s) = H(s, t), \qquad s \in [0, l], t \in [0, 1],$$
$$\tilde{H}(0, 0) = \tilde{\alpha}_0(0) = \tilde{p}.$$

Let us now prove that $\tilde{H}$ is continuous. Let $(s_0, t_0) \in [0, l] \times [0, 1]$. Since π is a local homeomorphism, there exists a neighborhood V of $\tilde{H}(s_0, t_0)$ such that the restriction π_0 of π to V is a homeomorphism onto a neighborhood U of $H(s_0, t_0)$. Let $Q_0 \subset H^{-1}(U) \subset [0, l] \times [0, 1]$ be an open square given by

$$s_0 - \epsilon < s < s_0 + \epsilon, \qquad t_0 - \epsilon < t < t_0 + \epsilon.$$

It suffices to prove that $\tilde{H}$ restricted to Q_0 may be written as $\tilde{H} = \pi_0^{-1} \circ H$ to conclude that $\tilde{H}$ is continuous at (s_0, t_0). Since (s_0, t_0) is arbitrary, $\tilde{H}$ is continuous in $[0, l] \times [0, 1]$, as desired.

For that, we observe that

$$\pi_0^{-1}(H(s_0, t)), \qquad t \in (t_0 - \epsilon, t_0 + \epsilon),$$

is a lifting of the arc $H(s_0, t)$ passing through $\tilde{H}(s_0, t_0)$. By uniqueness, $\pi_0^{-1}(H(s_0, t)) = \tilde{H}(s_0, t)$. Since Q_0 is a square, for every $(s_1, t_1) \in Q_0$ there exists an arc $H(s, t_1)$ in U, $s \in (s_0 - \epsilon, s_0 + \epsilon)$, which intersects the arc

$H(s_0, t)$. Since $\pi_0^{-1}(H(s_0, t_1)) = \tilde{H}(s_0, t_1)$, the arc $\pi_0^{-1}(H(s, t_1))$ is the lifting of $H(s, t_1)$ passing through $\tilde{H}(s_0, t_1)$. By uniqueness, $\pi_0^{-1}(H(s, t_1)) = \tilde{H}(s, t_1)$; hence, $\pi_0^{-1}(H(s_1, t_1)) = \tilde{H}(s_1, t_1)$. By the arbitrariness of $(s_1, t_1) \in Q_0$ we conclude that $\pi_0^{-1}(H(s, t)) = \tilde{H}(s, t)$, $(s, t) \in Q_0$, which ends the proof.

Q.E.D.

A consequence of Prop. 3 is the fact that if $\pi\colon \tilde{B} \longrightarrow B$ is a covering map, then homotopic arcs of B are lifted into homotopic arcs of $\tilde{B}$. This may be expressed in a more general and precise way as follows.

PROPOSITION 4. *Let $\pi\colon \tilde{B} \longrightarrow B$ be a local homeomorphism with the property of lifting arcs. Let $\alpha_0, \alpha_1\colon [0, l] \longrightarrow B$ be two arcs of B joining the points p and q and choose $\tilde{p} \in \tilde{B}$ such that $\pi(\tilde{p}) = p$. If α_0 and α_1 are homotopic, then the liftings $\tilde{\alpha}_0$ and $\tilde{\alpha}_1$ of α_0 and α_1, respectively, with origin $\tilde{p}$, are homotopic.*

Proof. Let H be the homotopy between α_0 and α_1 and let $\tilde{H}$ be its lifting, with origin at $\tilde{p}$. We shall prove that $\tilde{H}$ is a homotopy between $\tilde{\alpha}_0$ and $\tilde{\alpha}_1$ (see Fig. 5-28).

In fact, by the uniqueness of the lifting of arcs,

$$\tilde{H}(s, 0) = \tilde{\alpha}_0(s), \qquad \tilde{H}(s, 1) = \tilde{\alpha}_1(s), \qquad s \in [0, l],$$

which verifies condition 1 of Def. 2. Furthermore, $\tilde{H}(0, t)$ is the lifting of the "constant" arc $H(0, t) = p$, with origin at $\tilde{p}$. By uniqueness,

$$\tilde{H}(0, t) = \tilde{p}, \qquad t \in [0, 1].$$

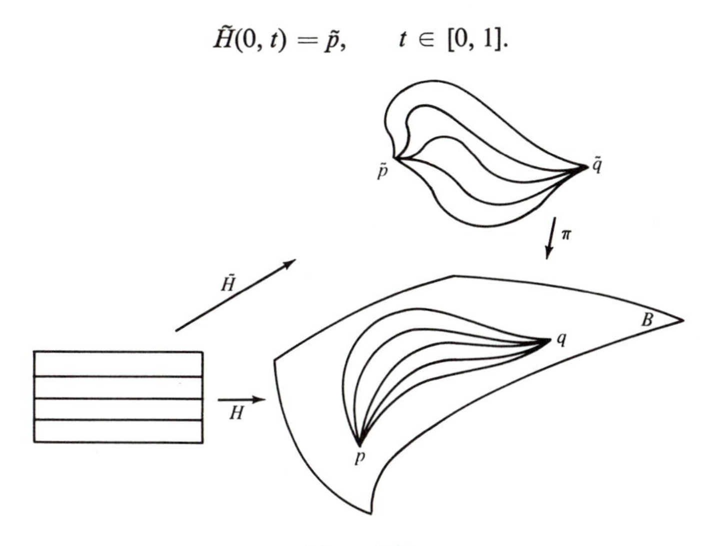

Figure 5-28

Similarly, $\tilde{H}(l, t)$ is the lifting of $H(l, t) = q$, with origin at $\tilde{\alpha}_0(l) = \tilde{q}$; hence,

$$\tilde{H}(l, t) = \tilde{q} = \alpha_1(l), \qquad t \in [0, 1].$$

Therefore, condition 2 of Def. 2 is verified, showing that $\tilde{H}$ is a homotopy between $\tilde{\alpha}_0$ and $\tilde{\alpha}_1$. **Q.E.D.**

Returning to the heuristic argument that led us to consider the concept of homotopy, we see that it still remains to explain what it is meant by a space without "holes." Of course we shall take as a definition of such a space precisely that property which was used in the heuristic argument.

DEFINITION 3. *An arcwise connected set* $B \subset R^3$ *is* simply connected *if given two points* $p, q \in B$ *and two arcs* $\alpha_0: [0, l] \to B$, $\alpha_1: [0, l] \to B$ *joining* p *to* q, *there exists a homotopy in* B *between* α_0 *and* α_1. *In particular, any closed arc of* B, $\alpha: [0, l] \to B$ *(closed means that* $\alpha(0) = \alpha(l) = p$*), is homotopic to the "constant" arc* $\alpha(s) = p$, $s \in [0, l]$ *(in Exercise* 5 *it is indicated that this last property is actually equivalent to the first one).*

Intuitively, an arcwise connected set B is simply connected if every closed arc in B can be continuously deformed into a point. It is possible to prove that the plane and the sphere are simply connected but that the cylinder and the torus are not simply connected (cf. Exercise 5).

We may now state and prove an answer to question 2 of this section. This will come out as a corollary of the following proposition.

PROPOSITION 5. *Let* $\pi: \tilde{B} \to B$ *be a local homeomorphism with the property of lifting arcs. Let* $\tilde{B}$ *be arcwise connected and* B *simply connected. Then* π *is a homeomorphism.*

Proof. The proof is essentially the same as that presented in the heuristic argument.

We need to prove that π is one-to-one. For this, let $\tilde{p}_1$ and $\tilde{p}_2$ be two points of $\tilde{B}$, with $\pi(\tilde{p}_1) = \pi(\tilde{p}_2) = p$. Since $\tilde{B}$ is arcwise connected, there exists an arc $\tilde{\alpha}_0$ of $\tilde{B}$, joining $\tilde{p}_1$ to $\tilde{p}_2$. Then $\pi \circ \tilde{\alpha}_0 = \alpha_0$ is a closed arc of B. Since B is simply connected, α_0 is homotopic to the constant arc $\alpha_1(s) = p$, $s \in [0, l]$. By Prop. 4, $\tilde{\alpha}_0$ is homotopic to the lifting $\tilde{\alpha}_1$ of α_1 which has origin in p. Since $\tilde{\alpha}_1$ is the constant arc joining the points $\tilde{p}_1$ and $\tilde{p}_2$, we conclude that $\tilde{p}_1 = \tilde{p}_2$. **Q.E.D.**

COROLLARY. *Let* $\pi: \tilde{B} \to B$ *be a covering map,* $\tilde{B}$ *arcwise connected, and* B *simply connected. Then* π *is a homeomorphism.*

The fact that we proved Props. 3, 4, and 5 with more generality than was strictly necessary will allow us to give another answer to question 1, as described below.

Let $\pi: \tilde{B} \longrightarrow B$ be a local homeomorphism with the property of lifting arcs, and assume that $\tilde{B}$ and B are locally "well-behaved" (to be made precise). Then π is, in fact, a covering map.

The required local properties are described as follows. Recall that $B \subset R^3$ is locally arcwise connected if any neighborhood of each point contains an arcwise connected neighborhood (appendix to Chap. 5, Def. 12).

DEFINITION 4. B *is* locally simply connected *if any neighborhood of each point contains a simply connected neighborhood.*

In other words, B is locally simply connected if each point has arbitrarily small simply connected neighborhoods. It is clear that if B is locally simply connected, then B is locally arcwise connected.

We remark that a regular surface S is locally simply connected, since $p \in S$ has arbitrarily small neighborhoods homeomorphic to the interior of a disk in the plane.

In the next proposition we shall need the following properties of a locally arcwise connected set $B \subset R^3$ (cf. the Appendix to Chap. 5, Part D). The union of all arcwise connected subsets of B which contain a point $p \in B$ is clearly an arcwise connected set A to be called the *arcwise connected component* of B containing p. Since B is locally arcwise connected, A is open in B. Thus, B can be written as a union $B = \bigcup_\alpha A_\alpha$ of its connected components A_α, which are open and pairwise disjoint.

We also remark that a regular surface is locally arcwise connected. Thus, in the proposition below, the hypotheses on B and $\tilde{B}$ are satisfied when both B and $\tilde{B}$ are regular surfaces.

PROPOSITION 6. *Let* $\pi: \tilde{B} \longrightarrow B$ *be a local homeomorphism with the property of lifting arcs. Assume that* B *is locally simply connected and that* $\tilde{B}$ *is locally arcwise connected. Then* π *is a covering map.*

Proof. Let $p \in B$ and let V be a simply connected neighborhood of p in B. The set $\pi^{-1}(V)$ is the union of its arcwise connected components; that is,

$$\pi^{-1}(V) = \bigcup_\alpha \tilde{V}_\alpha,$$

where the $\tilde{V}_\alpha$'s are open, arcwise connected, and pairwise disjoint sets. Consider the restriction $\pi: \tilde{V}_\alpha \longrightarrow V$. If we show that π is a homeomorphism of $\tilde{V}_\alpha$ onto V, π will satisfy the conditions of the definition of a covering map.

We first prove that $\pi(\tilde{V}_\alpha) = V$. In fact, $\pi(\tilde{V}_\alpha) \subset V$. Assume that there is a point $p \in V, p \notin \pi(\tilde{V}_\alpha)$. Then, since V is arcwise connected, there exists an arc $\alpha: [a, b] \longrightarrow V$ joining a point $q \in \pi(\tilde{V}_\alpha)$ to p. The lifting $\tilde{\alpha}: [a, b] \longrightarrow \tilde{B}$ of α with origin at $\tilde{q} \in \tilde{V}_\alpha$, where $\pi(\tilde{q}) = q$, is an arc in $\tilde{V}_\alpha$, since $\tilde{V}_\alpha$ is an arcwise connected component of B. Therefore,

$$\pi(\tilde{\alpha}(b)) = p \in \pi(\tilde{V}_\alpha),$$

which is a contradiction and shows that $\pi(\tilde{V}_\alpha) = V$.

Next, we observe that $\pi: \tilde{V}_\alpha \longrightarrow V$ is still a local homeomorphism, since $\tilde{V}_\alpha$ is open. Furthermore, by the above, the map $\pi: \tilde{V}_\alpha \longrightarrow V$ still has the property of lifting arcs. Therefore, we have satisfied the conditions of Prop. 5; hence, π is a homeomorphism. **Q.E.D.**

B. The Hadamard Theorems

We shall now return to the question posed in the beginning of this section, namely, under what conditions is the local diffeomorphism $\exp_p: T_p(S) \longrightarrow S$, where p is a point of a complete surface S of curvature $K \leq 0$, a global diffeomorphism of $T_p(S)$ onto S. The following propositions, which serve to "break up" the given question into questions 1 and 2, yield an answer to the problem.

We shall need the following lemma.

LEMMA 1. *Let* S *be a complete surface of curvature* $K \leq 0$. *Then* $exp_p: T_p(S) \longrightarrow S$, $p \in S$, *is length-increasing in the following sense: If* $u, w \in T_p(S)$, *we have*

$$\langle (d\, exp_p)_u(w), (d\, exp_p)_u(w) \rangle \geq \langle w, w \rangle,$$

where, as usual, w *denotes a vector in* $(T_p(S))_u$ *that is obtained from* w *by the translation* u.

Proof. For the case $u = 0$, the equality is trivially verified. Thus, let $v = u/|u|$, $u \neq 0$, and let $\gamma: [0, l] \longrightarrow S$, $l = |u|$, be the geodesic

$$\gamma(s) = \exp_p sv, \qquad s \in [0, l].$$

By the Gauss lemma, we may assume that $\langle w, v \rangle = 0$. Let $J(s) = s(d\exp_p)_{sv}(w)$ be the Jacobi field along γ given by Lemma 1 of Sec. 5-5. We know that $J(0) = 0$, $(DJ/ds)(0) = w$, and $\langle J(s), \gamma'(s) \rangle = 0$, $s \in [0, l]$.

Observe now that, since $K \leq 0$ (cf. Eq. (1), Sec. 5-5),

$$\frac{d}{ds}\left\langle J, \frac{DJ}{ds}\right\rangle = \left\langle \frac{DJ}{ds}, \frac{DJ}{ds}\right\rangle + \left\langle J, \frac{D^2J}{ds^2}\right\rangle = \left|\frac{DJ}{ds}\right|^2 - K|J|^2 \geq 0.$$

This implies that

$$\left\langle J, \frac{DJ}{ds} \right\rangle \geq 0;$$

hence,

$$\frac{d}{ds}\left\langle \frac{DJ}{ds}, \frac{DJ}{ds} \right\rangle = 2\left\langle \frac{DJ}{ds}, \frac{D^2J}{ds^2} \right\rangle = -2K\left\langle \frac{DJ}{ds}, J \right\rangle \geq 0. \quad \textbf{(1)}$$

It follows that

$$\left\langle \frac{DJ}{ds}, \frac{DJ}{ds} \right\rangle \geq \left\langle \frac{DJ}{ds}(0), \frac{DJ}{ds}(0) \right\rangle = \langle w, w \rangle = C; \quad \textbf{(2)}$$

hence,

$$\frac{d^2}{ds^2}\langle J, J \rangle = 2\left\langle \frac{DJ}{ds}, \frac{DJ}{ds} \right\rangle + 2\left\langle J, \frac{D^2J}{ds^2} \right\rangle \geq 2\left\langle \frac{DJ}{ds}, \frac{DJ}{ds} \right\rangle \geq 2C. \quad \textbf{(3)}$$

By integrating both sides of the above inequality, we obtain

$$\frac{d}{ds}\langle J, J \rangle \geq 2Cs + \left(\frac{d}{ds}\langle J, J \rangle\right)_{s=0} = 2Cs + 2\left\langle \frac{DJ}{ds}(0), J(0) \right\rangle = 2Cs.$$

Another integration yields

$$\langle J, J \rangle \geq Cs^2 + \langle J(0), J(0) \rangle = Cs^2.$$

By setting $s = l$ in the above expression and noticing that $C = \langle w, w \rangle$, we obtain

$$\langle J(l), J(l) \rangle \geq l^2 \langle w, w \rangle.$$

Since $J(l) = l(d\exp_p)_{lv}(w)$, we finally conclude that

$$\langle (d\exp_p)_{lv}(w), (d\exp)_{lv}(w) \rangle \geq \langle w, w \rangle. \quad \textbf{Q.E.D.}$$

For later use, it is convenient to establish the following consequence of the above proof.

COROLLARY (*of the proof*). *Let* $K \equiv 0$. *Then* $exp_p \colon T_p(S) \to S$, $p \in S$, *is a local isometry.*

It suffices to observe that if $K \equiv 0$, it is possible to substitute "≥ 0" by "$\equiv 0$" in Eqs. (1), (2), and (3) of the above proof.

PROPOSITION 7. *Let* S *be a complete surface with Gaussian curvature* $K \leq 0$. *Then the map* $exp_p \colon T_p(S) \to S$, $p \in S$, *is a covering map.*

Proof. Since we know that $\exp_p$ is a local diffeomorphism, it suffices (by Prop. 6) to show that $\exp_p$ has the property of lifting arcs.

Let $\alpha: [0, l] \longrightarrow S$ be an arc in S and also let $v \in T_p(S)$ be such that $\exp_p v = \alpha(0)$. Such a v exists since S is complete. Because $\exp_p$ is a local diffeomorphism, there exists a neighborhood U of v in $T_p(S)$ such that $\exp_p$ restricted to U is a diffeomorphism. By using $\exp_p^{-1}$ in $\exp_p(U)$, it is possible to define $\tilde{\alpha}$ in a neighborhood of 0.

Now let A be the set of $t \in [0, l]$ such that $\tilde{\alpha}$ is defined in $[0, t]$. A is non-empty, and if $\tilde{\alpha}(t_0)$ is defined, then $\tilde{\alpha}$ is defined in a neighborhood of t_0; that is, A is open in $[0, l]$. Once we prove that A is closed in $[0, l]$, we have, by connectedness of $[0, l]$, that $A = [0, l]$ and α may be entirely lifted.

The crucial point of the proof consists, therefore, in showing that A is closed in $[0, l]$. For this, let $t_0 \in [0, l]$ be an accumulation point of A and $\{t_n\}$ be a sequence with $\{t_n\} \longrightarrow t_0$, $t_n \in A$, $n = 1, 2, \ldots$. We shall first prove that $\tilde{\alpha}(t_n)$ has an accumulation point.

Assume that $\tilde{\alpha}(t_n)$ has no accumulation point in $T_p(S)$. Then, given a closed disk D of $T_p(S)$, with center $\tilde{\alpha}(0)$, there is an n_0 such that $\tilde{\alpha}(t_{n_0}) \notin D$. It follows that the distance, in $T_p(S)$, from $\tilde{\alpha}(0)$ to $\tilde{\alpha}(t_n)$ becomes arbitrarily large. Since, by Lemma 1, $\exp_p: T_p(S) \longrightarrow S$ increases lengths of the vectors, it is clear that the intrinsic distance in S from $\alpha(0)$ to $\alpha(t_n)$ becomes arbitrarily large. But that contradicts the fact that the intrinsic distance from $\alpha(0)$ to $\alpha(t_0) = \lim_{t_n \to t_0} \alpha(t_n)$ is finite, which proves our assertion.

We shall denote by q an accumulation point of $\tilde{\alpha}(t_n)$.

Now let V be a neighborhood of q in $T_p(S)$ such that the restriction of $\exp_p$ to V is a diffeomorphism. Since q is an accumulation point of $\{\tilde{\alpha}(t_n)\}$, there exists an n_1 such that $\tilde{\alpha}(t_{n_1}) \in V$. Moreover, since α is continuous, there exists an open interval $I \subset [0, l]$, $t_0 \in I$, such that $\alpha(I) \subset \exp_p(V) = U$. By using the restriction of $\exp_p^{-1}$ in U it is possible to define a lifting of α in I, with origin in $\tilde{\alpha}(t_{n_1})$. Since $\exp_p$ is a local diffeomorphism, this lifting coincides with $\tilde{\alpha}$ in $[0, t_0) \cap I$ and is therefore an extension of $\tilde{\alpha}$ to an interval containing t_0. Thus, the set A is closed, and this ends the proof of Prop. 7.

Q.E.D.

Remark 1. It should be noticed that the curvature condition $K \leq 0$ was used only to guarantee that $\exp_p: T_p(S) \longrightarrow S$ is a length-increasing local diffeomorphism. Therefore, we have actually proved that if *$\varphi: S_1 \longrightarrow S_2$ is a local diffeomorphism of a complete surface S_1 onto a surface S_2, which is length-increasing, then φ is a covering map.*

The following proposition, known as the Hadamard theorem, describes the topological structure of a complete surface with curvature $K \leq 0$.

THEOREM 1 (Hadamard). *Let* S *be a simply connected, complete surface, with Gaussian curvature* $K \leq 0$. *Then* $exp_p: T_p(S) \longrightarrow S$, $p \in S$, *is a diffeomorphism; that is,* S *is diffeomorphic to a plane.*

Proof. By Prop. 7, $\exp_p: T_p(S) \longrightarrow S$ is a covering map. By the corollary of Prop. 5, $\exp_p$ is a homeomorphism. Since $\exp_p$ is a local diffeomorphism, its inverse map is differentiable, and $\exp_p$ is a diffeomorphism. **Q.E.D.**

We shall now present another geometric application of the covering spaces, also known as the Hadamard theorem. Recall that a connected, compact, regular surface, with Gaussian curvature $K > 0$, is called an ovaloid (cf. Remark 1, Sec. 5-2).

THEOREM 2 (Hadamard). *Let* S *be an ovaloid. Then the Gauss map* $N: S \longrightarrow S^2$ *is a diffeomorphism. In particular,* S *is diffeomorphic to a sphere.*

Proof. Since for every $p \in S$ the Gaussian curvature of S, $K = \det(dN_p)$, is positive, N is a local diffeomorphism. By Prop. 1, N is a covering map. Since the sphere S^2 is simply connected, we conclude from the corollary of Prop. 5 that $N: S \longrightarrow S^2$ is a homeomorphism of S onto the unit sphere S^2. Since N is a local diffeomorphism, its inverse map is differentiable. Therefore, N is a diffeomorphism. **Q.E.D.**

Remark 2. Actually, we have proved somewhat more. Since the Gauss map N is a diffeomorphism, each unit vector v of R^3 appears exactly once as a unit normal vector to S. Taking a plane normal to v, away from the surface, and displacing it parallel to itself until it meets the surface, we conclude that S lies on one side of each of its tangent planes. This is expressed by saying that an ovaloid S is *locally convex*. It can be proved from this that S is actually the boundary of a convex set (that is, a set $K \subset R^3$ such that the line segment joining any two points $p, q \in K$ belongs entirely to K).

Remark 3. The fact that compact surfaces with $K > 0$ are homeomorphic to spheres was extended to compact surfaces with $K \geq 0$ by S. S. Chern and R. K. Lashof ("On the Total Curvature of Immersed Manifolds," *Michigan Math. J.* 5 (1958), 5–12). A generalization for complete surfaces was first obtained by J. J. Stoker ("Über die Gestalt der positiv gekrümnten offenen Fläche," *Compositio Math.* 3 (1936), 58–89), who proved, among other things, the following: *A complete surface with* $K > 0$ *is homeomorphic to a sphere or a plane.* This result still holds for $K \geq 0$ if one assumes that at some point $K > 0$ (for a proof and a survey of this problem, see M. do Carmo and E. Lima, "Isometric Immersions with Non-negative Sectional Curvatures," *Boletim da Soc. Bras. Mat.* 2 (1971), 9–22)

EXERCISES

1. Show that the map $\pi\colon R \longrightarrow S^1 = \{(x, y) \in R^2; x^2 + y^2 = 1\}$ that is given by $\pi(t) = (\cos t, \sin t)$, $t \in R$, is a covering map.

2. Show that the map $\pi\colon R^2 - \{0, 0\} \longrightarrow R^2 - \{0, 0\}$ given by

$$\pi(x, y) = (x^2 - y^2, 2xy), \qquad (x, y) \in R^2,$$

is a two-sheeted covering map.

3. Let S be the helicoid generated by the normals to the helix $(\cos t, \sin t, bt)$. Denote by L the z axis and let $\pi\colon S - L \longrightarrow R^2 - \{0, 0\}$ be the projection $\pi(x, y, z) = (x, y)$. Show that π is a covering map.

4. Those who are familiar with functions of a complex variable will have noticed that the map π in Exercise 2 is nothing but the map $\pi(z) = z^2$ from $\mathbb{C} - \{0\}$ onto $\mathbb{C} - \{0\}$; here $\mathbb{C}$ is the complex plane and $z \in \mathbb{C}$. Generalize that by proving that the map $\pi\colon \mathbb{C} - \{0\} \longrightarrow \mathbb{C} - \{0\}$ given by $\pi(z) = z^n$ is an n-sheeted covering map.

5. Let $B \subset R^3$ be an arcwise connected set. Show that the following two properties are equivalent (cf. Def. 3):

1. For any pair of points $p, q \in B$ and any pair of arcs $\alpha_0\colon [0, l] \longrightarrow B$, $\alpha_1\colon [0, l] \longrightarrow B$, there exists a homotopy in B joining α_0 to α_1.
2. For any $p \in B$ and any arc $\alpha\colon [0, l] \longrightarrow B$ with $\alpha(0) = \alpha(l) = p$ (that is, α is a closed arc with initial and end point p) there exists a homotopy joining α to the constant arc $\alpha(s) = p$, $s \in [0, l]$.

6. Fix a point $p_0 \in R^2$ and define a family of maps $\varphi_t\colon R^2 \longrightarrow R^2$, $t \in [0, 1]$, by $\varphi_t(p) = tp_0 + (1 - t)p$, $p \in R^2$. Notice that $\varphi_0(p) = p$, $\varphi_1(p) = p_0$. Thus, φ_t is a continuous family of maps which starts with the identity map and ends with the constant map p_0. Apply these considerations to prove that R^2 is simply connected.

7. **a.** Use stereographic projection and Exercise 6 to show that any closed arc on a sphere S^2 which omits at least one point of S^2 is homotopic to a constant arc.

b. Show that any closed arc on S^2 is homotopic to a closed arc in S^2 which omits at least one point.

c. Conclude from parts a and b that S^2 is simply connected. Why is part b necessary?

8. (*Klingenberg's Lemma.*) Let $S \subset R^3$ be a complete surface with Gaussian curvature $K \leq K_0$, where K_0 is a nonnegative constant. Let $p, q \in S$ and let γ_0 and γ_1 be two distinct geodesics joining p to q, with $l(\gamma_0) \leq l(\gamma_1)$; here $l(\ )$ denotes the length of the corresponding curve. Assume that γ_0 is homotopic to γ_1; i.e., there exists a continuous family of curves α_t, $t \in [0, 1]$, joining p to q with

$\alpha_0 = \gamma_0$, $\alpha_1 = \gamma_1$. The aim of this exercise is to prove that *there exists a* $t_0 \in [0, 1]$ *such that*

$$l(\gamma_0) + l(\alpha_{t_0}) > \frac{2\pi}{\sqrt{K_0}}.$$

(Thus, the homotopy has to pass through a "long" curve. See Fig. 5-29.) Assume that $l(\gamma_0) < \pi/\sqrt{K_0}$ (otherwise there is nothing to prove) and proceed as follows.

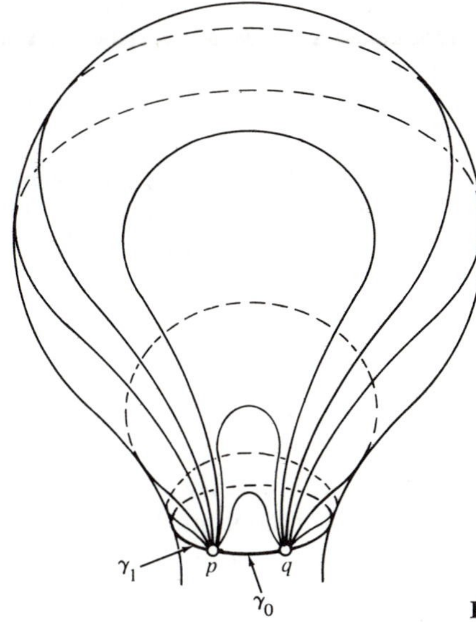

Figure 5-29. Klingenberg's lemma.

a. Use the first comparison theorem (cf. Exercise 3, Sec. 5-5) to prove that $\exp_p\colon T_p(S) \to S$ has no critical points in an open disk B of radius $\pi/\sqrt{K_0}$ about p.

b. Show that, for t small, it is possible to lift the curve α_t into the tangent plane $T_p(S)$; i.e., there exists a curve $\tilde{\alpha}_t$ joining $\exp_p^{-1}(p) = 0$ to $\exp_p^{-1}(q) = \tilde{q}$ and such that $\exp_p \circ \tilde{\alpha}_t = \alpha_t$.

c. Show that the lifting in part b cannot be defined for all $t \in [0, 1]$. Conclude that for every $\epsilon > 0$ there exists a $t(\epsilon)$ such that $\alpha_{t(\epsilon)}$ can be lifted into $\tilde{\alpha}_{t(\epsilon)}$ and $\tilde{\alpha}_{t(\epsilon)}$ contains points at a distance $<\epsilon$ from the boundary of B. Thus,

$$l(\gamma_0) + l(\alpha_{t(\epsilon)}) \geq \frac{2\pi}{\sqrt{K_0}} - 2\epsilon.$$

d. Choose in part c a sequence of ϵ's, $\{\epsilon_n\} \to 0$, and consider a converging subsequence of $\{t(\epsilon_n)\}$. Conclude the existence of a curve α_{t_0}, $t_0 \in [0, 1]$, such that

$$l(\gamma_0) + l(\alpha_{t_0}) \geq \frac{2\pi}{\sqrt{K_0}}.$$

9. a. Use Klingenberg's lemma to prove that if S is a complete, simply connected surface with $K \leq 0$, then $\exp_p\colon T_p(S) \longrightarrow S$ is one-to-one.

b. Use part a to give a simple proof of Hadamard's theorem (Theorem 1).

***10.** (*Synge's Lemma.*) We recall that a differentiable closed curve on a surface S is a differentiable map $\alpha\colon [0, l] \longrightarrow S$ such that α and all its derivatives agree at 0 and l. Two differentiable closed curves $\alpha_0, \alpha_1\colon [0, l] \longrightarrow S$ are *freely homotopic* if there exists a continuous map $H\colon [0, l] \times [0, 1] \longrightarrow S$ such that $H(s, 0) = \alpha_0(s)$, $H(s, 1) = \alpha_1(s)$, $s \in [0, l]$. The map H is called a *free homotopy* (the end points are not fixed) between α_0 and α_1. Assume that S is orientable and has positive Gaussian curvature. Prove that any simple closed geodesic on S is freely homotopic to a closed curve of smaller length.

11. Let S be a complete surface. A point $p \in S$ is called a *pole* if every geodesic $\gamma\colon [0, \infty) \longrightarrow S$ with $\gamma(0) = p$ contains no point conjugate to p relative to γ. Use the techniques of Klingenberg's lemma (Exercise 8) to prove that if S is simply connected and has a pole p, then $\exp_p\colon T_p(S) \longrightarrow S$ is a diffeomorphism.

5-7. Global Theorems for Curves; The Fary-Milnor Theorem

In this section, some global theorems for closed curves will be presented. The main tool used here is the degree theory for continuous maps of the circle. To introduce the notion of degree, we shall use some properties of covering maps developed in Sec. 5-6.

Let $S^1 = \{(x, y) \in R^2; x^2 + y^2 = 1\}$ and let $\pi\colon R \longrightarrow S^1$ be the covering of S^1 by the real line R given by

$$\pi(x) = (\cos x, \sin x), \qquad x \in R.$$

Let $\varphi\colon S^1 \longrightarrow S^1$ be a continuous map. The degree of φ is defined as follows. We can think of the first S^1 in the map $\varphi\colon S^1 \longrightarrow S^1$ as a closed interval $[0, l]$ with its end points 0 and l identified. Thus, φ can be thought of as a continuous map $\varphi\colon [0, l] \longrightarrow S^1$, with $\varphi(0) = \varphi(l) = p \in S^1$. Thus, φ is a closed arc at p in S^1 which, by Prop. 2 of Sec. 5-6, can be lifted into a unique arc $\tilde{\varphi}\colon [0, l] \longrightarrow R$, starting at a point $x \in R$ with $\pi(x) = p$. Since $\pi(\tilde{\varphi}(0)) = \pi(\tilde{\varphi}(l))$, the difference $\tilde{\varphi}(l) - \tilde{\varphi}(0)$ is an integral multiple of 2π. The integer $\deg \varphi$ given by

$$\tilde{\varphi}(l) - \tilde{\varphi}(0) = (\deg \varphi) 2\pi$$

is called the *degree* of φ.

Intuitively, $\deg \varphi$ is the number of times that $\varphi\colon [0, l] \longrightarrow S^1$ "wraps" $[0, l]$ around S^1 (Fig. 5-30). Notice that the function $\tilde{\varphi}\colon [0, l] \longrightarrow R$ is a con-

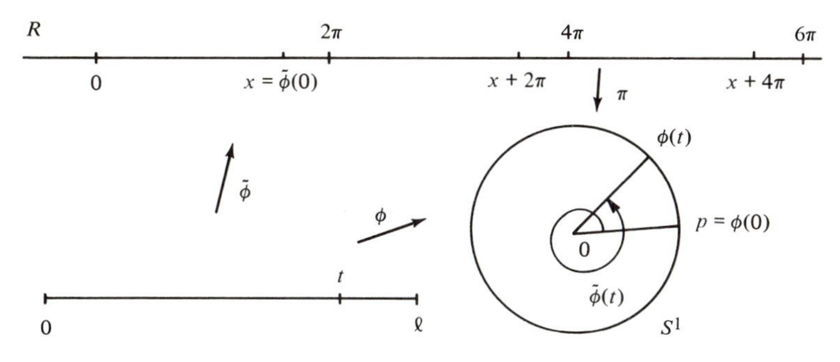

Figure 5-30

tinuous determination of the positive angle that the fixed vector $\varphi(0) - O$ makes with $\varphi(t) - O$, $t \in [0, l]$, $O = (0, 0)$—e. g., the map $\pi: S^1 \longrightarrow S^1$ described in Example 4 of Sec. 5-6, Part A, has degree k.

We must show that the definition of degree is independent of the choices of p and x.

First, deg φ is independent of the choice of x. In fact, let $x_1 > x$ be a point in R such that $\pi(x_1) = p$, and let $\tilde{\varphi}_1(t) = \tilde{\varphi}(t) + (x_1 - x)$, $t \in [0, l]$. Since $x_1 - x$ is an integral multiple of 2π, $\tilde{\varphi}_1$ is a lifting of φ starting at x_1. By the uniqueness part of Prop. 2 of Sec. 5-6, $\tilde{\varphi}_1$ is *the lifting* of φ starting at x_1. Since

$$\tilde{\varphi}_1(l) - \tilde{\varphi}_1(0) = \tilde{\varphi}(l) - \tilde{\varphi}(0) = (\deg \varphi)2\pi,$$

the degree of φ is the same whether computed with x or with x_1.

Second, deg φ is independent of the choice of $p \in S^1$. In fact, each point $p_1 \in S^1$, except the antipodal point of p, belongs to a distinguished neighborhood U_1 of p. Choose x_1, in the connected component of $\pi^{-1}(U_1)$ containing x, such that $\pi(x_1) = p_1$, and let $\tilde{\varphi}_1$ be the lifting of

$$\varphi: [0, l] \longrightarrow S^1, \varphi(0) = p_1,$$

starting at x_1. Clearly, $|\tilde{\varphi}_1(0) - \tilde{\varphi}(0)| < 2\pi$. It follows from the stepwise process through which liftings are constructed (cf. the proof of Prop. 2, Sec. 5-6) that $|\tilde{\varphi}_1(l) - \tilde{\varphi}(l)| < 2\pi$. Since both differences $\tilde{\varphi}(l) - \tilde{\varphi}(0)$, $\tilde{\varphi}_1(l) - \tilde{\varphi}_1(0)$ must be integral multiples of 2π, their values are actually equal. By continuity, the conclusion also holds for the antipodal point of p, and this proves our claim.

The most important property of degree is its invariance under homotopy. More precisely, let $\varphi_1, \varphi_2: S^1 \longrightarrow S^1$ be continuous maps. Fix a point $p \in S^1$, thus obtaining two closed arcs at p, $\varphi_1, \varphi_2: [0, l] \longrightarrow S^1$, $\varphi_1(0) = \varphi_2(0) = p$. If φ_1 and φ_2 are homotopic, then deg φ_1 = deg φ_2. This follows immediately from the fact that (Prop. 4, Sec. 5-6) the liftings of φ_1 and φ_2 starting from a fixed point $x \in R$ are homotopic, and hence have the same end points.

It should be remarked that if $\varphi\colon [0, l] \to S^1$ is differentiable, it determines differentiable functions $a = a(t)$, $b = b(t)$, given by $\varphi(t) = (a(t), b(t))$, which satisfy the condition $a^2 + b^2 = 1$. In this case, the lifting $\tilde{\varphi}$, starting at $\tilde{\varphi}_0 = x$, is precisely the differentiable function (cf. Lemma 1, Sec. 4-4)

$$\tilde{\varphi}(t) = \tilde{\varphi}_0 + \int_0^t (ab' - ba')\, dt.$$

This follows from the uniqueness of the lifting and the fact that $\cos \tilde{\varphi}(t) = a(t)$, $\sin \tilde{\varphi}(t) = b(t)$, $\tilde{\varphi}(0) = \tilde{\varphi}_0$. Thus, in the differentiable case, the degree of φ can be expressed by an integral,

$$\deg \varphi = \frac{1}{2\pi} \int_0^l \frac{d\tilde{\varphi}}{dt}\, dt.$$

In the latter form, the notion of degree has appeared repeatedly in this book. For instance, when $v\colon U \subset R^2 \to R^2$, $U \supset S^1$, is a vector field, and $(0, 0)$ is its only singularity, the index of v at $(0, 0)$ (cf. Sec. 4-5, Application 5) may be interpreted as the degree of the map $\varphi\colon S^1 \to S^1$ that is given by $\varphi(p) = v(p)/|v(p)|$, $p \in S^1$.

Before going into further examples, let us recall that a closed (differentiable) curve is a differentiable map $\alpha\colon [0, l] \to R^3$ (or R^2, if it is a plane curve) such that the components of α, together with all its derivatives, agree at 0 and l. The curve α is regular if $\alpha'(t) \neq 0$ for all $t \in [0, l]$, and α is simple if whenever $t_1 \neq t_2$, $t_1, t_2 \in [0, l)$, then $\alpha(t_1) \neq \alpha(t_2)$. Sometimes it is convenient to assume that α is merely continuous; in this case, we shall say explicitly that α is a *continuous closed curve*.

Example 1 (*The Winding Number of a Curve*). Let $\alpha\colon [0, l] \to R^2$ be a plane, continuous closed curve. Choose a point $p_0 \in R^2$, $p_0 \notin \alpha([0, l])$, and let $\varphi\colon [0, l] \to S^1$ be given by

$$\varphi(t) = \frac{\alpha(t) - p_0}{|\alpha(t) - p_0|}, \qquad t \in [0, l].$$

Clearly $\varphi(0) = \varphi(l)$, and φ may be thought of as a map of S^1 into S^1; it is called the *position map* of α relative to p_0. The degree of φ is called the *winding number* (or the *index*) of the curve α relative to p_0 (Fig. 5-31).

Notice that by moving p_0 along an arc β which does not meet $\alpha([0, l])$ the winding number remains unchanged. Indeed, the position maps of α relative to any two points of β can clearly be joined by a homotopy. It follows that the winding number of α relative to q is constant when q runs in a connected component of $R^2 - \alpha([0, l])$.

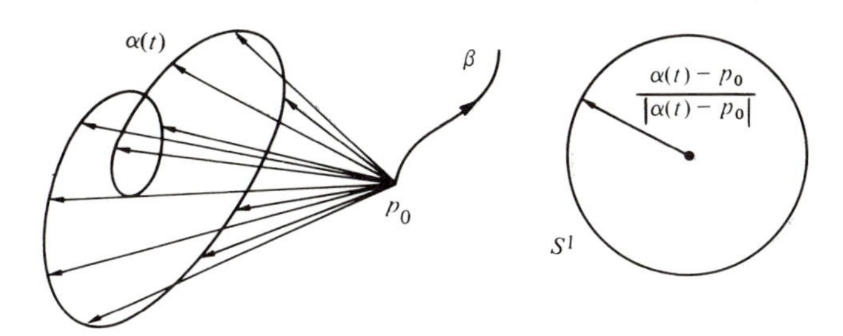

Figure 5-31

Example 2 (*The Rotation Index of a Curve*). Let $\alpha: [0, l] \longrightarrow R^2$ be a regular plane closed curve, and let $\varphi: [0, l] \longrightarrow S^1$ be given by

$$\varphi(t) = \frac{\alpha'(t)}{|\alpha'(t)|}, \qquad t \in [0, l].$$

Clearly φ is differentiable and $\varphi(0) = \varphi(l)$. φ is called the *tangent map* of α, and the degree of φ is called the *rotation index* of α. Intuitively, the rotation index of a closed curve is the number of complete turns given by the tangent vector field along the curve (Fig. 1-27, Sec. 1-7).

It is possible to extend the notion of rotation index to piecewise regular curves by using the angles at the vertices (see Sec. 4-5) and to prove that the rotation index of a simple, closed, piecewise regular curve is ± 1 (the theorem of turning tangents). This fact is used in the proof of the Gauss-Bonnet theorem. Later in this section we shall prove a differentiable version of the theorem of turning tangents.

Our first global theorem will be a differentiable version of the so-called Jordan curve theorem. For the proof we shall presume some familiarity with the material of Sec. 2-7.

THEOREM 1 (Differentiable Jordan Curve Theorem). *Let* $\alpha: [0, l] \longrightarrow R^2$ *be a plane, regular, closed, simple curve. Then* $R^2 - \alpha([0, l])$ *has exactly two connected components, and* $\alpha([0, l])$ *is their common boundary.*

Proof. Let $N_\epsilon(\alpha)$ be a tubular neighborhood of $\alpha([0, l])$. This is constructed in the same way as that used for the tubular neighborhood of a compact surface (cf. Sec. 2-7). We recall that $N_\epsilon(\alpha)$ is the union of open normal segments $I_\epsilon(t)$, with length 2ϵ and center in $\alpha(t)$. Clearly, $N_\epsilon(\alpha) - \alpha([0, l])$ has two connected components T_1 and T_2. Denote by $w(p)$ the winding number of α relative to $p \in R^2 - \alpha([0, l])$. The crucial point of the proof is to show that if both p_1 and p_2 belong to distinct connected components of $N_\epsilon(\alpha) - \alpha([0, l])$ and to the same $I_\epsilon(t_0)$, $t_0 \in [0, l]$, then $w(p_1) - w(p_2) = \pm 1$, the sign depending on the orientation of α.

Choose points $A = \alpha(t_1)$, $D = \alpha(t_2)$, $t_1 < t_0 < t_2$, so close to t_0 that the arc AD of α can be deformed homotopically onto the polygon $ABCD$ of Fig. 5-32. Here BC is a segment of the tangent line at $\alpha(t)$, and BA and CD are parallel to the normal line at $\alpha(t_0)$.

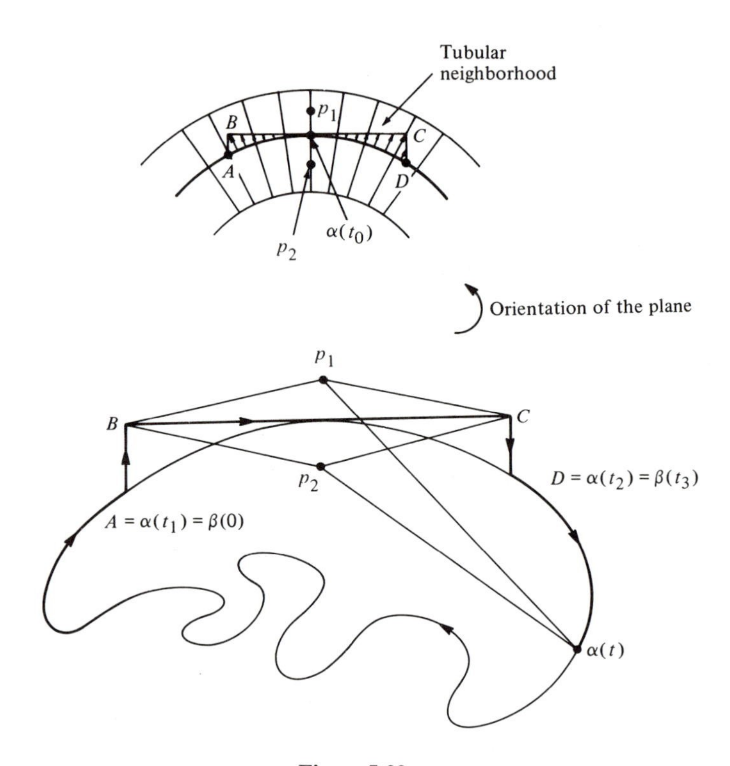

Figure 5-32

Let us denote by $\beta: [0, \bar{l}] \longrightarrow R^2$ the curve obtained from α by replacing the arc AD by the polygon $ABCD$, and let us assume that $\beta(0) = \beta(\bar{l}) = A$ and that $\beta(t_3) = D$. Clearly, $w(p_1)$ and $w(p_2)$ remain unchanged.

Let $\varphi_1, \varphi_2: [0, \bar{l}] \longrightarrow S^1$ be the position maps of β relative to p_1, p_2, respectively (cf. Example 1), and let $\tilde{\varphi}_1, \tilde{\varphi}_2: [0, \bar{l}] \longrightarrow R$ be their liftings from a fixed point, say $0 \in R$. For convenience, let us assume the orientation of β to be given as in Fig. 5-32.

We first remark that if $t \in [t_3, \bar{l}]$, the distances from $\alpha(t)$ to both p_1 and p_2 remain bounded below by a number independent of t, namely, the smallest of the two numbers dist$(p_1, \text{Bd } N_\epsilon(\alpha))$ and dist$(p_2, \text{Bd } N_\epsilon(\alpha))$. It follows that the angle of $\alpha(t) - p_1$ with $\alpha(t) - p_2$ tends uniformly to zero in $[t_3, \bar{l}]$ as p_1 approaches p_2.

Now, it is clearly possible to choose p_1 and p_2 so close to each other that

$\tilde{\varphi}_1(t_3) - \tilde{\varphi}_1(0) = \pi - \epsilon_1$, and $\tilde{\varphi}_2(t_3) - \tilde{\varphi}_2(0) = -(\pi + \epsilon_2)$, with ϵ_1 and ϵ_2 smaller than $\pi/3$. Furthermore,

$$\begin{aligned} 2\pi(w(p_1) - w(p_2)) &= (\tilde{\varphi}_1(\bar{l}) - \tilde{\varphi}_1(0)) - (\tilde{\varphi}_2(\bar{l}) - \tilde{\varphi}_2(0)) \\ &= \{(\tilde{\varphi}_1 - \tilde{\varphi}_2)(\bar{l}) - (\tilde{\varphi}_1 - \tilde{\varphi}_2)(t_3)\} \\ &\quad + \{(\tilde{\varphi}_1 - \tilde{\varphi}_2)(t_3) - (\tilde{\varphi}_1 - \tilde{\varphi}_2)(0)\}. \end{aligned}$$

By the above remark, the first term can be made arbitrarily small, say equal to $\epsilon_3 < \pi/3$, if p_1 is sufficiently close to p_2. Thus,

$$2\pi(w(p_1) - w(p_2)) = \epsilon_3 + \pi - \epsilon_1 - (-\pi - \epsilon_2) = 2\pi + \epsilon,$$

where $\epsilon < \pi$ if p_1 is sufficiently close to p_2. It follows that $w(p_1) - w(p_2) = 1$, as we had claimed.

It is now easy to complete the proof. Since $w(p)$ is constant in each connected component of $R^2 - \alpha([0, l]) = W$, it follows from the above that there are at least two connected components in W. We shall show that there are exactly two such components.

In fact, let C be a connected component of W. Clearly Bd $C \neq \phi$ and Bd $C \subset \alpha([0, l])$. On the other hand, if $p \in \alpha([0, l])$, there is a neighborhood of p that contains only points of $\alpha([0, l])$, points of T_1, and points of T_2 (T_1 and T_2 are the connected components of $N_\epsilon(\alpha) - \alpha([0, l])$). Thus, either T_1 or T_2 intersect C. Since C is a connected component, $C \supset T_1$ or $C \supset T_2$. Therefore, there are at most two (hence, exactly two) connected components of W. Denote them by C_1 and C_2. The argument also shows that Bd $C_1 = \alpha([0, l]) =$ Bd C_2. **Q.E.D.**

The two connected components given by Theorem 1 can easily be distinguished. One starts from the observation that if p_0 is outside a closed disk D containing $\alpha([0, l])$ (since $[0, l]$ is compact, such a disk exists), then the winding number of α relative to p_0 is zero. This comes from the fact that the lines joining p_0 to $\alpha(t)$, $t \in [0, l]$, are all within a region containing D and bounded by the two tangents from p_0 to the circle Bd D. Thus, the connected component with winding number zero is unbounded and contains all points outside a certain disk. Clearly the remaining connected component has winding number ± 1 and is bounded. It is usual to call them the *exterior* and the *interior* of α, respectively.

Remark 1. A useful complement to the above theorem, which was used in the applications of the Gauss-Bonnet theorem (Sec. 4-5), is the fact that the interior of α is homeomorphic to an open disk. A proof of that can be found in J. J. Stoker, *Differential Geometry*, Wiley-Interscience, New York, 1969, pp. 43–45.

We shall now prove a differentiable version of the theorem of turning tangents.

THEOREM 2. *Let $\beta: [0, l] \longrightarrow R^2$ be a plane, regular, simple, closed curve. Then the rotation index of β is ± 1 (depending on the orientation of β).*

Proof. Consider a line that does not meet the curve and displace it parallel to itself until it is tangent to the curve. Denote by l this position of the line and by p a point of tangency of the curve with l. Clearly the curve is entirely on one side of l (Fig. 5-33). Choose a new parametrization $\alpha: [0, l] \longrightarrow R^2$ for the curve so that $\alpha(0) = p$. Now let

$$T = \{(t_1, t_2) \in [0, l] \times [0, l]; 0 \leq t_1 \leq t_2 \leq l\}$$

be a triangle, and define a "secant map" $\psi: T \longrightarrow S^1$ by

$$\psi(t_1, t_2) = \frac{\alpha(t_2) - \alpha(t_1)}{|\alpha(t_2) - \alpha(t_1)|} \qquad \text{for } t_1 \neq t_2, (t_1, t_2) \in T - \{(0, l)\}$$

$$\psi(t, t) = \frac{\alpha'(t)}{|\alpha'(t)|}, \qquad \psi(0, l) = -\frac{\alpha'(0)}{|\alpha'(0)|}.$$

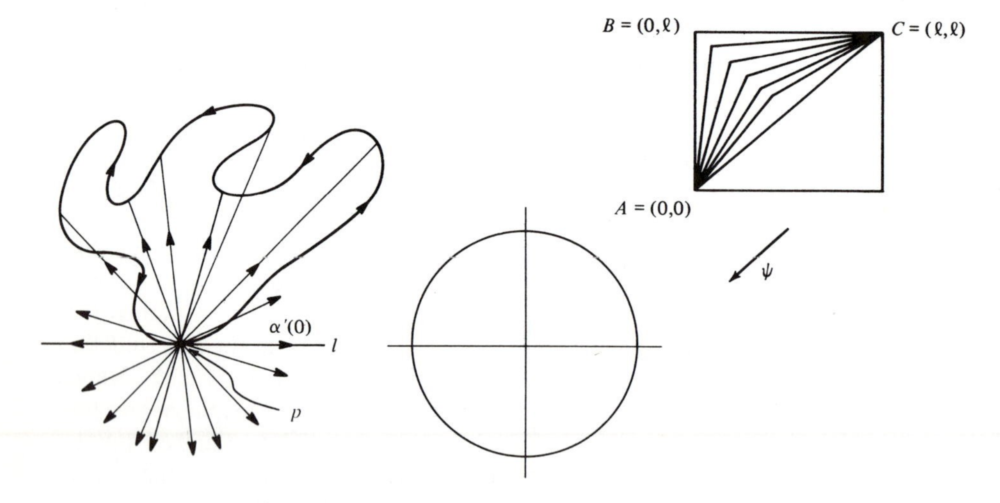

Figure 5-33

Since α is regular, ψ is easily seen to be continuous. Let $A = (0, 0)$, $B = (0, l)$, $C = (l, l)$ be the vertices of the triangle T. Notice that ψ restricted to the side AC is the tangent map of α, the degree of which is the rotation number of α. Clearly (Fig. 5-33), the tangent map is homotopic to the restriction of ψ to the remaining sides AB and BC. Thus, we are reduced to show that the degree of the latter map is ± 1.

Assume that the orientations of the plane and the curve are such that the oriented angle from $\alpha'(0)$ to $-\alpha'(0)$ is π. Then the restriction of ψ to AB covers half of S^1 in the positive direction, and the restriction of ψ to BC covers the remaining half also in the positive direction (Fig. 5-33). Thus, the degree of ψ restricted to AB and BC is $+1$. Reversing the orientation, we shall obtain -1 for this degree, and this completes the proof. **Q.E.D.**

The theorem of turning tangents can be used to give a characterization of an important class of curves, namely the convex curves.

A plane, regular, closed curve $\alpha: [0, l] \longrightarrow R^2$ is *convex* if, for each $t \in [0, l]$, the curve lies in one of the closed half-planes determined by the tangent line at t (Fig. 5-34; cf. also Sec. 1-7). If α is simple, convexity can be expressed in terms of curvature. We recall that for plane curves, curvature always means the signed curvature (Sec. 1-5, Remark 1).

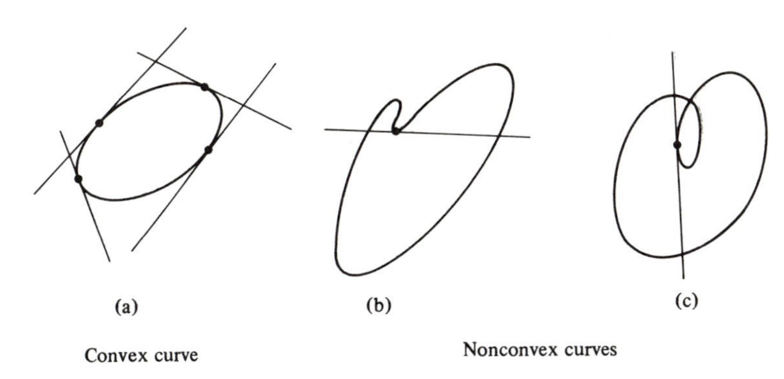

Figure 5-34

PROPOSITION 1. *A plane, regular, closed curve is convex if and only if it is simple and its curvature* k *does not change sign.*

Proof. Let $\varphi: [0, l] \longrightarrow S^1$ be the tangent map of α and $\tilde{\varphi}: [0, l] \longrightarrow R$ be the lifting of φ starting at $0 \in R$. We first remark that the condition that k does not change sign is equivalent to the condition that $\tilde{\varphi}$ is monotonic (nondecreasing if $k \geq 0$, or nonincreasing if $k \leq 0$).

Now, suppose that α is simple and that k does not change sign. We can orient the plane of the curve so that $k \geq 0$. Assume that α is not convex. Then there exists $t_0 \in [0, l]$ such that points of $\alpha([0, l])$ can be found on both sides of the tangent line T at $\alpha(t_0)$. Let $n = n(t_0)$ be the normal vector at t_0, and set

$$h_n(t) = \langle \alpha(t) - \alpha(t_0), n \rangle, \qquad t \in [0, l].$$

Since $[0, l]$ is compact and both sides of T contain points of the curve, the "height function" h_n has a maximum at $t_1 \neq t_0$ and a minimum at $t_2 \neq t_0$.

The tangent vectors at the points t_0, t_1, t_2 are all parallel, so two of them, say $\alpha'(t_0), \alpha'(t_1)$, have the same direction. It follows that $\varphi(t_0) = \varphi(t_1)$ and, by Theorem 2 (α is simple), $\tilde{\varphi}(t_0) = \tilde{\varphi}(t_1)$. Let us assume that $t_1 > t_0$. By the above remark, $\tilde{\varphi}$ is monotonic nondecreasing, and hence constant in $[t_0, t_1]$. This means that $\alpha([t_0, t_1]) \subset T$. But this contradicts the choice of T and shows that α is convex.

Conversely, assume that α is convex. We shall leave it as an exercise to show that if α is not simple, at a self-intersection point (Fig. 5-35(a)), or nearby it (Fig. 5-35(b)), the convexity condition is violated. Thus, α is simple.

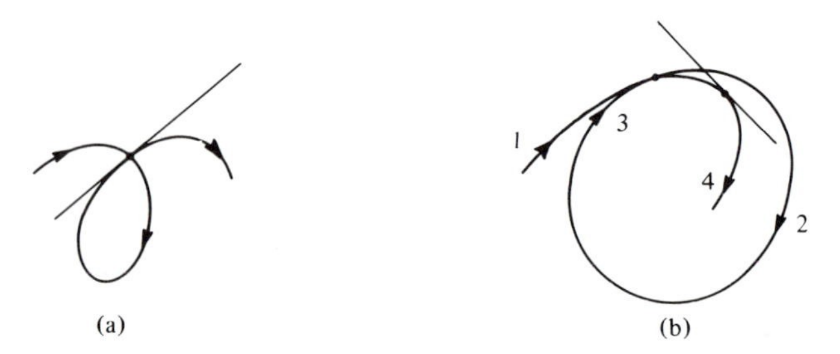

Figure 5-35

We now assume that α is convex and that k changes sign in $[0, l]$. Then there are points $t_1, t_2 \in [0, l]$, $t_1 < t_2$, with $\tilde{\varphi}(t_1) = \tilde{\varphi}(t_2)$ and $\tilde{\varphi}$ not constant in $[t_1, t_2]$.

We shall show that this leads to a contradiction, thereby concluding the proof. By Theorem 2, there exists $t_3 \in [0, l)$ with $\varphi(t_3) = -\varphi(t_1)$. By convexity, two of the three parallel tangent lines at $\alpha(t_1), \alpha(t_2), \alpha(t_3)$ must coincide. Assume this to be the case for $\alpha(t_1) = p$, $\alpha(t_3) = q$, $t_3 > t_1$. We claim that the arc of α between p and q is the linc segment pq.

In fact, assume that $r \neq q$ is the last point for which this arc is a line segment (r may agree with p). Since the curve lies in the same side of the line pq, it is easily seen that some tangent T near p will cross the segment $\overline{pq}$ in an interior point (Fig. 5-36). Then p and q lie on distinct sides of T. That is a contradiction and proves our claim.

It follows that the coincident tangent lines have the same directions; that

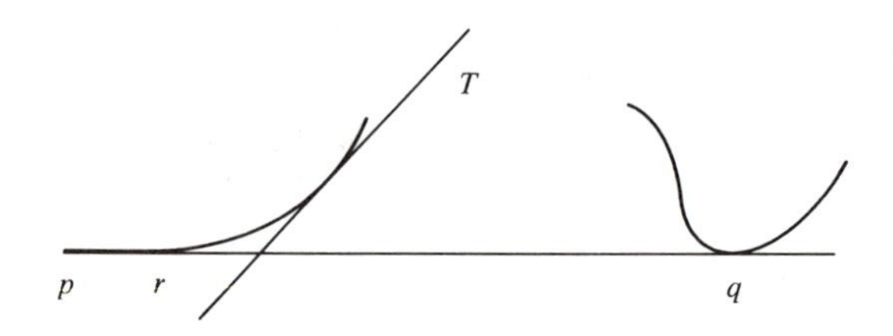

Figure 5-36

is, they are actually the tangent lines at $\alpha(t_1)$ and $\alpha(t_2)$. Thus, $\tilde{\varphi}$ is constant in $[t_1, t_2]$, and this contradiction proves that k does not change sign in $[0, l]$.

Q.E.D.

Remark 2. The condition that α is simple is essential to the proposition, as shown by the example of the curve in Fig. 5-34(c).

Remark 3. The proposition should be compared with Remarks 2 and 3 of Sec. 5-6; there it is stated that a similar situation holds for surfaces. It is to be noticed that, in the case of surfaces, the nonexistence of self-intersections is not an assumption but a consequence.

Remark 4. It can be proved that a plane, regular, closed curve is convex if and only if its interior is a convex set $K \subset R^2$ (cf. Exercise 4).

We shall now turn our attention to space curves. In what follows the word curve will mean a parametrized regular curve $\alpha: [0, l] \longrightarrow R^3$ with arc length s as parameter. If α is a plane curve, the curvature $k(s)$ is the signed curvature of α (cf. Sec. 1-5); otherwise, $k(s)$ is assumed to be positive for all $s \in [0, l]$. It is convenient to call

$$\int_0^l |k(s)|\, ds$$

the *total curvature* of α.

Probably the best-known global theorem on space curves is the so-called Fenchel's theorem.

THEOREM 3 (Fenchel's Theorem). *The total curvature of a simple closed curve is $\geq 2\pi$, and equality holds if and only if the curve is a plane convex curve.*

Before going into the proof, we shall introduce an auxiliary surface which is also useful for the proof of Theorem 4.

The *tube* of radius r around the curve α is the parametrized surface

$$\mathbf{x}(s, v) = \alpha(s) + r(n \cos v + b \sin v), \qquad s \in [0, l], v \in [0, 2\pi],$$

where $n = n(s)$ and $b = b(s)$ are the normal and the binormal vector of α, respectively. It is easily checked that

$$|\mathbf{x}_s \wedge \mathbf{x}_v| = EG - F^2 = r^2(1 - rk \cos v)^2.$$

We assume that r is so small that $rk_0 < 1$, where $k_0 < \max|k(s)|$, $s \in [0, l]$. Then $\mathbf{x}$ is regular, and a straightforward computation gives

$$N = -(n \cos v + b \sin v),$$
$$\mathbf{x}_s \wedge \mathbf{x}_v = r(1 - rk \cos v)N,$$
$$N_s \wedge N_v = k \cos v(n \cos v + b \sin v) = -k\, N \cos v$$
$$= -\frac{k \cos v}{r(1 - rk \cos v)} \mathbf{x}_v \wedge \mathbf{x}_s.$$

It follows that the Gaussian curvature $K = K(s, v)$ of the tube is given by

$$K(s, v) = -\frac{k \cos v}{r(1 - rk \cos v)}.$$

Notice that the trace T of $\mathbf{x}$ may have self-intersections. However, if α is simple, it is possible to choose r so small that this does not occur; we use the compactness of $[0, l]$ and proceed as in the case of a tubular neighborhood constructed in Sec. 2-7. If, in addition, α is closed, T is a regular surface homeomorphic to a torus, also called a *tube around* α. In what follows, we assume this to be the case.

Proof of Theorem 3. Let T be a tube around α, and let $R \subset T$ be the region of T where the Gaussian curvature of T is nonnegative. On the one hand,

$$\iint_R K\, d\sigma = \iint_R K \sqrt{EG - F^2}\, ds\, dv$$
$$= \int_0^l k\, ds \int_{\pi/2}^{3\pi/2} \cos v\, dv = 2 \int_0^l k(s)\, ds.$$

On the other hand, each half-line L through the origin of R^3 appears at least once as a normal direction of R. For if we take a plane P perpendicular to L such that $P \cap T = \phi$ and move P parallel to itself toward T (Fig. 5-37), it will meet T for the first time at a point where $K \geq 0$.

It follows that the Gauss map N of R covers the entire unit sphere S^2 at least once; hence, $\iint_R K\, d\sigma \geq 4\pi$. Therefore, the total curvature of α is $\geq 2\pi$, and we have proved the first part of Theorem 3.

Observe that the image of the Gauss map N restricted to each circle $s = \text{const.}$ is one-to-one and that its image is a great circle $\mathbf{\Gamma}_s \subset S^2$. We shall denote by $\mathbf{\Gamma}_s^+ \subset \mathbf{\Gamma}_s$ the closed half-circle corresponding to points where $K \geq 0$.

Assume that α is a plane convex curve. Then all $\mathbf{\Gamma}_s^+$ have the same end points p, q, and, by convexity, $\mathbf{\Gamma}_{s_1} \cap \mathbf{\Gamma}_{s_2} = \{p\} \cup \{q\}$ for $s_1 \neq s_2, s_1, s_2 \in [0, l)$. By the first part of the theorem, it follows that $\iint_R K\, d\sigma = 4\pi$; hence, the total curvature of α is equal to 2π.

Assume now that the total curvature of α is equal to 2π. By the first part

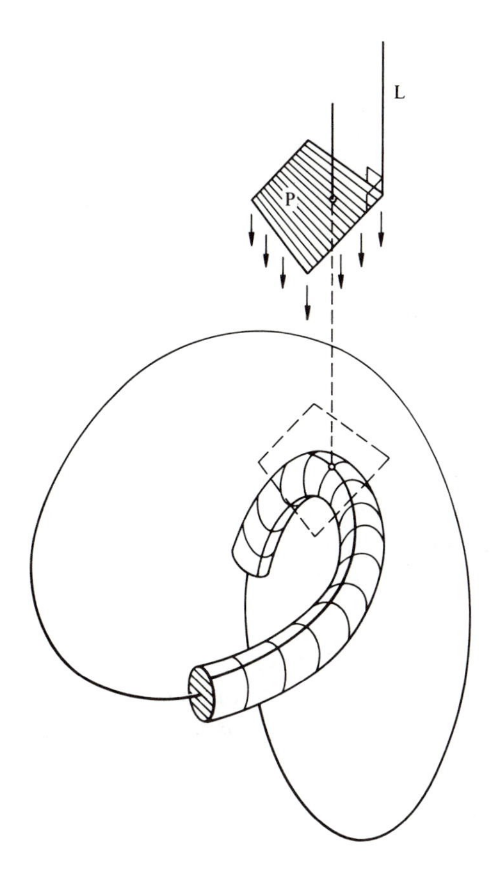

Figure 5-37

of the theorem, $\iint_R K\,d\sigma = 4\pi$. We claim that all Γ_s^+ have the same end points p and q. Otherwise, there are two distinct great circles $\Gamma_{s_1}, \Gamma_{s_2}$, s_1 arbitrarily close to s_2, that intersect in two antipodal points which are not in $N(R \cap Q)$, where Q is the set of points in T with nonpositive curvature. It follows that there are two points of positive curvature which are mapped by N into a single point of S^2. Since N is a local diffeomorphism at such points and each point of S^2 is the image of at least one point of R, we conclude that $\iint_R K\sigma > 4\pi$, a contradiction.

By observing that the points of zero Gaussian curvature in T are the intersections of the binormal of α with T, we see that the binormal vector of α is parallel to the line pq. Thus, α is contained in a plane normal to this line.

We finally prove that α is convex. We may assume that α is so oriented that its rotation number is positive. Since the total curvature of α is 2π, we have

$$2\pi = \int_0^l |k|\,ds \geq \int_0^l k\,ds.$$

On the other hand,

$$\int_J k\,ds \geq 2\pi,$$

where $J = \{s \in [0, l];\, k(s) \geq 0\}$. This holds for any plane closed curve and follows from an argument entirely similar to the one used for $R \subset T$ in the beginning of this proof. Thus,

$$\int_0^l k\,ds = \int_0^l |k|\,ds = 2\pi.$$

Therefore, $k \geq 0$, and α is a plane convex curve. **Q.E.D.**

Remark 5. It is not hard to see that the proof goes through even if α is not simple. The tube will then have self-intersections, but this is irrelevant to the argument. In the last step of the proof (the convexity of α), one has to observe that we have actually shown that α is nonnegatively curved and that its rotation index is equal to 1. Looking back at the first part of the proof of Prop. 1, one easily sees that this implies that α is convex.

We want to use the above method of proving Fenchel's theorem to obtain a sharpening of this theorem which states that if a space curve is knotted (a concept to be defined presently), then the total curvature is actually greater than or equal to 4π.

A simple closed *continuous* curve $C \subset R^3$ is *unknotted* if there exists a homotopy $H\colon S^1 \times I \to R^3$, $I = [0, 1]$, such that

$$H(S^1 \times \{0\}) = S^1$$
$$H(S^1 \times \{1\}) = C;$$
$$\text{and} \quad H(S^1 \times \{t\}) = C_t \subset R^3$$

is homeomorphic to S^1 for all $t \in [0, 1]$. Intuitively, this means that C can be deformed continuously onto the circle S^1 so that all intermediate positions are homeomorphic to S^1. Such a homotopy is called an *isotopy*; an unknotted curve is then a curve isotopic to S^1. When this is not the case, C is said to be *knotted* (Fig. 5-38).

THEOREM 4 (Fary-Milnor). *The total curvature of a knotted simple closed curve is greater than or equal to* 4π.

Proof. Let $C = \alpha([0, l])$, let T be a tube around α, and let $R \subset T$ be the region of T where $K \geq 0$. Let $b = b(s)$ be the binormal vector of α, and let

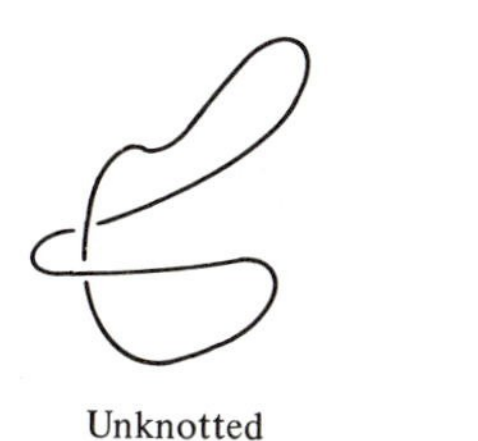

Unknotted

Knotted

Figure 5-38

$v \in R^3$ be a unit vector, $v \neq b(s)$, for all $s \in [0, l]$. Let $h_v: [0, l] \longrightarrow R$ be the height function of α in the direction of v; that is, $h_v(s) = \langle \alpha(s) - 0, v \rangle$, $s \in [0, l]$. Clearly, s is a critical point of h_v if and only if v is perpendicular to the tangent line at $\alpha(s)$. Furthermore, at a critical point,

$$\frac{d}{ds^2}(h_v) = \left\langle \frac{d^2\alpha}{ds^2}, v \right\rangle = k\langle n, v \rangle \neq 0,$$

since $v \neq b(s)$ for all s and $k > 0$. Thus, the critical points of h_v are either maxima or minima.

Now, assume the total curvature of α to be smaller than 4π. This means that

$$\iint_R K \, d\sigma = 2 \int k \, ds < 8\pi.$$

We claim that, for some $v_0 \notin b([0, l])$, h_{v_0} has exactly two critical points (since $[0, l]$ is compact, such points correspond to the maximum and minimum of h_{v_0}). Assume that the contrary is true. Then, for every $v \notin b([0, l])$, h_v has at least three critical points. We shall assume that two of them are points of minima, s_1 and s_2, the case of maxima being treated similarly.

Consider a plane P perpendicular to v such that $P \cap T = \phi$, and move it parallel to itself toward T. Either $h_v(s_1) = h_v(s_2)$ or, say, $h_v(s_1) < h_v(s_2)$. In the first case, P meets T at points $q_1 \neq q_2$, and since $v \notin b([0, l])$, $K(q_1)$ and $K(q_2)$ are positive. In the second case, before meeting $\alpha(s_1)$, P will meet T at a point q_1 with $K(q_1) > 0$. Consider a second plane $\bar{P}$, parallel to and at a distance r above P (r is the radius of the tube T). Move $\bar{P}$ further up until it reaches $\alpha(s_2)$; then P will meet T at a point $q_2 \neq q_1$ (Fig. 5-39). Since s_2 is a point of minimum and $v \notin b([0, l])$, $K(q_2) > 0$. In any case, there are two distinct points in T with $K > 0$ that are mapped by N into a single point of S^2. This contradicts the fact that $\iint_R K \, d\sigma < 8\pi$, and proves our claim.

Let s_1 and s_2 be the critical points of h_{v_0}, and let P_1 and P_2 be planes perpendicular to v_0 and passing through $\alpha(s_1)$ and $\alpha(s_2)$, respectively. Each plane parallel to v_0 and between P_1 and P_2 will meet C in exactly two points. Joining these pairs of points by line segments, we generate a surface bounded by C

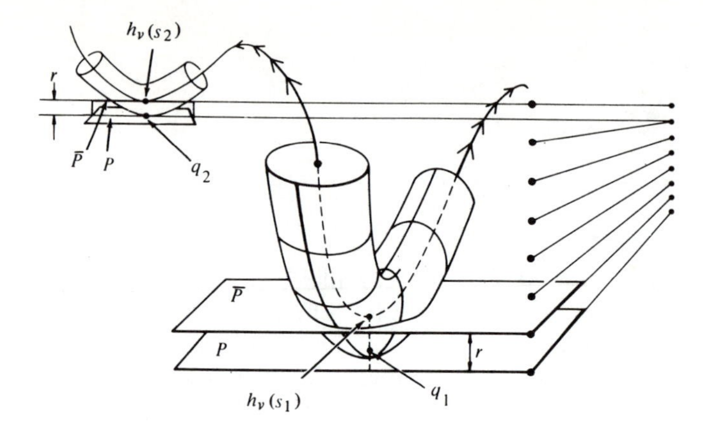

Figure 5-39

which is easily seen to be homeomorphic to a disk. Thus, C is unknotted, and this contradiction completes the proof. **Q.E.D.**

EXERCISES

1. Determine the rotation indices of curves (a), (b), (c), and (d) in Fig. 5-40.

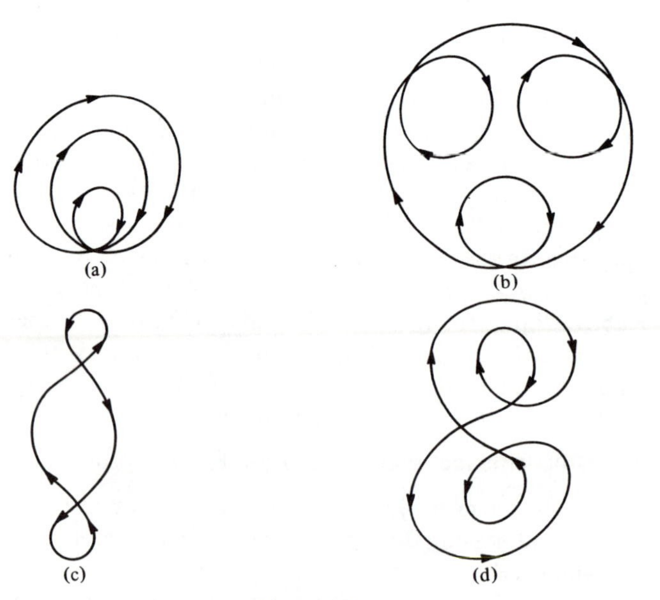

Figure 5-40

2. Let $\alpha(t) = (x(t), y(t))$, $t \in [0, l]$, be a differentiable plane closed curve. Let $p_0 = (x_0, y_0) \in R^2$, $(x_0, y_0) \notin \alpha([0, l])$, and define the functions

$$a(t) = \frac{x(t) - x_0}{\{(x(t) - x_0)^2 + (y(t) - y_0)^2\}^{1/2}},$$

$$b(t) = \frac{y(t) - y_0}{\{(x(t) - x_0)^2 + (y(t) - y_0)^2\}^{1/2}}.$$

a. Use Lemma 1 of Sec. 4-4 to show that the differentiable function

$$\varphi(t) = \varphi_0 + \int_0^t (ab' - ba')dt, \qquad a' = \frac{da}{dt}, b' = \frac{db}{dt},$$

is a determination of the angle that the x axis makes with the position vector $(\alpha(t) - p_0)/|\alpha(t) - p_0|$.

b. Use part a to show that when α is a differentiable closed plane curve, the winding number of α relative to p_0 is given by the integral

$$w = \frac{1}{2\pi} \int_0^l (ab' - ba')\, dt.$$

3. Let $\alpha: [0, l] \longrightarrow R^2$ and $\beta: [0, l] \longrightarrow R^2$ be two differentiable plane closed curves, and let $p_0 \in R^2$ be a point such that $p_0 \notin \alpha([0, l])$ and $p_0 \notin \beta([0, l])$. Assume that, for each $t \in [0, l]$, the points $\alpha(t)$ and $\beta(t)$ are closer than the points $\alpha(t)$ and p_0; i.e.,

$$|\alpha(t) - \beta(t)| < |\alpha(t) - p_0|.$$

Use Exercise 2 to prove that the winding number of α relative to p_0 is equal to the winding number of β relative to p_0.

4. a. Let C be a regular plane closed convex curve. Since C is simple, it determines, by the Jordan curve theorem, an interior region $K \subset R^2$. Prove that K is a convex set (i.e., given $p, q \in K$, the segment of straight line $\overline{pq}$ is contained in K; cf. Exercise 9, Sec. 1-7).

b. Conversely, let C be a regular plane curve (not necessarily closed), and assume that C is the boundary of a convex region. Prove that C is convex.

5. Let C be a regular plane, closed, convex curve. By Exercise 4, the interior of C is a convex set K. Let $p_0 \in K$, $p_0 \notin C$.

a. Show that the line which joins p_0 to an arbitrary point $q \in C$ is not tangent to C at q.

b. Conclude from part a that the rotation index of C is equal to the winding number of C relative to p_0.

c. Obtain from part b a simple proof for the fact that the rotation index of a closed convex curve is ± 1.

6. Let $\alpha\colon [0, l] \longrightarrow R^3$ be a regular closed curve parametrized by arc length. Assume that $0 \neq |k(s)| \leq 1$ for all $s \in [0, l]$. Prove that $l \geq 2\pi$ and that $l = 2\pi$ if and only if α is a plane convex curve.

7. (*Schur's Theorem for Plane Curves.*) Let $\alpha\colon [0, l] \longrightarrow R^2$ and $\tilde{\alpha}\colon [0, l] \longrightarrow R^2$ be two plane convex curves parametrized by arc length, both with the same length l. Denote by k and $\tilde{k}$ the curvatures of α and $\tilde{\alpha}$, respectively, and by d and $\tilde{d}$ the lengths of the chords of α and $\tilde{\alpha}$, respectively; i.e.,

$$d(s) = |\alpha(s) - \alpha(0)|, \qquad \tilde{d}(s) = |\tilde{\alpha}(s) - \tilde{\alpha}(0)|.$$

Assume that $k(s) \geq \tilde{k}(s)$, $s \in [0, l]$. We want to prove that $d(s) \leq \tilde{d}(s)$, $s \in [0, l]$ (i.e., if we stretch a curve, its chords become longer) and that equality holds for $s \in [0, l]$ if and only if the two curves differ by a rigid motion. We remark that the theorem can be extended to the case where $\tilde{\alpha}$ is a space curve and has a number of applications. Compare S. S. Chern [10].

The following outline may be helpful.

a. Fix a point $s = s_1$. Put both curves $\alpha(s) = (x(s), y(s))$, $\tilde{\alpha}(s) = (\tilde{x}(s), \tilde{y}(s))$ in the lower half-plane $y \leq 0$ so that $\alpha(0)$, $\alpha(s_1)$, $\tilde{\alpha}(0)$, and $\tilde{\alpha}(s_1)$ lie on the x axis and $x(s_1) > x(0)$, $\tilde{x}(s_1) > \tilde{x}(0)$ (see Fig. 5-41). Let $s_0 \in [0, s_1]$ be such that $\alpha'(s_0)$ is parallel to the x axis. Choose the function $\theta(s)$ which gives a differentiable determination of the angle that the x axis makes with $\alpha'(s)$ in such a way that $\theta(s_0) = 0$. Show that, by convexity, $-\pi \leq \theta \leq \pi$.

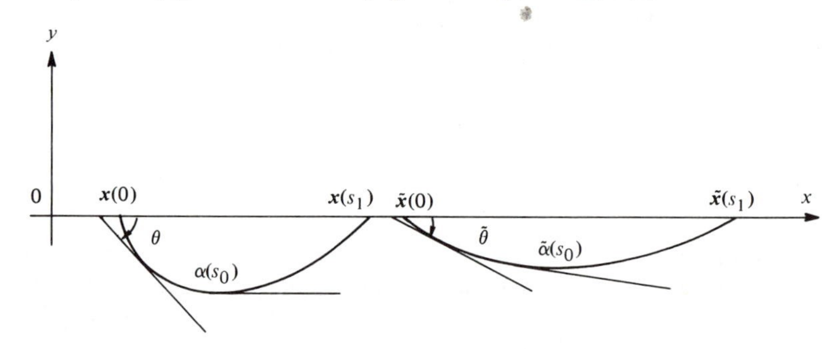

Figure 5-41

b. Let $\tilde{\theta}(s)$, $\tilde{\theta}(s_0) = 0$, be a differentiable determination of the angle that the x axis makes with $\alpha'(s)$. (Notice that $\tilde{\alpha}'(s_0)$ may no longer be parallel to the x axis.) Prove that $\tilde{\theta}(s) \leq \theta(s)$ and use part a to conclude that

$$d(s_1) = \int_0^{s_1} \cos\theta(s)\, ds \leq \int_0^{s_1} \cos\tilde{\theta}(s)\, ds \leq \tilde{d}(s_1).$$

For the equality case, just trace back your steps and apply the uniqueness theorem for plane curves.

8. (*Stoker's Theorem for Plane Curves.*) Let $\alpha\colon R \longrightarrow R^2$ be a regular plane curve parametrized by arc length. Assume that α satisfies the following conditions:

1. The curvature of α is strictly positive.
2. $\lim_{s \to \pm\infty} |\alpha(s)| = \infty$; that is, the curve extends to infinity in both directions.
3. α has no self-intersections.

The goal of the exercise is to prove that the total curvature of α is $\leq \pi$.

The following indications may be helpful. Assume that the total curvature is $> \pi$ and that α has no self-intersections. To obtain a contradiction, proceed as follows:

a. Prove that there exist points, say, $p = \alpha(0)$, $q = \alpha(s_1)$, $s_1 > 0$, such that the tangent lines T_p, T_q at the points p and q, respectively, are parallel and there exists no tangent line parallel to T_p in the arc $\alpha([0, s_1])$.

b. Show that as s increases, $\alpha(s)$ meets T_p at a point, say, r (Fig. 5-42).

c. The arc $\alpha((-\infty, 0))$ must meet T_p at a point t between p and r.

d. Complete the arc $tpqr$ of α with an arc β without self-intersections joining r to t, thus obtaining a closed curve C. Show that the rotation index of C is ≥ 2. Show that this implies that α has self-intersections, a contradiction.

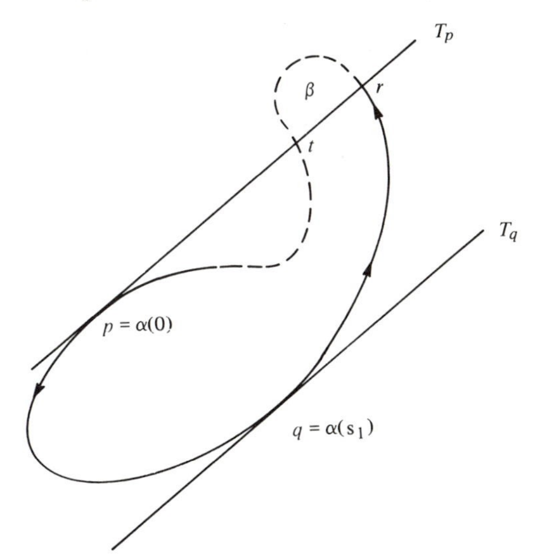

Figure 5-42

***9.** Let $\alpha: [0, l] \longrightarrow S^2$ be a regular closed curve on a sphere $S^2 = \{(x, y, z) \in R^3; x^2 + y^2 + z^2 = 1\}$. Assume that α is parametrized by arc length and that the curvature $k(s)$ is nowhere zero. Prove that

$$\int_0^l \tau(s)\, ds = 0.$$

(The above integral is actually a sufficient condition for a nonplanar curve to lie on the surface of a sphere. For this and related results, see H. Geppert, "Sopra una caractterizzazione della sfera," *Ann. di Mat. Pura ed App.* XX (1941), 59–66; and B. Segre, "Una nuova caracterizazzione della sfera," *Atti Accad. Naz. dei Lincei* 3 (1947), 420–422.)

5-8. Surfaces of Zero Gaussian Curvature

We have already seen (Sec. 4-6) that the regular surfaces with identically zero Gaussian curvature are locally isometric to the plane. In this section, we shall look upon such surfaces from the point of view of their position in R^3 and prove the following global theorem.

THEOREM. *Let* $S \subset R^3$ *be a complete surface with zero Gaussian curvature. Then* S *is a cylinder or a plane.*

By definition, a *cylinder* is a regular surface S such that through each point $p \in S$ there passes a unique line $R(p) \subset S$ (the generator through p) which satisfies the condition that if $q \neq p$, then the lines $R(p)$ and $R(q)$ are parallel or equal.

It is a strange fact in the history of differential geometry that such a theorem was proved only somewhat late in its development. The first proof came as a corollary of a theorem of P. Hartman and L. Nirenberg ("On Spherical Images Whose Jacobians Do Not Change Signs," *Amer. J. Math.* 81 (1959), 901–920) dealing with a situation much more general than ours. Later, W. S. Massey ("Surfaces of Gaussian Curvature Zero in Euclidean Space," *Tohoku Math. J.* 14 (1962), 73–79) and J. J. Stoker ("Developable Surfaces in the Large," *Comm. Pure and Appl. Math.* 14 (1961), 627–635) obtained elementary and direct proofs of the theorem. The proof we present here is a modification of Massey's proof. It should be remarked that Stoker's paper contains a slightly more general theorem.

We shall start with the study of some local properties of a surface of zero curvature.

Let $S \subset R^3$ be a regular surface with Gaussian curvature $K \equiv 0$. Since $K = k_1 k_2$, where k_1 and k_2 are the principal curvatures, the points of S are either parabolic or planar points. We denote by P the set of planar points and by $U = S - P$ the set of parabolic points of S.

P is closed in S. In fact, the points of P satisfy the condition that the mean curvature $H = \frac{1}{2}(k_1 + k_2)$ is zero. A point of accumulation of P has, by continuity of H, zero mean curvature; hence, it belongs to P. It follows that $U = S - P$ is open in S.

An instructive example of the relations between the sets P and U is given in the following example.

Example 1. Consider the open triangle ABC and add to each side a cylindrical surface, with generators parallel to the given side (see Fig. 5-43). It is possible to make this construction in such a way that the resulting surface is a regular surface. For instance, to ensure regularity along the open segment BC, it suffices that the section FG of the cylindrical band $BCDE$ by a plane normal to BC is a curve of the form

$$\exp\left(-\frac{1}{x^2}\right).$$

Observe that the vertices A, B, C of the triangle and the edges BE, CD, etc., of the cylindrical bands do not belong to S.

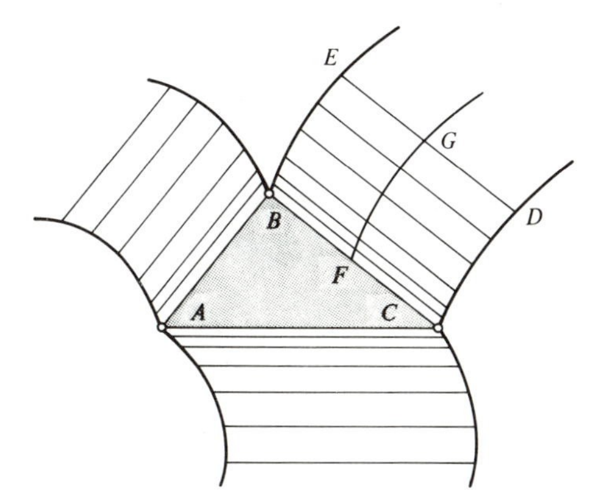

Figure 5-43

The surface S so constructed has curvature $K \equiv 0$. The set P is formed by the closed triangle ABC minus the vertices. Observe that P is closed in S but not in R^3. The set U is formed by the points which are interior to the cylindrical bands. Through each point of U there passes a unique line which will never meet P. The boundary of P is formed by the open segments AB, BC, and CA.

In the following, we shall prove that the relevant properties of this example appear in the general case.

First, let $p \in U$. Since p is a parabolic point, one of the principal directions at p is an asymptotic direction, and there is no other asymptotic direction at p. We shall prove that the unique asymptotic curve that passes through p is a segment of a line.

PROPOSITION 1. *The unique asymptotic line that passes through a parabolic point* $p \in U \subset S$ *of a surface* S *of curvature* $K \equiv 0$ *is an (open) segment of a (straight) line in* S.

Proof. Since p is not umbilical, it is possible to parametrize a neighborhood $V \subset U$ of p by $\mathbf{x}(u, v) = \mathbf{x}$ in such a way that the coordinate curves are lines of curvature. Suppose that $v =$ const. is an asymptotic curve; that is, it has zero normal curvature. Then, by the theorem of Olinde Rodrigues (Sec. 3-2, Prop. 4), $N_u = 0$ along $v =$ const. Since through each point of the neighborhood V there passes a curve $v =$ const., the relation $N_u = 0$ holds for every point of V.

It follows that in V

$$\langle \mathbf{x}, N \rangle_u = \langle \mathbf{x}_u, N \rangle + \langle \mathbf{x}, N_u \rangle = 0.$$

Therefore,

$$\langle \mathbf{x}, N \rangle = \varphi(v), \tag{1}$$

where $\varphi(v)$ is a differentiable function of v alone. By differentiating Eq. (1) with respect to v, we obtain

$$\langle \mathbf{x}, N_v \rangle = \varphi'(v). \tag{2}$$

On the other hand, N_v is normal to N and different from zero, since the points of V are parabolic. Therefore, N and N_v are linearly independent. Furthermore, $N_{vu} = N_{uv} = 0$ in V.

We now observe that along the curve $v =$ const. $= v_0$ the vector $N(u) = N_0$ and $N_v(u) = (N_v)_0 =$ const. Thus, Eq. (1) implies that the curve $\mathbf{x}(u, v_0)$ belongs to a plane normal to the constant vector N_0, and Eq. (2) implies that this curve belongs to a plane normal to the constant vector $(N_v)_0$. Therefore, the curve is contained in the intersection of two planes (the intersection exists since N_0 and $(N_v)_0$ are linearly independent); hence, it is a segment of a line. **Q.E.D.**

Remark. It is essential that $K \equiv 0$ in the above proposition. For instance, the upper parallel of a torus of revolution is an asymptotic curve formed by parabolic points and it is not a segment of a line.

We are now going to see what happens when we extend this segment of line. The following proposition shows that (cf. Example 1) the extended line never meets the set P; either it "ends" at a boundary point of S or stays indefinitely in U.

It is convenient to use the following terminology. An asymptotic curve passing through a point $p \in S$ is said to be *maximal* if it is not a proper subset of some asymptotic curve passing through p.

PROPOSITION 2 (Massey, *loc. cit.***).** *Let* r *be a maximal asymptotic line passing through a parabolic point* $p \in U \subset S$ *of a surface* S *of curvature* $K \equiv 0$ *and let* $P \subset S$ *be the set of planar points of* S. *Then* $r \cap P = \phi$.

The proof of Prop. 2 depends on the following local lemma, for which we use the Mainardi-Codazzi equations (cf. Sec. 4-3).

LEMMA 1. *Let* s *be the arc length of the asymptotic curve passing through a parabolic point* p *of a surface* S *of zero curvature and let* H = H(s) *be the mean curvature of* S *along this curve. Then, in* U,

$$\frac{d^2}{ds^2}\left(\frac{1}{H}\right) = 0.$$

Proof of Lemma 1. We introduce in a neighborhood $V \subset U$ of p a system of coordinates (u, v) such that the coordinate curves are lines of curvature and the curves $v =$ const. are the asymptotic curves of V. Let e, f, and g be the coefficients of the second fundamental form in this parametrization. Since $f = 0$ and the curve $v =$ const., $u = u(s)$ must satisfy the differential equation of the asymptotic curves

$$e\left(\frac{du}{ds}\right)^2 + 2f\frac{du}{ds}\frac{dv}{ds} + g\left(\frac{dv}{ds}\right)^2 = 0,$$

we conclude that $e = 0$. Under these conditions, the mean curvature H is given by

$$H = \frac{k_1 + k_2}{2} = \frac{1}{2}\left(\frac{e}{E} + \frac{g}{G}\right) = \frac{1}{2}\frac{g}{G}. \tag{3}$$

By introducing the values $F = f = e = 0$ in the Mainardi-Codazzi equations (Sec. 4-1, Eq. (7) and (7a)), we obtain

$$0 = \frac{1}{2}\frac{gE_v}{G}, \qquad g_u = \frac{1}{2}\frac{gG_u}{G}. \tag{4}$$

From the first equation of (4) it follows that $E_v = 0$. Thus, $E = E(u)$ is a function of u alone. Therefore, it is possible to make a change of parameters:

$$\bar{v} = v, \qquad \bar{u} = \int \sqrt{E(u)}\, du.$$

We shall still denote the new parameters by u and v. u now measures the arc length along $v =$ const., and thus $E = 1$.

In the new parametrization ($F = 0$, $E = 1$) the expression for the Gaussian curvature is

$$K = -\frac{1}{\sqrt{G}}(\sqrt{G})_{uu} = 0.$$

Therefore,

$$\sqrt{G} = c_1(v)u + c_2(v), \tag{5}$$

where $c_1(v)$ and $c_2(v)$ are functions of v alone.

On the other hand, the second equation of (4) may be written ($g \neq 0$)

$$\frac{g_u}{g} = \frac{1}{2}\frac{G_u}{\sqrt{G}\sqrt{G}} = \frac{(\sqrt{G})_u}{\sqrt{G}};$$

hence,

$$g = c_3(v)\sqrt{G}, \tag{6}$$

where $c_3(v)$ is a function of v. By introducing Eqs. (5) and (6) into Eq. (3) we obtain

$$H = \frac{1}{2}\frac{c_3(v)}{\sqrt{G}}\frac{\sqrt{G}}{\sqrt{G}} = \frac{1}{2}\frac{c_3(v)}{c_1(v)u + c_2(v)}.$$

Finally, by recalling that $u = s$ and differentiating the above expression with respect to s, we conclude that

$$\frac{d^2}{ds^2}\left(\frac{1}{H}\right) = 0,$$ **Q.E.D.**

Proof of Prop. 2. Assume that the maximal asymptotic line r passing through p and parametrized by arc length s contains a point $q \in P$. Since r is connected and U is open, there exists a point p_0 of r, corresponding to s_0, such that $p_0 \in P$ and the points of r with $s < s_0$ belong to U.

On the other hand, from Lemma 1, we conclude that along r and for $s < s_0$,

$$H(s) = \frac{1}{as + b},$$

where a and b are constants. Since the points of P have zero mean curvature, we obtain

$$H(p_0) = 0 = \lim_{s \to s_0} H(s) = \lim_{s \to s_0} \frac{1}{as + b},$$

which is a contradiction and concludes the proof. **Q.E.D.**

Let now $\mathrm{Bd}(U)$ be the *boundary* of U in S; that is, $\mathrm{Bd}(U)$ is the set of points $p \in S$ such that every neighborhood of p in S contains points of U and points of $S - U = P$. Since U is open in S, it follows that $\mathrm{Bd}(U) \subset P$. Furthermore, since the definition of a boundary point is symmetric in U

and P, we have that

$$\mathrm{Bd}(U) = \mathrm{Bd}(P).$$

The following proposition shows that (just as in Example 1) the set $\mathrm{Bd}(U) = \mathrm{Bd}(P)$ is formed by segments of straight lines.

PROPOSITION 3 (Massey). *Let* $p \in Bd(U) \subset S$ *be a point of the boundary of the set* U *of parabolic points of a surface* S *of curvature* $K \equiv 0$. *Then through* p *there passes a unique open segment of line* $C(p) \subset S$. *Furthermore,* $C(p) \subset Bd(U)$; *that is, the boundary of* U *is formed by segments of lines.*

Proof. Let $p \in \mathrm{Bd}(U)$. Since p is a limit point of U, it is possible to choose a sequence $\{p_n\}$, $p_n \in U$, with $\lim_{n\to\infty} p_n = p$. For every p_n, let $C(p_n)$ be the unique maximal asymptotic curve (open segment of a line) that passes through p_n (cf. Prop. 1). We shall prove that, as $n \longrightarrow \infty$, the directions of $C(p_n)$ converge to a certain direction that does not depend on the choice of the sequence $\{p_n\}$.

In fact, let $\Sigma \subset R^3$ be a sufficiently small sphere around p. Since the sphere Σ is compact, the points $\{q_n\}$ of intersection of $C(p_n)$ with Σ have at least one point of accumulation $q \in \Sigma$, which occurs simultaneously with its antipodal point. If there were another point of accumulation r besides q and its antipodal point, then through arbitrarily near points p_n and p_m there should pass asymptotic lines $C(p_n)$ and $C(p_m)$ making an angle greater than

$$\theta = \tfrac{1}{2}\,\mathrm{ang}(pq, pr),$$

thus contradicting the continuity of asymptotic lines. It follows that the lines $C(p_n)$ have a limiting direction. An analogous argument shows that this limiting direction does not depend on the chosen sequence $\{p_n\}$ with $\lim_{n\to\infty} p_n = p$, as previously asserted.

Since the directions of $C(p_n)$ converge and $p_n \longrightarrow p$, the open segments of lines $C(p_n)$ converge to a segment $C(p) \subset S$ that passes through p. The segment $C(p)$ does not reduce itself to the point p. Otherwise, since $C(p_n)$ is maximal, $p \in S$ would be a point of accumulation of the extremities of $C(p_n)$, which do not belong to S (cf. Prop. 2). By the same reasoning, the segment $C(p)$ does not contain its extreme points.

Finally, we shall prove that $C(p) \subset \mathrm{Bd}(U)$. In fact, if $q \in C(p)$, there exists a sequence

$$\{q_n\},\ q_n \in C(p_n) \subset U, \quad \text{with } \lim_{n\to\infty} q_n = q.$$

Then $q \in U \cup \mathrm{Bd}(U)$. Assume that $q \notin \mathrm{Bd}(U)$. Then $q \in U$, and, by the continuity of the asymptotic directions, $C(p)$ is the unique asymptotic line that passes through q. This implies, by Prop. 2, that $p \in U$, which is a contradiction. Therefore, $q \in \text{Bd}(U)$, that is, $C(p) \subset \text{Bd}(U)$, and this concludes the proof. **Q.E.D.**

We are now in a position to prove the global result stated in the beginning of this section.

Proof of the Theorem. Assume that S is not a plane. Then (Sec. 3-2, Prop. 5) S contains parabolic points. Let U be the (open) set of parabolic points of S and P be the (closed) set of planar points of S. We shall denote by int P, the *interior* of P, the set of points which have a neighborhood entirely contained in P. int P is an open set in S which contains only planar points. Therefore, each connected component of int P is contained in a plane (Sec. 3-2, Prop. 5).

We shall first prove that if $q \in S$ and $q \notin \text{int}\, P$, then through q there passes a unique line $R(q) \subset S$, and two such lines are either equal or do not intersect.

In fact, when $q \in U$, then there exists a unique maximal asymptotic line r passing through q. r is a segment of line (thus, a geodesic) and $r \cap P = \phi$ (cf. Props. 1 and 2). By parametrizing r by arc length we see that r is not a finite segment. Otherwise, there exists a geodesic which cannot be extended to all values of the parameter, which contradicts the completeness of S. Therefore, r is an entire line $R(q)$, and since $r \cap P = \phi$, we conclude that $R(q) \subset U$. It follows that when p is another point of U, $p \notin R(q)$, then $R(p) \cap R(q) = \phi$. Otherwise, through the intersection point there should pass two asymptotic lines, which contradicts the asserted uniqueness.

On the other hand, if $q \in \text{Bd}(U) = \text{Bd}(P)$, then (cf. Prop. 3) through q there passes a unique open segment of line which is contained in $\text{Bd}(U)$. By the previous argument, this segment may be extended into an entire line $R(q) \subset \text{Bd}(U)$, and if $p \in \text{Bd}(U)$, $p \notin R(q)$, then $R(p) \cap R(q) = \phi$.

Clearly, since U is open, if $q \in U$ and $p \in \text{Bd}(U)$, then $R(p) \cap R(q) = \phi$. In this way, through each point of $S - \text{int}\, P = U \cup \text{Bd}(U)$ there passes a unique line contained in $S - \text{int}\, P$, and two such lines are either equal or do not intersect, as we claimed. If we prove that these lines are parallel, we shall conclude that $\text{Bd}(U)$ ($= \text{Bd}(P)$) is formed by parallel lines and that each connected component of int P is an open set of a plane, bounded by two parallel lines. Thus, through each point $t \subset \text{int}\, P$ there passes a unique line $R(t) \subset \text{int}\, P$ parallel to the common direction. It follows that through each point of S there passes a unique generator and that the generators are parallel, that is, S is a cylinder, as we wish.

To prove that the lines passing through the points of $U \cup \text{Bd}(U)$ are parallel, we shall proceed in the following way. Let $q \in U \cup \text{Bd}(U)$ and $p \in U$. Since S is connected, there exists an arc $\alpha\colon [0, l] \longrightarrow S$, with $\alpha(0) = p$,

$\alpha(l) = q$. The map $\exp_p: T_p(S) \longrightarrow S$ is a covering map (Prop. 7, Sec. 5-6) and a local isometry (corollary of Lemma 2, Sec. 5-6). Let $\tilde{\alpha}: [0, l] \longrightarrow T_p(S)$ be the lifting of α, with origin at the origin $0 \in T_p(S)$. For each $\tilde{\alpha}(t)$, with $\exp_p \tilde{\alpha}(t) = \alpha(t) \in U \cup \text{Bd}(U)$, let r_t be the lifting of $R(\alpha(t))$ with origin at $\tilde{\alpha}(t)$. Since $\exp_p$ is a local isometry, r_t is a line in $T_p(S)$.

We shall prove that when $\alpha(t_1) \neq \alpha(t_2)$, $t_1, t_2 \in [0, l]$, the lines r_{t_1} and r_{t_2} are parallel. In fact, if $v \in r_{t_1} \cap r_{t_2}$, then

$$\exp_p(v) \in R(\alpha(t_1)) \cap R(\alpha(t_2)),$$

which is a contradiction.

So far we have not defined $R(\alpha(t))$ when $\alpha(t) \in \text{int}\, P$. This will now be done. When $\tilde{\alpha}(t)$ is such that $\exp_p \tilde{\alpha}(t) = \alpha(t) \in \text{int}\, P$, we draw through $\tilde{\alpha}(t)$ a line r in $T_p(S)$ parallel to the common direction we have just obtained. It is clear that $\exp_p(r) \subset \text{int}\, P$, and since $\exp_p(r)$ is a geodesic, $\exp_p(r)$ is an entire line contained in S. In this way, the line $R(\alpha(t))$ is defined for every $t \in [0, l]$.

We shall now prove that the lines $R(\alpha(t))$, $t \in [0, l]$, are parallel lines. In fact, by the usual compactness argument, it is possible to cover the interval $[0, l]$ with a finite number of open intervals $I_1, \ldots, I_n$ such that $\tilde{\alpha}(I_i)$ is contained in a neighborhood V_i of $\alpha(t_i)$, $t_i \in I_i$, where the restriction of $\exp_p$ is an isometry in V_i. Observe now that when $t_1, t_2 \in I_i$ and $\alpha(t_1) \neq \alpha(t_2)$ then $R(\alpha(t_1))$ is parallel to $R(\alpha(t_2))$. In fact, since r_{t_1} is parallel to r_{t_2} and $\exp_p$ is an isometry in V_i, the open segment $\exp_p(r_{t_1} \cap V_i)$ is parallel to $\exp_p(r_{t_2} \cap V_i)$; this means that the lines $\exp_p r_{t_1} = R(\alpha(t_1))$ and $\exp_p r_{t_2} = R(\alpha(t_2))$ have parallel open segments and are therefore parallel. By then using the decomposition of $[0, l]$ by $I_1, \ldots, I_n$, we shall prove, step by step, that the lines $R(\alpha(t))$ are parallel.

In particular, the line $R(q)$ is parallel to $R(p)$. If s is another point in $U \cup \text{Bd}(U)$, then, by the same argument, $R(s)$ is parallel to $R(p)$ and hence parallel to $R(q)$. In this way, it is proved that all the lines that pass through $U \cup \text{Bd}(U)$ are parallel, and this concludes the proof of the theorem.

Q.E.D.

5-9. Jacobi's Theorems

It is a fundamental property of a geodesic γ (Sec. 4-6, Prop. 4) that when two points p and q of γ are sufficiently close, then γ minimizes the arc length between p and q. This means that the arc length of γ between p and q is smaller than or equal to the arc length of any curve joining p to q. Suppose now that we follow a geodesic γ starting from a point p. It is then natural to ask how far the geodesic γ minimizes arc length. In the case of a sphere, for instance, a geodesic γ (a meridian) starting from a point p minimizes arc length up to the first conjugate point of p relative to γ (that is, up to the antipodal point of p). Past the antipodal point of p, the geodesic stops being minimal, as we may intuitively see by the following considerations.

A geodesic joining two points p and q of a sphere may be thought of as a thread stretched over the sphere and joining the two given points. When the arc $\widehat{pq}$ is smaller than a semimeridian and the points p and q are kept fixed, it is not possible to move the thread without increasing its length. On the other hand, when the arc $\widehat{pq}$ is greater than a semimeridian, a small displacement of the thread (with p and q fixed) "loosens" the thread (see Fig. 5-44).

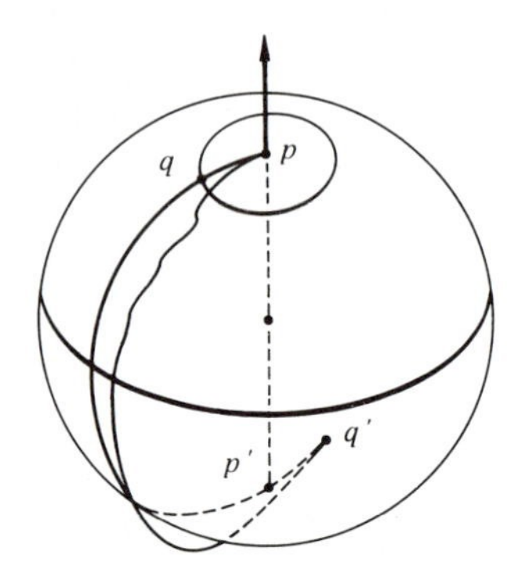

Figure 5-44

In other words, when q is farther away than the antipodal point of p, it is possible to obtain curves joining p to q that are close to the geodesic arc $\widehat{pq}$ and are shorter than this arc. Clearly, this is far from being a mathematical argument.

In this section we shall begin the study of this question and prove a result, due to Jacobi, which may be roughly described as follows. A geodesic γ starting from a point p minimizes arc length, relative to "neighboring" curves of γ, only up to the "first" conjugate point of p relative to γ (more precise statements will be given later; see Theorems 1 and 2).

For simplicity, the surfaces in this section are assumed to be complete and the geodesics are parametrized by arc length.

We need some preliminary results.

The following lemma shows that the image by $\exp_p\colon T_p(S) \to S$ of a segment of line of $T_p(S)$ with origin at $O \in T_p(S)$ (geodesic starting from p) is minimal relative to the images by $\exp_p$ curves of $T_p(S)$ which join the extremities of this segment.

More precisely, let

$$p \in S, \qquad u \in T_p(S), \qquad l = |u| \neq 0,$$

and let $\tilde{\gamma}\colon [0, l] \longrightarrow T_p(S)$ be the line of $T_p(S)$ given by

$$\tilde{\gamma}(s) = sv, \qquad s \in [0, l], \qquad v = \frac{u}{|u|}.$$

Let $\tilde{\alpha}\colon [0, l] \longrightarrow T_p(S)$ be a differentiable parametrized curve of $T_p(S)$, with $\tilde{\alpha}(0) = 0$, $\tilde{\alpha}(l) = u$, and $\tilde{\alpha}(s) \neq 0$ if $s \neq 0$. Furthermore, let (Fig. 5-45)

$$\alpha(s) = \exp_p \tilde{\alpha}(s) \quad \text{and} \quad \gamma(s) = \exp_p \tilde{\gamma}(s).$$

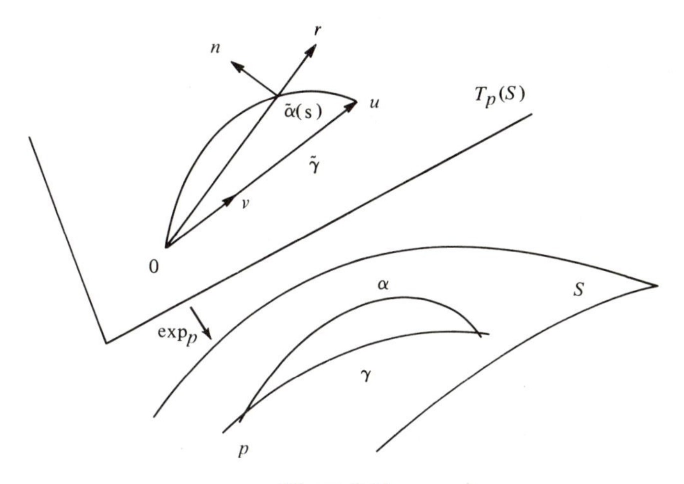

Figure 5-45

LEMMA 1. *With the above notation, we have*

1. $l(\alpha) \geq l(\gamma)$, *where* $l(\ )$ *denotes the arc length of the corresponding curve.*

In addition, if $\tilde{\alpha}(s)$ *is not a critical point of* $\exp_p$, $s \in [0, l)]$, *and if the traces of* α *and* γ *are distinct, then*

2. $l(\alpha) > l(\gamma)$.

Proof. Let $\tilde{\alpha}(s)/|\tilde{\alpha}(s)| = r$, and let n be a unit vector of $T_p(S)$, with $\langle r, n \rangle = 0$. In the basis $\{r, n\}$ of $T_p(S)$ we can write (Fig. 5-45)

$$\tilde{\alpha}'(s) = ar + bn,$$

where

$$a = \langle \tilde{\alpha}'(s), r \rangle,$$
$$b = \langle \tilde{\alpha}'(s), n \rangle.$$

By definition

$$\begin{aligned}\alpha'(s) &= (d\exp_p)_{\tilde{\alpha}(s)}(\tilde{\alpha}'(s)) \\ &= a(d\exp_p)_{\tilde{\alpha}(s)}(r) + b(d\exp_p)_{\tilde{\alpha}(s)}(n).\end{aligned}$$

Therefore, by using the Gauss lemma (cf. Sec. 5-5, Lemma 2) we obtain

$$\langle \alpha'(s), \alpha'(s) \rangle = a^2 + c^2,$$

where

$$c^2 = b^2\,|(d\exp_p)_{\tilde{\alpha}(s)}(n)|^2.$$

It follows that

$$\langle \alpha'(s), \alpha'(s) \rangle \geq a^2.$$

On the other hand,

$$\frac{d}{ds}\langle \tilde{\alpha}(s), \tilde{\alpha}(s)\rangle^{1/2} = \frac{\langle \tilde{\alpha}'(s), \tilde{\alpha}(s)\rangle}{\langle \tilde{\alpha}(s), \tilde{\alpha}(s)\rangle^{1/2}} = \langle \tilde{\alpha}'(s), r\rangle = a.$$

Therefore,

$$\begin{aligned}l(\alpha) &= \int_0^l \langle \alpha'(s), \alpha'(s)\rangle^{1/2}\,ds \geq \int_0^l a\,ds \\ &= \int_0^l \frac{d}{ds}\langle \tilde{\alpha}(s), \tilde{\alpha}(s)\rangle^{1/2}\,ds = |\tilde{\alpha}(l)| = l = l(\gamma),\end{aligned}$$

and this proves part 1.

To prove part 2, let us assume that $l(\alpha) = l(\gamma)$. Then

$$\int_0^l \langle \alpha'(s), \alpha'(s)\rangle^{1/2}\,ds = \int_0^l a\,ds,$$

and since

$$\langle \alpha'(s), \alpha'(s)\rangle^{1/2} \geq a,$$

the equality must hold in the last expression for every $s \in [0, l]$. Therefore,

$$c = |b|\,|(d\exp_p)_{\tilde{\alpha}(s)}(n)| = 0.$$

Since $\tilde{\alpha}(s)$ in not a critical point of $\exp_p$, we conclude that $b \equiv 0$. It follows that the tangent lines to the curve $\tilde{\alpha}$ all pass through the origin O of $T_p(s)$. Thus, $\tilde{\alpha}$ is a line of $T_p(S)$ which passes through O. Since $\tilde{\alpha}(l) = \tilde{\gamma}(l)$, the lines $\tilde{\alpha}$ and $\tilde{\gamma}$ coincide, thus contradicting the assumption that the traces of α and γ are distinct. From this contradiction it follows that $l(\alpha) > l(\gamma)$, which proves part 2 and ends the proof of the lemma. **Q.E.D.**

We are now in a position to prove that if a geodesic arc contains no conjugate points, it yields a local minimum for the arc length. More precisely, we have

THEOREM 1 (Jacobi.) *Let* $\gamma\colon [0, l] \to S$, $\gamma(0) = p$, *be a geodesic without conjugate points; that is,* $\exp_p\colon T_p(S) \to S$ *is regular at the points of the line* $\tilde{\gamma}(s) = s\gamma'(0)$ *of* $T_p(S)$, $s \in [0, l]$. *Let* $h\colon [0, l] \times (-\epsilon, \epsilon) \to S$ *be a proper variation of* γ. *Then*

1. *There exists a* $\delta > 0$, $\delta \leq \epsilon$, *such that if* $t \in (-\delta, \delta)$,

$$L(t) \geq L(0),$$

where $L(t)$ *is the length of the curve* $h_t\colon [0, l] \to S$ *that is given by* $h_t(s) = h(s, t)$.

2. *If, in addition, the trace of* h_t *is distinct from the trace of* γ, $L(t) > L(0)$.

Proof. The proof consists essentially of showing that it is possible, for every $t \in (-\delta, \delta)$, to lift the curve h_t into a curve $\tilde{h}_t$ of $T_p(S)$ such that $\tilde{h}_t(0) = 0$, $\tilde{h}_t(l) = \tilde{\gamma}(l)$ and then to apply Lemma 1.

Since $\exp_p$ is regular at the points of the line $\tilde{\gamma}$ of $T_p(S)$, for each $s \in [0, l]$ there exists a neighborhood U_s of $\tilde{\gamma}(s)$ such that $\exp_p$ restricted to U_s is a diffeomorphism. The family $\{U_s\}$, $s \in [0, l]$, covers $\tilde{\gamma}([0, l])$, and, by compactness, it is possible to obtain a finite subfamily, say, $U_1, \ldots, U_n$ which still covers $\tilde{\gamma}([0, l])$. It follows that we may divide the interval $[0, l]$ by points

$$0 = s_1 < s_2 < \cdots < s_n < s_{n+1} = l$$

in such a way that $\tilde{\gamma}([s_i, s_{i+1}]) \subset U_i$, $i = 1, \ldots, n$. Since h is continuous and $[s_i, s_{i+1}]$ is compact, there exists $\delta_i > 0$ such that

$$h([s_i, s_{k+1}] \times (-\delta_i, \delta_i)) \subset \exp_p(U_i) = V_i.$$

Let $\delta = \min(\delta_1, \ldots, \delta_n)$. For $t \in (-\delta, \delta)$, the curve $h_t\colon [0, l] \to S$ may be lifted into a curve $\tilde{h}_t\colon [0, l] \to T_p(S)$, with origin $\tilde{h}_t(0) = 0$, in the following way. Let $s \in [s_1, s_2]$. Then

$$\tilde{h}_t(s) = \exp_p^{-1}(h_t(s)),$$

where $\exp_p^{-1}$ is the inverse map of $\exp_p\colon U_1 \to V_1$. By applying the same technique we used for covering spaces (cf. Prop. 2, Sec. 5-6), we can extend $\tilde{h}_t$ for all $s \in [0, l]$ and obtain $\tilde{h}_t(l) = \tilde{\gamma}(l)$.

In this way, we conclude that $\gamma(s) = \exp_p \tilde{\gamma}(s)$ and that $h_t(s) = \exp_p \tilde{h}_t(s)$,

$t \in (-\delta, \delta)$, with $\tilde{h}_t(0) = 0$, $\tilde{h}_t(l) = \tilde{\gamma}(l)$. We then apply Lemma 1 to this situation and obtain the desired conclusions. **Q.E.D.**

Remark 1. A geodesic γ containing no conjugate points may well not be minimal relative to the curves which are not in a neighborhood of γ. Such a situation occurs, for instance, in the cylinder (which has no conjugate points), as the reader will easily verify by observing a closed geodesic of the cylinder.

This situation is related to the fact that conjugate points inform us only about the differential of the exponential map, that is, about the rate of "spreading out" of the geodesics nieghboring a given geodesic. On the other hand, the global behavior of the geodesics is controlled by the exponential map itself, which may not be globally one-to-one even when its differential is nonsingular everywhere.

Another example (this time simply connected) where the same fact occurs is in the ellipsoid, as the reader may verify by observing the figure of the ellipsoid in Sec. 5-5 (Fig. 5-19).

The study of the locus of the points for which the geodesics starting from p stop globally minimizing the arc length (called the *cut locus* of p) is of fundamental importance for certain global theorems of differential geometry, but it will not be considered in this book.

We shall proceed now to prove that a geodesic γ containing conjugate points is not a *local* minimum for the arc length; that is, "arbitrarily near" to γ there exists a curve, joining its extreme points, the length of which is smaller than that of γ.

We shall need some preliminaries, the first of which is an extension of the definition of variation of a geodesic to the case where piecewise differentiable functions are admitted.

DEFINITION 1. *Let* $\gamma\colon [0, l] \to S$ *be a geodesic of* S *and let*

$$h\colon [0, l] \times (-\epsilon, \epsilon) \to S$$

be a continuous map with

$$h(s, 0) = \gamma(s), \qquad s \in [0, l].$$

h *is said to be a* broken variation *of* γ *if there exists a partition*

$$0 = s_0 < s_1 < s_2 < \cdots < s_{n-1} < s_n = l$$

of $[0, l]$ *such that*

$$h\colon [s_i, s_{i+1}] \times (-\epsilon, \epsilon) \to S, \; i = 0, 1, \ldots, n-1,$$

is differentiable. The broken variation is said to be proper *if* $h(0, t) = \gamma(0)$, $h(l, t) = \gamma(l)$ *for every* $t \in (-\epsilon, \epsilon)$.

The curves $h_t(s)$, $s \in [0, l]$, of the variation are now piecewise differentiable curves. The variational vector field $V(s) = (\partial h/\partial t)(s, 0)$ is a piecewise differentiable vector field along γ; that is, $V: [0, l] \longrightarrow R^3$ is a continuous map, differentiable in each $[t_i, t_{i+1}]$. The broken variation h is said to be *orthogonal* if $\langle V(s), \gamma'(s)\rangle = 0$, $s \in [0, l]$.

In a way entirely analogous to that of Prop. 1 of Sec. 5-4, it is possible to prove that a piecewise differentiable vector field V along γ gives rise to a broken variation of γ, the variational field of which is V. Furthermore, if

$$V(0) = V(l) = 0,$$

the variation can be chosen to be proper.

Similarly, the function $L: (-\epsilon, \epsilon) \longrightarrow R$ (the arc length of a curve of the variation) is defined by

$$\begin{aligned} L(t) &= \sum_0^{n-1} \int_{t_i}^{t_{i+1}} \left|\frac{\partial h}{\partial s}(s, t)\right| ds \\ &= \int_0^l \left|\frac{\partial h}{\partial s}(s, t)\right| ds. \end{aligned}$$

By Lemma 1 of Sec. 5-4, each summand of this sum is differentiable in a neighborhood of 0. Therefore, L is differentiable in $(-\delta, \delta)$ if δ is sufficiently small.

The expression of the second variation of the arc length ($L''(0)$), for proper and orthogonal broken variations, is exactly the same as that obtained in Prop. 4 of Section 5-4, as may easily be verified. Thus, if V is a piecewise differentiable vector field along a geodesic $\gamma: [0, l] \longrightarrow S$ such that

$$\langle V(s), \gamma'(s)\rangle = 0, \qquad s \in [0, l], \qquad \text{and} \quad V(0) = V(l) = 0,$$

we have

$$L''_V(0) = \int_0^l \left(\left\langle \frac{DV}{ds}, \frac{DV}{ds}\right\rangle - K(s)\langle V(s), V(s)\rangle\right) ds.$$

Now let $\gamma: [0, l] \longrightarrow S$ be a geodesic and let us denote by $\mathcal{V}$ the set of piecewise differentiable vector fields along γ which are orthogonal to γ; that is, if $V \in \mathcal{V}$, then $\langle V(s), \gamma'(s)\rangle = 0$ for all $s \in [0, l]$. Observe that $\mathcal{V}$, with the natural operations of addition and multiplication by a real number, forms a vector space. Define a map $I: \mathcal{V} \times \mathcal{V} \longrightarrow R$ by

$$I(V, W) = \int_0^l \left(\left\langle \frac{DV}{ds}, \frac{DW}{ds} \right\rangle - K(s)\langle V(s), W(s)\rangle \right) ds,$$

where $V, W \in \mathcal{V}$.

It is immediate to verify that I is a symmetric bilinear map; that is, I is linear in each variable and $I(V, W) = I(W, V)$. Therefore, I determines a quadratic form in $\mathcal{V}$, given by $I(V, V)$. This quadratic form is called the *index form* of γ.

Remark 2. The index form of a geodesic was introduced by M. Morse, who proved the following result. Let $\gamma(s_0)$ be a conjugate point of $\gamma(0) = p$, relative to the geodesic $\gamma: [0, l] \to S$, $s_0 \in [0, l]$. The *multiplicity* of the conjugate point $\gamma(s_0)$ is the dimension of the largest subspace E of $T_p(S)$ such that $(d\exp_p)_{\gamma(s_0)}(u) = 0$ for every $u \in E$. The *index* of a quadratic form $Q: E \to R$ in a vector space E is the maximum dimension of a subspace L of E such that $Q(u) < 0$, $u \in L$. With this terminology, the *Morse index theorem* is stated as follows: *Let γ: [0, l] $\to$ S be a geodesic. Then the index of the quadratic form I of γ is finite, and it is equal to the number of conjugate points to $\gamma(0)$ in $\gamma((0, l])$, each one counted with its multiplicity.* A proof of this theorem may be found in J. Milnor, *Morse Theory*, *Annals of Mathematics Studies*, Vol. 51, Princeton University Press, Princeton, N. J., 1963.

For our purposes we need only the following lemma.

LEMMA 2. *Let* V $\in \mathcal{V}$ *be a Jacobi field along a geodesic* $\gamma: [0, l] \to$ S *and* W $\in \mathcal{V}$. *Then*

$$\mathrm{I(V, W)} = \left\langle \frac{\mathrm{DV}}{\mathrm{ds}}(l), \mathrm{W}(l) \right\rangle - \left\langle \frac{\mathrm{DV}}{\mathrm{ds}}(0), \mathrm{W}(0) \right\rangle.$$

Proof. By observing that

$$\frac{d}{ds}\left\langle \frac{DV}{ds}, W \right\rangle = \left\langle \frac{D^2V}{ds^2}, W \right\rangle + \left\langle \frac{DV}{ds}, \frac{DW}{ds} \right\rangle,$$

we may write I in the form (cf. Remark 4, Sec. 5-4)

$$I(V, W) = \left\langle \frac{DV}{ds}, W \right\rangle \Big|_0^l - \int_0^l \left(\left\langle \frac{D^2V}{ds^2} + K(s)V(s), W(s) \right\rangle \right) ds.$$

From the fact that V is a Jacobi field orthogonal to γ, we conclude that the integrand of the second term is zero. Therefore,

$$I(V, W) = \left\langle \frac{DV}{ds}(l), W(l) \right\rangle - \left\langle \frac{DV}{ds}(0), W(0) \right\rangle. \qquad \textbf{Q.E.D.}$$

We are now in a position to prove

THEOREM 2 (Jacobi). *If we let* $\gamma: [0, l] \to S$ *be a geodesic of* S *and we let* $\gamma(s_0) \in \gamma((0, l)$ *be a point conjugate to* $\gamma(0) = p$ *relative to* γ, *then there exists a proper broken variation* $h: [0, l] \times (-\epsilon, \epsilon) \to S$ *of* γ *and a real number* $\delta > 0$, $\delta \leq \epsilon$, *such that if* $t \in (-\delta, \delta)$ *we have* $L(t) < L(0)$.

Proof. Since $\gamma(s_0)$ is conjugate to p relative to γ, there exists a Jacobi field J along γ, not identically zero, with $J(0) = J(s_0) = 0$. By Prop. 4 of Sec. 5-5, it follows that $\langle J(s), \gamma'(s)\rangle = 0$, $s \in [0, l]$. Furthermore, $(DJ/ds)(s_0) \neq 0$; otherwise, $J(s) \equiv 0$.

Now let $\bar{Z}$ be a parallel vector field along γ, with $\bar{Z}(s_0) = -(DJ/ds)(s_0)$, and $f: [0, l] \to R$ be a differentiable function with $f(0) = f(l) = 0$, $f(s_0) = 1$. Define $Z(s) = f(s)\bar{Z}(s)$, $s \in [0, l]$.

For each real number $\eta > 0$, define a vector field Y_η along γ by

$$\begin{aligned} Y_\eta &= J(s) + \eta Z(s), \qquad & s \in [0, s_0], \\ &= \eta Z(s) & s \in [s_0, l]. \end{aligned}$$

Y_η is a piecewise differentiable vector field orthogonal to γ. Since $Y_\eta(0) = Y_\eta(l) = 0$, it gives rise to a proper, orthogonal, broken variation of γ. We shall compute $L''(0) = I(Y_\eta, Y_\eta)$.

For the segment of geodesic between 0 and s_0, we shall use the bilinearity of I and Lemma 2 to obtain

$$\begin{aligned} I_{s_0}(Y_\eta, Y_\eta) &= I_{s_0}(J + \eta Z, J + \eta Z) \\ &= I_{s_0}(J, J) + 2\eta I_{s_0}(J, Z) + \eta^2 I_{s_0}(Z, Z) \\ &= 2\eta\left\langle \frac{DJ}{ds}(s_0), Z(s_0)\right\rangle + \eta^2 I_{s_0}(Z, Z) \\ &= -2\eta\left|\frac{DJ}{ds}(s_0)\right|^2 + \eta^2 I_{s_0}(Z, Z), \end{aligned}$$

where I_{s_0} indicates that the coresponding integral is taken between 0 and s_0. By using I to denote the integral between 0 and l and noticing that the integral is additive, we have

$$I(Y_\eta, Y_\eta) = -2\eta\left|\frac{DJ}{ds}(s_0)\right|^2 + \eta^2 I(Z, Z).$$

Observe now that if $\eta = \eta_0$ is sufficiently small, the above expression is negative. Therefore, by taking Y_{η_0}, we shall obtain a proper broken variation, with $L''(0) < 0$. Since $L'(0) = 0$, this means that 0 is a point of local maximum for L; that is, there exists $\delta > 0$ such that if $t \in (-\delta, \delta)$, $t \neq 0$, then $L(t) < L(0)$. **Q.E.D.**

Remark 3. Jacobi's theorem is a particular case of the Morse index theorem, quoted in Remark 2. Actually, the crucial point of the proof of the index theorem is essentially an extension of the ideas presented in the proof of Theorem 2.

EXERCISES

1. (*Bonnet's Theorem.*) Let S be a complete surface with Gaussian curvature $K \geq \delta > 0$. By Exercise 5 of Sec. 5-5, every geodesic $\gamma: [0, \infty) \longrightarrow S$ has a point conjugate to $\gamma(0)$ in the interval $(0, \pi/\sqrt{\delta}\,]$. Use Jacobi's theorems to show that this implies that S is compact and that the diameter $\rho(S) \leq \pi/\sqrt{\delta}$ (*this gives a new proof of Bonnet's theorem of* Sec. 5-4).

2. (*Lines on Complete Surfaces.*) A geodesic $\gamma: [-\infty, \infty) \longrightarrow S$ is called a *line* if its length realizes the (intrinsic) distance between any two of its points.

a. Show that through each point of the complete cylinder $x^2 + y^2 = 1$ there passes a line.

b. Assume that S is a complete surface with Gaussian curvature $K > 0$. Let $\gamma:(-\infty, \infty) \longrightarrow S$ be a geodesic on S and let $J(s)$ be a Jacobi field along γ given by $\langle J(0), \gamma'(0)\rangle = 0$, $|J(0)| = 1$, $J'(0) = 0$. Choose an orthonormal basis $\{e_1(0) = \gamma'(0), e_2(0)\}$ at $T_{\gamma(0)}(S)$ and extend it by parallel transport along γ to obtain a basis $\{e_1(s), e_2(s)\}$ at each $T_{\gamma(s)}(S)$. Show that $J(s) = u(s)e_2(s)$ for some function $u(s)$ and that the Jacobi equation for J is

$$u'' + Ku = 0, \qquad u(0) = 1, \qquad u'(0) = 0. \qquad (*)$$

c. Extend to the present situation the comparison theorem of part b of Exercise 3, Sec. 5-5. Use the fact that $K > 0$ to show that it is possible to choose $\epsilon > 0$ sufficiently small so that

$$u(\epsilon) > 0, \qquad u(-\epsilon) > 0, \qquad u'(\epsilon) < 0, \qquad u'(-\epsilon) > 0,$$

where $u(s)$ is a solution of $(*)$. Compare $(*)$ with

$$v''(s) = 0, \qquad v(\epsilon) = u(\epsilon), \qquad v'(\epsilon) = u'(\epsilon) \qquad \text{for } s \in [\epsilon, \infty)$$

and with

$$w''(s) = 0, \qquad w(-\epsilon) = u(-\epsilon), \qquad w'(-\epsilon) = u'(-\epsilon) \qquad \text{for } s \in (-\infty, -\epsilon]$$

to conclude that if s_0 is sufficiently large, then $J(s)$ has two zeros in the interval $(-s_0, s_0)$.

d. Use the above to prove that a *complete surface with positive Gaussian curvature contains no lines.*

5-10. Abstract Surfaces; Further Generalizations

In Sec. 5-11, we shall prove a theorem, due to Hilbert, which asserts that there exists no complete regular surface in R^3 with constant negative Gaussian curvature.

Actually, the theorem is somewhat stronger. To understand the correct statement and the proof of Hilbert's theorem, it will be convenient to introduce the notion of an abstract geometric surface which arises from the following considerations.

So far the surfaces we have dealt with are subsets S of R^3 on which differentiable functions make sense. We defined a tangent plane $T_p(S)$ at each $p \in S$ and developed the differential geometry around p as the study of the variation of $T_p(S)$. We have, however, observed that all the notions of the intrinsic geometry (Gaussian curvature, geodesics, completeness, etc.) only depended on the choice of an inner product on each $T_p(S)$. If we are able to define abstractly (that is, with no reference to R^3) a set S on which differentiable functions make sense, we might eventually extend the intrinsic geometry to such sets.

The definition below is an outgrowth of our experience in Chap. 2. Historically, it took a long time to appear, probably due to the fact that the fundamental role of the change of parameters in the definition of a surface in R^3 was not clearly understood.

DEFINITION 1. *An* abstract surface (*differentiable manifold of dimension* 2) *is a set* S *together with a family of one-to-one maps* $\mathbf{x}_\alpha: U_\alpha \longrightarrow S$ *of open sets* $U_\alpha \subset R^2$ *into* S *such that*

1. $\bigcup_\alpha \mathbf{x}_\alpha(U_\alpha) = S$.
2. *For each pair* α, β *with* $\mathbf{x}_\alpha(U_\alpha) \cap \mathbf{x}_\beta(U_\beta) = W \neq \phi$, *we have that* $\mathbf{x}_\alpha^{-1}(W), \mathbf{x}_\beta^{-1}(W)$ *are open sets in* R^2, *and* $\mathbf{x}_\beta^{-1} \circ \mathbf{x}_\alpha, \mathbf{x}_\alpha^{-1} \circ \mathbf{x}_\beta$ *are differentiable maps* (*Fig.* 5-46)

The pair $(U_\alpha, \mathbf{x}_\alpha)$ with $p \in \mathbf{x}_\alpha(U_\alpha)$ is called a *parametrization* (or coordinate system) of S around p. $\mathbf{x}_\alpha(U_\alpha)$ is called a *coordinate neighborhood*, and if $q = \mathbf{x}_\alpha(u_\alpha, v_\alpha) \in S$, we say that (u_α, v_α) are the *coordinates* of q in this coordinate system. The family $\{U_\alpha, \mathbf{x}_\alpha\}$ is called a *differentiable structure* for S.

It follows immediately from condition 2 that the "change of parameters"

$$\mathbf{x}_\beta^{-1} \circ \mathbf{x}_\alpha : \mathbf{x}_\alpha^{-1}(W) \longrightarrow \mathbf{x}_\beta^{-1}(W)$$

is a diffeomorphism.

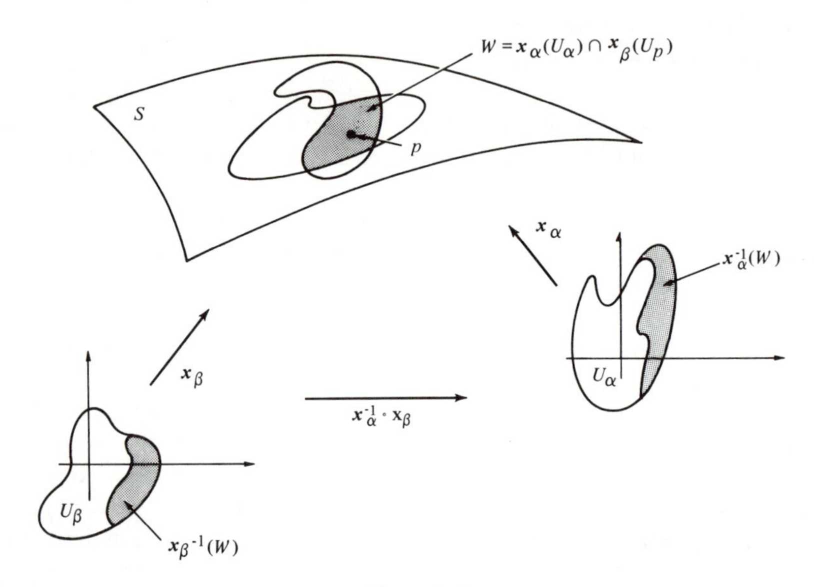

Figure 5-46

Remark 1. It is sometimes convenient to add a further axiom to Def. 1 and say that the differentiable structure should be *maximal* relative to conditions 1 and 2. This means that any other family satisfying conditions 1 and 2 is already contained in the family $\{U_\alpha, \mathbf{x}_\alpha\}$.

A comparison of the above definition with the definition of a regular surface in R^3 (Sec. 2-2, Def. 1) shows that the main point is to include the law of change of parameters (which is a theorem for surfaces in R^3, cf. Sec. 2-3, Prop. 1) in the definition of an abstract surface. Since this was the property which allowed us to define differentiable functions on surfaces in R^3 (Sec. 2-3, Def. 1), we may set

DEFINITION 2. *Let* S_1 *and* S_2 *be abstract surfaces. A map* $\varphi: S_1 \longrightarrow S_2$ *is* differentiable at $p \in S_1$ *if given a parametrization* $\mathbf{y}: V \subset R^2 \longrightarrow S_2$ *around* $\varphi(p)$ *there exists a parametrization* $\mathbf{x}: U \subset R^2 \longrightarrow S_1$ *around* p *such that* $\varphi(\mathbf{x}(U)) \subset \mathbf{y}(V)$ and the map

$$\mathbf{y}^{-1} \circ \varphi \circ \mathbf{x}: \mathbf{x}^{-1}(U) \subset R^2 \longrightarrow R^2 \tag{1}$$

is differentiable at $\mathbf{x}^{-1}(p)$. φ *is* differentiable *on* S_1 *if it is differentiable at every* $p \in S_1$ (*Fig.* 5-47).

It is clear, by condition 2, that this definition does not depend on the

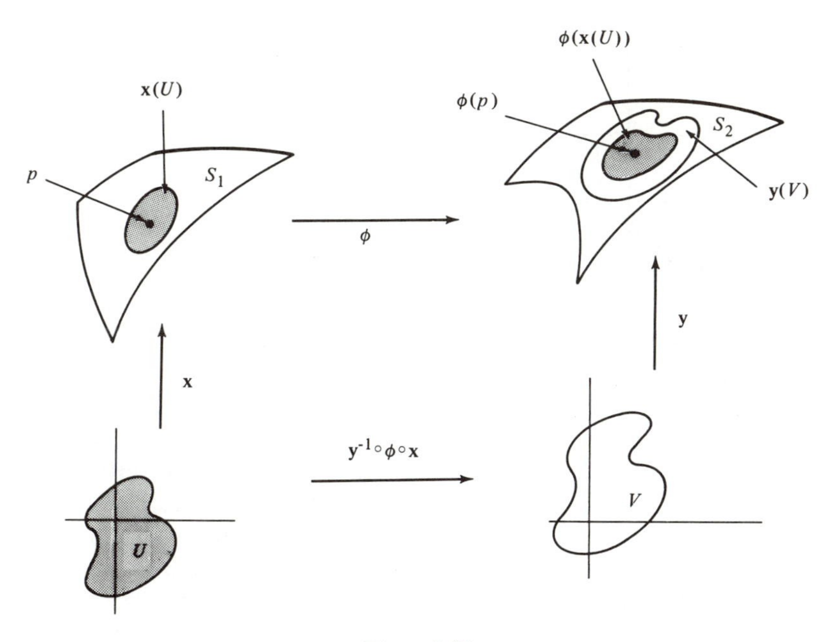

Figure 5-47

choices of the parametrizations. The map (1) is called the *expression* of φ in the parametrizations $\mathbf{x}$, $\mathbf{y}$.

Thus, on an abstract surface it makes sense to talk about differentiable functions, and we have given the first step toward the generalization of intrinsic geometry.

Example 1. Let $S^2 = \{(x, y, z) \in R^3;\ x^2 + y^2 + z^2 = 1\}$ be the unit sphere and let $A: S^2 \longrightarrow S^2$ be the antipodal map; i.e., $A(x, y, z) = (-x, -y, -z)$. Let P^2 be the set obtained from S^2 by identifying p with $A(p)$ and denote by $\pi: S^2 \longrightarrow P^2$ the natural map $\pi(p) = \{p, A(p)\}$. Cover S^2 with parametrizations $\mathbf{x}_\alpha: U_\alpha \longrightarrow S^2$ such that $\mathbf{x}_\alpha(U_\alpha) \cap A \circ \mathbf{x}_\alpha(U_\alpha) = \phi$. From the fact that S^2 is a regular surface and A is a diffeomorphism, it follows that P^2 together with the family $\{U_\alpha, \pi \circ \mathbf{x}_\alpha\}$ is an abstract surface, to be denoted again by P^2. P^2 is called the *real projective plane.*

Example 2. Let $T \subset R^3$ be a torus of revolution (Sec. 2-2, Example 4) with center in $(0, 0, 0) \in R^3$ and let $A: T \longrightarrow T$ be defined by $A(x, y, z) = (-x, -y, -z)$ (Fig. 5-48). Let K be the quotient space of T by the equivalence relation $p \sim A(p)$ and denote by $\pi: T \longrightarrow K$ the map $\pi(p) = \{p, A(p)\}$. Cover T with parametrizations $\mathbf{x}_\alpha: U_\alpha \longrightarrow T$ such that $\mathbf{x}_\alpha(U_\alpha) \cap A \circ \mathbf{x}_\alpha(U_\alpha) = \phi$. As before, it is possible to prove that K with the family $\{U_\alpha, \pi \circ \mathbf{x}_\alpha\}$ is an abstract surface, which is called the *Klein bottle.*

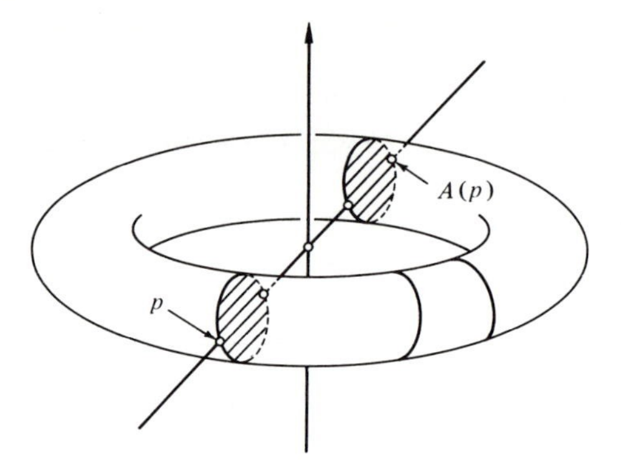

Figure 5-48

Now we need to associate a tangent plane to each point of an abstract surface S. It is again convenient to use our experience with surfaces in R^3 (Sec. 2-4). There the tangent plane was the set of tangent vectors at a point, a tangent vector at a point being defined as the velocity at that point of a curve on the surface. Thus, we must define what is the tangent vector of a curve on an abstract surface. Since we do not have the support of R^3, we must search for a characteristic property of tangent vectors to curves which is independent of R^3.

The following considerations will motivate the definition to be given below. Let $\alpha: (-\epsilon, \epsilon) \to R^2$ be a differentiable curve in R^2, with $\alpha(0) = p$. Write $\alpha(t) = (u(t), v(t))$, $t \in (-\epsilon, \epsilon)$, and $\alpha'(0) = (u'(0), v'(0)) = w$. Let f be a differentiable function defined in a neighborhood of p. We can restrict f to α and write the directional derivative of f relative to w as follows:

$$\left.\frac{d(f \circ \alpha)}{dt}\right|_{t=0} = \left.\left(\frac{\partial f}{\partial u}\frac{du}{dt} + \frac{\partial f}{\partial v}\frac{dv}{dt}\right)\right|_{t=0} = \left\{u'(0)\frac{\partial}{\partial u} + v'(0)\frac{\partial}{\partial v}\right\} f.$$

Thus, the directional derivative in the direction of the vector w is an operator on differentiable functions which depends only on w. This is the characteristic property of tangent vectors that we were looking for.

DEFINITION 3. *A differentiable map* $\alpha: (-\epsilon, \epsilon) \to S$ *is called a* curve *on* S. *Assume that* $\alpha(0) = p$ *and let* D *be the set of functions on* S *which are differentiable at* p. *The* tangent vector *to the curve* α *at* $t = 0$ *is the function* $\alpha'(0): D \to R$ *given by*

$$\alpha'(0)(f) = \left.\frac{d(f \circ \alpha)}{dt}\right|_{t=0}, \qquad f \in D.$$

A tangent vector *at a point* $p \in S$ *is the tangent vector at* $t = 0$ *of some curve* $\alpha: (-\epsilon, \epsilon) \to S$ *with* $\alpha(0) = p$.

By choosing a parametrization $\mathbf{x}: U \longrightarrow S$ around $p = \mathbf{x}(0, 0)$ we may express both the function f and the curve α in $\mathbf{x}$ by $f(u, v)$ and $(u(t), v(t))$, respectively. Therefore,

$$\alpha'(0)(f) = \frac{d}{dt}(f \circ \alpha)\bigg|_{t=0} = \frac{d}{dt}(f(u(t), v(t))\bigg|_{t=0}$$
$$= u'(0)\left(\frac{\partial f}{\partial u}\right)_0 + v'(0)\left(\frac{\partial f}{\partial v}\right)_0 = \left\{u'(0)\left(\frac{\partial}{\partial u}\right)_0 + v'(0)\left(\frac{\partial}{\partial v}\right)_0\right\}(f).$$

This suggests, given coordinates (u, v) around p, that we denote by $(\partial/\partial u)_0$ the tangent vector at p which maps a function f into $(\partial f/\partial u)_0$; a similar meaning will be attached to the symbol $(\partial/\partial v)_0$. We remark that $(\partial/\partial u)_0$, $(\partial/\partial v)_0$ may be interpreted as the tangent vectors at p of the "coordinate curves"

$$u \longrightarrow \mathbf{x}(u, 0), \qquad v \longrightarrow \mathbf{x}(0, v),$$

respectively (Fig. 5-49).

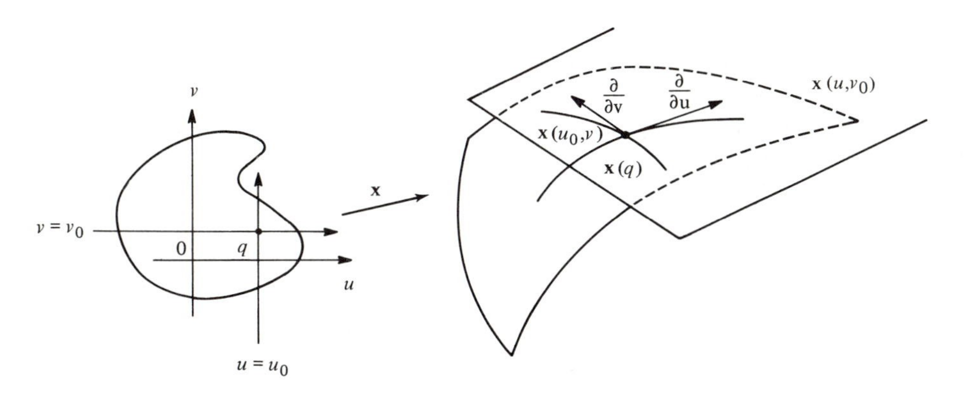

Figure 5-49

From the above, it follows that the set of tangent vectors at p, with the usual operations for functions, is a two-dimensional vector space $T_p(S)$ to be called the *tangent space* of S at p. It is also clear that the choice of a parametrization $\mathbf{x}: U \longrightarrow S$ around p determines an *associated basis* $\{(\partial/\partial u)_q, (\partial/\partial v)_q\}$ of $T_q(S)$ for any $q \in \mathbf{x}(U)$.

With the notion of tangent space, we can extend to abstract surfaces the definition of differential.

DEFINITION 4. *Let* S_1 *and* S_2 *be abstract surfaces and let* $\varphi: S_1 \longrightarrow S_2$ *be a differentiable map. For each* $p \in S_1$ *and each* $w \in T_p(S_1)$, *consider a differ-*

entiable curve $\alpha: (-\epsilon, \epsilon) \to S_1$, *with* $\alpha(0) = p$, $\alpha'(0) = w$. *Set* $\beta = \varphi \circ \alpha$. *The map* $d\varphi_p: T_p(S_1) \to T_p(S_2)$ *given by* $d\varphi_p(w) = \beta'(0)$ *is a well-defined linear map, called the* differential *of* φ *at* p.

The proof that $d\varphi_p$ is well defined and linear is exactly the same as the proof of Prop. 2 in Sec. 2-4.

We are now in a position to take the final step in our generalization of the intrinsic geometry.

DEFINITION 5. *A* geometric surface (*Riemannian manifold of dimension* 2) *is an abstract surface* S *together with the choice of an inner product* $\langle\ ,\ \rangle_p$ *at each* $T_p(S)$, $p \in S$, *which varies differentiably with* p *in the following sense. For some (and hence all) parametrization* $\mathbf{x}: U \to S$ *around* p, *the functions*

$$E(u, v) = \left\langle \frac{\partial}{\partial u}, \frac{\partial}{\partial u} \right\rangle, \qquad F(u, v) = \left\langle \frac{\partial}{\partial u}, \frac{\partial}{\partial v} \right\rangle, \qquad G(u, v) = \left\langle \frac{\partial}{\partial v}, \frac{\partial}{\partial v} \right\rangle$$

are differentiable functions in U. *The inner product* $\langle\ ,\ \rangle$ *is often called a* (Riemannian) metric *on* S.

It is now a simple matter to extend to geometric surfaces the notions of the intrinsic geometry. Indeed, with the functions E, F, G we define Christoffel symbols for S by system 2 of Sec. 4-3. Since the notions of intrinsic geometry were all defined in terms of the Christoffel symbols, they can now be defined in S.

Thus, covariant derivatives of vector fields along curves are given by Eq. (1) of Sec. 4-4. The existence of parallel transport follows from Prop. 2 of Sec. 4-4, and a geodesic is a curve such that the field of its tangent vectors has zero covariant derivative. Gaussian curvature can be either defined by Eq. (5) of Sec. 4-3 or in terms of the parallel transport, as in done in Sec. 4-5.

That this brings into play some new and interesting objects can be seen by the following considerations. We shall start with an example related to Hilbert's theorem.

Example 3. Let $S = R^2$ be a plane with coordinates (u, v) and define an inner product at each point $q = (u, v) \in R^2$ by setting

$$\left\langle \frac{\partial}{\partial u}, \frac{\partial}{\partial u} \right\rangle_q = E = 1, \qquad \left\langle \frac{\partial}{\partial u}, \frac{\partial}{\partial v} \right\rangle_q = F = 0,$$

$$\left\langle \frac{\partial}{\partial v}, \frac{\partial}{\partial v} \right\rangle_q = G = e^{2u}.$$

R^2 with this inner product is a geometric surface H called the *hyperbolic plane*. The geometry of H is different from the usual geometry of R^2. For instance, the curvature of H is (Sec. 4-3, Exercise 1)

$$K = -\frac{1}{2\sqrt{EG}}\left\{\left(\frac{E_v}{\sqrt{EG}}\right)_v + \left(\frac{G_u}{\sqrt{EG}}\right)_u\right\} = -\frac{1}{2e^u}\left(\frac{2e^{2u}}{e^u}\right)_u = -1.$$

Actually the geometry of H is an exact model for the non-euclidean geometry of Lobachewski, in which all the axioms of Euclid, except the axiom of parallels, are assumed (cf. Sec. 4-5). To make this point clear, we shall compute the geodesics of H.

If we look at the differential equations for the geodesics when $E = 1$, $F = 0$ (Sec. 4-6, Exercise 2), we see immediately that the curves $v = \text{const.}$ are geodesics. To find the other ones, it is convenient to define a map

$$\phi\colon H \longrightarrow R^2_+ = \{(x, y) \in R^2; y > 0\}$$

by $\phi(u, v) = (v, e^{-u})$. It is easily seen that ϕ is differentiable and, since $y > 0$, that it has a differentiable inverse. Thus, ϕ is a diffeomorphism, and we can induce an inner product in R^2_+ by setting

$$\langle d\phi(w_1), d\phi(w_2)\rangle_{\phi(q)} = \langle w_1, w_2\rangle_q.$$

To compute this inner product, we observe that

$$\frac{\partial}{\partial x} = \frac{\partial}{\partial v}, \qquad \frac{\partial}{\partial y} = -e^u\frac{\partial}{\partial u},$$

hence,

$$\left\langle\frac{\partial}{\partial x}, \frac{\partial}{\partial x}\right\rangle = e^{2u} = \frac{1}{y^2}, \qquad \left\langle\frac{\partial}{\partial x}, \frac{\partial}{\partial y}\right\rangle = 0, \qquad \left\langle\frac{\partial}{\partial y}, \frac{\partial}{\partial y}\right\rangle = \frac{1}{y^2}.$$

R^2_+ with this inner product is isometric to H, and it is sometimes called the *Poincaré half-plane*.

To determine the geodesics of H, we work with the Poincaré half-plane and make two further coordinate changes.

First, fix a point $(x_0, 0)$ and set (Fig. 5-50)

$$x - x_0 = \rho\cos\theta, \qquad y = \rho\sin\theta,$$

$0 < \theta < \pi, 0 < \rho < +\infty$. This is a diffeomorphism of R^2_+ into itself, and

$$\left\langle\frac{\partial}{\partial\rho}, \frac{\partial}{\partial\rho}\right\rangle = \frac{1}{\rho^2\sin^2\theta}, \qquad \left\langle\frac{\partial}{\partial\rho}, \frac{\partial}{\partial\theta}\right\rangle = 0, \qquad \left\langle\frac{\partial}{\partial\theta}, \frac{\partial}{\partial\theta}\right\rangle = \frac{1}{\sin^2\theta}.$$

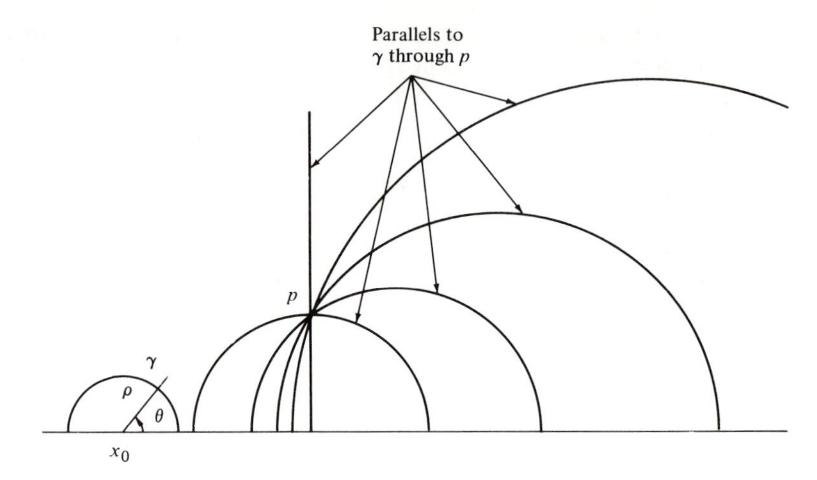

Figure 5-50

Next, consider the diffeomorphism of R^2_+ given by (we want to change θ into a parameter that measures the arc length along $\rho = \text{const.}$)

$$\rho_1 = \rho, \qquad \theta_1 = \int_0^\theta \frac{1}{\sin\theta}\,d\theta,$$

which yields

$$\left\langle \frac{\partial}{\partial\rho_1}, \frac{\partial}{\partial\rho_1} \right\rangle = \frac{1}{\rho_1^2 \sin^2\theta}, \qquad \left\langle \frac{\partial}{\partial\rho_1}, \frac{\partial}{\partial\theta_1} \right\rangle = 0, \qquad \left\langle \frac{\partial}{\partial\theta_1}, \frac{\partial}{\partial\theta_1} \right\rangle = 1.$$

By looking again at the differential equations for the geodesics ($F = 0$, $G = 1$), we see that $\rho_1 = \rho = \text{const.}$ are geodesics. (Another way of finding the geodesics of R^2_+ is given in Exercise 8.)

Collecting our observations, we conclude that the lines and the half-circles which are perpendicular to the axis $y > 0$ are geodesics of the Poincaré half-plane R^2_+. These are all the geodesics of R^2_+, since through each point $q \in R^2_+$ and each direction issuing from q there passes either a circle tangent to that line and normal to the axis $y = 0$ or a vertical line (when the direction is vertical).

The geometric surface R^2_+ is complete; that is, geodesics can be defined for all values of the parameter. The proof of this fact will be left as an exercise (Exercise 7; cf. also Exercise 6).

It is now easy to see, if we define a straight line of R^2_+ to be a geodesic, that all the axioms of Euclid but the axiom of parallels hold true in this geometry. The axiom of parallels in the Euclidean plane P asserts that from a point not in a straight line $r \subset P$ one can draw a unique straight line

$r' \subset P$ that does not meet r. Actually, in R^2_+, from a point not in a geodesic γ we can draw an infinite number of geodesics which do not meet γ.

The question then arises whether such a surface can be found as a regular surface in R^3. The natural context for this question is the following definition.

DEFINITION 6. *A differentiable map* $\varphi\colon S \longrightarrow R^3$ *of an abstract surface* S *into* R^3 *is an* immersion *if the differential* $d\varphi_p\colon T_p(S) \longrightarrow T_p(R^3)$ *is injective. If, in addition,* S *has a metric* $\langle\ ,\ \rangle$ *and*

$$\langle d\varphi_p(v), d\varphi_p(w)\rangle_{\varphi(p)} = \langle v, w\rangle_p, \qquad v, w \in T_p(S),$$

φ *is said to be an* isometric immersion.

Notice that the first inner product in the above relation is the usual inner product of R^3, whereas the second one is the given Riemannian metric on S. This means that in an isometric immersion, the metric "induced" by R^3 on S agrees with the given metric on S.

Hilbert's theorem, to be proved in Sec. 5-11, states that there is no isometric immersion into R^3 of the complete hyperbolic plane. In particular, one cannot find a model of the geometry of Lobachewski as a regular surface in R^3.

Actually, there is no need to restrict ourselves to R^3. The above definition of isometric immersion makes perfect sense when we replace R^3 by R^4 or, for that matter, by an arbitrary R^n. Thus, we can broaden our initial question, and ask: *For what values of* n *is there an isometric immersion of the complete hyperbolic plane into* R^n? Hilbert's theorem say that $n \geq 4$. As far as we know, the case $n = 4$ is still unsettled.

Thus, the introduction of abstract surfaces brings in new objects and illuminates our view of important questions.

In the rest of this section, we shall explore in more detail some of the ideas just introduced and shall show how they lead naturally to further important generalizations. This part will not be needed for the understanding of the next section.

Let us look into further examples.

Example 4. Let R^2 be a plane with coordinates (x, y) and $T_{m,n}\colon R^2 \longrightarrow R^2$ be the map (translation) $T_{m,n}(x, y) = (x + m, y + n)$, where m and n are integers. Define an equivalence relation in R^2 by $(x, y) \sim (x_1, y_1)$ if there exist integers m, n such that $T_{m,n}(x, y) = (x_1, y_1)$. Let T be the quotient space of R^2 by this equivalence relation, and let $\pi\colon R^2 \longrightarrow T$ be the natural projection map $\pi(x, y) = \{T_{m,n}(x, y)$; all integers $m, n\}$. Thus, in each open unit square whose vertices have integer coordinates, there is only one representative of

T, and T may be thought of as a closed square with opposite sides identified. (See Fig. 5-51. Notice that all points of R^2 denoted by x represent the same point p in T.)

Let $i_\alpha: U_\alpha \subset R^2 \longrightarrow R^2$ be a family of parametrizations of R^2, where i_α is the identity map, such that $U_\alpha \cap T_{m,n}(U_\alpha) = \phi$ for all m, n. Since $T_{m,n}$ is a diffeomorphism, it is easily checked that the family $(U_\alpha, \pi \circ i_\alpha)$ is a differentiable structure for T. T is called a (differentiable) *torus*. From the very definition of the differentiable structure on T, $\pi: R^2 \longrightarrow T$ is a differentiable map and a local diffeomorphism (the construction made in Fig. 5-51 indicates that T is diffeomorphic to the standard torus in R^3).

Now notice that $T_{m,n}$ is an isometry of R^2 and introduce a geometric (Riemannian) structure on T as follows. Let $p \in T$ and $v \in T_p(T)$. Let $q_1, q_2 \in R^2$ and $w_1, w_2 \in R^2$ be such that $\pi(q_1) = \pi(q_2) = p$ and $d\pi_{q_1}(w_1) = d\pi_{q_2}(w_2) = v$. Then $q_1 \sim q_2$; hence, there exists $T_{m,n}$ such that $T_{m,n}(q_1) = q_2$, $d(T_{m,n})_{q_1}(w_1) = w_2$. Since $T_{m,n}$ is an isometry, $|w_1| = |w_2|$. Now, define the length of v in $T_p(T)$ by $|v| = |d\pi_q(w_1)| = |w_1|$. By what we have seen, this is well defined. Clearly this gives rise to an inner product $\langle \ , \ \rangle_p$ on $T_p(T)$ for each $p \in T$. Since this is essentially the inner product of R^2 and π is a local diffeomorphism, $\langle \ , \ \rangle_p$ varies differentiably with p.

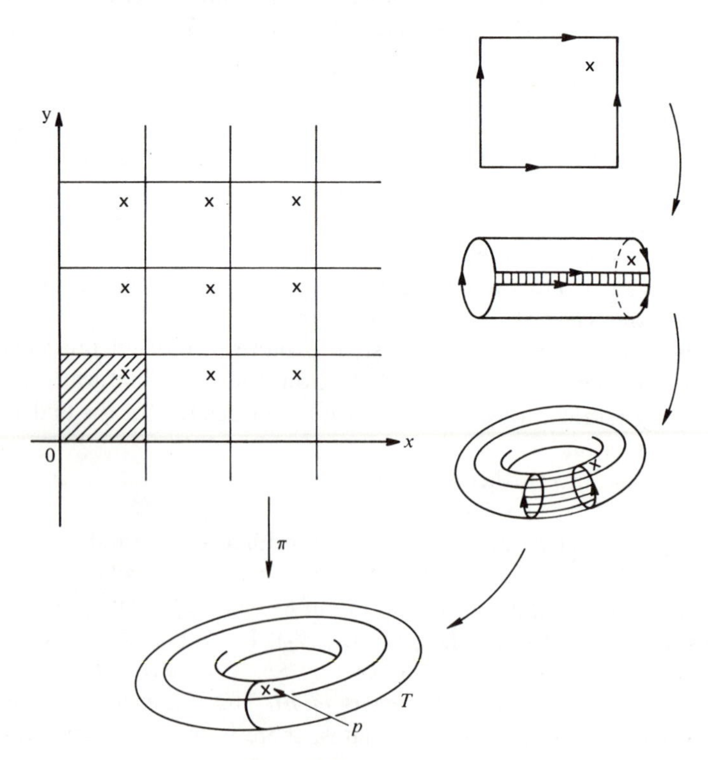

Figure 5-51 The torus.

Observe that the coefficients of the first fundamental form of T, in any of the parametrizations of the family $\{U_\alpha, \pi \circ i_\alpha\}$, are $E = G = 1$, $F = 0$. Thus, this torus behaves locally like a Euclidean space. For instance, its Gaussian curvature is identically zero (cf. Exercise 1, Sec. 4-3). This accounts for the name *flat torus*, which is usually given to T with the inner product just described.

Clearly the flat torus cannot be isometrically immersed in R^3, since, by compactness, it would have a point of positive curvature (cf. Exercise 16, Sec. 3-3, or Lemma 1, Sec. 5-2). However, it can be isometrically immersed in R^4.

In fact, let $F: R^2 \longrightarrow R^4$ be given by

$$F(x, y) = \frac{1}{2\pi}(\cos 2\pi x, \sin 2\pi x, \cos 2\pi y, \sin 2\pi y).$$

Since $F(x + m, y + n) = F(x, y)$ for all m, n, we can define a map $\varphi: T \longrightarrow R^4$ by $\varphi(p) = F(q)$, where $q \in \pi^{-1}(p)$. Clearly, $\varphi \circ \pi = F$, and since $\pi: R^2 \longrightarrow T$ is a local diffeomorphism, φ is differentiable. Furthermore, the rank of $d\varphi$ is equal to the rank of dF, which is easily computed to be 2. Thus, φ is an immersion. To see that the immersion is isometric, we first observe that if $e_1 = (1, 0)$, $e_2 = (0, 1)$ are the vectors of the canonical basis in R^2, the vectors $d\pi_q(e_1) = f_1$, $d\pi_q(e_2) = f_2$, $q \in R^2$, form a basis for $T_{\pi(q)}(T)$. By definition of the inner product on T, $\langle f_i, f_j \rangle = \langle e_i, e_j \rangle$, $i, j = 1, 2$. Next, we compute

$$\frac{\partial F}{\partial x} = dF(e_1) = (-\sin 2\pi x, \cos 2\pi x, 0, 0),$$

$$\frac{\partial F}{\partial y} = dF(e_2) = (0, 0, -\sin 2\pi y, \cos 2\pi y),$$

and obtain that

$$\langle dF(e_i), dF(e_j) \rangle = \langle e_i, e_j \rangle = \langle f_i, f_j \rangle.$$

Thus,

$$\langle d\varphi(f_i), d\varphi(f_j) \rangle = \langle d\varphi(d\pi(e_i)), d\varphi(d\pi(e_j)) \rangle = \langle f_i, f_j \rangle.$$

It follows that φ is an isometric immersion, as we had asserted.

It should be remarked that the image $\varphi(S)$ of an immersion $\varphi: S \longrightarrow R^n$ may have self-intersections. In the previous example, $\varphi: T \longrightarrow R^4$ is one-to-one, and furthermore φ is a homeomorphism onto its image. It is convenient to use the following terminology.

DEFINITION 7. *Let* S *be an abstract surface. A differentiable map* φ: S $\longrightarrow$ R^n *is an* embedding *if* φ *is an immersion and a homeomorphism onto its image.*

For instance, a regular surface in R^3 can be characterized as the image of an abstract surface S by an embedding $\varphi: S \longrightarrow R^3$. This means that only those abstract surfaces which can be embedded in R^3 could have been detected in our previous study of regular surfaces in R^3. That this is a serious restriction can be seen by the example below.

Example 5. We first remark that the definition of orientability (cf. Sec. 2-6, Def. 1) can be extended, without changing a single word, to abstract surfaces. Now consider the real projective plane P^2 of Example 1. We claim that P^2 is nonorientable.

To prove this, we first make the following general observation. Whenever an abstract surface S contains an open set M diffeomorphic to a Möbius strip (Sec. 2-6, Example 3), it is nonorientable. Otherwise, there exists a family of parametrizations covering S with the property that all coordinate changes have positive Jacobian; the restriction of such a family to M will induce an orientation on M which is a contradiction.

Now, P^2 is obtained from the sphere S^2 by identifying antipodal points. Consider on S^2 a thin strip B made up of open segments of meridians whose centers lay on half an equator (Fig. 5-52). Under identification of antipodal points, B clearly becomes an open Möbius strip in P^2. Thus, P^2 is nonorientable.

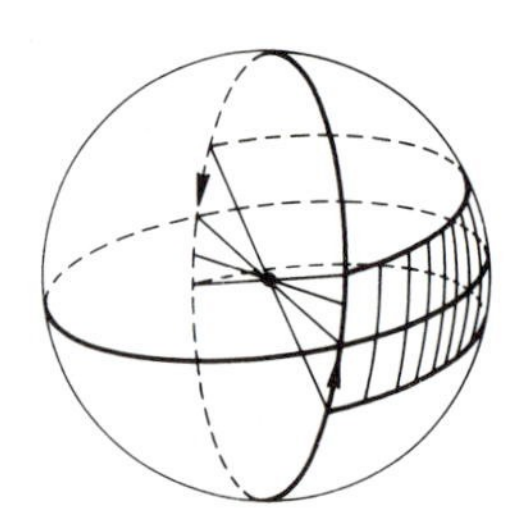

Figure 5-52. The projective plane contains a Möbius strip.

By a similar argument, it can be shown that the Klein bottle K of Example 2 is also nonorientable. In general, whenever a regular surface $S \subset R^3$ is symmetric relative to the origin of R^3, identification of symmetric points gives rise to a nonorientable abstract surface.

It can be proved that a compact regular surface in R^3 is orientable (cf. Remark 2, Sec. 2-7). Thus, P^2 and K cannot be embedded in R^3, and the same happens to the compact orientable surfaces generated as above. Thus, we miss quite a number of surfaces in R^3.

P^2 and K can, however, be embedded in R^4. For the Klein bottle K, consider the map $G: R^2 \longrightarrow R^4$ given by

$$G(u, v) = \Big((r \cos v + a) \cos u, (r \cos v + a) \sin u,$$
$$r \sin v \cos \frac{u}{2}, r \sin v \sin \frac{u}{2}\Big).$$

Notice that $G(u, v) = G(u + 2m\pi, 2n\pi - v)$, where m and n are integers. Thus, G induces a map ψ of the space obtained from the square

$$[0, 2\pi] \times [0, 2\pi] \subset R^2$$

by first reflecting one of its sides in the center of this side and then identifying opposite sides (see Fig. 5-53). That this is the Klein bottle, as defined in Example 2, can be seen by throwing away an open half of the torus in which antipodal points are being identified and observing that both processes lead to the same surface (Fig. 5-53).

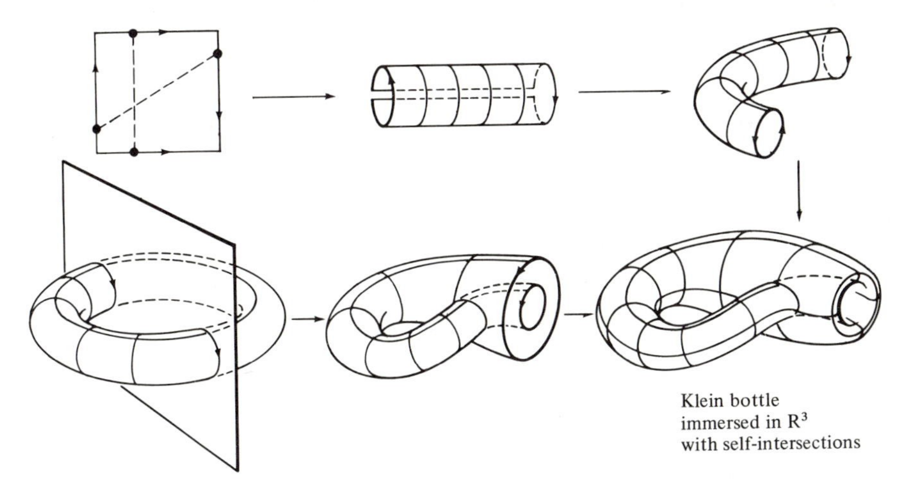

Figure 5-53

Thus, ψ is a map of K into R^4. Observe further that

$$G(u + 4m\pi, v + 2m\pi) = G(u, v).$$

It follows that $G = \psi \circ \pi_1 \circ \pi$, where $\pi\colon R^2 \to T$ is essentially the natural projection on the torus T (cf. Example 4) and $\pi_1\colon T \to K$ corresponds to identifying "antipodal" points in T. By the definition of the differentiable structures on T and K, π and π_1 are local diffeomorphisms. Thus, $\psi\colon K \to R^4$ is differentiable, and the rank of $d\psi$ is the same as the rank of dG. The latter is easily computed to be 2; hence, ψ is an immersion. Since K is compact and ψ is one-to-one, ψ^{-1} is easily seen to be continuous in $\varphi(K)$. Thus, ψ is an embedding, as we wished.

For the projective plane P^2, consider the map $F\colon R^3 \to R^4$ given by

$$F(x, y, z) = (x^2 - y^2, xy, xz, yz).$$

Let $S^2 \subset R^3$ be the unit sphere with center in the origin of R^3. It is clear that the restriction $\varphi = F/S^2$ is such that $\varphi(p) = \varphi(-p)$. Thus, φ induces a map

$$\tilde{\varphi}: P^2 \longrightarrow R^4 \quad \text{by} \quad \tilde{\varphi}(\{p, -p\}) = \varphi(p).$$

To see that φ (hence, $\tilde{\varphi}$) is an immersion, consider the parametrization $\mathbf{x}$ of S^2 given by $\mathbf{x}(x, y) = (x, y, +\sqrt{1 - x^2 - y^2})$, where $x^2 + y^2 \leq 1$. Then

$$\varphi \circ \mathbf{x}(x, y) = (x^2 - y^2, xy, xD, yD), \qquad D = \sqrt{1 - x^2 - y^2}.$$

It is easily checked that the matrix of $d(\varphi \circ \mathbf{x})$ has rank 2. Thus, $\tilde{\varphi}$ is an immersion.

To see that $\tilde{\varphi}$ is one-to-one, set

$$x^2 - y^2 = a, \qquad xy = b, \qquad xz = c, \qquad yz = d. \tag{2}$$

It suffices to show that, under the condition $x^2 + y^2 + z^2 = 1$, the above equations have only two solutions which are of the form (x, y, z) and $(-x, -y, -z)$. In fact, we can write

$$\begin{gathered} x^2 d = bc, \qquad y^2 c = bd, \\ z^2 b = cd, \qquad x^2 - y^2 = a, \\ x^2 + y^2 + z^2 = 1 \end{gathered} \tag{3}$$

where the first three equations come from the last three equations of (2).

Now, if one of the numbers b, c, d is nonzero, the equations in (3) will give x^2, y^2, and z^2, and the equations in (2) will determine the sign of two coordinates, once given the sign of the remaining one. If $b = c = d = 0$, the equations in (2) and the last equation of (3) show that exactly two coordinates will be zero, the remaining one being ± 1. In any case, the solutions have the required form, and $\tilde{\varphi}$ is one-to-one.

By compactness, φ is an embedding, and that concludes the example.

If we look back to the definition of abstract surface, we see that the number 2 has played no essential role. Thus, we can extend that definition to an arbitrary n and, as we shall see presently, this may be useful.

DEFINITION 1a. *A* differentiable manifold *of dimension* n *is a set* M *together with a family of one-to-one maps* $\mathbf{x}_\alpha: U_\alpha \longrightarrow M$ *of open sets* $U_\alpha \subset R^n$ *into* M *such that*

1. $\bigcup_\alpha \mathbf{x}_\alpha(U_\alpha) = M$.

2. *For each pair* α, β *with* $\mathbf{x}_\alpha(U_\alpha) \cap \mathbf{x}_\beta(U_\beta) = W \neq \phi$, *we have that* $\mathbf{x}_\alpha^{-1}(W)$, $\mathbf{x}_\beta^{-1}(W)$ *are open sets in* R^n *and that* $\mathbf{x}_\beta^{-1} \circ \mathbf{x}_\alpha$, $\mathbf{x}_\alpha^{-1} \circ \mathbf{x}_\beta$ *are differentiable maps.*

3. *The family* $\{U_\alpha, \mathbf{x}_\alpha\}$ *is maximal relative to conditions* 1 *and* 2.

A family $\{U_\alpha, \mathbf{x}_\alpha\}$ satisfying conditions 1 and 2 is called a *differentiable structure* on M. Given a differentiable structure on M we can easily complete it into a maximal one by adding to it all possible parametrizations that, together with some parametrization of the family $\{U_\alpha, \mathbf{x}_\alpha\}$, satisfy condition 2. Thus, with some abuse of language, we may say that a differentiable manifold is a set together with a differentiable structure.

Remark. A family of open sets can be defined in M by the following requirement: $V \subset M$ is an open set if for every α, $\mathbf{x}_\alpha^{-1}(V \cap \mathbf{x}_\alpha(U_\alpha))$ is an open set in R^n. The readers with some knowledge of point set topology will notice that such a family defines a natural topology on M. In this topology, the maps $\mathbf{x}_\alpha$ are continuous and the sets $\mathbf{x}_\alpha(U_\alpha)$ are open in M. In some deeper theorems on manifolds, it is necessary to impose some conditions on the natural topology of M.

The definitions of differentiable maps and tangent vector carry over, word by word, to differentiable manifolds. Of course, the tangent space is now an n-dimensional vector space. The definitions of differential and orientability also extend straightforwardly to the present situation.

In the following example we shall show how questions on two-dimensional manifolds lead naturally into the consideration of higher-dimensional manifolds.

Example 6 (*The Tangent Bundle*). Let S be an abstract surface and let $T(S) = \{(p, w), p \in S, w \in T_p(S)\}$. We shall show that the set $T(S)$ can be given a differentiable structure (of dimension 4) to be called the *tangent bundle* of S.

Let $\{U_\alpha, \mathbf{x}_\alpha\}$ be a differentiable structure for S. We shall denote by (u_α, v_α) the coordinates of U_α, and by $\{\partial/\partial u_\alpha, \partial/\partial v_\alpha\}$ the associated bases in the tangent planes of $\mathbf{x}_\alpha(U_\alpha)$. For each α, define a map $\mathbf{y}_\alpha\colon U_\alpha \times R^2 \longrightarrow T(S)$ by

$$\mathbf{y}_\alpha(u_\alpha, v_\alpha, x, y) = \left(\mathbf{x}_\alpha(u_\alpha, v_\alpha), x\frac{\partial}{\partial u_\alpha} + y\frac{\partial}{\partial v_\alpha}\right), \qquad (x, y) \in R^2.$$

Geometrically, this means that we shall take as coordinates of a point $(p, w) \in T(S)$ the coordinates u_α, v_α of p plus the coordinates of w in the basis $\{\partial/\partial u_\alpha, \partial/\partial v_\alpha\}$.

We shall show that $\{U_\alpha \times R^2, \mathbf{y}_\alpha\}$ is a differentiable structure for $T(S)$. Since $\bigcup_\alpha \mathbf{x}_\alpha(U_\alpha) = S$ and $(d\mathbf{x}_\alpha)_q(R^2) = T_{\mathbf{x}_\alpha(q)}(S)$, $q \in U_\alpha$, we have that

$$\bigcup_\alpha \mathbf{y}_\alpha(U_\alpha \times R^2) = T(S),$$

and that verifies condition 1 of Def. 1a. Now let

$$(p, w) \in \mathbf{y}_\alpha(U_\alpha \times R^2) \cap \mathbf{y}_\beta(U_\beta \times R^2).$$

Then

$$(p, w) = (\mathbf{x}_\alpha(q_\alpha), d\mathbf{x}_\alpha(w_\alpha)) = (\mathbf{y}_\beta(q_\beta), d\mathbf{x}_\beta(w_\beta)),$$

where $q_\alpha \in U_\alpha, q_\beta \in U_\beta, w_\alpha, w_\beta \in R^2$. Thus,

$$\begin{aligned}\mathbf{y}_\beta^{-1} \circ \mathbf{y}_\alpha(q_\alpha, w_\alpha) &= \mathbf{y}_\beta^{-1}(\mathbf{x}_\alpha(q_\alpha), d\mathbf{x}_\alpha(w_\alpha)) \\ &= ((\mathbf{x}_\beta^{-1} \circ \mathbf{x}_\alpha)(q_\alpha), d(\mathbf{x}_\beta^{-1} \circ \mathbf{x}_\alpha)(w_\alpha)).\end{aligned}$$

Since $\mathbf{x}_\beta^{-1} \circ \mathbf{x}_\alpha$ is differentiable, so is $d(\mathbf{x}_\beta^{-1} \circ \mathbf{x}_\alpha)$. It follows that $\mathbf{y}_\beta^{-1} \circ \mathbf{y}_\alpha$ is differentiable, and that verifies condition 2 of Def. 1a.

The tangent bundle of S is the natural space to work with when one is dealing with second-order differential equations on S. For instance, the equations of a geodesic on a geometric surface S can be written, in a coordinate neighborhood, as (cf. Sec. 4-7)

$$\begin{aligned}u'' &= f_1(u, v, u', v'), \\ v'' &= f_2(u, v, u', v').\end{aligned}$$

The classical "trick" of introducing new variables $x = u'$, $y = v'$ to reduce the above to the first-order system

$$\begin{aligned}x' &= f_1(u, v, x, y), \\ y' &= f_2(u, v, x, y), \\ u' &= f_3(u, v, x, y), \\ v' &= f_4(u, v, x, y)\end{aligned} \tag{4}$$

may be interpreted as bringing into consideration the tangent bundle $T(S)$, with coordinates (u, v, x, y) and as looking upon the geodesics as trajectories of a vector field given locally in $T(S)$ by (4). It can be shown that such a vector field is well defined in the entire $T(S)$; that is, in the intersection of two coordinate neighborhoods, the vector fields given by (4) agree. This field (or rather its trajectories) is called the *geodesic flow* on $T(S)$. It is a very natural object to work with when studying global properties of the geodesics on S.

By looking back to Sec. 4-7, it will be noticed that we have used, in a disguised form, the manifold $T(S)$. Since we were interested only in local properties, we could get along with a coordinate neighborhood (which is essentially an open set of R^4). However, even this local work becomes neater when the notion of tangent bundle is brought into consideration.

Of course, we can also define the tangent bundle of an arbitrary n-dimensional manifold. Except for notation, the details are the same and will be left as an exercise.

We can also extend the definition of a geometric surface to an arbitrary dimension.

DEFINITION 5a. *A* Riemannian manifold *is an* n*-dimensional differentiable manifold* M *together with a choice, for each* $p \in M$, *of an inner product* $\langle\ ,\ \rangle_p$ *in* $T_p(M)$ *that varies differentiably with* p *in the following sense. For some (hence, all) parametrization* $\mathbf{x}_\alpha: U_\alpha \to M$ *with* $p \in \mathbf{x}_\alpha(U_\alpha)$, *the functions*

$$g_{ij}(u_1, \ldots, u_n) = \left\langle \frac{\partial}{\partial u_i}, \frac{\partial}{\partial u_j} \right\rangle, \qquad i, j = 1, \ldots, n,$$

are differentiable at $\mathbf{x}_\alpha^{-1}(p)$; *here* $(u_1, \ldots, u_n)$ *are the coordinates of* $U_\alpha \subset R^n$.

The differentiable family $\{\langle\ \rangle_p, p \in M\}$ is called a *Riemannian structure* (or Riemannian metric) for M.

Notice that in the case of surfaces we have used the traditional notation $g_{11} = E, g_{12} = g_{21} = F, g_{22} = G$.

The extension of the notions of the intrinsic geometry to Riemannian manifolds is not so straightforward as in the case of differentiable manifolds.

First, we must define a notion of covariant derivative for Riemannian manifolds. For this, let $\mathbf{x}: U \to M$ be a parametrization with coordinates $(u_1, \ldots, u_n)$ and set $\mathbf{x}_i = \partial/\partial u_i$. Thus, $g_{ij} = \langle \mathbf{x}_i, \mathbf{x}_j \rangle$.

We want to define the covariant derivative $D_w v$ of a vector field v relative to a vector field w. We would like $D_w v$ to have the properties we are used to and that have shown themselves to be effective in the past. First, it should have the distributive properties of the old covariant derivative. Thus, if u, v, w are vector fields on M and f, g are differentiable functions on M, we want

$$D_{fu+gw}(v) = fD_u v + gD_w v, \tag{5}$$

$$D_u(fv + gw) = fD_u v + \frac{\partial f}{\partial u} v + gD_u w + \frac{\partial g}{\partial u} w, \tag{6}$$

where $\partial f/\partial u$, for instance, is a function whose value at $p \in M$ is the derivative $(f \circ \alpha)'(0)$ of the restriction of f to a curve $\alpha: (-\epsilon, \epsilon) \to M$, $\alpha(0) = p$, $\alpha'(0) = u$.

Equations (5) and (6) show that the covariant derivative D is entirely determined once we know its values on the basis vectors

$$D_{\mathbf{x}_i} \mathbf{x}_j = \sum_{k=1}^{n} \Gamma_{ij}^k \mathbf{x}_k, \qquad i, j, k = 1, \ldots, n,$$

where the coefficients Γ_{ij}^k are functions yet to be determined.

Second, we want the Γ_{ij}^k to be symmetric in i and $j(\Gamma_{ij}^k = \Gamma_{ji}^k)$; that is,

$$D_{\mathbf{x}_i}\mathbf{x}_j = D_{\mathbf{x}_j}\mathbf{x}_i \qquad \text{for all } i, j. \tag{7}$$

Third, we want the law of products to hold; that is,

$$\frac{\partial}{\partial u_k}\langle \mathbf{x}_i, \mathbf{x}_j\rangle = \langle D_{\mathbf{x}_k}\mathbf{x}_i, \mathbf{x}_j\rangle + \langle \mathbf{x}_i, D_{\mathbf{x}_k}\mathbf{x}_j\rangle. \tag{8}$$

From Eqs. (7) and (8), it follows that

$$\frac{\partial}{\partial u_k}\langle \mathbf{x}_i\, \mathbf{x}_j\rangle + \frac{\partial}{\partial u_i}\langle \mathbf{x}_j, \mathbf{x}_k\rangle - \frac{\partial}{\partial u_j}\langle \mathbf{x}_k, \mathbf{x}_i\rangle = 2\langle D_{\mathbf{x}_i}\mathbf{x}_k, \mathbf{x}_j\rangle,$$

or, equivalently,

$$\frac{\partial}{\partial u_k} g_{ij} + \frac{\partial}{\partial u_i} g_{jk} - \frac{\partial}{\partial u_j} g_{ki} = 2\sum_l \Gamma_{ik}^l\, g_{lj}.$$

Since $\det(g_{ij}) \neq 0$, we can solve the last system, and obtain the Γ_{ij}^k as functions of the Riemannian metric g_{ij} and its derivatives (the reader should compare the system above with system (2) of Sec. 4-3). If we think of g_{ij} as a matrix and write its inverse as g^{ij}, the solution of the above system is

$$\Gamma_{ij}^k = \frac{1}{2}\sum_l g^{kl}\left(\frac{\partial g_{il}}{\partial u_j} + \frac{\partial g_{jl}}{\partial u_i} - \frac{\partial u_{ij}}{\partial u_l}\right).$$

Thus, given a Riemannian structure for M, *there exists a unique covariant derivative on* M (also called the Levi-Civita *connection* of the given Riemannian structure) *satisfying* Eqs. (5)–(8).

Starting from the covariant derivative, we can define parallel transport, geodesics, geodesic curvature, the exponential map, completeness, etc. The definitions are exactly the same as those we have given previously. The notion of curvature, however, requires more elaboration. The following concept, due to Riemann, is probably the best analogue in Riemannian geometry of the Gaussian curvature.

Let $p \in M$ and let $\sigma \subset T_p(M)$ be a two-dimensional subspace of the tangent space $T_p(M)$. Consider all those geodesics of M that start from p and are tangent to σ. From the fact that the exponential map is a local diffeomorphism at the origin of $T_p(M)$, it can be shown that small segments of such geodesics make up an abstract surface S containing p. S has a natural geometric structure induced by the Riemannian structure of M. The Gaussian curvature of S at p is called the *sectional curvature* $K(p, \sigma)$ of M at p along σ.

It is possible to formalize the sectional curvature in terms of the Levi-Civita connection but that is too technical to be described here. We shall only mention that most of the theorems in this chapter can be posed as natural questions in Riemannian geometry. Some of them are true with little or no modification of the given proofs. (The Hopf-Rinow theorem, the Bonnet theorem, the first Hadamard theorem, and the Jacobi theorems are all in this class.) Some others, however, require further assumptions to hold true (the second Hadamard theorem, for instance) and were seeds for further developments.

A full development of the above ideas would lead us into the realm of Riemannian geometry. We must stop here and refer the reader to the bibliography at the end of the book.

EXERCISES

1. Introduce a metric on the projective plane P^2 (cf. Example 1) so that the natural projection $\pi: S^2 \longrightarrow P^2$ is a local isometry. What is the (Gaussian) curvature of such a metric?

2. (*The Infinite Möbius Strip.*) Let

$$C = \{(x, y, z) \in R^3; x^2 + y^2 = 1\}$$

be a cylinder and $A: C \longrightarrow C$ be the map (the antipodal map) $A(x, y, z) = (-x, -y, -z)$. Let M be the quotient of C by the equivalence relation $p \sim A(p)$, and let $\pi: C \longrightarrow M$ be the map $\pi(p) = \{p, A(p)\}, p \in C$.

a. Show that M can be given a differentiable structure so that π is a local diffeomorphism (M is then called the *infinite Möbius strip*).

b. Prove that M is nonorientable.

c. Introduce on M a Riemannian metric so that π is a local isometry. What is the curvature of such a metric?

3. a. Show that the projection $\pi: S^2 \longrightarrow P^2$ from the sphere onto the projective plane has the following properties: (1) π is continuous and $\pi(S^2) = P^2$; (2) each point $p \in P^2$ has a neighborhood U such that $\pi^{-1}(U) = V_1 \cup V_2$, where V_1 and V_2 are disjoint open subsets of S^2, and the restriction of π to each V_i, $i = 1, 2$, is a homeomorphism onto U. Thus, π satisfies formally the conditions for a covering map (see Sec. 5-6, Def. 1) with two sheets. Because of this, we say that S^2 is an *orientable double covering of* P^2.

b. Show that, in this sense, the torus T is an orientable double covering of the Klein bottle K (cf. Example 2) and that the cylinder is an orientable double covering of the infinite Möbius strip (cf. Exercise 2).

4. (*The Orientable Double Covering*). This exercise gives a general construction for the orientable double covering of a nonorientable surface. Let S be an abstract, connected, nonorientable surface. For each $p \in S$, consider the set B of all bases of $T_p(S)$ and call two bases *equivalent* if they are related by a matrix with positive determinant. This is clearly an equivalence relation and divides B into two disjoint sets (cf. Sec. 1-4). Let $\mathcal{O}_p$ be the quotient space of B by this equivalence relation. $\mathcal{O}_p$ has two elements, and each element $O_p \in \mathcal{O}_p$ is an orientation of $T_p(S)$ (cf. Sec. 1-4). Let $\tilde{S}$ be the set

$$\tilde{S} = \{(p, O_p); p \in S; O_p \in \mathcal{O}_p\}.$$

To give $\tilde{S}$ a differentiable structure, let $\{U_\alpha, \mathbf{x}_\alpha\}$ be the maximal differentiable structure of S and define $\tilde{\mathbf{x}}_\alpha: U_\alpha \longrightarrow \tilde{S}$ by

$$\tilde{\mathbf{x}}_\alpha(u_\alpha, v_\alpha) = \left(\mathbf{x}_\alpha(u_\alpha, v_\alpha), \left[\frac{\partial}{\partial u_\alpha}, \frac{\partial}{\partial v_\alpha}\right]\right),$$

where $(u_\alpha, v_\alpha) \in U_\alpha$ and $[\partial/\partial u_\alpha, \partial/\partial v_\alpha]$ denotes the element of $\mathcal{O}_p$ determined by the basis $\{\partial/\partial u_\alpha, \partial/\partial v_\alpha\}$. Show that

a. $\{U_\alpha, \tilde{\mathbf{x}}_\alpha\}$ is a differentiable structure on $\tilde{S}$ and that $\tilde{S}$ with such a differentiable structure is an orientable surface.

b. The map $\pi: \tilde{S} \longrightarrow S$ given by $\pi(p, O_p) = p$ is a differentiable surjective map. Furthermore, each point $p \in S$ has a neighborhood U such that $\pi^{-1}(U) = V_1 \cup V_2$, where V_1 and V_2 are disjoint open subsets of $\tilde{S}$ and π restricted to each V_i, $i = 1, 2$, is a diffeomorphism onto U. Because of this, $\tilde{S}$ is called an *orientable double covering of S*.

5. Extend the Gauss-Bonnet theorem (see Sec. 4-5) to orientable geometric surfaces and apply it to prove the following facts:

a. There is no Riemannian metric on an abstract surface T diffeomorphic to a torus such that its curvature is positive (or negative) at all points of T.

b. Let T and S^2 be abstract surfaces diffeomorphic to the torus and the sphere, respectively, and let $\varphi: T \longrightarrow S^2$ be a differentiable map. Then φ has at least one critical point, i.e., a point $p \in T$ such that $d\varphi_p = 0$.

6. Consider the upper half-plane R^2_+ (cf. Example 3) with the metric

$$E(x, y) = 1, \qquad F(x, y) = 0, \qquad G(x, y) = \frac{1}{y}, \qquad (x, y) \in R^2_+.$$

Show that the lengths of vectors become arbitrarily large as we approach the boundary of R^2_+ and yet the length of the vertical segment

$$x = 0, \qquad 0 < \epsilon \leq y \leq 1,$$

approaches 2 as $\epsilon \longrightarrow 0$. Conclude that such a metric is not complete.

***7.** Prove that the Poincaré half-plane (cf. Example 3) is a complete geometric surface. Conclude that the hyperbolic plane is complete.

8. Another way of finding the geodesics of the Poincaré half-plane (cf. Example 3) is to use the Euler-Lagrange equation for the corresponding variational problem (cf. Exercise 4, Sec. 5-4). Since we know that the vertical lines are geodesics, we can restrict ourselves to geodesics of the form $y = y(x)$. Thus, we must look for the critical points of the integral ($F = 0$)

$$\int \sqrt{E + G(y')^2}\, dx = \int \frac{\sqrt{1 + (y')^2}}{y}\, dx,$$

since $E = G = 1/y^2$. Use Exercise 4, Sec. 5-4, to show that the solution to this variational problem is a family of circles of the form

$$(x + k_1)^2 + y^2 = k_2^2, \qquad k_1, k_2 = \text{const.}$$

9. Let $\tilde{S}$ and S be connected geometric surfaces and let $\pi\colon \tilde{S} \longrightarrow S$ be a surjective differentiable map with the following property: For each $p \in S$, there exists a neighborhood U of p such that $\pi^{-1}(U) = \bigcup_\alpha V_\alpha$, where the V_α's are open disjoint subsets of $\tilde{S}$ and π restricted to each V_α is an isometry onto U (thus, π is essentially a covering map and a local isometry).

a. Prove that S is complete if and only if $\tilde{S}$ is complete.

b. Is the metric on the infinite Möbius strip, introduced in Exercise 2, part c, a complete metric?

10. (*Kazdan-Warner's Results.*)

a. Let a metric on R^2 be given by

$$E(x, y) = 1, \qquad F(x, y) = 0, \qquad G(x, y) > 0, \qquad (x, y) \in R^2.$$

Show that the curvature of this metric is given by

$$\frac{\partial^2(\sqrt{G})}{\partial x^2} + K(x, y)\sqrt{G} = 0. \tag{$*$}$$

b. Conversely, given a function $K(x, y)$ on R^2, regard y as a parameter and let $\sqrt{G}$ be the solution of ($*$) with the initial conditions

$$\sqrt{G}(x_0, y) = 1, \qquad \frac{\partial\sqrt{G}}{\partial x}(x_0, y) = 0.$$

Prove that G is positive in a neighborhood of (x_0, y) and thus defines a metric in this neighborhood. This shows that *every differentiable function is locally the curvature of some (abstract) metric.*

***c.** Assume that $K(x, y) \leq 0$ for all $(x, y) \in R^2$. Show that the solution of part b satisfies

$$\sqrt{G(x, y)} \geq \sqrt{G(x_0, y)} = 1 \qquad \text{for all } x.$$

Thus, $G(x, y)$ defines a metric on all of R^2. Prove also that this metric is complete. This shows that *any nonpositive differentiable function on* R^2 *is the curvature of some complete metric on* R^2. If we do not insist on the metric being complete, the result is true for any differentiable function K on R^2. Compare

J. Kazdan and F. Warner, "Curvature Functions for Open 2-Manifolds," *Ann. of Math.* 99 (1974), 203–219, where it is also proved that the condition on K given in Exercise 2 of Sec. 5-4 is necessary and sufficient for the metric to be complete.

5-11. Hilbert's Theorem

Hilbert's theorem can be stated as follows.

THEOREM. *A complete geometric surface* S *with constant negative curvature cannot be isometrically immersed in* R^3.

Remark 1. Hilbert's theorem was first treated in D. Hilbert, "Über Flächen von konstanter Gausscher Krümung," *Trans. Amer. Math. Soc.* 2 (1901), 87–99. A different proof was given shortly after by E. Holmgren, "Sur les surfaces à courbure constante negative," *C. R. Acad. Sci. Paris* 134 (1902), 740–743. The proof we shall present here follows Hilbert's original ideas. The local part is essentially the same as in Hilbert's paper; the global part, however, is substantially different. We want to thank J. A. Scheinkman for helping us to work out this proof and M. Spivak for suggesting Lemma 7 below.

We shall start with some observations. By multiplying the inner product by a constant factor, we may assume that the curvature $K \equiv -1$. Moreover, since $\exp_p: T_p(S) \longrightarrow S$ is a local diffeomorphism (corollary of the theorem of Sec. 5-5), it induces an inner product in $T_p(S)$. Denote by S' the geometric surface $T_p(S)$ with this inner product. If $\psi: S \longrightarrow R^3$ is an isometric immersion, the same holds for $\varphi = \psi \circ \exp_p: S' \longrightarrow R^3$. Thus, we are reduced to proving that there exists no isometric immersion $\varphi: S' \longrightarrow R^3$ of a plane S' with an inner product such that $K \equiv -1$.

LEMMA 1. *The area of* S′ *is infinite.*

Proof. We shall prove that S' is (globally) isometric to the hyperbolic plane H. Since the area of the latter is (cf. Example 3, Sec. 5-10)

$$\int_{-\infty}^{+\infty}\int_{-\infty}^{+\infty} e^u \, du \, dv = \infty,$$

this will prove the lemma.

Let $p \in H$, $p' \in S'$, and choose a linear isometry $\psi: T_p(H) \longrightarrow T_{p'}(S')$ between their tangent spaces. Define a map $\varphi: H \longrightarrow S'$ by $\varphi = \exp_p \circ \psi \circ \exp_p^{-1}$.

Since each point of H is joined to p by a unique minimal geodesic, φ is well defined.

We now use polar coordinates (ρ, θ) and (ρ', θ') around p and p', respectively, requiring that φ maps the axis $\theta = 0$ into the axis $\theta' = 0$. By the results of Sec. 4-6, φ preserves the first fundamental form; hence, it is locally an isometry. By using the remark made after Hadamard's theorem, we conclude that φ is a covering map. Since S' is simply connected, φ is a homeomorphism, and hence a (global) isometry. **Q.E.D.**

For the rest of this section we shall assume that there exists an isometric immersion $\varphi: S' \longrightarrow R^3$, where S' is a geometric surface homeomorphic to a plane and with $K \equiv -1$.

To avoid the difficulties associated with possible self-intersections of $\varphi(S')$, we shall work with S' and use the immersion φ to induce on S' the local extrinsic geometry of $\varphi(S') \subset R^3$. More precisely, since φ is an immersion, for each $p \in S'$ there exists a neighborhood $V' \subset S'$ of p such that the restriction $\varphi | V' = \bar{\varphi}$ is a diffeomorphism. At each $\bar{\varphi}(q) \in \bar{\varphi}(V')$, there exist, for instance, two asymptotic directions. Through $\bar{\varphi}$, these directions induce two directions at $q \in S'$, which will be called *the asymptotic directions on* S' at q. In this way, it makes sense to talk about asymptotic curves on S', and the *same procedure can be applied to any other local entity of* $\varphi(S')$.

We now recall that the coordinate curves of a parametrization constitute a *Tchebyshef net* if the opposite sides of any quadrilateral formed by them have equal length (cf. Exercise 7, Sec. 2-5). If this is the case, it is possible to reparametrize the coordinate neighborhood in such a way that $E = 1$, $F = \cos \theta$, $G = 1$, where θ is the angle formed by the coordinate curves, (Sec. 2-5, Exercise 8). Furthermore, in this situation, $K = -(\theta_{uv}/\sin \theta)$ (Sec. 4-3, Exercise 5).

LEMMA 2. *For* each $p \in S'$ *there is a parametrization* $\mathbf{x}: U \subset R^2 \longrightarrow S'$, $p \in \mathbf{x}(U)$, *such that the coordinate curves of* $\mathbf{x}$ *are the asymptotic curves of* $\mathbf{x}(U) = V'$ *and form a Tchebyshef net* (we shall express this by saying that the asymptotic curves of V' form a Tchebyshef net).

Proof. Since $K < 0$, a neighborhood $V' \subset S'$ of p can be parametrized by $\mathbf{x}(u, v)$ in such a way that the coordinate curves of $\mathbf{x}$ are the asymptotic curves of V'. Thus, if e, f, and g are the coefficients of the second fundamental form of S' in this parametrization, we have $e = g = 0$. Notice that we are using the above convention of referring to the second fundamental form of S' rather than the second fundamental form of $\varphi(S') \subset R^3$.

Now in $\varphi(V') \subset R^3$, we have

$$N_u \wedge N_v = K(\mathbf{x}_u \wedge \mathbf{x}_v);$$

hence, setting $D = \sqrt{EG - F^2}$,

$$(N \wedge N_v)_u - (N \wedge N_u)_v = 2(N_u \wedge N_v) = 2KDN.$$

Furthermore,

$$N \wedge N_u = \frac{1}{D}\{(\mathbf{x}_u \wedge \mathbf{x}_v) \wedge N_u\} = \frac{1}{D}\{\langle \mathbf{x}_u, N_u\rangle \mathbf{x}_v - \langle \mathbf{x}_v, N_u\rangle \mathbf{x}_u\}$$
$$= \frac{1}{D}(f\mathbf{x}_u - e\mathbf{x}_v),$$

and, similarly,

$$N \wedge N_v = \frac{1}{D}(g\mathbf{x}_u - f\mathbf{x}_v).$$

Since $K = -1 = -(f^2/D^2)$ and $e = g = 0$, we obtain

$$N \wedge N_u = \pm\mathbf{x}_u, \qquad N \wedge N_v = \pm\mathbf{x}_v;$$

hence,

$$2KDN = -2DN = \pm\mathbf{x}_{uv} \pm \mathbf{x}_{vu} = \pm 2\mathbf{x}_{uv}.$$

It follows that $\mathbf{x}_{uv}$ is parallel to N; hence, $E_v = 2\langle \mathbf{x}_{uv}, \mathbf{x}_u\rangle = 0$ and $G_u = 2\langle \mathbf{x}_{uv}, \mathbf{x}_v\rangle = 0$. But $E_v = G_u = 0$ implies (Sec. 2-5, Exercise 7) that the coordinate curves form a Tchebyshef net. **Q.E.D.**

LEMMA 3. *Let* $V' \subset S'$ *be a coordinate neighborhood of* S' *such that the coordinate curves are the asymptotic curves in* V'. *Then the area* A *of any quadrilateral formed by the coordinate curves is smaller than* 2π.

Proof. Let $(\bar{u}, \bar{v})$ be the coordinates of V'. By the argument of Lemma 1, the coordinate curves form a Tchebyshef net. Thus, it is possible to reparametrize V' by, say, (u, v) so that $E = G = 1$ and $F = \cos\theta$. Let R be a quadrilateral that is formed by the coordinate curves with vertices (u_1, v_1), (u_2, v_1), (u_2, v_2), (u_1, v_2) and interior angles $\alpha_1, \alpha_2, \alpha_3, \alpha_4$, respectively (Fig. 5-54). Since $E = G = 1$, $F = \cos\theta$, and $\theta_{uv} = \sin\theta$, we obtain

$$A = \int_R dA = \int_R \sin\theta \, du\, dv = \int_R \theta_{uv}\, du\, dv$$
$$= \theta(u_1, v_1) - \theta(u_2, v_1) + \theta(u_2, v_2) - \theta(u_1, v_2)$$
$$= \alpha_1 + \alpha_3 - (\pi - \alpha_2) - (\pi - \alpha_4) = \sum_{i=1}^{4} \alpha_i - 2\pi < 2\pi,$$

since $\alpha_i < \pi$. **Q.E.D.**

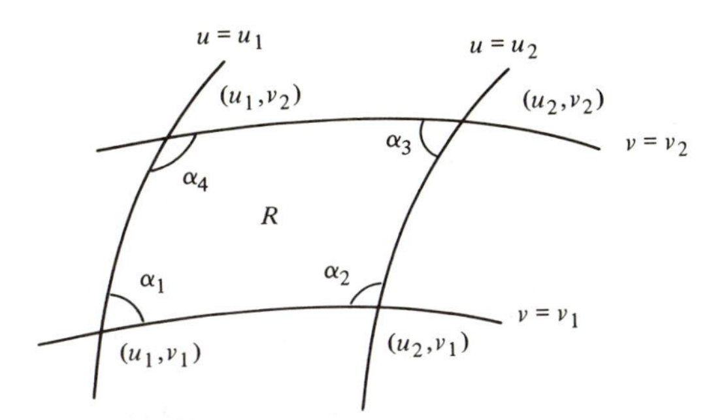

Figure 5-54

So far the considerations have been local. We shall now define a map $\mathbf{x}: R^2 \longrightarrow S'$ and show that $\mathbf{x}$ is a parametrization for the entire S'.

The map $\mathbf{x}$ is defined as follows (Fig. 5-55). Fix a point $O \in S'$ and choose orientations on the asymptotic curves passing through O. Make a definite choice of one of these asymptotic curves, to be called a_1, and denote the other one by a_2. For each $(s, t) \in R^2$, lay off on a_1 a length equal to s starting from O. Let p' be the point thus obtained. Through p' there pass two asymptotic curves, one of which is a_1. Choose the other one and give it the orientation obtained by the continuous extension, along a_1, of the orientation of a_2. Over this oriented asymptotic curve lay off a length equal to t starting from p'. The point so obtained is $\mathbf{x}(s, t)$.

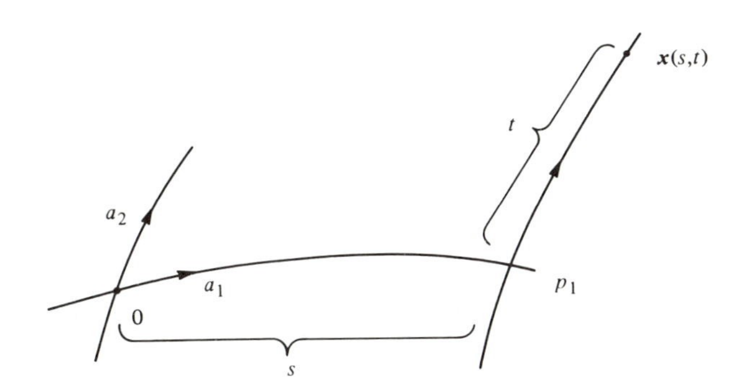

Figure 5-55

$\mathbf{x}(s, t)$ is well defined for all $(s, t) \in R^2$. In fact, if $\mathbf{x}(s, 0)$ is not defined, there exists s_1 such that $a_1(s)$ is defined for all $s < s_1$ but not for $s = s_1$. Let $q = \lim_{s \to s_1} a_1(s)$. By completeness, $q \in S'$. By using Lemma 2, we see that $a_1(s_1)$ is defined, which is a contradiction and shows that $\mathbf{x}(s, 0)$ is defined for all $s \in R$. With the same argument we show that $\mathbf{x}(s, t)$ is defined for all $t \in R$.

Now we must show that **x** is a parametrization of S'. This will be done through a series of lemmas.

LEMMA 4. *For a fixed* t, *the curve* **x**(s, t), $-\infty < s < \infty$, *is an asymptotic curve with* s *as arc length.*

Proof. For each point $\mathbf{x}(s', t') \in S'$, there exists by Lemma 2 a "rectangular" neighborhood (that is, of the form $t_a < t < t_b$, $s_a < s < s_b$) such that the asymptotic curves of this neighborhood form a Tchebyshef net. We first remark that if for some t_0, $t_a < t_0 < t_b$, the curve $\mathbf{x}(s, t_0)$, $s_a < s < s_b$, is an asymptotic curve, then we know the same holds for every curve $\mathbf{x}(s, \bar{t})$, $t_a < \bar{t} < t_b$. In fact, the point $\mathbf{x}(s, \bar{t})$ is obtained by laying off a segment of length $\bar{t}$ from $\mathbf{x}(s, 0)$ which is equivalent to laying off a segment of length $\bar{t} - t_0$ from $\mathbf{x}(s, t_0)$. Since the asymptotic curves form a Tchebyshef net in this neighborhood, the assertion follows.

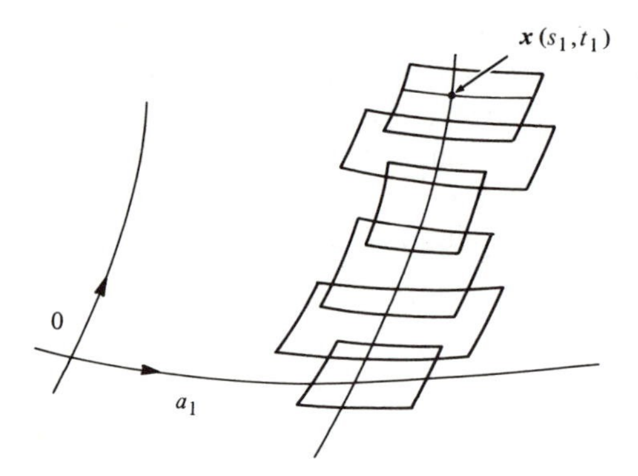

Figure 5-56

Now, let $\mathbf{x}(s_1, t_1) \in S'$ be an arbitrary point. By compactness of the segment $\mathbf{x}(s_1, t)$, $0 \leq t \leq t_1$, it is possible to cover it by a finite number of rectangular neighborhoods such that the asymptotic curves of each of them form a Tchebyshef net (Fig. 5-56). Since $\mathbf{x}(s, 0)$ is an asymptotic curve, we iterate the previous remark and show that $\mathbf{x}(s, t_1)$ is an asymptotic curve in a neighborhood of s_1. Since (s_1, t_1) was arbitrary, the assertion of the lemma follows. **Q.E.D.**

LEMMA 5. **x** *is a local diffeomorphism.*

Proof. This follows from the fact that on the one hand $\mathbf{x}(s_0, t)$, $\mathbf{x}(s, t_0)$ are asymptotic curves parametrized by arc length, and on the other hand S' can be locally parametrized in such a way that the coordinate curves are the asymptotic curves of S' and $E = G = 1$. Thus, **x** agrees locally with such a parametrization. **Q.E.D.**

LEMMA 6. $\mathbf{x}$ *is surjective.*

Proof. Let $Q = \mathbf{x}(R^2)$. Since $\mathbf{x}$ is a local diffeomorphism, Q is open in S'. We also remark that if $p' = \mathbf{x}(s_0, t_0)$, then the two asymptotic curves which pass through p' are entirely contained in Q.

Let us assume that $Q \neq S'$. Since S' is connected, the boundary Bd $Q \neq \phi$. Let $p \in$ Bd Q. Since Q is open in S', $p \notin Q$. Now consider a rectangular neighborhood R of p in which the asymptotic curves form a Tchebyshef net (Fig. 5-57). Let $q \in Q \cap R$. Then one of the asymptotic curves through q intersects one of the asymptotic curves through p. By the above remark, this is a contradiction. **Q.E.D.**

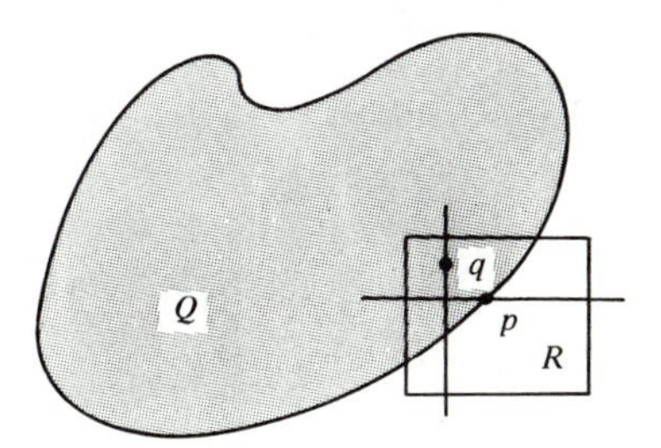

Figure 5-57

LEMMA 7. *On* S′ *there are two differentiable linearly independent vector fields which are tangent to the asymptotic curves of* S′.

Proof. Through each point of S' there pass two distinct asymptotic curves. Fix a point $p \in S'$ and choose two unit vectors, $v_1(p)$ and $v_2(p)$, tangent to the asymptotic curves through p. Let $q \in S'$ be an arbitrary point and let $\alpha_0\colon [0, l] \longrightarrow S'$ be an arc such that $\alpha_0(0) = p, \alpha_0(l) = q$. Define $v_1(\alpha_0(s))$, $s \in [0, l]$, as the (unique) continuous extension of $v_1(p)$ along α_0 which is tangent to an asymptotic curve. Define similarly $v_2(\alpha_0(s))$, $s \in [0, l]$. We claim that $v_1(q)$ and $v_2(q)$ do not depend on the choice of the arc joining p to q. Thus, v_1 and v_2 are well-defined continuous vector fields on S' which are tangent to the asymptotic curves. Hence, v_1 and v_2 are differentiable, and the lemma will be proved.

To prove our claim, let us work with v_1, the case of v_2 being similar. Let $\alpha_1 : [0, l] \longrightarrow S'$ be another arc with $\alpha_1(0) = p$, $\alpha_1(l) = q$. Since S' (which is homeomorphic to a plane) is simply connected (cf. Sec. 5-6, Def. 3), there exists a homotopy $\alpha_t(s) = H(s, t)$, $s \in [0, l]$, $t \in [0, 1]$, between α_0 and α_1 (cf. Sec. 5-6, Def. 2); that is, $\alpha_t(s)$ is a continuous family of arcs joining p to q. From the continuity of the asymptotic directions and the compactness of $[0, l]$, it follows that given $\epsilon > 0$ there exists $t_0 \in [0, 1]$ such that if $t < t_0$, then $|v_1(\alpha_t(l)) - v_1(\alpha_0(l))| < \epsilon$. Thus, if t_0 is small enough, we have $v_1(\alpha_t(l)) = v_1(\alpha_0(l))$ for $t < t_0$. Since $[0, 1]$ is compact, we can extend this argument stepwise to all $t \in [0, 1]$. Hence, $v_1(\alpha_1(l)) = v_1(\alpha_0(l))$.

This proves our claim and concludes the proof of the lemma. **Q.E.D.**

LEMMA 8. $\mathbf{x}$ *is injective.*

Proof. We want to show that $\mathbf{x}(s_0, t_0) = \mathbf{x}(s_1, t_1)$ implies that $(s_0, t_0) = (s_1, t_1)$.

We first assume that $\mathbf{x}(s_0, t_0) = \mathbf{x}(s_1, t_0)$, $s_1 > s_0$, and show that this leads to a contradiction. By Lemma 7, an asymptotic curve cannot intersect itself unless the tangent lines agree at the intersection point. Since $\mathbf{x}$ is a local diffeomorphism, there exists an $\epsilon > 0$ such that

$$\mathbf{x}(s_0, t) = \mathbf{x}(s_1, t), \; t_0 - \epsilon < t < t_0 + \epsilon.$$

By the same reason, the points of the curve $\mathbf{x}(s_0, t)$ for which

$$\mathbf{x}(s_0, t) = \mathbf{x}(s_1, t)$$

form an open and closed set of this curve; hence, $\mathbf{x}(s_0, t) = \mathbf{x}(s_1, t)$ for all t. Moreover, by the construction of the map $\mathbf{x}$, $\mathbf{x}(s_0 + a, t_0) = \mathbf{x}(s_1 + a, t_0)$, $0 \leq a \leq s_1 - s_0$; hence, $\mathbf{x}(s_0 + a, t) = \mathbf{x}(s_1 + a, t)$ for all t. Thus, either

1. $\mathbf{x}(s_0, t_0) \neq \mathbf{x}(s_0, t)$ for all $t > t_0$, or
2. There exists $t = t_1 > t_0$ such that $\mathbf{x}(s_0, t_0) = \mathbf{x}(s_0, t_1)$; by a similar argument, we shall prove that $\mathbf{x}(s, t_0 + b) = \mathbf{x}(s, t_1 + b)$ for all s, $0 \leq b \leq t_1 - t_0$.

In case 1, $\mathbf{x}$ maps each strip of R^2 between two vertical lines at a distance $s_1 - s_0$ onto S' and identifies points in these lines with the same t. This implies that S' is homeomorphic to a cylinder, and this is a contradiction (Fig. 5-58).

In case 2, $\mathbf{x}$ maps each square formed by two horizontal lines at a distance $s_1 - s_0$ and two vertical lines at a distance $t_1 - t_0$ onto S and identifies

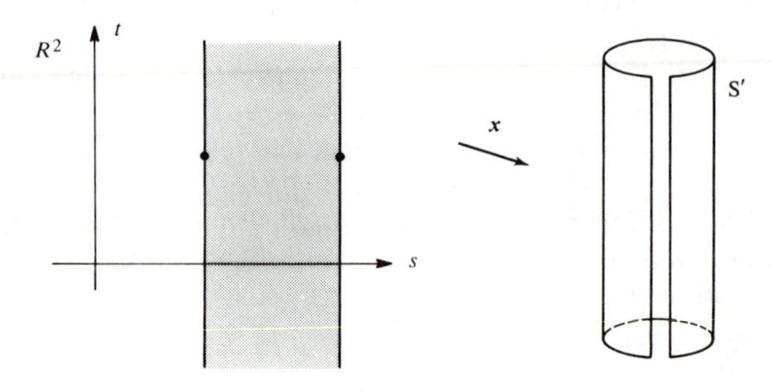

Figure 5-58

corresponding points on opposite sides of the boundary. This implies that S' is homeomorphic to a torus, and this is a contradiction (Fig. 5-59).

By a similar argument, we show that $\mathbf{x}(s_0, t_0) = \mathbf{x}(s_0, t_1)$, $t_1 > t_0$, leads to the same contradiction.

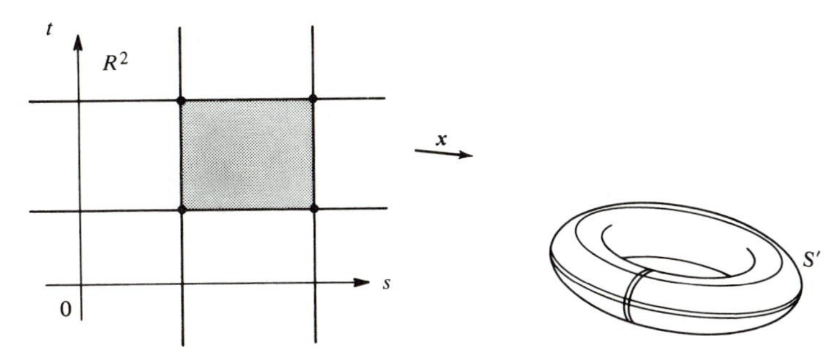

Figure 5-59

We now consider the case $\mathbf{x}(s_0, t_0) = \mathbf{x}(s_1, t_1)$, $s_1 > s_0$, $t_1 > t_0$. By using the fact that $\mathbf{x}$ is a local diffeomorphism and the connectedness of S', we see that $\mathbf{x}$ maps onto S' a strip of R^2 between two lines perpendicular to the vector $(s_1 - s_0, t_1 - t_0) \in R^2$ and at a distance $\sqrt{(s_1 - s_0)^2 + (t_1 - t_0)^2}$ apart. We can now consider cases 1 and 2 as in the previous argument and show that S' is then homeomorphic either to a cylinder or to a torus. In any case, this is a contradiction. **Q.E.D.**

The proof of Hilbert's theorem now follows easily.

Proof of the Theorem. Assume the existence of an isometric immersion $\psi: S \to R^3$, where S is a complete surface with $K \equiv -1$. Let $p \in S$ and denote by S' the tangent plane $T_p(S)$ endowed with the metric induced by $\exp_p: T_p(S) \to S$. Then $\varphi = \psi \circ \exp_p: S' \to R^3$ is an isometric immersion and Lemmas 5, 6, and 8 show the existence of a parametrization $\mathbf{x}: R^2 \to S'$ of the entire S' such that the coordinate curves of $\mathbf{x}$ are the asymptotic curves of S' (Lemma 4). Thus, we can cover S' by a union of "coordinate quadrilaterals" Q_n, with $Q_n \subset Q_{n+1}$. By Lemma 3, the area of each Q_n is smaller than 2π. On the other hand, by Lemma 1, the area of S' is unbounded. This is a contradiction and concludes the proof. **Q.E.D.**

Remark 2. Hilbert's theorem was generalized by N. V. Efimov, "Appearance of Singularities on Surfaces of Negative Curvature," *Math. Sb.* 106 (1954). A.M.S. Translations. Series 2, Vol. 66, 1968, 154–190, who proved the following conjecture of Cohn-Vossen: *Let* S *be a complete surface with curvature* K *satisfying* $K \leq \delta < 0$. *Then there exists no isometric immersion of* S *into* R^3. Efimov's proof is very long, and a shorter proof would be desirable.

An excellent exposition of Efimov's proof can be found in a paper by T. Klotz Milnor, "Efimov's Theorem About Complete Immersed Surfaces of Negative Curvature," *Advances in Mathematics* 8 (1972), 474–543. This paper also contains another proof of Hilbert's theorem which holds for surfaces of class C^2.

For further details on immersion of the hyperbolic plane see M. L. Gromov and V. A. Rokhlin, "Embeddings and Immersions in Riemannian Geometry," *Russian Math. Surveys* (1970), 1–57, especially p. 15.

EXERCISES

1. (*Stoker's Remark.*) Let S be a complete geometric surface. Assume that the Gaussian curvature K satisfies $K \leq \delta < 0$. Show that there is no isometric immersion $\varphi: S \longrightarrow R^3$ such that the absolute value of the mean curvature H is bounded. This proves Efimov's theorem quoted in Remark 2 with the additional condition on the mean curvature. The following outline may be useful:

a. Assume such a φ exists and consider the Gauss map $N: \varphi(S) \subset R^3 \longrightarrow S^2$, where S^2 is the unit sphere. Since $K \neq 0$ everywhere, N induces a new metric $(\, , \,)$ on S by requiring that $N \circ \varphi: S \longrightarrow S^2$ be a local isometry. Choose coordinates on S so that the images by φ of the coordinate curves are lines of curvature of $\varphi(S)$. Show that the coefficients of the new metric in this coordinate system are

$$g_{11} = (k_1)^2 E, \qquad g_{12} = 0, \qquad g_{22} = (k_2)^2 G,$$

where E, $F\,(=0)$, and G are the coefficients of the initial metric in the same system.

b. Show that there exists a constant $M > 0$ such that $k_1^2 \leq M$, $k_2^2 \leq M$. Use the fact that the initial metric is complete to conclude that the new metric is also complete.

c. Use part b to show that S is compact; hence, it has points with positive curvature, a contradiction.

2. The goal of this exercise is to prove that there is no regular complete surface of revolution S in R^3 with $K \leq \delta < 0$ (this proves Efimov's theorem for surfaces of revolution). Assume the existence of such an $S \subset R^3$.

a. Prove that the only possible forms for the generating curve of S are those in Fig. 5-60(a) and (b), where the meridian curve goes to infinity in both directions. Notice that in Fig. 5-60(b) the lower part of the meridian is asymptotic to the z axis.

b. Parametrize the generating curve $(\varphi(s), \psi(s))$ by arc length $s \in R$ so that $\psi(0) = 0$. Use the relations $\varphi'' + K\varphi = 0$ (cf. Example 4, Sec. 3-3, Eq. (9)) and $K \leq \delta < 0$ to conclude that there exists a point $s_0 \in [0, +\infty)$ such that $(\varphi'(s_0))^2 = 1$.

c. Show that each of the three possibilities to continue the meridian $(\varphi(s), \psi(s))$ of S past the point $p_0 = (\varphi(s_0), \psi(s_0))$ (described in Fig. 5-60(c) as I, II, and III) leads to a contradiction. Thus, S is not complete.

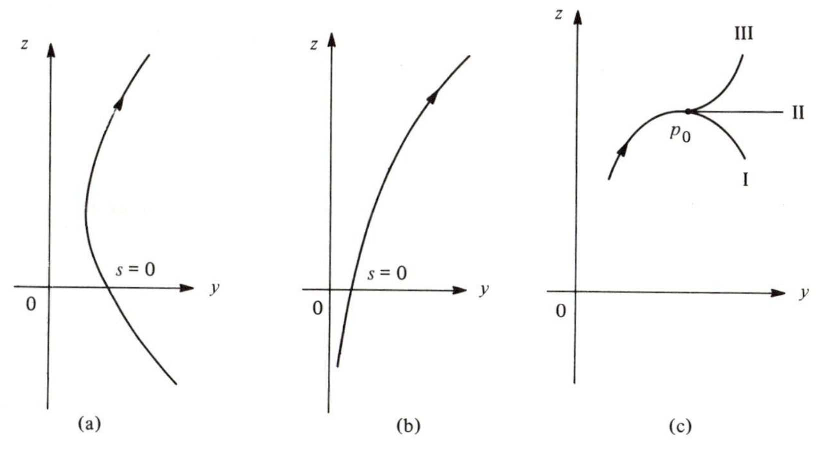

Figure 5-60

3. (*T. K. Milnor's Proof of Hilbert's Theorem.*) Let S be a plane with a complete metric g_1 such that its curvature $K \equiv -1$. Assume that there exists an isometric immersion $\varphi\colon S \longrightarrow R^3$. To obtain a contradiction, proceed as follows:

a. Consider the Gauss map $N\colon \varphi(S) \subset R^3 \longrightarrow S^2$ and let g_2 be the metric on S obtained by requiring that $N \circ \varphi\colon S \longrightarrow S^2$ be a local isometry. Choose local coordinates on S so that the images by φ of the coordinate curves are the asymptotic curves of $\varphi(S)$. Show that, in such a coordinate system, g_1 can be written as

$$du^2 + 2\cos\theta \, du\, dv + dv^2$$

and that g_2 can be written as

$$du^2 - 2\cos\theta \, du\, dv + dv^2.$$

b. Prove that $g_3 = \frac{1}{2}(g_1 + g_2)$ is a metric on S with vanishing curvature. Use the fact that g_1 is a complete metric and $3g_3 \geq g_1$ to conclude that the metric g_3 is complete.

c. Prove that the plane with the metric g_3 is globally isometric to the standard (Euclidean) plane R^2. Thus, there is an isometry $\varphi\colon S \longrightarrow R^2$. Prove further that φ maps the asymptotic curves of S, parametrized by arc length, into a rectangular system of straight lines in R^2, parametrized by arc length.

d. Use the global coordinate system on S given by part c, and obtain a contradiction as in the proof of Hilbert's theorem in the text.

Appendix Point-Set Topology of Euclidean Spaces

In Chap. 5 we have used more freely some elementary topological properties of R^n. The usual properties of compact and connected subsets of R^n, as they appear in courses of advanced calculus, are essentially all that is needed. For completeness, we shall make a brief presentation of this material here, with proofs. We shall assume the material of the appendix to Chap. 2, Part A, and the basic properties of real numbers.

A. Preliminaries

Here we shall complete in some points the material of the appendix to Chap. 2, Part A.

In what follows $U \subset R^n$ will denote an open set in R^n. The index i varies in the range $1, 2, \ldots, m, \ldots$, and if $p = (x_1, \ldots, x_n)$, $q = (y_1, \ldots, y_n)$, $|p - q|$ will denote the distance from p to q; that is,

$$|p - q|^2 = \sum_j (x_j - y_j)^2, \qquad j = 1, \ldots, n.$$

DEFINITION 1. *A sequence* $p_1, \ldots, p_i, \ldots \in R^n$ converges *to* $p_0 \in R^n$ *if given* $\epsilon > 0$, *there exists an index* i_0 *of the sequence such that* $p_i \in B_\epsilon(p_0)$ *for all* $i > i_0$. *In this situation,* p_0 *is the* limit *of the sequence* $\{p_i\}$, *and this is denoted by* $\{p_i\} \to p_0$.

Convergence is related to continuity by the following proposition.

PROPOSITION 1. *A map* $F: U \subset R^n \to R^m$ *is continuous at* $p_0 \in U$ *if and only if for each converging sequence* $\{p_i\} \to p_0$ *in* U, *the sequence* $\{F(p_i)\}$ *converges to* $F(p_0)$.

Proof. Assume F to be continuous at p_0 and let $\epsilon > 0$ be given. By continuity, there exists $\delta > 0$ such that $F(B_\delta(p_0)) \subset B_\epsilon(F(p_0))$. Let $\{p_i\}$ be a sequence in U, with $\{p_i\} \to p_0 \in U$. Then there exists in correspondence with δ an index i_0 such that $p_i \in B_\delta(p_0)$ for $i > i_0$. Thus, for $i > i_0$,

$$F(p_i) \in F(B_\delta(p_0)) \subset B_\epsilon(F(p_0)),$$

which implies that $\{F(p_i)\} \to F(p_0)$.

Suppose now that F is not continuous at p_0. Then there exists a number $\epsilon > 0$ such that for every $\delta > 0$ we can find a point $p \in B_\delta(p_0)$, with $F(p) \notin B_\epsilon(F(p_0))$. Fix this ϵ, and set $\delta = 1, 1/2, \ldots, 1/i, \ldots$, thus obtaining a sequence $\{p_i\}$ which converges to p_0. However, since $F(p_i) \notin B_\epsilon(F(p_0))$, the sequence $\{F(p_i)\}$ does not converge to $F(p_0)$. **Q.E.D.**

DEFINITION 2. *A point* $p \in R^n$ *is a* limit point *of a set* $A \subset R^n$ *if every neighborhood of* p *in* R^n *contains one point of* A *distinct from* p.

To avoid some confusion with the notion of limit of a sequence, a limit point is sometimes called a *cluster point* or an *accumulation point*.

Definition 2 is equivalent to saying that every neighborhood V of p contains infinitely many points of A. In fact, let $q_1 \neq p$ be the point of A given by the definition, and consider a ball $B_\epsilon(p) \subset V$ so that $q_1 \notin B_\epsilon(p)$. Then there is a point $q_2 \neq p, q_2 \in A \cap B_\epsilon(p)$. By repeating this process, we obtain a sequence $\{q_i\}$ in V, where the $q_i \in A$ are all distinct. Since $\{q_i\} \to p$, the argument also shows that p is a limit point of A if and only if p is the limit of some sequence of distinct points in A.

Example 1. The sequence $1, 1/2, 1/3, \ldots, 1/i, \ldots$ converges to 0. The sequence $3/2, 4/3, \ldots, i + 1/i, \ldots$ converges to 1. The "intertwined" sequence $1, 3/2, 1/2, 4/3, 1/3, \ldots, 1 + (1/i), 1/i, \ldots$ does not converge and has two limit points, namely 0 and 1 (Fig. A5-1).

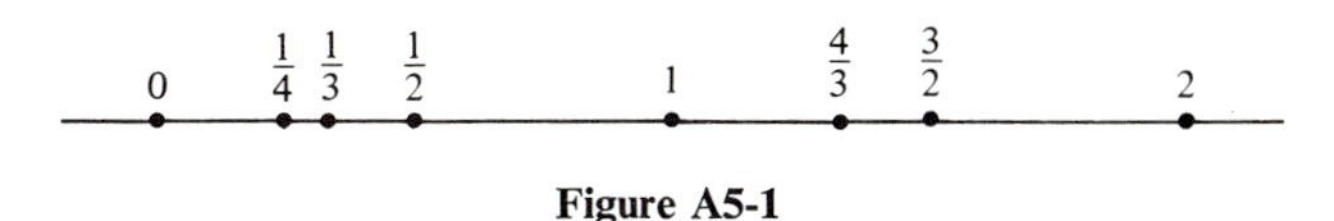

Figure A5-1

It should be observed that the limit p_0 of a converging sequence has the property that any neighborhood of p_0 contains all but a finite number of points of the sequence, whereas a limit point p of a set has the weaker property that any neighborhood of p contains infinitely many points of the set. Thus, a sequence which contains no constant subsequence is convergent if and only if, as a set, it contains only one limit point.

An interesting example is given by the rational numbers Q. It can be proved that Q is countable; that is, it can be made into a sequence. Since arbitrarily near any real number there are rational numbers, the set of limit points of the sequence Q is the real line R.

DEFINITION 3. *A set* $F \subset R^n$ *is* closed *if every limit point of* F *belongs to* F. *The* closure *of* $A \subset R^n$ *denoted by* $\bar{A}$, *is the union of* A *with its limit points.*

Intuitively, F is closed if it contains the limit of all its convergent sequences, or, in other words, it is invariant under the operation of passing to the limit.

It is obvious that the closure of a set is a closed set. It is convenient to make the convention that the empty set ϕ is both open and closed.

There is a very simple relation between open and closed sets.

PROPOSITION 2. $F \subset R^n$ *is closed if and only if the complement* $R^n - F$ *of* F *is open.*

Proof. Assume F to be closed and let $p \in R^n - F$. Since p is not a limit point of F, there exists a ball $B_\epsilon(p)$ which contains no points of F. Thus, $B_\epsilon \subset R^n - F$; hence $R^n - F$ is open.

Conversely, suppose that $R^n - F$ is open and that p is a limit point of F. We want to prove that $p \in F$. Assume the contrary. Then there is a ball $B_\epsilon(p) \subset R^n - F$. This implies that $B_\epsilon(p)$ contains no point of F and contradicts the fact that p is a limit point of F. **Q.E.D.**

Continuity can also be expressed in terms of closed sets, which is a consequence of the following fact.

PROPOSITION 3. *A map* $F: U \subset R^n \to R^m$ *is continuous if and only if for each open set* $V \subset R^m$, $F^{-1}(V)$ *is an open set.*

Proof. Assume F to be continuous and let $V \subset R^m$ be an open set in R^m. If $F^{-1}(V) = \phi$, there is nothing to prove, since we have set the convention that the empty set is open. If $F^{-1}(V) \neq \phi$, let $p \in F^{-1}(V)$. Then $F(p) \in V$, and since V is open, there exists a ball $B_\epsilon(F(p)) \subset V$. By continuity of F, there exists a ball $B_\delta(p)$ such that

$$F(B_\delta(p)) \subset B_\epsilon(F(p)) \subset V.$$

Thus, $B_\delta(p) \subset F^{-1}(V)$; hence, $F^{-1}(V)$ is open.

Assume now that $F^{-1}(V)$ is open for every open set $V \subset R^m$. Let $p \in U$ and $\epsilon > 0$ be given. Then $A = F^{-1}(B_\epsilon(F(p)))$ is open. Thus, there exists $\delta > 0$ such that $B_\delta(p) \subset A$. Therefore,

$$F(B_\delta(p)) \subset F(A) \subset B_\epsilon(F(p));$$

hence, F is continuous in p. **Q.E.D.**

COROLLARY. $F: U \subset R^n \to R^m$ *is continuous if and only if for every closed set* $A \subset R^m$, $F^{-1}(A)$ *is a closed set.*

Example 2. Proposition 3 and its corollary give what is probably the best way of describing open and closed subsets of R^n. For instance, let $f: R^2 \to R$ be given by $f(x, y) = (x^2/a^2) - (y^2/b^2) - 1$. Observe that f is continuous, $0 \in R$ is a closed set in R, and $(0, +\infty)$ is an open set in R. Thus, the set

$$F_1 = \{(x, y);\ f(x, y) = 0\} = f^{-1}(0)$$

is closed in R^2, and the sets

$$U_1 = \{(x, y);\ f(x, y) > 0\},$$
$$U_2 = \{(x, y);\ f(x, y) < 0\}$$

are open in R^2. On the other hand, the set

$$A = \{(x, y) \in R^2,\ x^2 + y^2 < 1\} \cup \{(x, y) \in R^2;\ x^2 + y^2 = 1, x > 0, y > 0\}$$

is neither open nor closed (Fig. A5-2).

The last example suggests the following definition.

DEFINITION 4. *Let* $A \subset R^n$. *The* boundary $\operatorname{Bd} A$ *of* A *is the set of points* p *in* R^n *such that every neighborhood of* p *contains points in* A *and points in* $R^n - A$.

Thus, if A is the set of Example 2, $\operatorname{Bd} A$ is the circle $x^2 + y^2 = 1$. Clearly, $A \subset R^n$ is open if and only if no point of $\operatorname{Bd} A$ belongs to A, and $B \subset R^n$ is closed if and only if all points of $\operatorname{Bd} B$ belong to B.

A final remark on these preliminary notions: Here, as in the appendix to Chap. 2, definitions were given under the assumption that R^n was the "am-

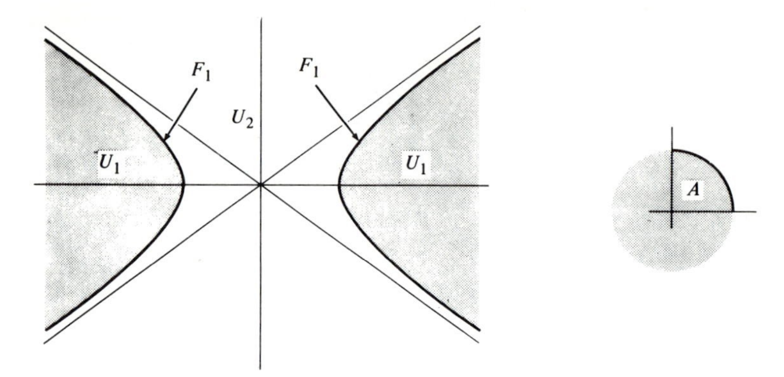

Figure A5-2

bient" space. It is often convenient, as already remarked in the appendix to Chap. 2, to extend such definitions to subsets of an arbitrary set $A \subset R^n$. To do that, we adopt the following definition.

DEFINITION 5. *Let* $A \subset R^n$. *We say that* $V \subset A$ *is an* open set in A *if there exists an open set* U *in* R^n *such that* $V = U \cap A$. *A* neighborhood of $p \in A$ in A *is an open set in* A *containing* p.

With this notion of "proximity" in A, it is a simple matter to extend the previous definitions to subsets of A and to check that the propositions already proved still hold with the new definitions.

Now we want to recall a basic property of the real numbers. We need some definitions.

DEFINITION 6. *A subset* $A \subset R$ *of the real line* R *is* bounded above *if there exists* $M \in R$ *such that* $M \geq a$ *for all* $a \in A$. *The number* M *is called an* upper bound *for* A. *When* A *is bounded above, a* supremum *or a* least upper bound *of* A, sup A (*or l.u.b.* A) *is an upper bound* M *which satisfies the following condition: Given* $\epsilon > 0$, *there exists* $a \in A$ *such that* $M - \epsilon < a$. *By changing the sign of the above inequalities, we define similarly a* lower bound *for* A *and an* infimum (*or a* greatest lower bound) *of* A, *inf A* (*or g.l.b.* A).

AXIOM OF COMPLETENESS OF REAL NUMBERS. *Let* $A \subset R$ *be nonempty and bounded above* (*below*). *Then there exists* sup A (*inf* A).

There are several equivalent ways of expressing the basic property of completeness of the real-number system. We have chosen the above, which, although not the most intuitive, is probably the most effective one.

It is convenient to set the following convention. If $A \subset R$ is not bounded above (below), we say that sup $A = +\infty$ (inf $A = -\infty$). With this convention the above axiom can be stated as follows: *Every nonempty set of real numbers has a* sup *and an* inf.

Example 3. The sup of the set (0, 1) is 1, which does not belong to the set. The sup of the set

$$B = \{x \in R; 0 < x < 1\} \cup \{2\}$$

is 2. The point 2 is an *isolated* point of B; that is, it belongs to B but is not a limit point of B. Observe that the greatest limit point of B is 1, which is not sup B. However, if a bounded set has no isolated points, its sup is certainly a limit point of the set.

One important consequence of the completeness of the real numbers is the following "intrinsic" characterization of convergence, which is actually equivalent to completeness (however, we shall not prove that).

LEMMA 1. *Call a sequence* $\{x_i\}$ *of real numbers a Cauchy sequence if given* $\epsilon < 0$, *there exists* i_0 *such that* $|x_i - x_j| < \epsilon$ *for all* $i, j > i_0$. *A sequence is convergent if and only if it is a Cauchy sequence.*

Proof. Let $\{x_i\} \to x_0$. Then, if $\epsilon > 0$ is given, there exists i_0 such that $|x_i - x_0| < \epsilon/2$ for $i > i_0$. Thus, for $i, j > i_0$, we have

$$|x_i - x_j| \leq |x_i - x_0| + |x_j - x_0| < \epsilon;$$

hence, $\{x_i\}$ is a Cauchy sequence.

Conversely, let $\{x_i\}$ be a Cauchy sequence. The set $\{x_i\}$ is clearly a bounded set. Let $a_1 = \inf\{x_i\}$, $b_1 = \sup\{x_i\}$. Either, one of these points is a limit point of $\{x_i\}$ and then $\{x_i\}$ converges to this point, or both are isolated points of the set $\{x_i\}$. In the latter case, consider the set of points in the open interval (a_1, b_1), and let a_2 and b_2 be its inf and sup, respectively. Proceeding in this way, we obtain that either $\{x_i\}$ converges or there are two bounded sequences $a_1 < a_2 < \cdots$ and $b_1 > b_2 > \cdots$. Let $a = \sup\{a_i\}$ and $b = \inf\{b_i\}$. Since $\{x_i\}$ is a Cauchy sequence, $a = b$, and this common value x_0 is the unique limit point of $\{x_i\}$. Thus, $\{x_i\} \to x_0$. **Q.E.D.**

This form of completness extends naturally to Euclidean spaces.

DEFINITION 7. *A sequence* $\{p_i\}$, $p_i \in R^n$, *is a* Cauchy sequence *if given* $\epsilon > 0$, *there exists an index* i_0 *such that the distance* $|p_i - p_j| < \epsilon$ *for all* $i, j > i_0$.

PROPOSITION 4. *A sequence* $\{p_i\}$, $p_i \in R^n$, *converges if and only if it is a Cauchy sequence.*

Proof. A convergent sequence is clearly a Cauchy sequence (see the argument in Lemma 1). Conversely, let $\{p_i\}$ be a Cauchy sequence, and consider its projection on the j axis of R^n, $j = 1, \ldots, n$. This gives a sequence of real numbers $\{x_{ji}\}$ which, since the projection decreases distances, is again a Cauchy sequence. By Lemma 1, $\{x_{ji}\} \longrightarrow x_{j0}$. It follows that $\{p_i\} \longrightarrow p_0 = \{x_{10}, x_{20}, \ldots, x_{n0}\}$. **Q.E.D.**

B. Connected Sets

DEFINITION 8. *A continuous curve* $\alpha: [a, b] \longrightarrow A \subset R^n$ *is called an* arc in A joining $\alpha(a)$ to $\alpha(b)$.

DEFINITION 9. $A \subset R^n$ is arcwise connected *if, given two points* $p, q \in A$, *there exists an arc in* A *joining* p *to* q.

Earlier in the book we have used the word connected to mean arcwise connected (Sec. 2-2). Since we were considering only regular surfaces, this can be justified, as will be done presently. For a general subset of R^n, however, the notion of arcwise connectedness is much too restrictive, and it is more convenient to use the following definition.

DEFINITION 10. $A \subset R^n$ *is* connected *when it is not possible to write* $A = U_1 \cup U_2$, *where* U_1 *and* U_2 *are nonempty open sets in* A *and* $U_1 \cap U_2 = \phi$.

Intuitively, this means that it is impossible to decompose A into disjoint pieces. For instance, the sets U_1 and F_1 in Example 2 are not connected. By taking the complements of U_1 and U_2, we see that we can replace the word "open" by "closed" in Def. 10.

PROPOSITION 5. *Let* $A \subset R^n$ *be connected and let* $B \subset A$ *be simultaneously open and closed in* A. *Then either* $B = \phi$ *or* $B = A$.

Proof. Suppose that $B \neq \phi$ and $B \neq A$ and write $A = B \cup (A - B)$. Since B is closed in A, $A - B$ is open in A. Thus, A is a union of disjoint, nonvoid, open sets, namely B and $A - B$. This contradicts the connectedness of A. **Q.E.D.**

The next proposition shows that the continuous image of a connected set is connected.

PROPOSITION 6. *Let* $F: A \subset R^n \longrightarrow R^m$ *be continuous and* A *be connected. Then* $F(A)$ *is connected.*

Proof. Assume that $F(A)$ is not connected. Then $F(A) = U_1 \cup U_2$, where U_1 and U_2 are disjoint, nonvoid, open sets in $F(A)$. Since F is continuous, $F^{-1}(U_1)$, $F^{-1}(U_2)$ are also disjoint, nonvoid, open sets in A. Since $A = F^{-1}(U_1) \cup F^{-1}(U_2)$, this contradicts the connectedness of A. **Q.E.D.**

For the purposes of this section, it is convenient to extend the definition of interval as follows:

DEFINITION 11. *An interval of the real line* R *is any of the sets* $a < x < b$, $a \leq x \leq b$, $a < x \leq b$, $a \leq x < b$, $x \in R$. *The cases* $a = b$, $a = -\infty$, $b = +\infty$ *are not excluded, so that an interval may be a point, a half-line, or* R *itself.*

PROPOSITION 7. $A \subset R$ *is connected if and only if* A *is an interval.*

Proof. Let $A \subset R$ be an interval and assume that A is not connected. We shall arrive at a contradiction.

Since A is not connected, $A = U_1 \cup U_2$, where U_1 and U_2 are nonvoid, disjoint, and open in A. Let $a_1 \in U_1, b_1 \in U_2$ and assume that $a_1 < b_1$. By dividing the closed interval $[a_1, b_1] = I_1$ by the midpoint $(a_1 + b_1)/2$, we obtain two intervals, one of which, to be denoted by I_2, has one of its end points in U_1 and the other end point in U_2. Considering the midpoint of I_2 and proceeding as before, we obtain an interval $I_3 \subset I_2 \subset I_1$. Thus, we obtain a family of closed intervals $I_1 \supset I_2 \supset \cdots \supset I_n \supset \cdots$ whose lengths approach zero. Let us rewrite $I_i = [c_i, d_i]$. Then $c_1 \leq c_2 \leq \cdots \leq c_n \leq \cdots$, and $d_1 \geq d_2 \geq \cdots \geq d_n \geq \cdots$. Let $c = \sup\{c_1\}$ and $d = \inf\{d_i\}$. Since $d_i - c_i$ is arbitrarily small, $c = d$. Furthermore, any neighborhood of c contains some I_i for i sufficiently large. Thus, c is a limit point of both U_1 and U_2. Since U_1 and U_2 are closed, $c \in U_1 \cap U_2$, and that contradicts the disjointness of U_1 and U_2.

Conversely, assume that A is connected. If A has a single element, A is trivially an interval. Suppose that A has at least two elements, and let $a = \inf A$, $b = \sup A$, $a \neq b$. Clearly, $A \subset [a, b]$. We shall show that $(a, b) \subset A$, and that implies that A is an interval. Assume the contrary; that is, there exists t, $a < t < b$, such that $t \notin A$. The sets $A \cap (-\infty, t) = V_1$, $A \cap (t, +\infty) = V_2$ are open in $A = V_1 \cup V_2$. Since A is connected, one of these sets, say, V_2, is empty. Since $b \in (t, +\infty)$, this implies both that $b \notin A$ and b is not a limit point of A. This contradicts the fact that $b = \sup A$. In the same way, if $V_1 = \phi$, we obtain a contradiction with the fact that $a = \inf A$. **Q.E.D.**

PROPOSITION 8. *Let* f: $A \subset R^n \to R$ *be continuous and* A *be connected. Assume that* $f(q) \neq 0$ *for all* $q \in A$. *Then* f *does not change sign in* A.

Proof. By Prop. 5, $f(A) \subset R$ is connected. By Prop. 7, $f(A)$ is an interval. By hypothesis, $f(A)$ does not contain zero. Thus, the points in $f(A)$ all have the same sign. **Q.E.D.**

PROPOSITION 9. *Let* $A \subset R^n$ *be arcwise connected. Then* A *is connected.*

Proof. Assume that A is not connected. Then $A = U_1 \cup U_2$, where U_1, U_2 are nonvoid, disjoint, open sets in A. Let $p \in U_1, q \in U_2$. Since A is arcwise connected, there is an arc $\alpha: [a, b] \to A$ joining p to q. Since α is continuous, $B = \alpha([a, b]) \subset A$ is connected. Set $V_1 = B \cap U_1, V_2 = B \cap U_2$. Then $B = V_1 \cup V_2$, where V_1 and V_2 are nonvoid, disjoint, open sets in B, and that is a contradiction. **Q.E.D.**

The converse is, in general, not true. However, there is an important special case where the converse holds.

DEFINITION 12. *A set* $A \subset R^n$ *is* locally arcwise *connected if for each* $p \in A$ *and each neighborhood* V *of* p *in* A *there exists an arcwise connected neighborhood* $U \subset V$ *of* p *in* A.

Intuitively, this means that each point of A has arbitrarily small arcwise connected neighborhoods. A simple example of a locally arcwise connected set in R^3 is a regular surface. In fact, for each $p \in S$ and each neighborhood W of p in R^3, there exists a neighborhood $V \subset W$ of p in R^3 such that $V \cap S$ is homeomorphic to an open disk in R^2; since open disks are arcwise connected, each neighborhood $W \cap S$ of $p \in S$ contains an arcwise connected neighborhood.

The next proposition shows that our usage of the word connected for arcwise connected surfaces was entirely justified.

PROPOSITION 10. *Let* $A \subset R^n$ *be a locally arcwise connected set. Then* A *is connected if and only if it is arcwise connected.*

Proof. Half of the statement has already been proved in Prop. 9. Now assume that A is connected. Let $p \in A$ and let A_1 be the set of points in A that can be joined to p by some arc in A. We claim that A_1 is open in A.

In fact, let $q \in A_1$ and let $\alpha: [a, b] \to A$ be the arc joining p to q. Since A is locally arcwise connected, there is a neighborhood V of q in A such that q can be joined to any point $r \in V$ by an arc $\beta: [b, c] \to V$ (Fig. A5-3).

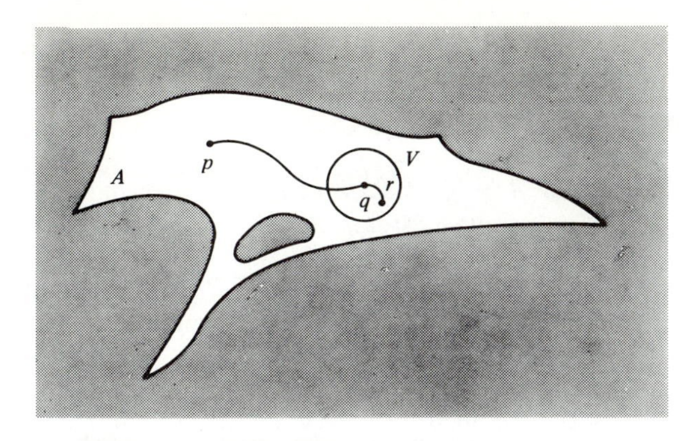

Figure A5-3

It follows that the arc in A,

$$\alpha \circ \beta = \begin{cases} \alpha(t), & t \in [a, b], \\ \beta(t), & t \in [b, c], \end{cases}$$

joins q to r, and this proves our claim.

By a similar argument, we prove that the complement of A_1 is also open in A. Thus, A_1 is both open and closed in A. Since A is locally arcwise connected, A_1 is not empty. Since A is connected, $A_1 = A$. **Q.E.D.**

Example 4. A set may be arcwise connected and yet fail to be locally arcwise connected. For instance, let $A \subset R^2$ be the set made up of vertical lines passing through $(1/n, 0)$, $n = 1, \ldots,$ plus the x and y axis. A is clearly arcwise connected, but a small neighborhood of $(0, y)$, $y \neq 0$, is not arcwise connected. This comes from the fact that although there is a "long" arc joining any two points $p, q \in A$, there may be no short arc joining these points (Fig. A5-4).

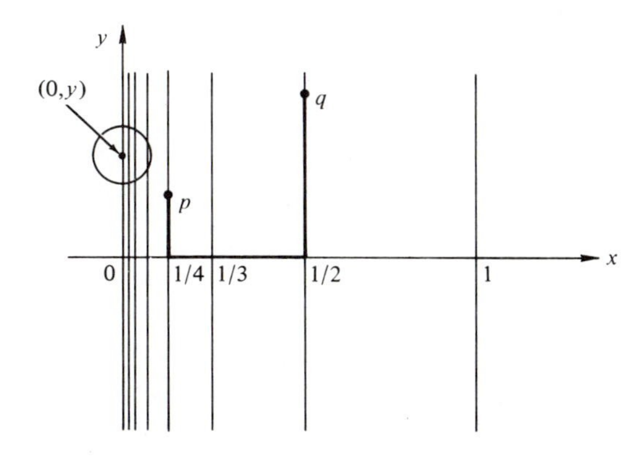

Figure A5-4

C. Compact Sets

DEFINITION 13. *A set* $\mathrm{A} \subset \mathrm{R}^n$ *is bounded if it is contained in some ball of* R^n. *A set* $\mathrm{K} \subset \mathrm{R}^n$ *is* compact *if it is closed and bounded.*

We have already met compact sets in Sec. 2-7. For completness, we shall prove here properties 1 and 2 of compact sets, which were assumed in Sec. 2-7.

DEFINITION 14. *An* open cover *of a set* $\mathrm{A} \subset \mathrm{R}^n$ *is a family of open sets* $\{\mathrm{U}_\alpha\}$, $\alpha \in \mathfrak{A}$ *such that* $\bigcup_\alpha \mathrm{U}_\alpha = \mathrm{A}$. *When there are only finitely many* U_α *in the family, we say that the cover is* finite. *If the subfamily* $\{\mathrm{U}_\beta\}$, $\beta \in \mathfrak{B} \subset \mathfrak{A}$, *still covers* A, *that is,* $\bigcup_\beta \mathrm{U}_\beta = \mathrm{A}$, *we say that* $\{\mathrm{U}_\beta\}$ *is a* subcover *of* $\{\mathrm{U}_\alpha\}$.

PROPOSITION 11. *For a set* $\mathrm{K} \subset \mathrm{R}^n$ *the following assertions are equivalent:*

1. K *is compact.*
2. (Heine-Borel). *Every open cover of* K *has a finite subcover.*
3. (Bolzano-Weierstrass). *Every infinite subset of* K *has a limit point in* K.

Proof. We shall prove $1 \Rightarrow 2 \Rightarrow 3 \Rightarrow 1$.

$1 \Rightarrow 2$: Let $\{U_\alpha\}$, $\alpha \in \mathfrak{A}$, be an open cover of the compact K, and assume that $\{U_\alpha\}$ has no finite subcover. We shall show that this leads to a contradiction.

Since K is compact, it is contained in a closed rectangular region

$$B = \{(x_1, \dots, x_n) \in R^n;\ a_j \leq x_j \leq b_j, \qquad j = 1, \dots, n\}.$$

Let us divide B by the hyperplanes $x_j = (a_j + b_j)/2$ (for instance, if $K \subset R^2$, B is a rectangle, and we are dividing B into $2^2 = 4$ rectangles). We thus obtain 2^n smaller closed rectangular regions. By hypothesis, at least one of these regions, to be denoted by B_1, is such that $B_1 \cap K$ is not covered by a finite number of open sets of $\{U_\alpha\}$. We now divide B_1 in a similar way, and, by repeating the process, we obtain a sequence of closed rectangular regions (Fig. A5-5)

$$B_1 \supset B_2 \supset \cdots \supset B_i \supset \cdots$$

which is such that no $B_i \cap K$ is covered by a finite number of open sets of $\{U_\alpha\}$ and the length of the largest side of B_i converges to zero.

We claim that there exists $p \in \cap B_i$. In fact, by projecting each B_i on the j axis of R^n, $j = 1, \dots, n$, we obtain a sequence of closed intervals

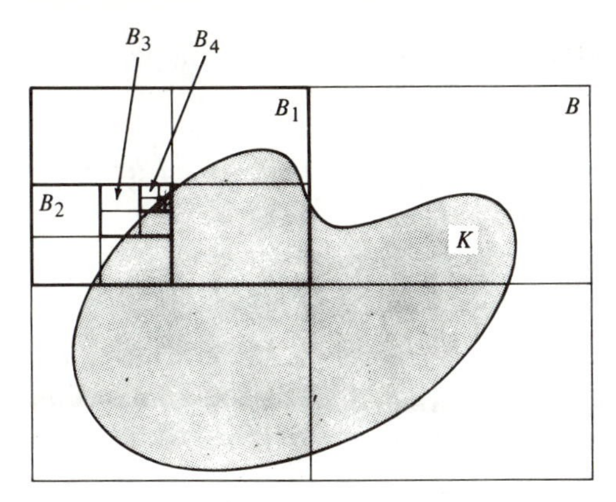

Figure A5-5

$$[a_{j1}, b_{j1}] \supset [a_{j2}, b_{j2}] \supset \cdots \supset [a_{ji}, b_{ji}] \supset \cdots.$$

Since $(b_{ji} - a_{ji})$ is arbitrarily small, we see that

$$a_j = \sup\{a_{ji}\} = \inf\{b_{ji}\} = b_j;$$

hence,

$$a_j \in \bigcap_i [a_{ji}, b_{ji}].$$

Thus, $p = (a_1, \ldots, a_n) \in \bigcap_i B_i$, as we claimed.

Now, any neighborhood of p contains some B_i for i sufficiently large; hence, it contains infinitely many points of K. Thus, p is a limit point of K, and since K is closed, $p \in K$. Let U_0 be an element of the family $\{U_\alpha\}$ which contains p. Since U_0 is open, there exists a ball $B_\epsilon(p) \subset U_0$. On the other hand, for i sufficiently large, $B_i \subset B_\epsilon(p) \subset U_0$. This contradicts the fact that no $B_i \cap K$ can be covered by a finite number of U_α's and proves that $1 \Rightarrow 2$.

$2 \Rightarrow 3$. Assume that $A \subset K$ is an infinite subset of K and that no point of K is a limit point of A. Then it is possible, for each $p \in K$, $p \notin A$, to choose a neighborhood V_p of p such that $V_p \cap A = \phi$ and for each $q \in A$ to choose a neighborhood W_q of q such that $W_q \cap A = q$. Thus, the family $\{V_p, W_q\}$, $p \in K - A, q \in A$, is an open cover of K. Since A is infinite and the omission of any W_q of the family leaves the point q uncovered, the family $\{V_p, W_q\}$ has no finite subcover. This contradicts assertion 2.

$3 \Rightarrow 1$: We have to show that K is closed and bounded. K is closed, because if p is a limit point of K, by considering concentric balls $B_{1/i}(p) = B_i$, we obtain a sequence $p_1 \in B_1 - B_2$, $p_2 \in B_2 - B_3, \ldots, p_i \in B_i - B_{i+1}, \ldots$ which has p as a limit point. By assertion 3, $p \in K$.

K is bounded. Otherwise, by considering concentric balls $B_i(p)$, of radius $1, 2, \ldots, i, \ldots$, we will obtain a sequence $p_1 \in B_1$, $p_2 \in B_2 - B_1, \ldots$, $p_i \in B_i - B_{i-1}, \ldots$ with no limit point. This proves that $3 \Rightarrow 1$. **Q.E.D.**

The next proposition shows that a continuous image of a compact set is compact.

PROPOSITION 12. *Let* $F: K \subset R^n \to R^m$ *be continuous and let* K *be compact. Then* F(K) *is compact.*

Proof. If $F(K)$ is finite, it is trivially compact. Assume that $F(K)$ is not finite and consider an infinite subset $\{F(p_\alpha)\} \subset F(K)$, $p_\alpha \in K$. Clearly the set $\{p_\alpha\} \subset K$ is infinite and has, by compactness, a limit point $q \in K$. Thus, there exists a sequence $p_1, \ldots, p_i, \ldots, \to q, p_i \in \{p_\alpha\}$. By the continuity of F, the sequence $F(p_i) \to F(q) \in F(K)$ (Prop. 1). Thus, $\{F(p_\alpha)\}$ has a limit point $F(q) \in F(K)$; hence, $F(K)$ is compact. **Q.E.D.**

The following is probably the most important property of compact sets.

PROPOSITION 13. *Let* $f: K \subset R^n \to R$ *be a continuous function defined on a compact set* K. *Then there exists* $p_1, p_2 \in K$ *such that*

$$f(p_2) \leq f(p) \leq f(p_1) \qquad \textit{for all } p \in K;$$

that is, f *reaches a maximum at* p_1 *and a minimum at* p_2.

Proof. We shall prove the existence of p_1; the case of minimum can be treated similarly.

By Prop. 12, $f(K)$ is compact, and hence closed and bounded. Thus, there exists sup $f(K) = x_1$. Since $f(K)$ is closed, $x_1 \in f(K)$. It follows that there exists $p_1 \in K$ with $x_1 = f(p_1)$. Clearly, $f(p) \leq f(p_1) = x_1$ for all $p \in K$. **Q.E.D.**

Although we shall make no use of it, the notion of uniform continuity fits so naturally in the present context that we should say a few words about it.

A map $F: A \subset R^n \to R^m$ is *uniformly continuous* in A if given $\epsilon > 0$, there exists $\delta > 0$ such that $F(B_\delta(p)) \subset B_\epsilon(F(p))$ *for all* $p \in A$.

Formally, the difference between this definition and that of (simple) continuity is the fact that here, given ϵ, the number δ is the same for all $p \in B$, whereas in simple continuity, given ϵ, the number δ may vary with p. Thus, uniform continuity is a global, rather than a local, notion.

It is an important fact that on compact sets the two notions agree. More precisely, *let* $F: K \subset R^n \to R^m$ *be continuous and* K *be compact. Then* F *is uniformly continuous in* K.

The proof of this fact is simple if we recall the notion of the Lebesgue number of an open cover, introduced in Sec. 2-7. In fact, given $\epsilon > 0$, there exists for each $p \in K$ a number $\delta(p) > 0$ such that $F(B_{\delta(p)}(p)) \subset B_{\epsilon/2}(F(p))$.

The family $\{B_{\delta(p)}(p), p \in K\}$ is an open cover of K. Let $\delta > 0$ be the Lebesgue number of this family (Sec. 2-7, property 3). If $q \in B_\delta(p)$, $p \in K$, then q and p belong to some element of the open cover. Thus, $|F(p) - F(q)| < \epsilon$. Since q is arbitrary, $F(B_\delta(p)) \subset B_\epsilon(F(p))$. This shows that δ satisfies the definition of uniform continuity, as we wished.

D. Connected Components

When a set is not connected, it may be split into its connected components. To make this idea precise, we shall first prove the following proposition.

PROPOSITION 14. *Let* $C_\alpha \subset R^n$ *be a family of connected sets such that*

$$\bigcap_\alpha C_\alpha \neq \phi.$$

Then $\bigcup_\alpha C_\alpha = C$ *is a connected set.*

Proof. Assume that $C = U_1 \cup U_2$, where U_1 and U_2 are nonvoid, disjoint, open sets in C, and that some point $q \in \bigcap_\alpha C_\alpha$ belongs to U_1. Let $q \in U_2$. Since $C = \bigcup_\alpha C_\alpha$ and $p \in \bigcap_\alpha C_\alpha$, there exists some C_α such that p, $q \in C_\alpha$. Then $C_\alpha \cap U_1$ and $C_\alpha \cap U_2$ are nonvoid, disjoint, open sets in C_α. This contradicts the connectedness of C_α and shows that C is connected. **Q.E.D.**

DEFINITION 15. *Let* $A \subset R^n$ *and* $p \in A$. *The union of all connected subsets of* A *which contain* p *is called the* connected component *of* A *containing* p.

By Prop. 14, a connected component is a connected set. Intuitively the connected component of A containing $p \in A$ is the largest connected subset of A (that is, it is contained in no other connected subset of A that contains p).

A connected component of a set A is always closed in A. This is a consequence of the following proposition.

PROPOSITION 15. *Let* $C \subset A \subset R^n$ *be a connected set. Then the closure* $\bar{C}$ *of* C *in* A *is connected.*

Proof. Let us suppose that $\bar{C} = U_1 \cup U_2$, where U_1, U_2 are nonvoid, disjoint, open sets in $\bar{C}$. Since $\bar{C} \supset C$, the sets $C \cap U_1 = V_1$, $C \cap U_2 = V_2$ are open in C, disjoint, and $V_1 \cup V_2 = C$. We shall show that V_1 and V_2 are nonvoid, thus reaching a contradiction with the connectedness of C.

Let $p \in U_1$. Since U_1 is open in $\bar{C}$, there exists a neighborhood W of p in A such that $W \cap \bar{C} \subset U_1$. Since p is a limit of C, there exists $q \in W \cap C \subset W \cap \bar{C} \subset U_1$. Thus, $q \in C \cap U_1 = V_1$, and V_1 is not empty. In a similar way, it can be shown that V_2 is not empty. **Q.E.D.**

COROLLARY. *A connected component* $C \subset A \subset R^n$ *of a set* A *is closed in* A.

In fact, if $\bar{C} \neq C$, there exists a connected subset of A, namely $\bar{C}$, which contains C properly. This contradicts the maximality of the connected component C.

In some special cases, a connected component of set A is also an open set in A.

PROPOSITION 16. *Let* $C \subset A \subset R^n$ *be a connected component of a locally arcwise connected set* A. *Then* C *is open in* A.

Proof. Let $p \in C \subset A$. Since A is locally arcwise connected, there exists an arcwise connected neighborhood V of p in A. By Prop. 9, V is connected. Since C is maximal, $C \supset V$; hence, C is open in A. **Q.E.D.**

Bibliography and Comments

The basic work of differential geometry of surfaces is Gauss' paper "Disquisitiones generales circa superficies curvas," *Comm. Soc. Göttingen* Bd 6, 1823–1827. There are translations into several languages, for instance,

1. Gauss, K. F., *General Investigations of Curved Surfaces,* Raven Press, New York, 1965.

We believe that the reader of this book is now in a position to try to understand that paper. Patience and open-mindedness will be required, but the experience is most rewarding.

The classical source of differential geometry of surfaces is the four-volume treatise of Darboux:

2. Darboux, G., *Théorie des Surfaces,* Gauthier-Villars, Paris, 1887, 1889, 1894, 1896. There exists a reprint published by Chelsea Publishing Co., Inc., New York.

This is a hard reading for beginners. However, beyond the wealth of information, there are still many unexplored ideas in this book that make it worthwhile to come to it from time to time.

The most influential classical text in the English language was probably

3. Eisenhart, L. P., *A Treatise on the Differential Geometry of Curves and Surfaces,* Ginn and Company, Boston, 1909, reprinted by Dover, New York, 1960.

An excellent presentation of some intuitive ideas of classical differential geometry can be found in Chap. 4 of

4. Hilbert, H., and S. Cohn-Vossen, *Geometry and Imagination*, Chelsea Publishing Company, Inc., New York, 1962 (translation of a book in German, first published in 1932).

Below we shall present, in chronological order, a few other textbooks. They are more or less pitched at about the level of the present book. A more complete list can be found in [9], which, in addition, contains quite a number of global theorems.

5. Struik, D. J., *Lectures on Classical Differential Geometry*, Addison-Wesley, Reading, Mass., 1950.

6. Pogorelov, A. V., *Differential Geometry*, Noordhoff, Groningen, Netherlands, 1958.

7. Willmore, T. J., *An Introduction to Differential Geometry*, Oxford University Press, Inc., London 1959.

8. O'Neill, B., *Elementary Differential Geometry*, Academic Press, New York, 1966.

9. Stoker, J. J., *Differential Geometry*, Wiley-Interscience, New York, 1969.

A clear and elementary exposition of the method of moving frames, not treated in the present book, can be found in [8]. Also, more details on the theory of curves, treated briefly here, can be found in [5], [6], and [9].

Although not textbooks, the following references should be included. Reference [10] is a beautiful presentation of some global theorems on curves and surfaces, and [11] is a set of notes which became a classic on the subject.

10. Chern, S. S., *Curves and Surfaces in Euclidean Spaces*, Studies in Global Geometry and Analysis, MAA Studies in Mathematics, The Mathematical Association of America, 1967.

11. Hopf, H., *Lectures on Differential Geometry in the Large*, notes published by Stanford University, 1955.

For more advanced reading, one should probably start by learning something of differentiable manifolds and Lie groups. For instance,

12. Spivak, M., *A Comprehensive Introduction to Differential Geometry*, Vol. 1, Brandeis University, 1970.

13. Warner, F., *Foundations of Differentiable Manifolds and Lie Groups*, Scott, Foresman, Glenview, Ill., 1971.

Reference [12] is a delightful-reading. Chapters 1–4 of [13] provide a short and efficient account of the basics of the subject.

After that, there is a wide choice of reading material, depending on the reader's tastes and interests. Below we include a possible choice, by no means unique. In [16] and [17] one can find extensive lists of books and papers.

14. Berger, M., P. Gauduchon, and E. Mazet, *Le Spectre d'une Variété Riemannienne*, Lecture Notes 194, Springer, Berlin, 1971.

15. Bishop, R. L., and R. J. Crittenden, *Geometry of Manifolds*, Academic Press, New York, 1964.

16. Cheeger, J., and D. Ebin, *Comparison Theorems in Riemannian Geometry*, North-Holland, Amsterdam, 1974.

17. Helgason, S., *Differential Geometry and Symmetric Spaces*, Academic Press, New York, 1963.

18. Kobayashi, S., and K. Nomizu, *Foundations of Differential Geometry*, Vols. I and II, Wiley-Interscience, New York, 1963 and 1969.

19. Klingenberg, W., D. Gromoll, and W. Meyer, *Riemannsche Geometrie im Grossen*, Lecture Notes 55, Springer-Verlag, Berlin, 1968.

20. Lawson, B., *Lectures on Minimal Submanifolds*, Monografias de Matemática, IMPA, Rio de Janeiro, 1973.

21. Milnor, J., *Morse Theory*, Princeton University Press, Princeton, N. J., 1963.

22. Spivak, M., *A Comprehensive Introduction to Differential Geometry*, Vol. II, Brandeis University, 1970.

The theory of minimal submanifolds, [20] and the references therein; the problems associated with the spectrum, [14]; and the topological behavior of positively curved manifolds, [16] and [19], are only three of the many interesting topics of present-day differential geometry.

Hints and Answers

SECTION 1-3

2. a. $\alpha(t) = (t - \sin t, 1 - \cos t)$; see Fig. 1-7. Singular points: $t = 2\pi n$, where n is any integer.

7. b. Apply the mean value theorem to each of the functions x, y, z to prove that the vector $(\alpha(t + h) - \alpha(t + k))/(h - k)$ converges to the vector $\alpha'(t)$ as $h, k \longrightarrow 0$. Since $\alpha'(t) \neq 0$, the line determined by $\alpha(t + h), \alpha(t + k)$ converges to the line determined by $\alpha'(t)$.

8. By the definition of integral, given $\epsilon > 0$, there exists a $\delta' > 0$ such that if $|P| < \delta'$, then

$$\left|\left(\int_a^b |\alpha'(b)|\, dt\right) - \sum (t_i - t_{i-1})|\alpha'(t_i)|\right| < \frac{\epsilon}{2}.$$

On the other hand, since α' is uniformly continuous in $[a, b]$, given $\epsilon > 0$, there exists $\delta'' > 0$ such that if $t, s \in [a, b]$ with $|t - s| < \delta''$, then

$$|\alpha'(t) - \alpha'(s)| < \epsilon/2(b - a).$$

Set $\delta = \min(\delta', \delta'')$. Then if $|P| < \delta$, we obtain, by using the mean value theorem for vector functions,

$$\begin{aligned} &|\textstyle\sum |\alpha(t_{i-1}) - \alpha(t_i)| - \sum (t_{i-1} - t_i)|\alpha'(t_i)|| \\ &\qquad \leq |\textstyle\sum (t_{i-1} - t_i) \sup\limits_{s_i} |\alpha'(s_i)| - \sum (t_{i-1} - t_i)|\alpha'(t_i)|| \\ &\qquad \leq |\textstyle\sum (t_{i-1} - t_i) \sup\limits_{s_i} |\alpha'(s_i) - \alpha'(t_i)|| \leq \dfrac{\epsilon}{2}, \end{aligned}$$

where $t_{i-1} \leq s_i \leq t_i$. Together with the above, this gives the required inequality.

SECTION 1-4

2. Let the points $p_0 = (x_0, y_0, z_0)$ and $p = (x, y, z)$ belong to the plane P. Then $ax_0 + by_0 + cz_0 + d = 0 = ax + by + cz + d$. Thus, $a(x - x_0) + b(y - y_0) + c(z - z_0) = 0$. Since the vector $(x - x_0, y - y_0, z - z_0)$ is parallel to P, the vector (a, b, c) is normal to P. Given a point $p = (x, y, z) \in P$, the distance ρ from the plane P to the origin O is given by $\rho = |p| \cos \theta = (p \cdot v)/|v|$, where θ is the angle of Op with the normal vector v. Since $p \cdot v = -d$,

$$\rho = \frac{p \cdot v}{|v|} = -\frac{d}{|v|}.$$

3. This is the angle of their normal vectors.

4. Two planes are parallel if and only if their normal vectors are parallel.

6. v_1 and v_2 are both perpendicular to the line of intersection. Thus, $v_1 \wedge v_2$ is parallel to this line.

7. A plane and a line are parallel when a normal vector to the plane is perpendicular to the direction of the line.

8. The direction of the common perpendicular to the given lines is the direction of $u \wedge v$. The distance between these lines is obtained by projecting the vector $r = (x_0 - x_1, y_0 - y_1, z_0 - z_1)$ onto the common perpendicular. Such a projection is clearly the inner product of r with the unit vector $(u \wedge v)/|u \wedge v|$.

SECTION 1-5

2. Use the fact that $\alpha' = t$, $\alpha'' = kn$, $\alpha''' = kn' + k'n = -k^2 t + k'n - k\tau b$.

4. Differentiate $\alpha(s) + \lambda(s)n(s) = \text{const.}$, obtaining

$$(1 - \lambda k)t + \lambda' n - \lambda \tau b = 0.$$

It follows that $\tau = 0$ (the curve is contained in a plane) and that $\lambda = \text{const.} = 1/k$.

7. a. Parametrize α by arc length.

b. Parametrize α by arc length s. The normal lines at s_1 and s_2 are

$$\beta_1(t) = \alpha(s_1) + tn(s_1), \qquad \beta_2(\tau) = \alpha(s_2) + \tau n(s_2), \qquad t \in R, \tau \in R,$$

respectively. Their point of intersection will be given by values of t and τ such that

$$\frac{\alpha(s_2) - \alpha(s_1)}{s_2 - s_1} = \frac{tn(s_1) - \tau n(s_2)}{s_2 - s_1}.$$

Use Taylor's formula $n(s_2) = n(s_1) + (s_2 - s_1)n'(s_1) + R$, and let $s_2 \longrightarrow s_1$ to conclude that $\alpha'(s_1) = -\bar{t}n'(s_1)$, where $\bar{t}$ is the limiting common value of t and τ as $s_2 \longrightarrow s_1$. It follows that $\bar{t} = 1/k$.

13. To prove that the condition is necessary, differentiate three times $|\alpha(s)|^2 =$ const., obtaining $\alpha(s) = -Rn + R'Tb$. For the sufficiency, differentiate $\beta(s) = \alpha(s) + Rn - R'Tb$, obtaining

$$\beta'(s) = t + R(-kt - \tau b) + R'n - (TR')'b - R'n = -(R\tau + (TR')')b.$$

On the other hand, by differentiating $R^2 + (TR')^2 =$ const., one obtains

$$0 = 2RR' + 2(TR')(TR')' = \frac{2R'}{\tau}(R\tau + (TR')'),$$

since $k' \neq 0$ and $\tau \neq 0$. Hence, $\beta(s)$ is a constant p_0, and

$$|\alpha(s) - p_0|^2 = R^2 + (TR')^2 = \text{const.}$$

15. Since $b' = \tau n$ is known, $|\tau| = |b'|$. Then, up to a sign, n is determined. Since $t = n \wedge b$ and the curvature is positive and given by $t' = kn$, the curvature can also be determined.

16. First show that

$$\frac{n \wedge n' \cdot n''}{|n'|^2} = \frac{\dfrac{k}{\tau}}{\left(\dfrac{k}{\tau}\right)^2 + 1} = a(s).$$

Thus, $\int a(s)\, ds = \arctan (k/\tau)$; hence, k/τ can be determined; since k is positive, this also gives the sign of τ. Furthermore, $|n'|^2 = |-kt - \tau b|^2 = k^2 + \tau^2$ is also known. Together with k/τ, this suffices to determine k^2 and τ^2.

17. a. Let a be the unit vector of the fixed direction and let θ be the constant angle. Then $t \cdot a = \cos\theta =$ const., which differentiated gives $n \cdot a = 0$. Thus, $a = t\cos\theta + b\sin\theta$, which differentiated gives $k\cos\theta + \tau\sin\theta = 0$, or $k/\tau = -\tan\theta =$ const. Conversely, if $k/\tau =$ const. $= -\tan\theta = -(\sin\theta/\cos\theta)$, we can retrace our steps, obtaining that $t\cos\theta + b\sin\theta$ is a constant vector a. Thus, $t \cdot a = \cos\theta =$ const.

b. From the argument of part a, it follows immediately that $t \cdot a =$ const. implies that $n \cdot a = 0$; the last condition means that n is parallel to a plane normal to a. Conversely, if $n \cdot a = 0$, then $(dt/ds) \cdot a = 0$; hence, $t \cdot a =$ const.

c. From the argument of part a, it follows that $t \cdot a =$ const. implies that $b \cdot a =$ const. Conversely, if $b \cdot a =$ const., by differentiation we find that $n \cdot a = 0$.

18. a. Parametrize α by arc length s and differentiate $\bar{\alpha} = \alpha + rn$ with respect to s, obtaining

$$\frac{d\bar{\alpha}}{ds} = (1 - rk)t + r'n - r\tau b.$$

Since $d\bar{\alpha}/ds$ is tangent to $\bar{\alpha}$, $(d\bar{\alpha}/ds)\cdot n = 0$; hence, $r' = 0$.

b. Parametrize α by arc length s, and denote by $\bar{s}$ and $\bar{t}$ the arc length and the unit tangent vector of $\bar{\alpha}$. Since $d\bar{t}/ds = (d\bar{t}/d\bar{s})(d\bar{s}/ds)$, we obtain that

$$\frac{d}{ds}(t \cdot \bar{t}) = t \cdot \frac{d\bar{t}}{ds} + \frac{dt}{ds} \cdot \bar{t} = 0;$$

hence, $t \cdot \bar{t} = \text{const.} = \cos\theta$. Thus, by using that $\bar{\alpha} = \alpha + rn$, we have

$$\cos\theta = \bar{t} \cdot t = \frac{d\bar{\alpha}}{ds}\frac{ds}{d\bar{s}} \cdot t = \frac{ds}{d\bar{s}}(1 - rk),$$

$$|\sin\theta| = |\bar{t} \wedge t| = \left|\frac{ds}{d\bar{s}}((t + rn') \wedge t\right| = \left|\frac{ds}{d\bar{s}}r\tau\right|.$$

From these two relations, it follows that

$$\frac{1 - rk}{r\tau} = \text{const.} = \frac{B}{r}.$$

Thus, setting $r = A$, we finally obtain that $Ak + B\tau = 1$.

Conversely, let this last relation hold, set $A = r$, and define $\bar{\alpha} = \alpha + rn$. Then, by again using the relation, we obtain

$$\frac{d\bar{\alpha}}{ds} = (1 - rk)t - r\tau b = \tau(Bt - rb).$$

Thus, a unit vector $\bar{t}$ of $\bar{\alpha}$ is $(Bt - rb)/\sqrt{B^2 + r^2} = \bar{t}$. It follows that $d\bar{t}/ds = ((Bk - r\tau)/\sqrt{B^2 + r^2})n$. Therefore, $\bar{n}(s) = \pm n(s)$, and the normal lines of $\bar{\alpha}$ and α at s agree. Thus, α is a Bertrand curve.

c. Assume the existence of two distinct Bertrand mates $\bar{\alpha} = \alpha + \bar{r}n$, $\tilde{\alpha} = \alpha + \tilde{r}n$. By part b there exist constants c_1 and c_2 so that $1 - \bar{r}k = c_1(\bar{r}\tau)$, $1 - \tilde{r}k = c_2(\tilde{r}\tau)$. Clearly, $c_1 \neq c_2$. Differentiating these expressions, we obtain $k' = \tau' c_1$, $k' = \tau' c_2$, respectively. This implies that $k' = \tau' = 0$. Using the uniqueness part of the fundamental theorem of the local theory of curves, it is easy to see that the circular helix is the only such curve.

SECTION 1-6

1. Assume that $s = 0$, and consider the canonical form around $s = 0$. By condition 1, P must be of the form $z = cy$, or $y = 0$. The plane $y = 0$ is the rectifying plane, which does not satisfy condition 2. Observe now that if $|s|$ is sufficiently small, $y(s) > 0$, and $z(s)$ has the same sign as s. By condition 2, $c = z/y$ is simultaneously positive and negative. Thus, P is the plane $z = 0$.

2. a. Consider the canonical form of $\alpha(s) = (x(s), y(s), z(s))$ in a neighborhood of $s = 0$. Let $ax + by + cz = 0$ be the plane that passes through $\alpha(0)$, $\alpha(0 + h_1)$, $\alpha(0 + h_2)$. Define a function $F(s) = ax(s) + by(s) + cz(s)$ and notice that $F(0) = F(h_1) = F(h_2) = 0$. Use the canonical form to show that $F'(0) = a$,

$F''(0) = bk$. Use the mean value theorem to show that as $h_1, h_2 \longrightarrow 0$, then $a \longrightarrow 0$ and $b \longrightarrow 0$. Thus, as $h_1, h_2 \longrightarrow 0$ the plane $ax + by = cz = 0$ approaches the plane $z = 0$, that is, the osculating plane.

SECTION 1-7

1. No. Use the isoperimetric inequality.

2. Let S^1 be a circle such that $\overline{AB}$ is a chord of S^1 and one of the two arcs α and β determined by A and B on S^1, say α, has length l. Consider the piecewise C^1 closed curve (see Remark 2 after Theorem 1) formed by β and C. Let β be fixed and C vary in the family of all curves joining A to B with length l. By the isoperimetric inequality for piecewise C^1 curves, the curve of the family that bounds the largest area is S^1. Since β is fixed, the arc of circle α is the solution to our problem.

4. Choose coordinates such that the center O is at p and the x and y axes are directed along the tangent and normal vectors at p, respectively. Parametrize C by arc length, $\alpha(s) = (x(s), y(s))$, and assume that $\alpha(0) = p$. Consider the (finite) Taylor's expansion

$$\alpha(s) = \alpha(0) + \alpha'(0)s + \alpha''(0)\frac{s^2}{2} + R,$$

where $\lim_{s \to 0} R/s^2 = 0$. Let k be the curvature of α at $s = 0$, and obtain

$$x(s) = s + R_x, \qquad y(s) = \pm\frac{ks^2}{2} + R_y,$$

where $R = (R_x, R_y)$ and the sign depends on the orientation of α. Thus,

$$|k| = \lim_{s \to 0} \frac{2|y(s)|}{s^2} = \lim_{d \to 0} \frac{2h}{d^2}.$$

5. Let O be the center of the disk D. Shrink the boundary of D through a family of concentric circles until it meets the curve C at a point p. Use Exercise 4 to show that the curvature k of C at p satisfies $|k| \geq 1/r$.

8. Since α is simple, we have, by the theorem of turning tangents,

$$\int_0^l k(s)\, ds = \theta(l) - \theta(0) = 2\pi.$$

Since $k(s) \leq c$, we obtain

$$2\pi = \int_0^l k(s)\, ds \leq c\int_0^l ds = cl.$$

9. By the Jordan curve theorem, a simple closed curve C bounds a set K. If K is not convex, there are points $p, q \in K$ such that the segment $\overline{pq}$ contains points that do not belong to K, and $\overline{pq}$ meets C at a point r, $r \neq p, q$. Use the argument given in the middle of the proof of the four-vertex theorem to show that the line L determined by p and q is tangent to C at the points p, q, r and that the segment $\overline{pq}$ is contained in $C \subset K$. This is a contradiction.

11. Observe that the area bounded by H is greater than or equal to the area bounded by C and that the length of H is smaller than or equal to the length of C. Expand H through a family of curves parallel to H (Exercise 6) until its length reaches the length of C. Since the area either remains the same or has been further increased in this process, we obtain a convex curve H' with the same length as C but bounding an area greater than or equal to the area of C.

12.

$$M_1 = \int_0^{2\pi} \left(\int_0^{1/3} dp \right) d\theta = \frac{2\pi}{3},$$

$$M_2 = \int_0^{2\pi} \left(\int_0^{1} dp \right) d\theta = 2\pi.$$

(See Fig. 1-40.)

SECTION 2-2

5. No. $\mathbf{x}$ is not one-to-one.

11. b. To see that $\mathbf{x}$ is one-to-one, observe that from z one obtains $\pm u$. Since $\cosh v > 0$, the sign of u is the same as the sign of x. Thus, $\sinh v$ (and hence v) is determined.

13. $\mathbf{x}(u, v) = (\sinh u \cos v, \sinh u \sin v, \cosh v)$.

15. Eliminate t in the equations $x/a = y/t = -(z - t)/t$ of the line joining $p(t) = (0, 0, t)$ to $q(t) = (a, t, 0)$.

17. c. Extend Prop. 3 for plane curves and apply the argument of Example 5.

18. For the first part, use the inverse function theorem. To determine F, set $u = \rho^2$, $v = \tan \varphi$, $w = \tan^2 \theta$. Write $x = f(\rho, \theta) \cos \varphi$, $y = f(\rho, \theta) \sin \varphi$, where f is to be determined. Then

$$x^2 + y^2 + z^2 = f^2 + z^2 = \rho^2, \qquad \frac{f^2}{z^2} = \tan^2 \theta.$$

It follows that $f = \rho \cos \theta$, $z = \rho \sin \theta$. Therefore,

$$F(u, v, w) = \left(\frac{\sqrt{u}}{\sqrt{(1 + w)(1 + v^2)}}, \frac{v\sqrt{u}}{\sqrt{(1 + w)(1 + v^2)}}, \frac{\sqrt{uw}}{\sqrt{1 + w}} \right).$$

19. No. For C, observe that no neighborhood in R^2 of a point in the vertical arc can be written as the graph of a differentiable function. The same argument applies to S.

SECTION 2-3

1. Since $A^2 =$ identity, $A = A^{-1}$.

5. d is the restriction to S of a function $d\colon R^3 \longrightarrow R$:

$$d(x, y, z) = \{(x - x_0)^2 + (y - y_0)^2 + (z - z_0)^2\}^{1/2}, \qquad (x, y, z) \neq (x_0, y_0, z_0).$$

8. If $p = (x, y, z)$, $F(p)$ lies in the intersection with H of the line $t \longrightarrow (tx, ty, z)$, $t > 0$. Thus,

$$F(p) = \left(\frac{\sqrt{1 + z^2}}{\sqrt{x^2 + y^2}}x, \frac{\sqrt{1 + z^2}}{\sqrt{x^2 + y^2}}y, z\right).$$

Let U be R^3 minus the z axis. Then $F\colon U \subset R^3 \longrightarrow R^3$ as defined above is differentiable.

13. If f is such a restriction, f is differentiable (Example 1). To prove the converse, let $\mathbf{x}\colon U \longrightarrow R^3$ be a parametrization of S in p. As in Prop. 1, extend $\mathbf{x}$ to $F\colon U \times R \longrightarrow R^3$. Let W be a neighborhood of p in R^3 on which F^{-1} is a diffeomorphism. Define $g\colon W \longrightarrow R$ by $g(q) = f \circ \mathbf{x} \circ \pi \circ F^{-1}(q)$, $q \in W$, where $\pi\colon U \times R \longrightarrow U$ is the natural projection. Then g is differentiable, and the restriction $g \mid W \cap S = f$.

16. F is differentiable in $S^2 - \{N\}$ as a composition of differentiable maps. To prove that F is differentiable at N, consider the stereographic projection π_S from the south pole $S = (0, 0, -1)$ and set $Q = \pi_S \circ F \circ \pi_S^{-1}\colon \mathbb{C} \longrightarrow \mathbb{C}$ (of course, we are identifying the plane $z = 1$ with $\mathbb{C}$). Show that $\pi_N \circ \pi_S^{-1}\colon \mathbb{C} - \{0\} \longrightarrow \mathbb{C}$ is given by $\pi_N \circ \pi_S^{-1}(\zeta) = 4/\bar{\zeta}$. Conclude that

$$Q(\zeta) = \frac{\zeta^n}{\bar{a}_0 + \bar{a}_1\zeta + \cdots + \bar{a}_n\zeta^n};$$

hence, Q is differentiable at $\zeta = 0$. Thus, $F = \pi_S^{-1} \circ Q \circ \pi_S$ is differentiable at N.

SECTION 2-4

1. Let $\alpha(t) = (x(t), y(t), z(t))$ be a curve on the surface passing through $p_0 = (x_0, y_0, z_0)$ for $t = 0$. Thus, $f(x(t), y(t), z(t)) = 0$; hence, $f_x x'(0) + f_y y'(0) + f_z z'(0) = 0$, where all derivatives are computed at p_0. This means that all tangent vectors at p_0 are perpendicular to the vector (f_x, f_y, f_z), and hence the desired equation.

4. Denote by f' the derivative of $f(y/x)$ with respect to $t = y/x$. Then $z_x = f - (y/x)f'$, $z_y = f'$. Thus, the equation of the tangent plane at (x_0, y_0) is $z = x_0 f + (f - (y_0/x_0)f')(x - x_0) + f'(y - y_0)$, where the functions are computed at (x_0, y_0). It follows that if $x = 0$, $y = 0$, then $z = 0$.

12. For the orthogonality, consider, for instance, the first two surfaces. Their normals are parallel to the vectors $(2x - a, 2y, 2z)$, $(2x, 2y - b, 2z)$. In the intersection of these surfaces, $ax = by$; introduce this relation in the inner product of the above vectors to show that this inner product is zero.

13. a. Let $\alpha(t)$ be a curve on S with $\alpha(0) = p$, $\alpha'(0) = w$. Then

$$df_p(w) = \frac{d}{dt}(\langle \alpha(t) - p_0, \alpha(t) - p_0 \rangle^{1/2})|_{t=0} = \frac{\langle w, p - p_0 \rangle}{|p - p_0|}.$$

It follows that p is a critical point of f if and only if $\langle w, p - p_0 \rangle = 0$ for all $w \in T_p(S)$.

14. a. $f(t)$ is continuous in the interval $(-\infty, c)$, and $\lim_{t \to -\infty} f(t) = 0$, $\lim_{t \to c, t<c} f(t) = +\infty$. Thus, for some $t_1 \in (-\infty, c)$, $f(t_1) = 1$. By similar arguments, we find real roots $t_2 \in (c, b)$, $t_3 \in (b, a)$.

b. The condition for the surfaces $f(t_1) = 1$, $f(t_2) = 1$ to be orthogonal is

$$f_x(t_1)f_x(t_2) + f_y(t_1)f_y(t_2) + f_z(t_1)f_z(t_2) = 0.$$

This reduces to

$$\frac{x^2}{(a - t_1)(a - t_2)} + \frac{y^2}{(b - t_1)(b - t_2)} + \frac{z^2}{(c - t_1)(c - t_2)} = 0,$$

which follows from the fact that $t_1 \neq t_2$ and $f(t_1) - f(t_2) = 0$.

17. Since every surface is locally the graph of a differentiable function, S_1 is given by $f(x, y, z) = 0$ and S_2 by $g(x, y, z) = 0$ in a neighborhood of p; here 0 is a regular value of the differentiable functions f and g. In this neighborhood of p, $S_1 \cap S_2$ is given as the inverse image of $(0, 0)$ of the map $F: R^3 \to R^2: F(q) = (f(q), g(q))$. Since S_1 and S_2 intersect transversally, the normal vectors (f_x, f_y, f_z) and (g_x, g_y, g_z) are linearly independent. Thus, $(0, 0)$ is a regular value of F and $S_1 \cap S_2$ is a regular curve (cf. Exercise 17, Sec. 2-2).

20. The equation of the tangent plane at (x_0, y_0, z_0) is

$$\frac{xx_0}{a^2} + \frac{yy_0}{b^2} + \frac{zz_0}{c^2} = 1.$$

The line through O and perpendicular to the tangent plane is given by

$$\frac{xa^2}{x_0} = \frac{yb^2}{y_0} = \frac{zc^2}{z_0}.$$

From the last expression, we obtain

$$\frac{x^2a^2}{xx_0} = \frac{y^2b^2}{yy_0} = \frac{z^2c^2}{zz_0} = \frac{a^2x^2 + b^2y^2 + c^2z^2}{xx_0 + yy_0 + zz_0}.$$

From the same expression, and taking into account the equation of the ellipsoid, we obtain

$$\frac{xx_0}{x_0^2/a^2} = \frac{yy_0}{y_0^2/b^2} = \frac{zz_0}{z_0^2/c^2} = \frac{xx_0 + yy_0 + zz_0}{1}.$$

Again from the same expression and using the equation of the tangent plane, we obtain

$$\frac{x^2}{(x_0x)/a^2} = \frac{y^2}{(y_0y)/b^2} = \frac{z^2}{(z_0z)/c^2} = \frac{x^2 + y^2 + z^2}{1}.$$

The right-hand sides of the three last equations are therefore equal, and hence the asserted equation.

21. Imitate the proof of Prop. 9 of the appendix to Chap. 2.

22. Let r be the fixed line which is met by the normals of S and let $p \in S$. The plane P_1, which contains p and r, contains all the normals to S at the points of $P_1 \cap S$. Consider a plane P_2 passing through p and perpendicular to r. Since the normal through p meets r, P_2 is transversal to $T_p(S)$; hence, $P_2 \cap S$ is a regular plane curve C in a neighborhood of p (cf. Exercise 17, Sec. 2-4). Furthermore $P_1 \cap P_2$ is perpendicular to $T_p(S) \cap P_2$; hence, $P_1 \cap P_2$ is normal to C. It follows that the normals of C all pass through a fixed point $q = r \cap P_2$; hence, C is contained in a circle (cf. Exercise 4, Sec. 1-5). Thus, every $p \in S$ has a neighborhood contained in some surface of revolution with axis r. By connectedness, S is contained in a fixed one of these surfaces.

SECTION 2-5

8. Since $\partial E/\partial v = 0$, $E = E(u)$ is a function of u alone. Set $\bar{u} = \int \sqrt{E}\, du$. Similarly, $G = G(v)$ is a function of v alone, and we can set $\bar{v} = \int \sqrt{G}\, dv$. Thus, $\bar{u}$ and $\bar{v}$ measure arc lengths along the coordinate curves, whence $\bar{E} = \bar{G} = 1$, $\bar{F} = \cos\theta$.

9. Parametrize the generating curve by arc length.

SECTION 3-2

13. Since the osculating plane is normal to N, $N' = \tau n$ and, therefore, $\tau^2 = |N'|^2 = k_1^2 \cos^2\theta + k_2^2 \sin^2\theta$, where θ is the angle of e_1 with the tangent to the curve. Since the direction is asymptotic, we obtain $\cos^2\theta$ and $\sin^2\theta$ as functions of k_1 and k_2, which substituted in the expression above yields $\tau^2 = -k_1k_2$.

14. By setting $\boldsymbol{\lambda}_1 = \lambda_1 N_2$ and $\boldsymbol{\lambda}_2 = \lambda_2 N_1$ we have that

$$\begin{aligned} |\boldsymbol{\lambda}_1 - \boldsymbol{\lambda}_2| &= k\,|\langle n, N_1\rangle N_2 - \langle n, N_2\rangle N_1| \\ &= \sqrt{\lambda_1^2 + \lambda_2^2 - 2\lambda_1\lambda_2\cos\theta}. \end{aligned}$$

On the other hand,

$$|\sin \theta| = |N_1 \wedge N_2| = |n \wedge (N_1 \wedge N_2)|$$
$$= |\langle n, N_2\rangle N_1 - \langle n, N_1\rangle N_2|.$$

16. Intersect the torus by a plane containing its axis and use Exercise 15.

18. Use the fact that if $\theta = 2\pi/m$, then

$$\sigma(\theta) = 1 + \cos^2\theta + \cdots + \cos^2(m-1)\theta = \frac{m}{2},$$

which may be proved by observing that

$$\sigma(\theta) = \frac{1}{4}\left(\sum_{\nu=-(m-1)}^{\nu=m-1} e^{2\nu i\theta} + 2m + 1\right)$$

and that the expression under the summation sign is the sum of a geometric progression, which yields

$$\frac{\sin(2m\theta - \theta)}{\sin\theta} = -1.$$

19. a. Express t and h in the basis $\{e_1, e_2\}$ given by the principal directions, and compute $\langle dN(t), h\rangle$.

b. Differentiate $\cos\theta = \langle N, n\rangle$, use that $dN(t) = -k_n t + \tau_g h$, and observe that $\langle N, b\rangle = \langle h, N\rangle = \sin\theta$, where b is the binormal vector.

20. Let S_1, S_2, and S_3 be the surfaces that pass through p. Show that the geodesic torsions of $C_1 = S_2 \cap S_3$ relative to S_2 and S_3 are equal; it will be denoted by τ_1. Similarly, τ_2 denotes the geodesic torsion of $C_2 = S_1 \cap S_3$ and τ_3 that of $S_1 \cap S_2$. Use the definition of τ_g to show that, since C_1, C_2, C_3 are pairwise orthogonal, $\tau_1 + \tau_2 = 0$, $\tau_2 + \tau_3 = 0$, $\tau_3 + \tau_1 = 0$. It follows that $\tau_1 = \tau_2 = \tau_3 = 0$.

SECTION 3-3

2. Asymptotic curves: $u =$ const., $v =$ const. Lines of curvature:

$$\log(v + \sqrt{v^2 + c^2}) \pm u = \text{const.}$$

3. $u + v =$ const. $u - v =$ const.

6. a. By taking the line r as the z axis and a normal to r as the x axis, we have that

$$z' = \frac{\sqrt{1 - x^2}}{x}.$$

By setting $x = \sin\theta$, we obtain

$$z(\theta) = \int \frac{\cos^2\theta}{\sin\theta}\,d\theta = \log\tan\frac{\theta}{2} + \cos\theta + C.$$

If $z(\pi/2) = 0$, then $C = 0$.

8. a. The assertion is clearly true if $\mathbf{x} = \mathbf{x}_1$ and $\bar{\mathbf{x}} = \bar{\mathbf{x}}_1$ are parametrizations that satisfy the definition of contact. If $\mathbf{x}$ and $\bar{\mathbf{x}}$ are arbitrary, observe that $\mathbf{x} = \mathbf{x}_1 \circ h$, where h is the change of coordinates. It follows that the partial derivatives of $f \circ \mathbf{x} = f \circ \mathbf{x}_1 \circ h$ are linear combinations of the partial derivatives of $f \circ \mathbf{x}_1$. Therefore, they become zero with the latter ones.

b. Introduce parametrizations $\mathbf{x}(x, y) = (x, y, f(x, y))$ and $\bar{\mathbf{x}}(x, y) = (x, y, \bar{f}(x, y))$, and define a function $h(x, y, z) = f(x, y) - z$. Observe that $h \circ \mathbf{x} = 0$ and $h \circ \bar{\mathbf{x}} = f - \bar{f}$. It follows from part a, applied the function h, that $f - \bar{f}$ has partial derivatives of order ≤ 2 equal to zero at $(0, 0)$.

d. Since contact of order ≥ 2 implies contact of order ≥ 1, the paraboloid passes through p and is tangent to the surface at p. By taking the plane $T_p(S)$ as the xy plane, the equation of the paraboloid becomes

$$\bar{f}(x, y) = ax^2 + 2bxy + cy^2 + dx + ey.$$

Let $z = f(x, y)$ be the representation of the surface in the plane $T_p(S)$. By using part b, we obtain that $d = c = 0$, $a = \frac{1}{2}f_{xx}$, $b = f_{xy}$, $c = \frac{1}{2}f_{yy}$.

15. If there exists such an example, it may locally be written in the form $z = f(x, y)$, with $f(0, 0) = 0$, $f_x(0, 0) = f_y(0, 0) = 0$. The given conditions require that $f_{xx}^2 + f_{yy}^2 \neq 0$ at $(0, 0)$ and that $f_{xx}f_{yy} - f_{xy}^2 = 0$ if and only if $(x, y) = (0, 0)$.

By setting, tentatively, $f(x, y) = \alpha(x) + \beta(y) + xy$, where $\alpha(x)$ is a function of x alone and $\beta(y)$ is a function of y alone, we verify that $\alpha_{xx} = \cos x$, $\beta_{yy} = \cos y$ satisfy the conditions above. It follows that

$$f(x, y) = \cos x + \cos y + xy - 2$$

is such an example.

16. Take a sphere containing the surface and decrease its radius continuously. Study the normal sections at the point (or points) where the sphere meets the surface for the first time.

19. Show that the hyperboloid contains two one-parameter families of lines which are necessarily the asymptotic lines. To find such families of lines, write the equation of the hyperboloid as

$$(x + z)(x - z) = (1 - y)(1 + y)$$

and show that, for each $k \neq 0$, the line $x + z = k(1 + y)$, $x - z = (1/k)(1 - y)$ belongs to the surface.

20. Observe that $(x/a^2, y/b^2, z/c^2) = fN$ for some function f and that an umbilical point satisfies the equation

$$\left\langle \frac{d(fN)}{dt} \wedge \frac{d\alpha}{dt}, N \right\rangle = 0$$

for every curve $\alpha(t) = (x(t), y(t), z(t))$ on the surface. Assume that $z \neq 0$, multiply this equation by z/c^2, and eliminate z and dz/dt (observe that the equation holds for every tangent vector on the surface). Four umbilical points are found, namely,

$$y = 0, \qquad x^2 = a^2\frac{a^2 - b^2}{a^2 - c^2}, \qquad z^2 = c^2\frac{b^2 - c^2}{a^2 - c^2}.$$

The hypothesis $z = 0$ does not yield any further umbilical points.

21. a. Let $dN(v_1) = av_1 + bv_2$, $dN(v_2) = cv_1 + dv_2$. A direct computation yields

$$\langle d(fN)(v_1) \wedge d(fN)(v_2), fN \rangle = f^3 \det(dN).$$

b. Show that $fN = (x/a^2, y/b^2, z/c^2) = W$, and observe that

$$d(fN)(v_i) = \left(\frac{\alpha_i}{a^2}, \frac{\beta_i}{b^2}, \frac{\gamma_i}{c^2}\right), \qquad \text{where } v_i = (\alpha_i, \beta_i, \gamma_i),$$

$i = 1, 2$. By choosing v_i so that $v_1 \wedge v_2 = N$, conclude that

$$\langle d(fN)(v_1) \wedge df(N)(v_2), fN \rangle = \frac{\langle W, X \rangle}{a^2b^2c^2}\frac{1}{f},$$

where $X = (x, y, z)$, and therefore $\langle W, X \rangle = 1$.

24. d. Choose a coordinate system in R^3 so that the origin O is at $p \in S$, the xy plane agrees with $T_p(S)$, and the positive direction of the z axis agrees with the orientation of S at p. Furthermore, choose the x and y axes in $T_p(S)$ along the principal directions at p. If V is sufficiently small, it can then be represented as the graph of a differentiable function

$$z = f(x, y), \qquad (x, y) \in D \subset R^2,$$

where D is an open disk in R^2 and

$$f_x(0, 0) = f_y(0, 0) = f_{xy}(0, 0) = 0, \qquad f_{xx}(0, 0) = k_1, \qquad f_{yy}(0, 0) = k_2.$$

We can assume, without loss of generality, that $k_1 \geq 0$ and $k_2 \geq 0$ on D, and we want to prove that $f(x, y) \geq 0$ on D.

Assume that, for some $(\bar{x}, \bar{y}) \in D$, $f(\bar{x}, \bar{y}) < 0$. Consider the function $h_0(t) = f(t\bar{x}, t\bar{y})$, $0 \leq t \leq 1$. Since $h_0'(0) = 0$, there exists a t_1, $0 \leq t_1 \leq 1$, such that $h_0''(t_1) < 0$, Let $p_1 = (t_1\bar{x}, t_1\bar{y}, f(t_1\bar{x}, t_1\bar{y})) \in S$, and consider the height function h_1 of V relative to the tangent plane $T_{p_1}(S)$ at p_1. Restricted to the curve $\alpha(t)=(t\bar{x}, t\bar{y}, f(t\bar{x}, t\bar{y}))$, this height function is $h_1(t)=\langle\alpha(t)-p_1, N_1\rangle$, where N_1 is the unit normal vector at p_1. Thus, $h_1''(t) = \langle\alpha''(t), N_1\rangle$, and, at $t = t_1$,

$$h_1''(t_1) = \langle(0, 0, h_0''(t_1)), (-f_x(p_1), -f_y(p_1), 1)\rangle = h_0''(t_1) < 0.$$

But $h_1''(t_1) = \langle\alpha''(t_1), N_1\rangle$ is, up to a positive factor, the normal curvature at p in the direction of $\dot{\alpha}'(t_1)$. This is a contradiction.

SECTION 3-4

10. c. Reduce the problem to the fact that if λ is an irrational number and m and n run through the integers, the set $\{\lambda m + n\}$ is dense in the real line. To prove the last assertion, it suffices to show that the set $\{\lambda m + n\}$ has arbitrarily small positive elements. Assume the contrary, show that the greatest lower bound of the positive elements of $\{\lambda m + n\}$ still belongs to that set, and obtain a contradiction.

11. Consider the set $\{\alpha_i : I_i \longrightarrow I\}$ of trajectories of w, with $\alpha_i(0) = p$, and set $I = \bigcup_i I_i$. By uniqueness, the maximal trajectory $\alpha : I \longrightarrow U$ may be defined by setting $\alpha(t) = \alpha_i(t)$, where $t \in I_i$.

12. For every $q \in S$, there exist a neighborhood U of q and an interval $(-\epsilon, \epsilon)$, $\epsilon > 0$, such that the trajectory $\alpha(t)$, with $\alpha(0) = q$, is defined in $(-\epsilon, \epsilon)$. By compactness, it is possible to cover S with a finite number of such neighborhoods. Let ϵ_0 = minimum of the corresponding ϵ's. If $\alpha(t)$ is defined for $t < t_0$ and is not defined for t_0, take $t_1 \in (0, t_0)$, with $|t_0 - t_1| < \epsilon_0/2$. Consider the trajectory $\beta(t)$ of w, with $\beta(t_1) = \alpha(t_1)$, and obtain a contradiction.

SECTION 4-2

3. The "only if" part is immediate. To prove the "if" part, let $p \in S$ and $v \in T_p(S)$, $v \neq 0$. Consider a curve $\alpha : (-\epsilon, \epsilon) \longrightarrow S$, with $\alpha'(0) = v$. We claim that $|d\varphi_p(\alpha'(0))| = |\alpha'(0)|$. Otherwise, say, $|d\varphi_p(\alpha'(0))| > |\alpha'(0)|$, and in a neighborhood J of 0 in $(-\epsilon, \epsilon)$, we have $|d\varphi_p(\alpha'(t))| > |\alpha'(t)|$. This implies that the length of $\alpha(J)$ is greater than the length of $\varphi \circ \alpha(J)$, a contradiction.

6. Parametrize α by arc length s in a neighborhood of t_0. Construct in the plane a curve with curvature $k = k(s)$ and apply Exercise 5.

8. Set $0 = (0, 0, 0)$, $G(0) = p_0$, and $G(p) - p_0 = F(p)$. Then $F: R^3 \longrightarrow R^3$ is a map such that $F(0) = 0$ and $|F(p)| = |G(p) - G(0)| = |p|$. This implies that F preserves the inner product of R^3. Thus, it maps the basis

$$\{(1, 0, 0) = f_1, (0, 1, 0) = f_2, (0, 0, 1) = f_3\}$$

onto an orthonormal basis, and if $p = \sum a_i f_i$, $i = 1, 2, 3$, then $F(p) = \sum a_i F(f_i)$. Therefore, F is linear.

11. a. Since F is distance-preserving and the arc length of a differentiable curve is the limit of the lengths of inscribed polygons, the restriction $F|S$ preserves the arc length of a curve in S.

c. Consider the isometry of an open strip of the plane onto a cylinder minus a generator.

12. The restriction of $F(x, y, z) = (x, -y, -z)$ to C is an isometry of C (cf. Exercise 11), the fixed points of which are $(1, 0, 0)$ and $(-1, 0, 0)$.

17. The loxodromes make a constant angle with the meridians of the sphere. Under Mercator's projection (see Exercise 16) the meridians go into parallel straight lines in the plane. Since Mercator's projection is conformal, the loxodromes also go into straight lines. Thus, the sum of the interior angles of the triangle in the sphere is the same as the sum of the interior angles of a rectilinear plane triangle.

SECTION 4-4

6. Use the fact that the absolute value of the geodesic curvature is the absolute value of the projection onto the tangent plane of the usual curvature.

8. Use Exercise 1, part b, and Prop. 5 of Sec. 3-2.

9. Use the fact that the meridians are geodesics and that the parallel transport preserves angles.

10. Apply the relation $k_g^2 + k_n^2 = k^2$ and the Meusnier theorem to the projecting cylinder.

12. Parametrize a neighborhood of $p \in S$ in such a way that the two families of geodesics are coordinate curves (Corollary 1, Sec. 3-4). Show that this implies that $F = 0$, $E_v = 0 = G_u$. Make a change of parameters to obtain that $\bar{F} = 0$, $\bar{E} = \bar{G} = 1$.

13. Fix two orthogonal unit vectors $v(p)$ and $w(p)$ in $T_p(S)$ and parallel transport them to each point of V. Two differentiable, orthogonal, unit vector fields are thus obtained. Parametrize V in such a way that the directions of these vectors are tangent to the coordinate curves, which are then geodesics. Apply Exercise 12.

16. Parametrize a neighborhood of $p \in S$ in such a way that the lines of curvature are the coordinate curves and that $v =$ const. are the asymptotic curves. It follows that $e_v = 0$, and from the Mainardi-Codazzi equations, we conclude that $E_v = 0$. This implies that the geodesic curvature of $v =$ const. is zero. For the example, look at the upper parallel of the torus.

18. Use Clairaut's relation (cf. Example 5).

19. Substitute in Eq. (4) the Christoffel symbols by their values as functions of E, F, and G and differentiate the expression of the first fundamental form:

$$1 = E(u')^2 + 2Fu'v' + G(v')^2.$$

20. Use Clairaut's relation.

SECTION 4-5

4. b. Observe that the map $x = \bar{x}$, $y = (\bar{y})^2$, $z = (\bar{z})^3$ gives a homeomorphism of the sphere $x^2 + y^2 + z^2 = 1$ onto the surface $(\bar{x})^2 + (\bar{y})^4 + (\bar{z})^6 = 1$.

6. a. Restrict v to the curve $\alpha(t) = (\cos t, \sin t)$, $t \in [0, 2\pi]$. The angle that $v(t)$ forms with the x axis is t. Thus, $2\pi I = 2\pi$; hence, $I = 1$.

d. By restricting v to the curve $\alpha(t) = (\cos t, \sin t)$, $t \in [0, 2\pi]$, we obtain $v(t) = (\cos^2 t - \sin^2 t, -2 \cos t \sin t) = (\cos 2t, -\sin 2t)$. Thus, $I = -2$.

SECTION 4-6

8. Let (ρ, θ) be a system of geodesic polar coordinates such that its pole is one of the vertices of Δ and one of the sides of Δ corresponds to $\theta = 0$. Let the two other sides be given by $\theta = \theta_0$ and $\rho = h(\theta)$. Since the vertex that corresponds to the pole does not belong to the coordinate neighborhood, take a small circle of radius ϵ around the pole. Then

$$\iint_{\Delta} K\sqrt{G}\, d\rho\, d\theta = \int_0^{\theta_0} d\theta \left(\lim_{\epsilon \to 0} \int_{\epsilon}^{h(\theta)} K\sqrt{G}\, d\rho \right) \cdot$$

Observing that $K\sqrt{G} = -(\sqrt{G})_{\rho\rho}$ and that $\lim_{\epsilon \to 0} (\sqrt{G})_\rho = 1$, we have that the limit enclosed in parentheses is given by

$$1 - \frac{\partial(\sqrt{G})}{\partial \rho}(h(\theta), \theta).$$

By using Exercise 7, we obtain

$$\iint_{\Delta} K\sqrt{G}\, d\rho\, d\theta = \int_0^{\theta_0} d\theta - \int_0^{\theta_0} d\varphi = \alpha_3 - (\pi - \alpha_2 - \alpha_1) = \sum_1^3 \alpha_i - \pi.$$

12. c. For $K \equiv 0$, the problem is trivial. For $K > 0$, use part b. For $K < 0$, consider a coordinate neighborhood V of the pseudosphere (cf. Exercise 6, part b, Sec. 3-3), parametrized by polar coordinates (ρ, θ); that is, $E = 1$, $F = 0$, $G = \sinh^2 \rho$. Compute the geodesics of V; it is convenient to use the change of coordinates $\tanh \rho = 1/w$, $\rho \neq 0$, $\theta = \theta$, so that

$$E = \frac{1}{(w^2 - 1)^2}, \qquad G = \frac{1}{w^2 - 1}, \qquad F = 0,$$

$$\Gamma^1_{11} = -\frac{2w}{w^2 - 1}, \qquad \Gamma^1_{12} = -\frac{w}{w^2 - 1}, \qquad \Gamma^1_{22} = w,$$

and the other Christoffel symbols are zero. It follows that the nonradial geodesics satisfy the equation $(d^2w/d\theta^2) + w = 0$, where $w = w(\theta)$. Thus,

$w = A\cos\theta + B\sin\theta$; that is

$$A\tanh\rho\cos\theta + B\tanh\rho\sin\theta = 1.$$

Therefore, the map of V into R^2 given by

$$\xi = \tanh\rho\cos\theta, \qquad \eta = \tanh\rho\sin\theta,$$

$(\xi, \eta) \in R^2$, is a geodesic mapping.

13. b. Define $\mathbf{x} = \varphi^{-1}: \varphi(U) \subset R^2 \longrightarrow S$. Let $v = v(u)$ be a geodesic in U. Since φ is a geodesic mapping and the geodesics of R^2 are lines, then $d^2v/du^2 \equiv 0$. By bringing this condition into part a, the required result is obtained.

c. Equation (a) is obtained from Eq. (5) of Sec. 4-3 using part b. From Eq. (5a) of Sec. 4-3 together with part b we have

$$KF = (\Gamma^1_{12})_u - 2(\Gamma^2_{12})_v + \Gamma^2_{12}\Gamma^1_{12}.$$

By interchanging u and v in the expression above and subtracting the results, we obtain $(\Gamma^1_{12})_u = (\Gamma^2_{12})_v$, whence Eq. (b). Finally, Eqs. (c) and (d) are obtained from Eqs. (a) and (b), respectively, by interchanging u and v.

d. By differentiating Eq. (a) with respect to v, Eq. (b) with respect to u, and subtracting the results, we obtain

$$EK_v - FK_u = -K(E_v - F_u) + K(-F\Gamma^2_{12} + E\Gamma^1_{12}).$$

By taking into account the values of Γ^k_{ij}, the expression above yields

$$EK_v - FK_u = -K(E_v - F_u) + K(E_v - F_u) = 0.$$

Similarly, from Eqs. (c) and (d) we obtain $FK_v - GK_u = 0$, whence $K_v = K_u = 0$.

SECTION 4-7

1. Consider an orthonormal basis $\{e_1, e_2\}$ at $T_{\alpha(0)}(S)$ and take the parallel transport of e_1 and e_2 along α, obtaining an orthonormal basis $\{e_1(t), e_2(t)\}$ at each $T_{\alpha(t)}(S)$. Set $w(\alpha(t)) = w_1(t)e_1(t) + w_2(t)e_2(t)$. Then $D_y w = w_1'(0)e_1 + w_2'(0)e_2$ and the second member is the velocity of the curve $w_1(t)e_1 + w_2(t)e_2$ in $T_p(S)$ at $t = 0$.

2. b. Show that if $(t_1, t_2) \subset I$ is small and does not contain "break points of α," then the tangent vector field of $\alpha((t_1, t_2))$ can be extended to a vector field y in a neighborhood of $\alpha((t_1, t_2))$. Thus, by restricting v and w to α, property 3 becomes

$$\frac{d}{dt}\langle v(t), w(t)\rangle = \left\langle \frac{Dv}{dt}, w\right\rangle + \left\langle v, \frac{Dw}{dt}\right\rangle,$$

which implies that parallel transport in $\alpha \mid (t_1, t_2)$ is an isometry. By compactness, this can be extended to the entire I. Conversely, assume that parallel transport is an isometry. Let α be the trajectory of y through a point $p \in S$. Restrict v and w to α. Choose orthonormal basis $\{e_1(t), e_2(t)\}$ as in the solution of Exercise 1, and set $v(t) = v_1 e_1 + v_2 e_2$, $w(t) = w_1 e_1 + w_2 e_2$. Then property 3 becomes the "product rule":

$$\frac{d}{dt}\Big(\sum_i v_i w_i\Big) = \sum_i \frac{dv_i}{dt} w_i + \sum_i v_i \frac{dw_i}{dt}, \qquad i = 1, 2.$$

c. Let D be given and choose an orthogonal parametrization $\mathbf{x}(u, v)$. Let $y = y_1 \mathbf{x}_u + y_2 \mathbf{x}_v$, $w = w_1 \mathbf{x}_u + w_2 \mathbf{x}_v$. From properties 1, 2, and 3, it follows that $D_y w$ is determined by the knowledge of $D_{\mathbf{x}_u}\mathbf{x}_u$, $D_{\mathbf{x}_u}\mathbf{x}_v$, $D_{\mathbf{x}_v}\mathbf{x}_v$. Set $D_{\mathbf{x}_u}\mathbf{x}_u = A_{11}^1 \mathbf{x}_u + A_{11}^2 \mathbf{x}_v$, $D_{\mathbf{x}_u}\mathbf{x}_v = A_{12}^1 \mathbf{x}_u + A_{12}^2 \mathbf{x}_v$, $D_{\mathbf{x}_v}\mathbf{x}_v = A_{22}^1 \mathbf{x}_u + A_{22}^2 \mathbf{x}_v$. From property 3 it follows that the A_{ij}^k satisfy the same equations as the Γ_{ij}^k (cf. Eq. (2), Sec. 4-3). Thus, $A_{ij}^k = \Gamma_{ij}^k$, which proves that $D_y v$ agrees with the operation "Take the usual derivative and project it onto the tangent plane."

3. a. Observe that

$$d\mathbf{x}_{(0,t)}(1, 0) = \left(\frac{\partial \mathbf{x}}{\partial s}\right)_{s=0} = \frac{d}{ds}\gamma(s, \alpha(t), v(t))\Big|_{s=0} = v(t),$$

$$d\mathbf{x}_{(0,t)}(0, 1) = \left(\frac{\partial \mathbf{x}}{\partial t}\right)_{s=0} = \alpha'(t).$$

b. Use the fact that $\mathbf{x}$ is a local diffeomorphism to cover the compact set I with a family of open intervals in which $\mathbf{x}$ is one-to-one. Use the Heine-Borel theorem and the Lebesgue number of the covering (cf. Sec. 2-7) to globalize the result.

c. To show that $F = 0$, we compute (cf. property 4 of Exercise 2)

$$\frac{d}{ds}F = \frac{d}{ds}\left\langle \frac{\partial \mathbf{x}}{\partial s}, \frac{\partial \mathbf{x}}{\partial t}\right\rangle = \left\langle \frac{D}{\partial s}\frac{\partial \mathbf{x}}{\partial s}, \frac{\partial \mathbf{x}}{\partial t}\right\rangle + \left\langle \frac{\partial \mathbf{x}}{\partial s}, \frac{D}{\partial s}\frac{\partial \mathbf{x}}{\partial t}\right\rangle = \left\langle \frac{\partial \mathbf{x}}{\partial s}, \frac{D}{\partial t}\frac{\partial \mathbf{x}}{\partial s}\right\rangle,$$

because the vector field $\partial \mathbf{x}/\partial s$ is parallel along $t = \text{const}$. Since

$$0 = \frac{d}{dt}\left\langle \frac{\partial \mathbf{x}}{\partial s}, \frac{\partial \mathbf{x}}{\partial s}\right\rangle = 2\left\langle \frac{D}{\partial t}\frac{\partial \mathbf{x}}{\partial s}, \frac{\partial \mathbf{x}}{\partial s}\right\rangle,$$

F does not depend on s. Since $F(0, t) = 0$, we have $F = 0$.

d. This is a consequence of the fact that $F = 0$.

4. a. Use Schwarz's inequality,

$$\left(\int_a^b fg\, dt\right)^2 \leq \int_a^b f^2\, dt \int_a^b g^2\, dt,$$

with $f \equiv 1$ and $g = |d\alpha/dt|$.

5. a. By noticing that $E(t) = \int_0^l \{(\partial u/\partial v)^2 + G(\gamma(v, t), v)\}\, dv$, we obtain (we write $\gamma(v, t) = u(v, t)$, for convenience)

$$E'(t) = \int_0^l \left\{ 2 \frac{\partial u}{\partial v} \frac{\partial^2 u}{\partial v \, \partial t} + \frac{\partial G}{\partial u} u' \right\} dv.$$

Since, for $t = 0$, $\partial u/\partial v = 0$ and $\partial G/\partial u = 0$, we have proved the first part. Furthermore,

$$E''(t) = \int_0^l \left\{ 2\left(\frac{\partial^2 u}{\partial v \, \partial t}\right)^2 + 2 \frac{\partial u}{\partial v} \frac{\partial^3 u}{\partial v \, \partial^2 t} + \frac{\partial^2 G}{\partial u^2}(u')^2 + \frac{\partial G}{\partial u} u'' \right\} dv.$$

Hence, by using $G_{uu} = -2K\sqrt{G}$ and noting that $\sqrt{G} = 1$ for $t = 0$, we obtain

$$E''(0) = 2 \int_0^l \left\{ \left(\frac{d\eta}{dv}\right)^2 - K\eta^2 \right\} dv.$$

6. b. Choose $\epsilon > 0$ and coordinates in $R^3 \supset S$ so that $\varphi(\rho, \epsilon) = q$. Consider the points $(\rho, \epsilon) = r_0$, $(\rho, \epsilon + 2\pi \sin \beta) = r_1, \ldots, (\rho, \epsilon + 2\pi k \sin \beta) = r_k$. Taking ϵ sufficiently small, we see that the line segments $\overline{r_0 r_1}, \ldots, \overline{r_0 r_k}$ belong to V if $2\pi k \sin \beta < \pi$ (Fig. 4-49). Since φ is a local isometry, the images of these segments will be geodesics joining q to q, which are clearly broken at q (Fig. 4-49).

c. It must be proved that each geodesic $\gamma\colon [0, l] \longrightarrow S$ with $\gamma(0) = \gamma(l) = q$ is the image by φ of one of the line segments $\overline{r_0 r_1}, \ldots, \overline{r_0 r_k}$ referred to in part b. For some neighborhood $U \subset V$ of r_0, the restriction $\varphi \mid U = \tilde{\varphi}$ is an isometry. Thus, $\tilde{\varphi}^{-1} \circ \gamma$ is a segment of a half-line L starting at r_0. Since $\varphi(L)$ is a geodesic which agrees with $\gamma([0, l])$ in an open interval, it agrees with γ where γ is defined. Since $\gamma(l) = q$, L passes through one of the points r_i, $i = 1, \ldots, k$, say r_j, and so γ is the image of $\overline{r_0 r_j}$.

SECTION 5-2

3. a. Use the relation $\varphi'' = -K\varphi$ to obtain $(\varphi'^2 + K\varphi^2)' = K'\varphi^2$. Integrate both sides of the last relation and use the boundary conditions of the statement.

SECTION 5-3

5. Assume that every Cauchy sequence in d converges and let $\gamma(s)$ be a geodesic parametrized by arc length. Suppose, by contradiction, that $\gamma(s)$ is defined for $s < s_0$ but not for $s = s_0$. Choose a sequence $\{s_n\} \longrightarrow s_0$. Thus, given $\epsilon > 0$, there exists n_0 such that if $n, m > n_0$, $|s_n - s_m| < \epsilon$. Therefore,

$$d(\gamma(s_m), \gamma(s_n)) \leq |s_n - s_m| < \epsilon$$

and $\{\gamma(s_n)\}$ is a Cauchy sequence in d. Let $\{\gamma(s_n)\} \longrightarrow p_0 \in S$ and let W be a neighborhood of p_0 as given by Prop. 1 of Sec. 4-7. If m, n are sufficiently large, the small geodesic joining $\gamma(s_m)$ to $\gamma(s_n)$ clearly agrees with γ. Thus, γ can be extended through p_0, a contradiction.

Conversely, assume that S is complete and let $\{p_n\}$ be a Cauchy sequence in d of points on S. Since d is greater than or equal to the Euclidean distance $\bar{d}$, $\{p_n\}$ is a Cauchy sequence in $\bar{d}$. Thus, $\{p_n\}$ converges to $p_0 \in R^3$. Assume, by contradiction, that $p_0 \notin S$. Since a Cauchy sequence is bounded, given $\epsilon > 0$ there exists an index n_0 such that, for all $n > n_0$, the distance $d(p_{n_0}, p_n) < \epsilon$. By the Hopf-Rinow theorem, there is a minimal geodesic γ_n joining p_{n_0} to p_n with length $< \epsilon$. As $n \longrightarrow \infty$, γ_n tends to a minimal geodesic γ with length $\leq \epsilon$. Parametrize γ by arc length s. Then, since $p_0 \notin S$, γ is not defined for $s = \epsilon$. This contradicts the completeness of S.

6. Let $\{p_n\}$ be a sequence of points on S such that $d(p, p_n) \longrightarrow \infty$. Since S is complete, there is a minimal geodesic $\gamma_n(s)$ (parametrized by arc length)joining p to p_n with $\gamma_n(0) = p$. The unit vectors $\gamma_n'(0)$ have a limit point v on the (compact) unit sphere of $T_p(S)$. Let $\gamma(s) = \exp_p sv$, $s \geq 0$. Then $\gamma(s)$ is a ray issuing from p. To see this, notice that, for a fixed s_0 and n sufficiently large, $\lim_{n\to\infty} \gamma_n(s_0) = \gamma(s_0)$. This follows from the continuous dependence of geodesics from the initial conditions. Furthermore, since d is continuous,

$$\lim_{n\to\infty} d(p, \gamma_n(s_0)) = d(p, \gamma(s_0)).$$

But if n is large enough, $d(p, \gamma_n(s_0)) = s_0$. Thus, $d(p, \gamma(s_0)) = s_0$, and γ is a ray.

8. First show that if d and $\bar{d}$ denote the intrinsic distances of S and $\bar{S}$, respectively, then $d(p,q) \geq c\bar{d}(\varphi(p), \varphi(q))$ for all $p, q \in S$. Now let $\{p_n\}$ be a Cauchy sequence in d of points on S. By the initial remark, $\{\varphi(p_n)\}$ is a Cauchy sequence in $\bar{d}$. Since $\bar{S}$ is complete, $\{\varphi(p_n)\} \longrightarrow \varphi(p_0)$. Since φ^{-1} is continuous, $\{p_n\} \longrightarrow p_0$. Thus, every Cauchy sequence in d converges; hence S is complete (cf. Exercise 5).

9. *φ is one-to-one*: Assume, by contradiction, that $p_1 \neq p_2 \in S_1$ are such that $\varphi(p_1) = \varphi(p_2) = q$. Since S_1 is complete, there is a minimal geodesic γ joining p_1 to p_2. Since φ is a local isometry, $\varphi \circ \gamma$ is a geodesic joining q to itself with the same length as γ. Any point distinct from q on $\varphi \circ \gamma$ can be joined to q by two geodesics, a contradiction.

φ is onto: Since φ is a local diffeomorphism, $\varphi(S_1) \subset S_2$ is an open set in S_2. We shall prove that $\varphi(S_1)$ is also closed in S_2; since S_2 is connected, this will imply that $\varphi(S_1) = S_2$. If $\varphi(S_1)$ is not closed in S_2, there exists a sequence $\{\varphi(p_n)\}$, $p_n \in S_1$, such that $\{\varphi(p_n)\} \longrightarrow p_0 \notin \varphi(S_1)$. Thus, $\{\varphi(p_n)\}$ is a nonconverging Cauchy sequence in $\varphi(S_1)$. Since φ is a one-to-one local isometry, $\{p_n\}$ is a nonconverging Cauchy sequence in S_1, a contradiction to the completeness of S_1.

10. a. Since

$$\frac{d}{dt}(h \circ \varphi(t)) = \frac{d}{dt}\langle \varphi(t), v\rangle = \langle \varphi'(t), v\rangle = \langle \text{grad } h, v\rangle$$

and

$$\frac{d}{dt}(h \circ \varphi(t)) = dh(\varphi'(t)) = dh(\text{grad } h) = \langle \text{grad } h, \text{grad } h\rangle,$$

we conclude, by equating the last members of the above relations, that $|\text{grad } h| \leq 1$.

b. Assume that $\varphi(t)$ is defined for $t < t_0$ but not for $t = t_0$. Then there exists a sequence $\{t_n\} \longrightarrow t_0$ such that the sequence $\{\varphi(t_n)\}$ does not converge. If m and n are sufficiently large, we use part a to obtain

$$d(\varphi(t_m), \varphi(t_m)) \leq \int_{t_n}^{t_m} |\text{grad } h(\varphi(t))|\, dt \leq |t_m - t_n|,$$

where d is the intrinsic distance of S. This implies that $\{\varphi(t_n)\}$ is a nonconverging Cauchy sequence in d, a contradiction to the completeness of S.

SECTION 5-4

2. Assume that

$$\lim_{r\to\infty}\left(\inf_{x^2+y^2\geq r} K(x, y)\right) = 2c > 0.$$

Then there exists $R > 0$ such that if $(x, y) \notin D$, where

$$D = \{(x, y) \in R^2;\, x^2 + y^2 < R^2\},$$

then $K(x, y) \geq c$. Thus, by taking points outside the disk D, we can obtain arbitrarily large disks where $K(x, y) \geq c > 0$. This is easily seen to contradict Bonnet's theorem.

SECTION 5-5

3. b. Assume that $a > b$ and set $s = b$ in relation $(*)$. Use the initial conditions and the facts $v'(b) < 0$, $u(b) > 0$, $uv \geq 0$ in $[0, b]$ to obtain a contradiction.

c. From $[uv' - vu']_0^s \geq 0$, one obtains $v'/v \geq u'/u$; that is, $(\log v)' \geq (\log u)'$. Now, let $0 < s_0 \leq s \leq a$, and integrate the last inequality between s_0 and s to obtain

$$\log v(s) - \log v(s_0) \geq \log u(s) - \log u(s_0);$$

that is, $v(s)/u(s) \geq v(s_0)/u(s_0)$. Next, observe that

$$\lim_{s_0\to 0}\frac{v(s_0)}{u(s_0)} = \lim_{s_0\to 0}\frac{v'(s_0)}{u'(s_0)} = 1.$$

Thus, $v(s) \geq u(s)$ for all $s \in [0, a)$.

6. Suppose, by contradiction, that $u(s) \neq 0$ for all $s \in (0, s_0]$. By using Eq. $(*)$ of Exercise 3, part b (with $\tilde{K} = L$ and $s = s_0$), we obtain

$$\int_0^{s_0} (K - L)uv\, ds + u(s_0)v'(s_0) - u(0)v'(0) = 0.$$

Assume, for instance, that $u(s) > 0$ and $v(s) < 0$ on $(0, s_0]$. Then $v'(0) < 0$ and $v'(s_0) > 0$. Thus, the first term of the above sum is ≥ 0 and the two remaining terms are > 0, a contradiction. All the other cases can be treated similarly.

8. Let $\mathcal{V}$ be the vector space of Jacobi fields J along γ with the property that $J(l) = 0$. $\mathcal{V}$ is a two-dimensional vector space. Since $\gamma(l)$ is not conjugate to $\gamma(0)$, the linear map $\theta\colon \mathcal{V} \longrightarrow T_{\gamma(0)}(S)$ given by $\theta(J) = J(0)$ is injective, and hence, for dimensional reasons, an isomorphism. Thus, there exists $J \in \mathcal{V}$ with $J(0) = w_0$. By the same token, there exists a Jacobi field $\bar{J}$ along γ with $\bar{J}(0) = 0$, $\bar{J}(l) = w_1$. The required Jacobi field is given by $J + \bar{J}$.

SECTION 5-6

10. Let $\gamma\colon [0, l] \longrightarrow S$ be a simple closed geodesic on S and let $v(0) \in T_{\gamma(0)}(S)$ be such that $|v(0)| = 1$, $\langle v(0), \gamma'(0)\rangle = 0$. Take the parallel transport $v(s)$ of $v(0)$ along γ. Since S is orientable, $v(l) = v(0)$ and v defines a differentiable vector field along γ. Notice that v is orthogonal to γ and that $Dv/ds = 0$, $s \in [0, l]$. Define a variation (with free end points) $h\colon [0, l] \times (-\epsilon, \epsilon) \longrightarrow S$ by

$$h(s, t) = \exp_{\gamma(s)} tv(s).$$

Check that, for t small, the curves of the variation $h_t(s) = h(s, t)$ are closed. Extend the formula for the second variation of arc length to the present case, and show that

$$L_v''(0) = -\int_0^l K\, ds < 0.$$

Thus, $\gamma(s)$ is longer than all curves $h_t(s)$ for t small, say, $|t| < \delta \leq \epsilon$. By changing the parameter t into t/δ, we obtain the required homotopy.

SECTION 5-7

9. Use the notion of geodesic torsion τ_g of a curve on a surface (cf. Exercise 19, Sec. 3-2). Since

$$\frac{d\theta}{ds} = \tau - \tau_g,$$

where $\cos\theta = \langle N, n\rangle$ and the curve is closed and smooth, we obtain

$$\int_0^l \tau\, ds - \int_0^l \tau_g\, ds = 2\pi n,$$

where n is an integer. But on the sphere, all curves are lines of curvature. Since the lines of curvature are characterized by having vanishing geodesic torsion (cf.

Exercise 19, Sec. 3-2), we have

$$\int_0^l \tau\, ds = 2\pi n.$$

Since every closed curve on a sphere is homotopic to zero, the integer n is easily seen to be zero.

SECTION 5-10

7. We have only to show that the geodesics $\gamma(s)$ parametrized by arc length which approach the boundary of R^2_+ are defined for all values of the parameter s. If the contrary were true, such a geodesic would have a finite length l, say, from a fixed point p_0. But for the circles of R^2_+ that are geodesics, we have

$$l = \left| \lim_{\epsilon \to 0} \int_{\theta_0 > \pi/2}^{\epsilon} \frac{d\theta}{\sin\theta} \right| \geq \left| \lim_{\epsilon \to 0} \int_{\theta_0 > \pi/2}^{\epsilon} \frac{\cos\theta\, d\theta}{\sin\theta} \right| = \infty,$$

and the same holds for the vertical lines of R^2_+.

10. c. To prove that the metric is complete, notice first that it dominates the Euclidean metric on R^2. Thus, if a sequence is a Cauchy sequence in the given metric, it is also a Cauchy sequence in the Euclidean metric. Since the Euclidean metric is complete, such a sequence converges. It follows that the given metric is complete (cf. Exercise 1, Sec. 5-3).

Index

Acceleration vector, 345
Accumulation point, 457
Angle:
  between two surfaces, 87
  external, 266
  interior, 274
Antipodal map, 80
Arc, 462
  regular, 266
Arc length, 6
  in polar coordinates, 25 (Ex. 11)
  reparametrization by, 22
Area, 98
  geometric definition of, 115
  of a graph, 100 (Ex. 5)
  oriented, 15 (Ex. 10), 166
  of surface of revolution, 101 (Ex. 11)
Area-preserving diffeomorphisms, 230 (Ex. 18), 231 (Ex. 20)
Asymptotic curve, 148
Asymptotic direction, 148

Ball, 118
Beltrami-Enneper, theorem of, 152 (Ex. 13)
Beltrami's theorem on geodesic mappings, 296 (Ex. 13)
Bertrand curve, 26
Bertrand mate, 26
Binormal line, 19
Binormal vector, 18
Bolzano-Weierstrass theorem, 112, 124, 466
Bonnet, O., 535
Bonnet's theorem, 352, 424
Boundary of a set, 459
Braunmühl, A., 363
Buck, R. C., 43, 97, 132

Calabi, E., 354
Catenary, 23 (Ex. 8)
Catenoid, 221
  asymptotic curves of, 168 (Ex. 3)
  local isometry of, with a helicoid, 213 (Ex. 14), 223
  as a minimal surface, 202
Cauchy-Crofton formula, 41
Cauchy sequence, 461
  in the intrinsic distance, 336 (Ex. 5)
Chain rule, 91 (Ex. 24), 126, 129
Chern, S. S., 318
  and Lashof, R., 387
Christoffel symbols, 232
  in normal coordinates, 295 (Ex. 4)
  for a surface of revolution, 232
Cissoid of Diocles, 7 (Ex. 3)
Clairaut's relation, 257
Closed plane curve, 30
Closed set, 458
Closure of a set, 458
Compact set, 112, 465
Comparison theorems, 369 (Ex. 3)
Compatibility equations, 235
Complete surface, 325
Cone, 64, 65 (Ex. 3), 327
  geodesics of, 307 (Ex. 6)
  local isometry of, with plane, 223
  as a ruled surface, 189
Conformal map, 226
  linear, 229 (Ex. 13)
  local, 226, 229 (Ex. 14)
    of planes, 229 (Ex. 15)
    of spheres into planes, 230 (Ex. 16)
Conjugate directions, 150
Conjugate locus, 363

Conjugate minimal surfaces, 213 (Ex. 14)
Conjugate points, 362
  Kneser criterion for, 370 (Ex. 7)
Connected, 462
  arcwise, 462
    locally, 464
  component, 469
  simply, 382
    locally, 383
Connection, 306 (Ex. 2), 442
Conoid, 210 (Ex. 5)
Contact of curves, 171 (Ex. 9)
Contact of curves and surfaces, 171 (Ex. 10)
Contact of surfaces, 91 (Ex. 27), 170 (Ex. 8)
Continuous map, 120
  uniformly, 468
Convergence, 456
  in the intrinsic distance, 336 (Ex. 4)
Convex curve, 37
Convex hull, 48 (Ex. 11)
Convex neighborhood, 303
  existence of, 305
Convex set, 48 (Ex. 9)
Convexity and curvature, 40, 174 (Ex. 24), 387, 397
Coordinate curves, 53
Coordinate neighborhood, 53
Coordinate system, 52
Courant, R., 115
Covariant derivative, 238
  algebraic value of, 248
  along a curve, 240
  expression of, 239
  properties of, 306 (Ex. 2)
  in terms of parallel transport, 305 (Ex. 1)
Covering space, 371
  number of sheets of, 377
  orientable double, 443 (Exs. 3, 4)
Critical point, 58, 89 (Ex. 13)
  nondegenerate, 173 (Ex. 23)
Critical value, 58
Cross product, 12
Curvature:
  Gaussian, 146, 155 (*see also* Gaussian curvature)
  geodesic, 248, 253
  lines of, 145
    differential equations of, 161
  mean, 146, 156, 163
    vector, 201
  normal, 141
  of a plane curve, 21
  principal, 144
  radius of, 19
  sectional, 442
  of a space curve, 16
    in arbitrary parameters, 25 (Ex. 12)
Curve:
  asymptotic, 148
    differential equations for, 160
    maximal, 410
  of class C^k, 10 (Ex. 7)
  closed, 30
Curve (*Cont.*)
  closed (*Cont.*)
    continuous, 392
    piecewise regular, 266
    simple, 30
  coordinate, 53
  divergent, 336 (Ex. 7)
  knotted, 402
  level, 102 (Ex. 14)
  parametrized, 3
    piecewise differentiable, 329
    piecewise regular, 244
    regular, 6
  piecewise C^1, 35
  simple, 10 (Ex. 7)
Cut locus, 420
Cycloid, 7
Cylinder, 65 (Ex. 1)
  first fundamental form of, 93
  isometries of, 229 (Ex. 12)
  local isometry of, with plane, 219
  normal sections of, 144
  as a ruled surface, 188

Darboux trihedron, 261 (Ex. 14)
Degree of a map, 390
Developable surface, 194, 210 (Ex. 3)
  classification of, 194
  as the envelope of a family of tangent planes, 195
  tangent plane of a, 210 (Ex. 6)
Diffeomorphism, 74
  area-preserving, 230 (Exs. 18, 19)
  local, 86
    orientation-preserving, 165
    orientation-reversing, 166
Differentiable function, 72, 80 (Ex. 9), 82 (Ex. 13), 125
Differentiable manifold, 438
Differentiable map, 73, 126, 426
Differentiable structure, 425, 439
Differential of a map, 86, 127, 430
Direction:
  asymptotic, 148
  principal, 144
Directions:
  conjugate, 150
  field of, 178
Directrix of a ruled surface, 188
Distance on a surface, 329
Distribution parameter, 192
do Carmo, M. and E. Lima, 387
Domain, 97
Dot product, 4
Dupin indicatrix, 148
  geometric interpretation of, 164
Dupin's theorem on triply orthogonal systems, 153

Edges of a triangulation, 271
Efimov, N. V., 453
Eigenvalue, 216
Eigenvector, 216
Ellipsoid, 61, 80 (Ex. 4), 90 (Ex. 20)

Ellipsoid (*Cont.*)
conjugate locus of, 263
first fundamental form of, 99 (Ex. 1)
Gaussian curvature of, 173 (Ex. 21)
parametrization of, 66 (Ex. 12)
umbilical points of, 172 (Ex. 20)
Embedding, 435
of the Klein bottle into R^4, 436
of the projective plane into R^4, 437
of the torus into R^4, 435
Energy of a curve, 307 (Ex. 4)
Enneper's surface, 168 (Ex. 5)
as a minimal surface, 205
Envelope of a family of tangent planes, 195, 210 (Ex. 8), 212 (Ex. 10), 244, 308 (Ex. 7)
Euclid's fifth axiom, 279, 431, 432
Euler formula, 145
Euler-Lagrange equation, 365
Euler-Poincaré characteristic, 272
Evolute, 23 (Ex. 7)
Exponential map, 284
differentiability of, 285

Faces of a triangulation, 271
Fary-Milnor Theorem, 402
Fenchel's theorem, 399
Fermi coordinates, 306 (Ex. 3)
Field of directions, 178
differential equation of, 179
integral curves of, 178
Field of unit normal vectors, 104
First fundamental form, 92
Flat torus, 435
Focal surfaces, 210 (Ex. 9)
Folium of Descartes, 8 (Ex. 5)
Frenet formulas, 19
Frenet trihedron, 19
Function:
analytic, 207
component, 120
continuous, 119
differentiable, 72, 125
harmonic, 201
height, 72
Morse, 173 (Ex. 23)
Fundamental theorem for the local theory of curves, 19, 309
Fundamental theorem for the local theory of surfaces, 236, 311

Gauss-Bonnet theorem (global), 274
application of, 276
Gauss-Bonnet theorem (local), 268
Gauss formula, 234
in orthogonal coordinates, 237 (Ex. 1)
Gauss lemma, 288
Gauss map, 136
Gauss theorem egregium, 234
Gaussian curvature, 146, 155
geometric interpretation of, 167
for graphs of differentiable functions, 163
in terms of parallel transport, 270, 271
Genus of a surface, 273
Geodesic:
circles, 287
coordinates, 306 (Ex. 3)
curvature, 248, 253
flow, 440
mapping, 296 (Ex. 12)
parallels, 306 (Ex. 3)
polar coordinates, 286
first fundamental form in, 287
Gaussian curvature in, 288
geodesics in, 295 (Ex. 7)
torsion, 153 (Ex. 19), 261 (Ex. 14)
Geodesics, 307
of a cone, 306 (Ex. 6)
of a cylinder, 246, 247
differential equations of, 254
existence of, 255
minimal, 303, 332
minimizing properties of, 292
of a paraboloid of revolution, 258–260
of the Poincaré half-plane, 431, 432, 444 (Ex. 8)
radial, 287
as solutions to a variational problem, 345
of a sphere, 246
of surfaces of revolution, 255–258, 356 (Ex. 5)
Geppert, H., 407
Gluck, H., 41
Gradient on surfaces, 101 (Ex. 14)
Graph of a differentiable function, 58
area of, 100 (Ex. 5)
Gaussian curvature of, 163
mean curvature of, 163
second fundamental form of, 163, 164
tangent plane of, 88 (Ex. 3)
Green, L., 363
Gromov, M. L., and V. A. Rokhlin, 454
Group of isometries, 229 (Ex. 9)

Hadamard's theorem on complete surfaces with $K \leq O$, 387, 390 (Ex. 9)
Hadamard's theorem on ovaloids, 387
Hartman, P. and L. Nirenberg, 408
Heine-Borel theorem, 112, 124
Helicoid, 94
asymptotic curves of, 168 (Ex. 2)
distribution parameter of, 209 (Ex. 1)
generalized, 101 (Ex. 13), 186 (Ex. 6)
line of striction of, 209 (Ex. 1)
lines of curvature of, 168 (Ex. 2)
local isometry of, with a catenoid, 213 (Ex. 14), 223
as a minimal surface, 204
as the only minimal ruled surface, 204
tangent plane of, 89 (Ex. 9)
Helix, 3, 22 (Ex. 1)
generalized, 26 (Ex. 17)
Hessian, 164, 173 (Ex. 22)
Hilbert, D., 318, 446
Hilbert's theorem, 446
Holmgren, E., 446
Holonomy group, 297 (Ex. 14)
Homeomorphism, 123

Homotopic arcs, 378
Homotopy of arcs, 378
free, 390 (Ex. 10)
lifting of, 379
Hopf, H. and W. Rinow, 326, 354
Hopf-Rinow's theorem, 333
Hopf's theorem on surfaces with H = const., 234 (Ex. 4)
Hurewicz, W., 177
Hyperbolic paraboloid (saddle surface), 66 (Ex. 11), Fig. 3–7
asymptotic curves of, 184
first fundamental form of, 99 (Ex. 1)
Gauss map of, 139
parametrization of, 66 (Ex. 11)
as a ruled surface, 193
Hyperbolic plane, 431
Hyperboloid of one sheet, 88 (Ex. 2), Fig. 3–34
Gauss map of, 151 (Ex. 8)
as a ruled surface, 189, 209 (Ex. 2)
Hyperboloid of two sheets, 61
first fundamental form of, 99 (Ex. 1)
parametrization of, 67 (Ex. 13), 99 (Ex. 1)

Immersion, 433
isometric, 433
Index form of a geodesic, 422
Index of a vector field, 280
Infimum (g.l.b.), 460
Integral curve, 178
Intermediate value theorem, 124
Intrinsic distance, 225, 329
Intrinsic geometry, 217, 235, 238
Inverse function theorem, 131
Inversion, 121
Isometry, 218
linear, 228 (Ex. 7)
local, 219
in local coordinates, 220, 228 (Ex. 2)
of tangent surfaces to planes, 228 (Ex. 6)
Isoperimetric inequality, 33
for geodesic circles, 295 (Ex. 9)
Isothermal coordinates, 201, 227
for minimal surfaces, 213 (Ex. 13(b))

Jacobi equation, 357
Jacobi field, 357
on a sphere, 362
Jacobian determinant, 128
Jacobian matrix, 128
Jacobi's theorem on the normal indicatrix, 278
Jacobi's theorems on conjugate points, 419, 423
Joachimstahl, theorem of, 152 (Ex. 15)
Jordan curve theorem, 393

Kazdan, J. and F. Warner, 446
Klein bottle, 427
embedding of, into R^4, 436, 437
non-orientability of, 436
Klingenberg's lemma, 388 (Ex. 8)
Kneser criterion for conjugate points, 370 (Ex. 7)
Knotted curve, 402

Lashof, R. and S. S. Chern, 387
Lebesgue number of a family, 113
Levi–Civita connection, 442
Lifting:
of an arc, 376
of a homotopy, 379
property of, arcs, 380
Lima, E. and M. do Carmo, 387
Limit point, 457
Limit of a sequence, 456
Line of curvature, 145
Liouville:
formula of, 253
surfaces of, 263
Local canonical form of a curve, 27
Locally convex, 174 (Ex. 24), 387
strictly, 174 (Ex. 24)
Logarithmic spiral, 9
Loxodromes of a sphere, 96, 230

Mainardi–Codazzi equations, 235
Mangoldt, H., 363
Map:
antipodal, 80 (Ex. 1)
conformal, 226
linear, 229 (Ex. 13)
continuous, 120
covering, 371
differentiable, 73, 126, 426
distance-preserving, 228 (Ex. 8)
exponential, 284
Gauss, 136
geodesic, 296 (Ex. 12)
self-adjoint linear, 214
Massey, W., 408
Mean curvature, 146, 156, 163
Mean curvature vector, 201
Mercator projection, 230 (Ex. 16), 231 (Ex. 20)
Meridian, 76
Meusnier theorem, 142
Milnor, T. Klotz, 454
Minding's theorem, 288
Minimal surfaces, 197
conjugate, 213 (Ex. 14)
Gauss map of, 212 (Ex. 13)
isothermal parameters on, 202, 213 (Ex. 13(b))
of revolution, 202
ruled, 204
as solutions to a variational problem, 199
Möbius strip, 106
Gaussian curvature of, 172 (Ex. 18)
infinite, 443 (Ex. 2)
nonorientability of, 107, 109 (Exs. 1, 7)
parametrization of, 106
Monkey saddle, 159, 171 (Ex. 11)
Morse index theorem, 422

Neighborhood, 119, 123
  convex, 303
  coordinate, 53
  distinguished, 371
  normal, 285
Nirenberg, L. and P. Hartman, 408
Norm of a vector, 4
Normal:
  coordinates, 286
  curvature, 141
  indicatrix, 278
  line, 87
  plane to a curve, 19
  principal, 19
  section, 142
  vector to a curve, 17
  vector to a surface, 87

Olinde Rodrigues, theorem of, 145
Open set, 118
Orientation:
  change of, for curves, 6
  for curves, 109 (Ex. 6)
  positive, of R^n, 12
  for surfaces, 103, 136
  of a vector space, 12
Oriented:
  area in R^2, 15 (Ex. 10)
  positively, boundary of a simple region, 267
  positively, simple closed plane curve, 31
  surface, 103, 106
  volume in R^3, 16 (Ex. 11)
Orthogonal:
  families of curves, 102 (Ex. 15), 181, 186 (Ex. 6)
  fields of directions, 181, 185 (Ex. 4), 186 (Ex. 5)
  parametrization, 95, 183
  projection, 80 (Ex. 2), 121
  transformation, 23 (Ex. 6), 228 (Ex. 7)
Osculating:
  circle to a curve, 30 (Ex. 2(b))
  paraboloid to a surface, 170 (Ex. 8(c))
  plane to a curve, 17, 29, 29 (Ex. 1), 30 (Ex. 2)
  sphere to a curve, 171 (Ex. 10(c))
Osserman's theorem, 208, 337 (Ex. 11)
Ovaloid, 322, 387

Paraboloid of revolution, 80 (Ex. 3)
  conjugate points on, 368 (Ex. 2)
  Gauss map of, 140
  geodesics of, 258
Parallel:
  curves, 47 (Ex. 6)
  surfaces, 212 (Ex. 11)
  transport, 242
    existence and uniqueness of, 242, 253
    geometric construction of, 244
  vector field, 241
Parallels:
  geodesic, 306 (Ex. 3(d))
  of a surface of revolution, 76
Parameter:
  of a curve, 3
  distribution, 192
Parameters:
  change of, for curves, 82 (Ex. 15)
  change of, for surfaces, 70
  isothermal, 227
    existence of, 227
    existence of, for minimal surfaces, 213 (Ex. 13(b))
Parametrization of a surface, 52
  by asymptotic curves, 184
  by lines of curvature, 185
  orthogonal, 95
    existence of, 183
Partition, 10 (Ex. 8), 114
Plane:
  hyperbolic, 431
  normal, 19
  osculating, 17, 29, 29 (Ex. 1), 30 (Ex. 2)
  real projective, 427
  rectifying, 19
  tangent, 84
Planes, one-parameter family of tangent, 212 (Ex. 10), 308 (Ex. 7)
Plateau's problem, 200
Poincaré half-plane, 431
  completeness of, 444 (Ex. 7)
  geodesics of, 432, 444 (Ex. 8)
Poincaré's theorem on indices of a vector field, 282
Point:
  accumulation, 457
  central, 191
  conjugate, 362
  critical, 58, 89 (Ex. 13), 364
  elliptic, 146
  hyperbolic, 146
  isolated, 461
  limit, 457
  parabolic, 146
  umbilical, 147
Pole, 390 (Ex. 11)
Principal:
  curvature, 144
  direction, 144
  normal, 19
Product:
  cross, 12
  dot, 4
  inner, 4
  vector, 12
Projection, 80 (Ex. 2), 121
  Mercator, 230 (Ex. 16), 231 (Ex. 20)
  stereographic, 67 (Ex. 16), 228 (Ex. 4)
Projective plane, 427
  embedding of, into R^4, 437
  nonorientability of, 436
  orientable double covering of, 443 (Ex. 3)
Pseudo-sphere, 168 (Ex. 6)

Radius of curvature, 19
Ray, 336 (Ex. 6)
Rectifying plane, 19
  envelope of, 308 (Ex. 7(b))
Region, 97
  bounded, 97
  regular, 271
  simple, 267
Regular:
  curve, 68 (Ex. 17), 75
  parametrized curve, 6
  parametrized surface, 78
  surface, 52
  value, 58, 92 (Ex. 28)
    inverse image of, 59, 92 (Ex. 28)
Reparametrization by arc length, 22
Riemannian:
  manifold, 441
    covariant derivative on, 442
  metric, 441
    on abstract surfaces, 430
  structure, 442
Rigid motion, 23 (Ex. 6), 42
Rigidity of the sphere, 317
Rinow, W. and H. Hopf, 326, 354
Rokhlin, V. A. and M. L. Gromov, 454
Rotation, 74, 86
Rotation axis, 76
Rotation index of a curve, 37, 393
Ruled surface, 188
  central points of, 191
  directrix of, 188
  distribution parameter of, 192
  Gaussian curvature of, 192
  line of striction of, 191
  noncylindrical, 190
  rulings of, 188
Ruling, 188

Samelson, H., 114
Santaló, L., 45
Scherk's minimal surface, 207
Schneider, R., 54
Schur's theorem for plane curves, 406 (Ex. 8)
Second fundamental form, 141
Segre, B., 408
Set:
  arcwise connected, 462
  bounded, 112
  closed, 458
  compact, 112, 466
  connected, 462
  convex, 48 (Ex. 9)
  locally simply connected, 383
  open, 118
  simply connected, 382
Similarity, 296 (Ex. 12)
Similitude, 187 (Ex. 9), 229 (Ex. 13)
Simple region, 267
Singular point:
  of a parametrized curve, 6
  of a parametrized surface, 78
  of a vector field, 279
Smooth function, 2
Soap films, 199
Sphere, 55
  conjugate locus on, 362–363
  as double covering of projective plane, 443 (Ex. 2)
  first fundamental form of, 95
  Gauss map of, 137
  geodesics of, 246
  isometries of, 229 (Ex. 11), 264 (Ex. 23)
  isothermal parameters on, 228 (Ex. 4)
  Jacobi field on, 362
  orientability of, 104
  parametrizations of, 55–58, 67 (Ex. 16)
  rigidity of, 317
  stereographic projection of, 67 (Ex. 16)
Spherical image, 152 (Ex. 9), 279
Stereographic projection, 67 (Ex. 16), 228 (Ex. 4)
Stoker, J. J., 387, 408
Stoker's remark on Efimov's theorem, 454 (Ex. 1)
Stoker's theorem for plane curves, 406 (Ex. 8)
Striction, line of, 191
Sturm's oscillation theorem, 370 (Ex. 6)
Supremum (l.u.b.), 460
Surface:
  abstract, 425
  complete, 325
  connected, 61
  developable, 194, 210 (Ex. 3)
  focal, 210 (Ex. 9)
  geometric, 430
  of Liouville, 263
  minimal, 197
  parametrized, 78
    regular, 78
  regular, 52
  of revolution (*see* Surfaces of revolution)
  rigid, 317
  ruled (*see* Ruled surface)
  tangent, 78
Surfaces of revolution, 76
  area of, 101 (Ex. 11)
  area-preserving maps of, 231 (Ex. 20)
  Christoffel symbols, 232
  conformal maps of, 231 (Ex. 20)
  with constant curvature, 169 (Ex. 7), 320
  extended, 78
  Gaussian curvature of, 162
  geodesics of, 255–258
  isometries of, 229 (Ex. 10)
  mean curvature of, 162
  minimal, 202–203
  parametrization of, 76
  principal curvatures of, 162
Symmetry, 74, 121
Synge's lemma, 390 (Ex. 10)

Tangent:
  bundle, 439
  indicatrix, 23 (Ex. 3), 36
  line to a curve, 5

Tangent (*Cont.*)
  map of a curve, 393
  plane, 84, 88 (Exs. 1, 3)
    of abstract surfaces, 429
  strong, 10 (Ex. 7)
  surface, 78
  vector to an abstract surface, 428
  vector to a curve, 2
  vector to a regular surface, 83
  weak, 10 (Ex. 7)
Tangents, theorem of turning, 267, 396
Tchebyshef net, 100 (Exs. 3, 4), 237 (Ex. 5), 447
Tissot's theorem, 187 (Ex. 9)
Topological properties of surfaces, 271–273
Torsion:
  in an arbitrary parametrization, 25 (Ex. 12)
  geodesic, 153 (Ex. 19), 261 (Ex. 14)
  in a parametrization by arc length, 22 (Ex. 12)
  sign of, 28
Torus, 61
  abstract, 434
  area of, 98
  flat, 435
  Gaussian curvature of, 157
  implicit equation of, 62
  as orientable double covering of Klein bottle, 443 (Ex. 3)
  parametrization of, 65
Total curvature, 399
Trace of a parametrized curve, 2
Trace of a parametrized surface, 78
Tractrix, 7 (Ex. 4)
Translation, 23 (Ex. 6)
Transversal intersection, 90 (Ex. 17)
Triangle on a surface, 271
  geodesic, 264, 278
    free mobility of small, 295 (Ex. 8)
Triangulation, 271
Trihedron:
  Darboux, 261 (Ex. 14)
  Frenet, 19
Tubular:
  neighborhood, 110, 400
  surfaces, 89 (Ex. 10), 399

Umbilical point, 147
Uniformly continuous map, 468
Unit normal vector, 87

Variation:
  first, of arc length, 345
  second, of arc length, 351
  second, of energy for simple geodesics, 307 (Ex. 5)
Variations:
  broken, 420
  calculus of, 354–356 (Exs. 4, 5)
  of curves, 339
  orthogonal, 346
  proper, 339
  of simple geodesics, 307
  of surfaces, 197
Vector:
  acceleration, 345
  length of, 4
  norm of, 4
  tangent (*see* Tangent, vector)
  velocity, 2
Vector field along a curve, 240
  covariant derivative of, 240
  parallel, 241
  variational, 340
Vector field along a map, 343
Vector field on a plane, 175
  local first integral of, 178
  local flow of, 177
  trajectories of, 175
Vector field on a surface, 179, 238
  covariant derivative of, 238
  derivative of a function relative to, 186 (Ex. 7)
  maximal trajectory of, 187 (Ex. 11)
  singular point of, 279
Vertex:
  the four, theorem, 37
  of a plane curve, 37
Vertices of a piecewise regular curve, 266
Vertices of a triangulation, 271

Warner, F. and J. Kazdan, 446
Weingarten, equations of, 155
Winding number, 392